ARILD STUBHAUG

The Mathematician Sophus Lie

Springer
Berlin
Heidelberg
New York
Barcelona
Hong Kong
London
Milan
Paris
Tokyo

ARILD STUBHAUG

The Mathematician Sophus Lie

It was the Audacity of My Thinking

Translated from the Norwegian
by Richard H. Daly

Springer

ARILD STUBHAUG
Department of Mathematics
University of Oslo
P. O. Box 1053, Blindern
0316 Oslo, Norway
arild.stubhaug@online.no

Translator:

RICHARD H. DALY
Stensrudlia 1
1294 Oslo, Norway
daly@powertech.no

Title of the Norwegian original edition 2000:
Det var mine tankers djervhet – Matematikeren Sophus Lie
Publisher: H. Aschehoug & Co. (W. Nygaard), Oslo
ISBN 82-03-22297-8

Publisher's note: Springer-Verlag thanks Elisabet W. Middlethon of MUNIN (Marketing Unit for Norwegian International Non-fiction), Oslo, for the excellent collaboration and financial support for the translation, which made the publication of this book possible. Thanks go also to Harald Engelstad of H. Aschehoug & Co., Oslo.

Library of Congress Cataloging-in-Publication Data
Stubhaug, Arild, 1948–. [Det var mine tankers djervhet. English]
The Mathematician Sophus Lie: it was the audacity of my thinking / Arild Stubhaug;
translated from the Norwegian by Richard H. Daly.
p. cm. Includes bibliographical references and index. ISBN 3540421378 (alk. paper)
1. Lie, Sophus, 1842–1899. 2. Mathematicians–Norway–Biography. I. Title
QA29.L.29 S8813 2002 510'.92–dc21. [B] 2001049574

Mathematics Subject Classification (2000):
01A55, 01A70

ISBN 3-540-42137-8 Springer-Verlag Berlin Heidelberg New York

With 105 figures, 8 in color

Springer-Verlag Berlin Heidelberg New York
a member of BertelsmannSpringer Science+Business Media GmbH
http://www.springer.de

Printed in Germany

Typeset in Word by the translator and edited by PublicationService Gisela Koch, Wiesenbach, using
a modified Springer LaTeX macro-package.
Cover design: *design & production GmbH*, Heidelberg
Printed on acid-free paper SPIN 10699495 46/3142Ko – 5 4 3 2 1 0

Contents

PART VI

Professor in Leipzig

PART VII

At the End of the Road

Appendices

List of Illustrations

PART I

The Measure of a Life

Sophus Lie and his wife Anna: late in the 1890s at Åsgårdstrand

Tracking Him Down: A Torrent of Stories

According to most accounts of Sophus Lie, he was the embodiment of an archetypical character in a theatrical drama – with his forceful beard, his sparkling green-blue eyes magnified by the stout lenses of his spectacles – the blonde Nordic prototype, as it was called across Europe – the Germanic gigantic being – a primal force, a titan replete with the lust for life, with audacious goals and an indomitable will. These descriptions of his physical and mental strength also contained a subtext, an embryonic notion, not only about this brilliant man of science, the prophet who intuitively conceived new mathematical truths, but also about the colossus who, in his constant zeal for new knowledge, might push others aside, and inadvertently trample them underfoot. He was described as highly committed and richly innovative, someone with unusual physical strength, and the stamina to overcome the majority of obstacles, but also, a man who afterwards had to pay for this with correspondingly great swings of mood and temperament.

Everyone who met him immediately had something to tell. All of those who knew him seem to have had their own unique version of this man who created whole new fields of mathematics. They cheerfully sought to popularise their own image of Sophus Lie, expanding their accounts with assorted stories. Long before Sophus Lie received world renown he had become the object of his friends' conversation, and a centre of their discussions. Embellished with local stories and anecdotes, these accounts flourished and nobody seems to have been able to pass him by in silence.

What, in detail, did the stories recount, and what, on the other hand, do we know with certainty? What was imaginary and what was real?

Even as a two- or three year-old, little Sophus was described as a boy with extreme traits of character. His father, the parish vicar of Nordfjordeid[1], caused explosions of laughter whenever, in social gatherings, he told about how his little son simply fell asleep one day with a half-eaten pancake in his mouth – it was as though two instincts, hunger and sleep, were competing for the last word. The child himself, the arena for these impulses, had gaped, open-mouthed, at what was happening, and had done so with unusual persistence and stamina until hunger had been vanquished by sleep.

[1] A seaside settlement at the junction of Eidsfjord and Nordfjord in the fjord country of western Norway's Sogn og Fjordane County; trade and administrative centre for the middle of Nordfjord.

Once upon a time, in the middle of the 1840s at Stranda, in a convivial gathering at his father's pastoral colleague, Aage Schavland, little Sophus had been described in the above-mentioned manner. Otherwise, the conversation and discussion between the two vicars had to do with new technical innovations in boating. One of them felt that the paddlewheel was the best way to propel a boat through the water, and the other was a supporter of screws or propellers. They jested that in order to settle the matter they should meet on the open sea, each in his own boat. That meeting never took place, but the two vicars each had a son, and the sons would come to meet and become friends in the university milieu in Christiania. These sons were Sophus and Olaf, who was four years older and wrote his name "Skavlan". Olaf Skavlan became Norway's first professor of European Comparative Literature.

At this time, during the 1840s, Sophus Lie's father, together with his wife and seven children, occupied the vicarage at Nordfjordeid. Sophus, the penultimate child, was born here, and the world of his first nine years was this hamlet, spread out between the fjord and the icy blue hues of the mountain vastness. The Lie family had lived here since 1836. The vicar was well regarded and he became the first mayor of the community. The pastor's wife was particularly well-liked by the ordinary people as well, both for her accommodation and concern for others, as well as for her competence and her helping hand in the many duties on the vicarage's farm. Mrs. Lie was described as a strong personality who protected her home and created a generous and hospitable spirit around the farm.

In 1851 the family moved to Moss, a small city on the east shore of what today is Oslofjord. Father Lie had been appointed to the position of city pastor, and here he remained until his death, more than twenty years later. But Mrs. Lie, following a brief illness, died only a year after they had moved from Nordfjordeid. The atmosphere around the home now changed. The father became both more austere and more attentive, as, simultaneously, Sophus' eldest sisters, aged sixteen and fifteen, gradually took over the work of running the home, together with a housekeeper.

For Sophus, who was now ten years old, early childhood was definitely over. In addition, rapidly following his arrival in the city he was placed in the city's middle-class, science-oriented Moss [Realskole] Grammar School. This was a school that after operating for twenty years, now had about eighty pupils, and where modern languages such as English and French had their place, at the cost of Latin and Greek. In keeping with its unflagging faith in progress, the school was also introducing instruction in physics and chemistry. In every way, young Sophus was a bright and conscientious pupil at the Moss Grammar School. He attended the school for five years and gradually worked himself up from the fifth place in his class to first by the time of his school graduation in the summer of 1856. Fellow pupils later recounted that he always enthusiastically helped them with assignments when this was needed, and that when it came to mathematics he was considered to be a sort of "oracle".

The course of studies at the school in Moss did not go through to the *examen artium*, as the school matriculation exam was called. Therefore, in the autumn of 1857, at the age of fifteen, Sophus Lie journeyed to Christiania to continue his schooling there.

Apart from his priestly duties, Sophus Lie's father was interested in, and concerned with, teaching. He himself had been a teacher for a number of years following his public service examination in theology, and he eagerly engaged himself in all sorts of schooling and enlightenment work. He also arranged public lectures on themes drawn from the natural sciences. These lectures were directed particularly at the workers, and in this, Vicar Lie was regarded as a forerunner of what later came to be known as the Workers' Academy[2]. It is possible that he was also inspired in this enterprise by his eldest son Fredrik, who at this time was studying the sciences at the university in Christiania. It might also be assumed that the father thought that this was the natural direction that Sophus' education should take. In any case, the fifteen-year-old Sophus began at Nissen's School of Latin and Science in Christiania[3] in the autumn of 1857. This was a school that had been founded fourteen years earlier to give greater place to modern languages and the natural sciences than was currently available in the well-established Latin schools.

At Nissen's School, Sophus Lie met Ernst Motzfeldt, who was his own age, and would come to be his steadfast friend and an important support throughout his life. Moreover, the exchange of letters between these close comrades yields important information about Sophus Lie's life. Ernst Motzfeldt, later a Supreme Court lawyer, held a series of public positions of trust. He helped Sophus Lie both with accommodations and money. They undertook long hiking tours together, and Ernst Motzfeldt always stepped forward to help whenever difficulties piled up around his friend.

After two years at Nissen's School, Sophus Lie wrote an extremely good *examen artium*. He became a university student and prepared for the obligatory secondary exam, a rather comprehensive preparatory examination at the university. He took this examination in the autumn of 1860 and was one of the two best in his year – there were about a hundred students from across the country in his year's cohort. But Sophus remained extremely uncertain about where he should direct his further studies, and toward which career. If he continued at the university he stood to choose between languages and the sciences; however, he was also interested in a military career, such as his middle brother, two years his senior, had chosen. Besides, his mother's kinfolk were full of military men whose lives were something to live up to. In any case, it seems to have been a disappointment to him when it was ascertained that his eyesight was not good enough. The idea of a military career was subsequently abandoned since he suffered from oblique corneas.

---

[2] The people's or workers' academy movement began in England as a popular extension of cultural awakening for the general public and reached Sweden in 1880 and Christiania by 1885. Its dissemination of knowledge excluded matters of politics and religion.

[3] Oslo, founded about A.D. 1000, was called Christiania (after the Danish king, Christian IV) between 1624 and 1924. The spelling was made consistent with Norwegian orthography in the 1870s, and until 1924 it was known as Kristiania. In 1925 the capital city reverted to the old name, Oslo.

The only thing with which he had been dissatisfied in his *examen artium* was the mark he obtained in Greek. In this subject he had received "very good" instead of "exceedingly good", and it was said that this was the reason he withdrew from philology, which was known as one of the elite courses of studies. At any rate, Sophus Lie now chose to study the sciences. In this choice he was quite certainly influenced by his eldest brother, Fredrik. His big brother was more than halfway through the course of studies in the sciences, or the *reallærereksamen*, the science teachers' examination, as this particular examination to qualify for the public service was called, a course of studies that had only been instituted at the University of Christiania a few years earlier, in 1851.

The studies that Sophus Lie now undertook were divided into three sections: mathematics, physics-chemistry, and natural history. The mathematical section was taught by Professors Carl Anton Bjerknes and Ole Jacob Broch, with Senior Teacher Ludvig Sylow as substitute for a couple of semesters while Broch was a cabinet minister. Three years later, in the autumn of 1863, Sophus Lie took the examination in mathematics with the top mark, *præ ceteris*. In the two next sections, each of one year's duration, his results were not of equal excellence; it seems in particular that he had little interest in the botany, zoology and geology that composed the final section. Consequently he did not obtain the result that he had hoped for, namely, the public service examination awarded with distinction. But he came very close to it. The fact that he did not master these subjects in their entirety affected him so very strongly that twenty years later he still felt his last year of studies had made him "melancholic and eccentric", and he feared that he had lost his "mental powers".

Many of his fellow students had stories to tell about their time at the university with Sophus Lie. One of Lie's friends from these years was the medical student, Armauer Hansen, the man who later discovered the leprosy bacillus and became world famous as a fighter against leprosy. Armauer Hansen later wrote a book of memoirs in which he told about his student years and experiences together with Sophus Lie and Ernst Motzfeldt. The three students often relaxed from their studies by exercising with the gymnastic rings, and when they were at the highest point they used to execute a backward somersault with their legs outside the suspension ropes, and so to a landing on a mat placed on the floor below. On one occasion, Sophus Lie, while executing this manoeuvre, let go too late, flew too far through the air, and landed with a thump on his backside, out beyond the mat. He sat there with such a remote and startled expression on his face that the other two broke into loud laughter. But when Sophus heard this he flew into an uncontrollable rage and flailed them with his curses because they laughed when he had almost killed himself. But then in the middle of his tirade, he broke off suddenly and mumbled, almost to himself, "It's strange though, because I even calculated precisely when I should let go!"

Otherwise, Sophus Lie was known to be a master at leaping over the wooden horse, and he would jump over whatever he encountered. There were many who maintained they had witnessed him spring upon a moving horse, place his hands on the horse's back and jump right over it while the beast continued to trot.

This was a time when many students were avid gymnasts and sportsmen. On weekends, small groups of close friends often undertook long hikes in the Nordmarka hills behind Christiania, or out along the fjord. On these excursions they usually camped for the night under the open sky. Wandering on foot through the mountain fastness of Norway was becoming a popular activity, particularly among students. Like the natural scientists, the painters and the writers before them, other groups – such as university students – now began to pay attention to the Norwegian landscape. This led to the founding of Den Norske Turistforening (The Norwegian Mountain Touring Association) some years later. Its work of establishing proper facilities and conditions for foot tourists in the mountains was an endeavour in which Sophus Lie too would participate. He became a member of the Mountain Touring Association's executive and was one of the first to direct attention to the hitherto little-known alpine landscape in the interior of Nordmøre County. In addition, during these years the ski jumper, Sondre Nordheim appeared in Christiania with his skis and bindings, and made a great impression with his aerial leaps, something that laid the basis for a ski association and the now famous Holmenkollen ski jump, together with its cross-country runs in the hills behind the capital city. Sophus Lie was part of this right from the beginning. He was especially known for the power and force of his alpine trekking; some even felt that he would have become famous as a foot "tourist" if it had not been for science. His normal day's march was usually said to be three or four Norwegian miles [4] (20–27 English miles), but Sophus Lie happily put behind him seven or eight miles (49–54 English miles). A song about him stated:

Fifty miles in one shake,
With rucksack and its ache
Is to him a piece of cake.

While he was a student in Christiania he walked home to Moss, a distance of about thirty-six English miles, to satisfy his desire to visit his father and his siblings. One particular story has him walking from Christiania to Moss one Saturday, and back again. The errand on that occasion had been to fetch a book from home, and he had headed back again from Moss without even greeting his father, who himself was out on an errand.

During the summer of 1863, on the way into the Jotunheimen Mountains with Ernst Motzfeldt, Sophus also got to display his other skills when, in Svatsum in Vestre Gausdal, they were invited to a farm where a wedding was being celebrated. They danced, ate and drank, and enjoyed themselves "excellently", he wrote in a letter to his father. He continued: "In the course of the day I often had occasion to show that I was a fellow of some prowess. For example, up under the roof beams there hung a ring. In the interval some boys began to pull themselves up and bonk their heads on the underside of the roof. The lads appeared to be quite stalwart but they probably lacked practice in such gymnastic moves. They did not manage to do it more than once or twice. For me, a trained gymnast, who moreover has a

[4] One Norwegian mile is 11,295 m., or approximately 6.8 English miles, or 11.3 kilometres.

forehead that can tolerate a lot, it was a small matter to pull myself up and bonk violently against the roof 10 or 12 times. Later they began to leap the wooden horse, and here as well, I possessed a definite superiority." In the same letter to his father, Sophus otherwise said that he had a couple of bottles of hard spirits in his rucksack, and that these were empty before they had reached Espedalsvannet, a good mile (11.3 km.) from Svatsum. A little later they met "3 pretty and quick alpine milkmaids", who, according to the words of Sophus Lie, were "free from all kinds of superfluous reticence."

Beyond these activities there was the Student Society in Christiania – a large and natural gathering point where both new and old students met, and on festive occasions both professors and academics from all professions were in attendance – those taking the leadership were usually persons who had long since completed their university education. The Student Society was a signficant forum for social debate, particularly around the end of the 1860s and thereafter, when great problems were "submitted for debate".

As a student in the first part of the 1860s, Sophus Lie was never a "Societies Man". Like many others he probably thought that the Student Society's social gatherings were by and large a waste of time. Every Saturday evening, members joined one another around a long table, drank bad punch, and listened to boring speeches. One was expected to be convivial and spirited, and the discussions most frequently were among the younger and older jurists who sought to outshine one another in sophistry and debate as they disputed small sums on the bill. Sophus Lie found most joy and benefit listening to lectures at the newly instituted Realistforeningen, the Science Students Association, where scientific lectures were combined with social gatherings. In addition, other subjects and disciplines began creating their own associations that would allow students to delve more deeply into their respective fields of study.

During the spring of 1864 there were stormy discussions about how Norwegians, in the name of Scandinavianism, should involve themselves in the Danish-German War. Some thought that one ought to take part actively in the battles along the Danish-German border, and an "Instruction School" was set up, with the participation of up to two hundred aspiring reserve officers. Many students joined in, including Sophus Lie, who underwent such a form of officer-training during the summer of 1864. Nonetheless, they saw no active service. In the coming years Sophus Lie was on the military rolls as a reserve lieutenant of the Trondheim Infantry Brigade, and second lieutenant seconded to a battalion in the Trondheim region. His greatest exploit as a military man seems to have been when, during a short-lived exercise, he marched the more than fifty kilometres into Trondheim and the same distance back again, and the next day got up at six o'clock without the least sign of fatigue.

Another group that would have significance for Sophus Lie was a circle of radical students and mutual friends who, following dinner at the Student Society, took their coffee in a little room next to the dining hall. Every day they spent several hours together in animated discussion on all the world's current issues. Because both chairs and sofas had green upholstery, and the heavy window curtains were also green, the circle gradually acquired the designation, "The Green Room". Most of the fifteen to twenty members were the same age as Sophus Lie, a few younger, and a number older, but it was open to all, including guests and guests of guests. For example, the writer and incisive journalist, A. O. Vinje, came here and read aloud from his articles, as well as commenting upon events and personalities with an astonishing wit, and on other occasions raucous laughter broke out when, for instance, the historian Ernst Sars presented his perceptions concerning Ontology and the State of Being.

His last year as a student, 1865, seems to have been a hard and difficult time for Sophus Lie. He was profoundly uncomfortable with the subjects of the final section of his studies: botany, zoology and geology, and he suffered from insomnia and depression. Armauer Hansen believed that Sophus Lie would perhaps have run aground completely if Ernst Motzfeldt had not taken him in hand at that point. That is to say, Motzfeldt took his friend out on long hiking tours in Nordmarka, the hills and forests along the north side of Christiania, with the aim of tiring him enough physically so that he could begin to sleep again. But Sophus Lie was not easy to tire out; he was, as Armauer Hansen observed, "exceedingly strong".

After examinations in December, 1865, Sophus Lie went home to his father in Moss, and was both extremely depressed and of many minds concerning what would become of him. He considered himself, in his own words, to be "a bogus Subject". He lacked what was much talked about in intellectual circles, namely a *calling*, or a vocation. Then, in March, 1866, from home in Moss, he wrote a letter to his friend Motzfeldt: "When I bid you Farewell before Christmas I believed that it was for Now and all Eternity; for it was my Intention to become a Suicide. But I do not have the Strength for it. So in consequence I get another Chance to try to live."

From the family home in Moss he took the summer boat across what today is called Oslofjord, to Horten and travelled south, overland on foot, to visit his sister and brother-in-law at Tvedestrand, a town in Sørlandet, the region along the Norwegian shore of the Skagerrak. His eldest sister, Mathilde, was married to the doctor, Fredrik Vogt, and with their children they lived at the coastal town of Tvedestrand, where Vogt conducted a large medical practice. Vogt was a well-known doctor and leader of a national health commission, and on his travels to and from the capital he had probably taken his brother-in-law, Sophus Lie, back home with him several times in recent years for rest and relaxation at Tvedestrand. In any case, Sophus Lie was well-known in that town, and had already been there on many holidays before he arrived during this particular summer, and found himself entangled in local scandal of this tiny society. According to his sister's son, Johan Herman Lie Vogt, who later became professor of mineralogy and geology, Sophus Lie used to set the town on its head with a steady stream of escapades. That summer of 1866, from the account of the nephew, who at the time was scarcely eight years old, Sophus Lie initiated a swimming school for Johan Herman and his friends. When they rowed out into the fjord in the doctor's distinctive boat, which was rowed with three pairs of oars – black on the outside and red inside – Uncle

Sophus always sat in the stern with a bailer full of water ready to throw at whichever boy dared to break the unison of the stroke. One Sunday as they rowed out, Uncle Sophus got it into his head to fasten a flotation belt to his nephew and throw him overboard. There was a more-than-fresh breeze blowing down the fjord, and the moment little Johan Herman was in the water his small, slender form was borne away among the wave tops. The boat drifted there among the waves for they had lost sight of the boy. But the residents along the shore had been keeping an eye on the doctor's boat because, it was said, one never knew what was going to happen when Sophus Lie was visiting. And when people saw that the doctor's son had been thrown overboard, they set out with their own boats, located the lad, pulled him blue and freezing from the sea, and wrapped him in an overcoat. But when Sophus Lie, whose boat had drifted off several hundred metres, came back alongside, a furore without equal broke out. Sophus wanted to have his nephew back aboard and to get him dressed, but the rescue team refused, and instead, wanted to have the boy's clothes passed over to them. Sophus Lie was said to have cursed them up and down, taunted them and threatened to smash in their skulls if the boy did not come over to his boat. In the end, however, the clothes were passed over, and the whole town of Tvedestrand greeted them when they came ashore. Later it was said that an expression circulated in the district, with which mothers used to threaten their children: "You behave now, or Sophus Lie is going to get you!"

The public service examination that Sophus Lie had passed was called the *science teacher examination*. When the curriculum was established in 1851, the view and plans were that the candidates, at least the majority of them, would find positions in the school system and thereby help alleviate the lack of decent teaching in the sciences – something from which Latin schools across the country suffered. Indeed, many did follow this programme. For example, Sophus Lie's big brother, Fredrik Gill Lie, following the science teacher examination, first became a teacher at Nissen's Latin and Science School, and then science teacher at the Learned School of Drammen in 1866, before, six years later, becoming a head teacher at the Kristiansand Cathedral School. But Sophus Lie had little desire to enter the school system, even though he liked to teach, and all through his student days, had been superbly successful as a private tutor. As a qualified candidate in 1866, he had been offered a teaching position at Nissen's School to replace his big brother, but he did not accept.

Instead, it appears that during the autumn of 1866 Sophus Lie expanded his activity as private tutor, mainly for students facing the university's comprehensive set of preparatory exams, called the secondary examination, and for students at the military college. This gave him some income to live on; besides, he lived cheaply in a garret at Mrs. Motzfeldt's, at Grottebakken Nr. 1 (across from Grotten, the venerable artists' residence that faces what today is Oslo's Royal Palace Park) where he gradually came to be counted as almost a member of the family. Perhaps he turned down the offer from Nissen's School because he had applied for an assistant's position at the astronomical observatory in Chrisitiania. His favourite subject at this time seems to have been astronomy, for, from the lending records of the University Library one finds he borrowed many textbooks and works on that subject. That autumn he also began to work as a teaching assistant at the observatory. Carl Fredrik Fearnley was Professor of Astronomy, and it was decided that a new position was to be announced after Fearnley's former assistant was appointed director of the newly-established Meteorological Institute in Christiania. But Professor Fearnley did not want to have Sophus Lie as his assistant. This was a new blow to the young Lie, who was still perplexed as to what his calling in life was to be. It was said that Fearnley and Lie had completely different temperaments, that they did not get along with one another – and it was said that the cautious and careful Fearnley had been outraged when Lie, in order to keep warm on cold days, practiced his gymnastic leaps over the instruments. Some who were party to the knowledge recounted the time that Fearnley had, wittingly or unwittingly, locked Lie into the observatory, and that the boisterous graduate was said to have leaped out of an upstairs window.

But the disappointment that was probably occasioned by his dismissal from the observatory did not cause him to stop his studies in astronomy. Eager to bring his astronomical insights to wider circles during the spring of 1867, he requested permission from the Collegium to use university premises for "popular Lectures". When this was rejected, he approached the Student Society, and received permission, for free, to hold astronomical lectures in the Society's small hall, "to which any Student would have free Admission". The only stipulation was that the lectures must be finished before seven thirty in the evening. And it was here that Science Candidate Sophus Lie now held a series of lectures for, on average, fifty students, and it was said he awoke people's interest with "the unusually lively, eccentric, and almost grotesque means by which he illustrated his lectures".

If one can judge from the lending records of the University Library, during this autumn, little by little and in a fresh way, Sophus Lie was also turning toward mathematics. In any case, he borrowed Euclid's *Elements*, and other studies on the foundations of geometry, and a little later he borrowed Newton's *Arithmetica*, and books on the philosophy of science. He continued his lectures on astronomy at the Student Society through the autumn; he also arranged courses in trigonometry and stereometry for students who were facing the secondary examination, and in this connection he also wrote a little textbook on trigonometry. Among those who sought out Sophus Lie in order to achieve better examination results was Alexander Kielland, who was to become one of Norway's leading nineteenth century writers. Lie was later to have commented that judging from the poor quality of Kielland's mathematical abilities, one would not have suspected he would become famous, even as a writer of literary prose.

All this lecturing and teaching activity, and the fact that he was continuing his mathematical studies after completing the science teacher examination, led Sophus Lie to believe he was now sufficiently qualified to apply to the Academic Collegium for a lecturer's stipend. However, the faculty felt that he had "provided too little information pertaining to his studies", and could not recommend acceptance of his application. A few days after the refusal, right before Christmas 1867, Sophus Lie once more petitioned the Collegium for the use of the University's auditorium for his astronomical lectures, and mentioned as well that in the coming semester he would be publishing a textbook in astronomy. The Collegium refused once again. Sophus Lie never came to write such a textbook, but he continued his lectures on astronomical issues, as he had before, on the Student Society's premises. And it seems that from New Year's 1868, and thereafter, he had very seriously begun to borrow and read a great amount of mathematical literature. He worked with issues from Euclid to Abel, from the concepts of geometry and real numbers, all the way to the subject of elliptic functions.

The theory of elliptic functions was a popular mathematical theme at the time, but the shift of direction in Sophus Lie's mathematical interest may also have been linked to the fact that during the summer of 1868 a large Scandinavian meeting of the natural sciences was organised in Christiania. Sophus Lie seemed to have wanted to be *à jour* with these developments, and he probably viewed this meeting of the sciences as an opportunity to distinguish himself. He participated on the theoretical side of the conference, but his views as yet were without any basis in research. Many years later Sophus Lie wrote down his recollections from this period: "To get involved in original scientific work never occurred to me. Above all, I was thinking of improving our mathematical pedagogy. I was preoccupied considerably with this."

The Tenth Meeting of the Scandinavian Natural Sciences, with 368 participants, was held in Christiania in July 1868, and it probably constituted a turning point in the life of Sophus Lie. He himself made no contributions to the proceedings, but he listened to a number of lectures and seems to have been particularly inspired by his encounter with the Danish mathematician, Hieronymus Zeuthen. Zeuthen was three years older than Lie; he had studied for two years in Paris under one of the period's great teachers of geometry, Michel Chasles, and he had received his doctorate upon the submission of a treatise on "The Science of Systems of Conic Sections". At the meetings in Christiania, Zeuthen lectured "On a new Space Coordinate System", and he referred to works of Chasles and the German mathematicians, August Ferdinand Möbius and Julius Plücker. It seems that the approaching season, the autumn of 1868, became one long continuing period of work for Sophus Lie, with the frequent and extensive borrowing of books from the library. In addition to Chasles, Möbius and Plücker, he discovered the Frenchman, Poncelet, the Englishman, Hamilton, and the Italian, Cremona, as well as others who had made important contributions to algebraic and analytic geometry.

In the course of that fall, he ploughed through many volumes of the leading mathematical journals from Paris and Berlin, and in Realistforeningen (The Science Students Association) he gave several lectures during the spring of 1869, about what he called his "Theory of the Imaginaries", about how information on real geometric objects could be transferred to his "imaginary objects" – that is, how to move properties back and forth between the one real material sphere and the corresponding imaginary one. For example, he asked, in what ways do the properties of a circle explain the properties of a hyperboloid? Sophus Lie now managed to illustrate what mathematics called imaginary entities. He transcended the boundaries of reality that limited conceptions to a space extending in only three dimensions. He operated in a four-dimensional space.

Sophus Lie wrote a paper on his discovery. The paper was four pages long and published at his personal expense. At the request of Professor Ole Jacob Broch, the work was translated into German and published with money that Lie borrowed from his friend, Ernst Motzfeldt. The work was published the same autumn under the title "Repräsentation der Imaginären der Plangeometrie" in *Crelle's Journal* in Berlin. A somewhat expanded version was published in the Proceedings of the Christiania Academy of Sciences. On this basis he applied to the Collegium for a travel grant, and on this occasion the Collegium said yes to Sophus Lie. He received his travel grant.

Setting course for Berlin, he left Christiania in September 1869. He was away for a little more than a year. In the course of this period he established an acquaintance with several influential mathematicians and he came to have great faith in his own powers and abilities. His meeting in Berlin with the twenty-year-old Felix Klein was particularly important. Klein had been a student of Plücker, and was the publisher of the master's final work, a work that Lie had read in Christiania and had been inspired by. A close friendship grew between Lie and Klein and they had a common approach to geometric problem-solving, something that would become highly significant for them both. Apart from this, Sophus Lie met the prominent mathematicians, Leopold Kronecker and Karl Weierstrass in Berlin, and in the midst of lectures by Professor Ernst E. Kummer, Lie created quite an impression by going up to the blackboard and presenting complete solutions to problems the professor was working on.

At the new year, 1870, Lie travelled to Paris via Göttingen where he visited Alfred Clebsch, one of the great specialists in the relationship between algebra and geometry, between Abelian integrals and algebraic curve theory. In Paris, Lie made contact with prominent mathematicians, particularly Gaston Darboux and Camille Jordan, and a few months later Klein also joined him in Paris, staying at the same hotel. A stream of new mathematical ideas seem to have come into the light of day in this inspiring ambiance.

But then war broke out between France and Germany in July, and the German Klein had quickly to return to Berlin. Whereupon Lie, who now found it less useful to remain in Paris, wanted very much to visit the Italian, Luigi Cremona, and he planned to journey on foot from Paris to Milan. He had not got beyond Fountainbleau however, before he was stopped, suspected of being a German spy, and imprisoned. It was said that Lie talked aloud to himself as he walked, and everyone could hear that he was not French. Other versions maintained that Lie, on top of everything, used to sketch the landscape in his notebook, and he was said to have been sitting and drawing French defense works when he was apprehended by the police. Moreover, the mathematical symbols in his notebook were taken to be secret cypher script and suspicious codes. Lie protested his innocence, showed his Norwegian passport, explained that he was a mathematician, and tried to explain his mathematical symbols and expressions as well and as simply as he could – all to no avail. When Lie asked one of his guards what they did with prisoners, he was said to have replied, "We usually shoot them at six in the morning."

Lie sat in prison for a month, and when the item – "Norwegian Scientist Jailed as German Spy" – appeared in Norwegian newspapers, it made his name widely known in Norway.

Later he himself recounted that his stay in Fountainbleau gave him the best possible peace in which to work out an idea that had occurred to him in Paris, an idea that a year later became the subject of his doctoral dissertation. Working in four-space, he had already established a relationship between such dissimilar geometric objects as straight lines and spheres, and now he was in a position to demonstrate the relationship. Apart from this, he recalled, there were many dark prison hours spent reading novels by Sir Walter Scott, in a French translation.

Lie was released from prison when his mathematical friend Darboux came from Paris with a letter from the Ministry of the Interior, the powerful Gambetta. Once released, Lie took the train, the quickest means of transport, through Switzerland to Italy. On the way home again he met Klein in Berlin, and the two of them agreed to publish a joint work in the Berlin Academy's *Monatsberichte*.

When Sophus Lie returned to Christiania before Christmas of 1870 he was known, and his name was "on everybody's lips", as the man of science who had resided in a French prison, mistaken for a German spy. And Lie himself willingly recounted his experiences. From this time on in public opinion there was always some anecdote associated with Sophus Lie. And this spurious "fame", together with stories of the fabled hiker and the mistaken spy, caused him later to become more cautious in talking about himself, about things that in the popular consciousness might denigrate his reputation as a mathematician.

One of those that Lie met for the first time during this period was Elling Holst. Holst quickly became Lie's pupil, and later, a university teacher as well. Holst would also come to write several articles about Sophus Lie. About their first meeting, Holst wrote: "Right from the first moment he made a deep impression on me. Tall, erect and slim, and of quite an athletic build, blonde, with an open, friendly face: obliging and courteous, but on closer inspection, expressing quite a manly determination of will, warm and engaging, which manifested itself in his scientific enthusiasm or his personal sympathy."

Now, in January 1871, as a visible result of his "imaginary theory", Lie became a research fellow in mathematics. It was said that his teaching and lecturing "were uncommonly animated and provoking", that he had few or no notes with him, and that the lectures consequently, according to Holst, became "a series of inspiring improvisations, filled with the brilliant ingenuity of the moment. He had a masterful understanding of how to present a great, ground-breaking idea."

It was a busy time. From New Year's 1871, Lie was also teaching at Nissen's School; he was chairman of Realistforeningen, the Science Students Association, and was more active in the Student Society as well, being rather frequently present in "The Green Room". The social atmosphere and the quality of discussion in the whole milieu had now changed character, becoming more radical and free, and the honour for this was largely ascribed to the Norwegian poet, Bjørnstjerne Bjørnson, who for a short time had been leader of the Student Society. Apart from his social activity, Lie was now writing his doctorate, and defending it (in June 1871), with the thesis "Over en Classe geometriske Transformationer" (Concerning a Class of Geometric Transformations). Here he presented his latest insights, and for his listeners it was a totally unknown reasoning expressed in a radically new language. But since a doctoral defense was a rarity, many had gathered to listen. The experience was difficult for his audience; and worse for the three professors, C. A. Bjerknes, C. M. Gullberg, and E. B. Münster, who were appointed to evaluate the content of Lie's work. The latter two did not make any contribution to Lie's subject matter; Münster sat in total silence, and Bjerknes' classical mathematics was far from the modernity of Lie's. Nonetheless, Lie was created a doctor with distinction. Elling Holst commented on this event: "Not one mother's son had understood a single word of it!"

Sophus Lie's doctoral work resonated out in Europe. Darboux, a trend-setting mathematician in Paris, characterised the work as "one of the most handsome discoveries in modern geometry." As conditions in Norway were not the best for Lie's scientific work, he submitted an application when a professorship became vacant at the University of Lund during the autumn of 1871. It was maintained in radical circles that this would be a defeat if now the country's great mathematical talent had to leave the fatherland and move to Sweden. This was protested from many circles, not least, "The Green Room" and the Science Students Association. Thus, in the new year, 1872, a proposal was put before Parliament to establish an extraordinary professorship for Sophus Lie. Testaments concerning Lie's work had been solicited from European mathematicians, and several Members of Parliament reminded their audience that the nation of Norway must not now make the same mistake that, in its time, had been committed against Niels Henrik Abel, the country's earlier mathematical genius, who never received a permanent appointment in his fatherland.

When the matter was to come before Parliament, Sophus Lie wanted very much to be in the parliamentary gallery to hear the debate. But at the very moment that the matter was being dealt with from the rostrum, he was said to have been pushing and shoving his way forward, to the detriment of others, and had quite certainly been grabbed by the neck and tossed out by one of the constables. But down on the floor of the parliament things went well. Following a rather intense exchange of words, where the fledgling political "right" and "left" opposed one another, Parliament voted by 85 to 16 to award an annual salary of 800 speciedaler[5]. Sophus Lie had become a professor, barely thirty years of age.

The same autumn he had the opportunity to travel to Germany again. In Erlangen he met his friend Felix Klein, who was now in the process of working out what would later be his so famous Erlangen Programme, a milestone in the develop-

[5] When the speciedaler was replaced by the crown (krone) in 1874 it had the value of four kroner.

ment of geometry. The two friends conducted intense mathematical discussions, and this exchange of ideas in relation to the Erlangen Programme would later come to create a deep split between them.

Following this trip and his safe return to Christiania, Sophus became engaged to the eighteen year-old Anna Sophie Birch. How the two young people met for the first time is not known for sure, but it probably occurred at the Motzfeldt family home. Indeed, Anna Sophie Birch was Ernst Motzfeldt's cousin. In any case, the proposal itself was conducted with the utmost care. In the form of a letter Lie formulated his proposal, and he received her assent, also in writing, on Christmas Eve 1872. A short time later, he told her that his friends had long since lost all hope of him ever becoming engaged to marry, and he confessed that for many years he had "never had any Sense of young Ladies". Of course, as an eighteen- or twenty-year-old he had been "as youthful and lusty a Man as any", but since then, as it were, he had lost all interest in "the Social Life of Balls and Divertissements", and he continued, "Thanks to You, my dearest Friend, all this has changed. And it is my constant Hope that You shall see there, in my strong nature, that a not inconsiderable Fund of youthful Desire, of youthful Power has lodged itself. Dear Anna, You get to take care of me, and correct me when You think I am beginning to become old and boring, and so You see, it will work out."

Anna Sophie Birch came from Risør where her father had been a customs officer. On her mother's side she was related to Niels Henrik Abel. Her mother's paternal grandfather was actually Abel's mother's father, the merchant and shipowner, Saxild Simonsen of Risør. But when she was only three years old, Anna Sophie Birch had lost her mother, and a few years before she now became engaged, her father had been removed and awarded a pension due to mental illness. There rapidly developed an extensive exchange of letters between Anna and Sophus, and this persisted through the years during periods when he was away from home – travelling abroad or hiking in the alpine regions. *À propos* of this, a short time after their betrothal, he commented upon his passion for long hiking tours. He wrote: "People think I have strange Taste: but it strengthens the Muscles and Nerves, and in addition I feel so well roaming through the Country like a Vagabond. On those occasions I tend always to be deep in my own Thoughts and building Castles in the Air for the Future."

There are various accounts that indicate Sophus Lie did look like a vagabond, and worse. According to rumour, he often walked in white linen garments, and so as not to dirty them he rolled his trouser legs up to his thighs, and his shirtsleeves to his shoulders. But nevertheless his clothes were frequently dirty and tattered. It is said that on one hiking tour in Trondheim's Trøndelag County, he was even taken for a dangerous criminal. There had been a murder in the vicinity, and when Sophus Lie was observed crossing the landscape at a rapid pace, he was taken for the murderer. The police sergeant was summoned with all haste. He jumped quickly into his cariole and drove off as fast as he could after this terrible fiend. But the story ends with the sergeant never managing to catch up to the speedy wanderer.

Much of the exchange of letters between Anna and Sophus has been preserved, and it obviously gives many insights about their life together. In August 1874, they married. They had two daughters and a son, the first daughter was born in 1877, and their last child, in 1884.

In its external features, the life of Sophus Lie can now be viewed as falling roughly into three parts; first, his period as professor in Christiania, marked by hard work without a particularly exciting milieu around him, up to the spring of 1886; the second, his time as professor in Leipzig as a leading mathematician with enormous production, students and eager collaborators, until the autumn of 1898; and third, his last, very short year in Norway, his fatal illness and world fame – a year which ended with his death in February 1899.

The first period, as professor in Christiania, or *Kristiania*, as it was now commonly written (with the shift from Danish to Norwegian orthography), was also a period when scientific circles were fighting over new ideas and power positions. There was talk of "the modern breakthrough" in literature and the cultural life; there was a growing realism and the will to submit "problems to debate" that would result in "the emancipation of the individual and the liberation of thought." In Norway these trends led to a golden age of literature with writers such as Bjørnson, Ibsen, Kielland, Jonas Lie and many more. (There were, by the way, many who wrongly thought that Sophus Lie and Jonas Lie[6] were related.) The new ways of thinking also found a trenchant organ in *Nyt norsk Tidsskrift* (The New Norwegian Journal) and later in *Nyt Tidsskrift* (New Journal). At the centre of the editorial board stood Ernst Sars and Olaf Skavlan, both members of "The Green Room", and both, following tough struggles, appointed extraordinary professors by Parliament. Sophus Lie was the first to be appointed by Parliament in this way, but "Parliamentary Professor" gradually became a term of ridicule in conservative circles, as such appointments were seen as the popularly elected Parliament taking over a jurisdiction traditionally exercised by Government. A politicisation spread among large sections of the population, the powers of liberalism were on the march; new times were emerging. Sophus Lie very definitely stood on the anti-Swedish[7] and radical wing in the acute political debates that raged.

A great change was also taking place in people's everyday life. There was also a total overhaul of the country's currency and its system of weights and measures. In 1875 the *speciedaler* or *daler* was replaced with the *krone* and *øre*, and *meter, kilo,* and *litre*, with all their decimals, became new unaccustomed words replacing the well-known concepts like *foot, alen* (0.627m.), and *fathom, barrel, cask, jug, pot, Bismer pound* (12 English pounds), *mark* (234 grams), and *lodd* (15.57 grams), etc.

[6] Jonas Lie (1833–1908) was one of Norway's "Golden Age" writers. Among other things, he wrote about the far north, life at sea, and the struggle for survival in the new, urban industrial way of life.

[7] Between Nov. 4, 1814, and June 7, 1905, there existed a political union between Sweden and Norway in which Norway was under the suzerainty of Sweden and its head of state, the king.

In the natural sciences, Darwinism and positivism were on the agenda, and the conflicts between old and new views were in the process of bursting the learned world asunder. Sophus Lie took part in this situation by establishing a new periodical, *Archiv for Mathematik og Naturvidenskab* (Archive for Mathematics and the Natural Sciences), in obvious opposition to the already well-established journal of the natural sciences, *Nyt Magazin for Naturvidenskaberne*. Sophus Lie was on the editorial board of this periodical until the end of his days, and he published many articles in *Archiv for Mathematik og Naturvidenskab*, often first drafts of what, in later versions, would appear in better known foreign publications.

Lie's work, fundamentally new in form and content, initiated a conceptual work that was path-breaking. Beginning with a geometric study of differential equations, Lie developed a symmetry concept of such differential equations. This concept was capable of determining when and how they were capable of solution. Concepts that emerged from this work were *transformations* and *groups of transformations, infinitesimal transformations*, and *continuous groups*. Today, *Lie Groups* and *Lie Algebra* are indispensable tools in mathematical research and in studies of the more intricate structures of nature.

During the period that he was Parliamentary Professor in Christiania, between 1872 and 1886, Sophus Lie travelled abroad twice. In the autumn of 1874 he was in Paris (on his honeymoon tour), and he met his friend Felix Klein in Düsseldorf, together with another prominent German mathematician, Adolph Mayer, who was a rich and prominent man at the University of Leipzig. During the autumn of 1882, Lie spend two months in Paris, and French mathematicians like Henri Poincaré, Emile Picard and Georges Halphèn were able to acquaint themselves with Lie's theories and his work, particularly his integration methods of differential equations. Lie also wove himself personal contacts with these French mathematicians – his fame grew and almost no foreign mathematician was followed with greater attention in Paris than Sophus Lie.

Already by 1876, as one of twelve foreign members, Lie had been made honorary member of the Mathematical Society in London, which had been the scientific home ground of Issac Newton. He was a corresponding member of the academy of sciences in Göttingen, and the academy of sciences in Christiania, but it was still, to use Lie's own words "lonesome", "frightfully lonesome here in Christiania where there are no human beings who understand my work and interests." He was restless and impatient – some called it nervousness – in any case it was due to an expressed conviction that he would "accomplish ten times as much" if his work and ideas were known to a still greater extent, were used and understood in the correct manner: "I am certain, absolutely certain, that in my case these theories will be recognized as fundamental at some point in the future."

Klein and Mayer in Germany tried to accommodate Lie. In the autumn of 1884 they sent one of their promising mathematicians, Friedrich Engel, to Christiania to help, and to assist Lie with the working out and writing up of the new ideas and theories. The plan was to work out a fundamental work, a systematic theoretical account, and to this end, they began "mit voller Musik" [with full music] as Lie expressed it. They sat down together twice a day. Lie ran through his way of thinking, orally, for each of the chapters and he supplied Engel with what Engel later characterized as "a short sketch, that is to say, a skeleton that I was to supply with flesh and blood". Engel stayed in Christiania for nine months and about those days at Lie's home he later wrote: "I consider them to have been the happiest times of my whole life."

This comprehensive work that they initiated would eventually result in the three-volume *Theorie der Transformationsgruppen*, Lie's major work – in all, 2000 pages long – published in the years 1888, 1890 and 1893.

Then, in 1886, an offer came for a professorship in a rich and pulsating centre, a milieu of science and scholarship. He moved with his wife and children to Leipzig, and remained in that city for twelve years.

In the course of the annual festive dinner for all the new professors, Lie immediately drew attention. When, in a state of ardour and cordiality during his welcome toast, the rector of the university came to say that he could well understand that Lie wanted to be away from Norway where the peasant group among the Members of Parliament had treated the noble King Oscar so abominably, Lie then stood up, protested vociferously, and left the hall, to the consternation of all.

The Lie family were otherwise well-received in all circles in Leipzig, and Mrs. Anna and the children came to enjoy themselves so well, at school and among friends, that in the end it was almost difficult to return to Norway. Lie tried to give the children everything that would serve their growth and development; they should be healthy in both body and soul. Every Sunday the family had to go out on a long expedition on foot. In the wintertime they would go skating, and at the end of the 1880s he had skis sent from Norway. Skiing[8] was something completely new to Leipzig, and when they went skiing in the parks, people pointed at them and shouted, "Seht da sind Eskimos!" ["Look at the Eskimos!"] The children were anything but happy to be called Eskimos, but nonetheless they remembered their father as always considerate and touching in relation to them and their mother. After dinner he always had a coffee hour with his wife, and this the children were not allowed to disturb. He would then drink a cup of coffee and Anna would dip a sugar lump in it.

For the first while after Sophus Lie settled down in Leipzig, the number of mathematical students at the university fell drastically. The reason was that in previous years there had been an unusually high number studying mathematics – such that the demand in the German market for science teachers in schools and general teaching was well on the way to being met. At the outset, Lie was hurt by the fact that his lectures on common mathematical questions attracted smaller audiences than expected, but the special lectures in which he dealt with his own theories went better, and here he soon had an extremely gifted student in Georg Scheffers, who

[8] Snorre's sagas mention transport by means of cross-country skis almost a millennium ago in the Nordic regions. Skis have long been part of winter transport in Scandinavia, but modern ski touring only became a popular sport, and particularly across other parts of Europe, toward the end of the nineteenth century.

later assisted Lie in the preparation for publication of three volumes of lectures. As well, from all over Europe and North America promising young students came to sit in front of Lie's lectern. Sophus Lie was now a central character in international mathematics.

But the teaching and guiding of students gradually took more time than he liked, and he became tired of always having to be at the disposal of examination candidates, especially since they were not always of an equal competence. Nor did his personal plans for the stay on "the Leipzig plains" develop quite as he might have hoped. He had problems speaking good German, and the German university system had its partisan divisions and personal rivalries that he was not immediately able to cope with; so too, were the forms of social comportment and etiquette new to him and different from the Norwegian milieu. He began to miss friends in Christiania, and above all, he missed Norwegian nature. From abroad, his homeland's natural landscape seemed to him to be doubly radiant.

In one of his memoirs, the writer Theodor Caspari tells of a mountain journey he made with Sophus Lie in Norway at this time. It was so late in the autumn that those touring on foot had left the mountains, and the haymaking had been finished at the alpine camps. (These "dairy" encampments in the high altitude summer pastures were, and are known as *sœters.*) On that occasion the "message stick"[9] had gone its rounds, inviting all the sæter folk to the autumn dance at Sikkilsdalgård, where forty or fifty hay-cutters from all the surrounding sæters stood against the walls or sat on the edge of beds in the farm's large living rooms. The fiddler seized his fiddle and began to play his music for the customary roundel dances, but the atmosphere was grudging and sluggish. And thus, after about half an hour, when there were still only one or two couples on the dance floor, Sophus Lie sprang to his feet. "No," he said, "this won't do, Caspari! Come along now, and we'll show these country folk how it should be danced!" Whereupon he grabbed the nearest sœter girl and began to dance as best he could. Caspari did not like the idea of playing the leader, he admitted, but he felt nevertheless that he had to follow Lie's example, even though he knew that this inertia was indeed conventional good form, and was normal at the beginning of such dance parties, and in consequence of this, it was useless to try to force the tempo. It went as Caspari expected – there was much grinning and whispering along the walls and edges of the beds, without anyone following the example of these funny tramps from the city. Then Lie, who by this time was well known both as professor and mountain hiker, suddenly stopped short and rushed over to the entrance way where he met the farm's proprietor, Sjur Sande, who was almost an institution for the first generation of wayfarers in the Jotunheimen Mountains. Lie burst out, "There's nothin' more t'do, Sjur! I'm stumped, and so I'm going to bed. There's going to be no dance here at Sikkilsdalgård tonight!" The stalwart old Sjur was said to have had facial ticks, and with almost a hiccup of laughter, he replied, "You, Professor,

[9] The *budstikk* was traditionally sent around from man to man, and farm to farm, to summon people for military, political or social gatherings. The custom continued to the end of the 1800s.

better come back in twelve hours, and then we'll talk about it!" And at eight or nine the next morning when Lie and Caspari stuck their heads into the dance hall, the walls were shaking with the rhythmic beat of the dancers' footsteps, as boys and girls threw themselves back and forth across the floor.

Apart from this, Lie seems to have been an avid dancer. In Leipzig he danced for the sake of exercise, and he gladdened the heart of many a professorial wife with his social skills at parties. But the problems of cooperating with colleagues began to emerge, and these would prove to be difficult to surmount. Much of the source of conflict seems to have been due, on one hand, to Lie's colossal will to work away at, and strive for insights into fields, whose vast scope no others could see; and on the other hand, to the anxiety of being misunderstood and misused. The disagreements made him sleepless and depressed, and the experiences had a dissolving effect on his ego. Taciturn and apathetic, he was admitted to a psychiatric clinic near Hanover in November 1889. Although Lie made few allusions to his "nervous breakdown", he did report that he had been "seized by a boundless despair".

He was a patient at this psychiatric clinic for over seven months, where he was treated with opium and other medicines to allow him to sleep. However, he himself maintained that in the end it was through hiking tours that he quite literally walked the illness out of his system. Be that as it may, when he returned home, he was very insecure about the future and his own health. But he took up his work with mathematics again, and at New Year 1891, he resumed his lecturing. Later on in life, he would never speak about the reasons for his depression. In the autumn of 1891 he wrote to Elling Holst: "Things are going badly as before. Of course I am sleeping and have absolutely no bodily inconveniences but my spirit has been broken. Pessimism and dark despair reign over my soul."

In the autumn of 1892 the depression seems to have released its grip for the first time. Lie was then able to take delight over the interest that French mathematicians were taking in his work. He had become a corresponding member of The Academy of the Sciences of France, and with full honours he was twice invited to Paris. In Paris, they frequently gathered – when not engaged in more august meetings and lectures – at Café de la Source on Boulevard St. Michel, and it was there that Lie often covered the marble-topped table with symbol after symbol to illustrate the development of his ideas.

Lie was a central figure in the work of making contacts and connections between the German and French scientific milieus. But in Leipzig, and around Germany, his colleagues and friends felt that a great change had occurred in Lie's attitude and behaviour since his illness. It was said that Lie had become suspicious and almost paranoid in his accusations and allegations that others were furnishing themselves with – and stealing – his ideas; then publishing them as their own. The boundary lines between his ideas being used (and credited to him), and their being stolen, (or used without attribution), then, as today, were not always straightforward.

Lie's stay in Germany became more difficult after this; mistrust and problems of collaboration seeped over everything. The worst was the break with Felix Klein. Indeed, Lie's posthumous reputation also came to suffer when, in his memorial to his former friend, Klein characterized him in terms that followed the customary romantic wedding of "genius to madness".

But Lie continued his mathematical work, and it was published, disseminated, and appreciated to an increasing degree. He received prizes, and he became a member of the scientific academies of a series of countries. There was no doubt in Norway about Lie's international stature, and throughout the 1890s his discomfort at living abroad also became known and discussed in his homeland. Many Norwegians felt that it was high time that the country's great son be brought back into their midst, and Lie himself expressed the desire to work in the land of his birth. In principle he was only on a leave of absence from his professorship in Kristiania, and he had only to come home whenever he liked. But obligations regarding work, students and family in Leipzig held him back. It was, moreover, an economic question; in Germany he earned substantially more than what could accrue to him from the position in Norway.

Therefore, both personal and national political motives lay behind the large-scale action that now began to mobilise support for getting Norway's most famous man of science back home. Lie's discontent there "on the Leipzig plains" was brought to light by Fridtjof Nansen, whereupon Elling Holst and Bjørnstjerne Bjørnson, among others, worked effectively on the campaign to repatriate Lie. Their work was so extremely effective that as early as the summer of 1894, Parliament undertook to change the title of Lie's position to "Professor of Transformation Group Theory", and offer him a salary of 10,000 kroner – almost double that of an ordinary professor at that time.

But despite all the goodwill and the unanimity of desire to get Lie to Norway, it would be almost four years before he returned on a permanent basis. The fact that his return home was postponed for so long created certain problems in relation to friends and colleagues in Kristiania. As often as possible, and for as long as it was possible, he meanwhile spent holiday time in Norway, and as a rule he had his family with him. And in addition to this, he participated, to the best of his ability, in ongoing debates in Norwegian society. He wrote newspaper articles about mathematical instruction in the Norwegian school system; he engaged himself in university matters; and he participated in the preparations for celebrating the one-hundredth anniversary of the birth of Niels Henrik Abel.

In September, 1898, Sophus Lie finally returned to Norway for good. Everyone could see that he was ill. He suffered from pernicious anæmia, a disorder of the blood that was incurable at the time. Normally this illness attacks hard and with brutal force, but it can also lie dormant, and perhaps a creeping anæmia had been corroding his powers over a number of years. Over the course of the autumn Lie's condition rapidly deteriorated. He only managed to get through his lectures with great effort – and at the last, from his sickbed. Then, on the 18th of February 1899, he died in his sleep.

Laudatory words from the Prime Minister, professors and foreign experts expanded and embellished the loss to the nation and to science. Sophus Lie passed away before the nation could properly celebrate and demonstrate that he was one of their great men. He died before he could gather Norwegian mathematicians around his ideas such that they might follow in his tracks – and he took with him to the grave a corpus of ideas and knowledge that only he possessed.

Nevertheless, he had provided an ideological foundation that, to a great degree, came to mark the development of modern mathematics and mathematical model-building.

Lie's method of solving differential equations was important for the calculations behind Einstein's Theory of General Relativity. Lie Theory is also an indispensable tool in the fundamental formulation of our laws of nature and in the understanding of the innermost structures of the atom.

The great French mathematician, Élie Cartan, wrote in 1948: "The culturally interested public are familiar with the great writers of novels, the great dramatists and the great musicians of Norway; but science also owes a debt to this nation, in particular for two extremely great mathematicians: the first, Niels Henrik Abel, born in 1802 and died at the age of twenty-six, has bequeathed us a body of work of an admirable depth; the second is Sophus Lie."

PART II

Family Background and Upbringing

The old vicarage at Eid where Sophus Lie was born in 1842

The Family Tree

On the paternal side, Sophus Lie's family tree stems from a certain Peder Lauridtzen of Strømsø in what is today the city of Drammen, west of Oslo. The sixty year-old Peder Lauridtzen is described in the census for 1665 as a beach dweller; in other words, he was far from being a member of the merchant bourgeoisie, and one might surmise, was a type of crofter who fished the fjord with a small boat.

His two sons, aged fourteen and sixteen are also mentioned in the census. The older of the two, Lars Pedersen, appears again, about thirty years later, in the church records of Bragernes/Drammen, described as a tailor by occupation and a horse handler on the Lie Farm at Åssiden in the Drammen area. This association with the Lie Farm gradually led to the adoption of the Lie name. In several of the tax rolls for Bragernes, Lars Pedersen Lie is entered as *capitaine des armes*, that is, the highest grade of non-commissioned officer. He lived at Øvre Sund, that much-frequented ferrying place on the Drammen River, where the wagon road from Kongsberg reached Bragernes. His second marriage, to a certain Marthe Christensdatter, produced three daughters and two sons.

When Lars Pedersen Lie died in 1738 his youngest son, Andreas, was seventeen years old, and this same Andreas Lie was to climb high on the social ladder – rising from modest circumstances, and without even a degree in law, he worked his way up into the civil servant class, and became a bailiff – first for eleven years in the County of Agder, at Råbyggelaget, and later for six years in the township of Rakkestad. On the family tree, Bailiff Andreas Lie figures as the great grandfather of Sophus Lie.

Andreas Lie no doubt owed his rise in social status both to talent and to the dint of hard work, but perhaps as well, to some degree, to the fact that in 1754 he married Sidsel Maria Leerberg. She was the daughter of an innkeeper and merchant in Christiania, Mads Jensen Leerberg and his wife, Sidsel Margrethe Petersen, who in turn was the daughter of a bailiff in the Ringerike District. When this merchant/innkeeper, Mads Leerberg, died two years after his daughter's marriage, it seems that Andreas Lie took over his father-in-law's business activities. In addition to this, Andreas Lie got practical juridical experience by being in service both to the bailiff and the prefect in the county of Oppland. With his appointment to the post of bailiff in Råbyggelaget in 1765, Andreas Lie's "Experience and Capability" was given emphasis, and his certificate spoke of his "Fidelity, Diligence and other Qualities".

Bailiff Andreas Lie and his wife had one daughter and two sons, and the younger son, Lars Lie, who was born in 1770, was to become Sophus Lie's grandfather.

When Lars Lie was twelve years of age, he lost his father, and presumably his mother had passed away even earlier. In any case, it was said that Lars Lie was orphaned at an early age, and he consequently went to sea. However, on one of his first sea voyages he fell from the top mast and lost the sight of one eye. He thereupon left the sea and went into service in the establishment of a juridical public servant in the southern part of Akershus Diocese[10]. At some point in the 1790s he took the public service examination in jurisprudence and obtained the mark of "suitable". He now went into service as a solicitor's clerk for a magistrate in Ringerike, north of Christiania, and later worked for Stipendiary Magistrate Heyerdahl in Høland. For a period while he was in Ringerike he served as acting magistrate, and here he met his wife. Lars Lie married Caspara Fredrikke Gill in 1798, and she gave birth to eleven children, seven of whom reached maturity and one of them subsequently became the father of Sophus Lie – and the *"Gillian"* physical and mental inheritance was often mentioned by later members of the family, and was said to have stamped many of the descendants for several generations.

Caspara Fredrikke Gill was the daughter of the vicar of Høland Parish, "the mighty" Jonas Gill, as he was frequently called. Jonas Gill was said to be extremely strong and robust, with a powerful physique. He was an attractive man, certainly of a rather vehement temperament, but nonetheless hearty and forthright of thought and manner. He was a fifth-generation descendant of a certain Christen from Jutland in Denmark who had been in service to the administrative head of the Kristiansand Bishopric in southern Norway and his descendants bought the Gill Farm. Here at the Gill Farm near Kristiansand, Jonas Gill was born in 1738. He was enrolled in the city's Cathedral School, took his *examen artium*, and later the public service exam in theology. At the age of twenty-seven he journeyed to Greenland as a missionary "where he worked with great Conscientiousness for 7 Years." After this he spent a year in Copenhagen in search of a posting, and in November, 1773, he was appointed parish vicar of Nes in Hallingdal, on the eastern side of the Hardangervidda plateau, between Bergen and Christiania. As soon as he received this posting, he set off on his homeward journey. Then, when, shortly before Christmas, upon disembarking from his ship's passage to Lillesand, up the coast from Kristiansand, he immediately began the journey on foot, homewards towards the Gill farm. However he was tired, and when he was passing the vicarage at Tvedt he decided to ask his fellow cleric if he might borrow a horse. As it turned out, however, the vicar's wife was home alone with her three daughters. She invited the newly appointed and handsome vicar to come in. And while the daughters milled around and waited expectantly, the vicar's wife was said to have asked Gill if he did not need a housekeeper now that he had been appointed to a calling. Three months later, Jonas Gill married one of the vicar's daughters, Rebekka Dorthea Pettersen, before journeying on to Nes in Hallingdal. Nine years later the priest and

[10] In the 1700s and 1800s, Akershus Diocese and Akershus County surrounded Christiania and extended mainly into areas north, east and southeast of the city.

his wife moved to Høland, a parish east of Christiania near the Swedish border, and here, for more than twenty years, Jonas Gill presided with great authority over his parish. By force of will he seems to have imposed order and obedience all around him, be it in the parish, the home or the vicarage. Some considered him stiff-necked and obstinate. According to hearsay, he usually had many legal proceedings on the go, claiming that many things were missing from the vicarage when he took it over. It is quite certain as well that his wife frequently intervened when Gill became too heated and vehement with one or other of their fourteen children. There is documentary evidence about how Vicar Gill acted when, one day in 1804, he found out that "a Rabble" of Hans Nielsen Hauge's followers[11] were holding a meeting in his parish. Gill himself marched in and interrupted the speaker who, according to the vicar, had stood there "with darting Eyes and hypocritical Gestures and a smug and self-important Voice". The powerful Gill threw the speaker out the door, and drove the gathering out into the waiting arms of an officer, who conducted them to the sheriff. The next day, four of them, who were known Haugians – and whom Vicar Gill described as "Hypocrites and Rabble" – were sentenced and sent to prison in Christiania.

Vicar Gill, who died three years after this event, had spent the last six years of his life as a widower. Twelve of the fourteen Gill children survived into maturity – eight daughters and four sons, and in turn eleven of these twelve each had flocks of children. The four sons all became officers, and the daughters were all well-married – and indeed, in 1798 one of them married Lars Lie, who then had a position with the magistrate at Høland.

Lars Lie was a temporarily appointed stipendiary magistrate when he and his wife, Caspara Fredrikke Gill, left Høland for Copenhagen in 1804, probably so that he could obtain a more remunerative position. By then, the couple had three sons to care for, but they took only the eldest son John with them to Copenhagen. The other two boys were left with relatives and "brave Peasants" in Høland. The younger of the two left in Høland was christened Johan Herman. This Johan Herman, usually simply called Johan, would in time become a priest and as well, the father of Sophus Lie. Johan's big brother, John, later became an unusual personality and much talked about doctor in Christiania. He was doctor to the famous writer, Henrik Wergeland during the final year of the poet's life, and since the doctor himself lived to be eighty-two, Sophus Lie in his time frequently partook of his father's brother's hospitable home on Øvre Vollgade in Christiania.

One way or another, whether they were among the "brave Peasants" of Høland or in regal Copenhagen, Johan and John Lie received a certain upbringing and schooling. After three meagre and difficult years in Copenhagen, Lars Lie obtained a magistrate's position in Stjør- og Værdalen, near Trondheim. In 1808 the family moved to Levanger, a town north of Trondheim, and bought the Gilstad farm at

[11] Hans Nielsen Hauge (1771–1824) was an evangelical populist who developed a following in much of rural Norway. He castigated churchmen in the Lutheran state church for being lazy and vain. He lauded "the honourable work of the hands" on the land and advocated peasant engagement in trade and commerce. His ideas affected Norwegian Lutherism even after his death.

Skogn. The magistrate's farm at Skogn, with its outlook over the settlement and the fjord proved to be a good home in which to grow up. Both Sophus Lie's father, Johan, and his father's brother, John, later told about their rich years of childhood and their considerable social intercourse with those neighbours and families who constituted "the cream of society" in the district. Magistrate Lie gradually came to have seven children; in addition, Mrs. Lie gave birth to four who died young. Apart from this, it was said that Mrs. Caspara Gill Lie was a strong, dynamic woman who ran her large house with great zeal. She was kind and helpful to everyone, and they said that whenever the poor came to the farm she willingly gave away her husband's clothes. Magistrate Lie had a manner that commanded respect, and besides, he was somewhat stout and stocky. Of his sons, it was certainly Johan, Sophus Lie's father, who resembled him the most physically. Magistrate Lie was said to have played the violin well, and he played as often as he could. This was an activity that his son, Johan, also took up. In later years, when the family's musical gifts were discussed, it was said that these stemmed from Magistrate Lie – however, this was a gift that Sophus did not seem to have, although his nephew, Sigurd Lie came to achieve fame as a composer. Magistrate Lie was also an avid marksman and there were holes all over the farm's buildings and equipment, from the buckshot and bullets of his guns.

At the tender age of nine, in 1808, the eldest son, John, was enrolled in the Trondheim Cathedral School. In 1816, at the age of thirteen, Johan was enrolled in the same school, and a third brother, Fredrik, was also a disciple (as the pupils were called) at the cathedral school in Trondheim, Norway's cathedral city. For a time, all three of the Lie boys had lodgings at the home of a certain Lieutenant-Colonel Dons. Mrs. Dons' sister was married to a Major Stabell, and it so happened that, beginning in 1820, Johan lived at the home of this Major Stabell for four years. He fell in love with the youngest daughter of the house, Mette Maren, whom he later married, and who subsequently became the mother of Sophus Lie.

Johan was sent to Christiania in 1824 to take the *examen artium* with a letter of attestation from Rector Boye of the Trondheim Cathedral School that Johan Lie possessed more than commonly good abilities, that he was an attentive listener and that his behaviour was, in the best sense, modest and seemly. He subsequently wrote a good examination and was awarded the overall mark of *laud*, which he also received in the preparatory exams – called the secondary examination – a year later. Thereafter he began to study theology. For a period of three years he received free lodgings at the University's main building, Regentsen. As the result of an economic loss, which was said to have been caused either by a great robbery or an act of fraud at the magistrate's farm at Skogn, there were no funds to be had from home. In order to earn money to support himself, Johan gave private lessons, and for a time he also had to take care of a seven-year younger brother, Mads Severin, who was known as "the root puller" due to his physical strength, and who had come to Christiania to study, but he gave up long before facing any public service examination, ending his days as a police sergeant in Levanger.

Magistrate Lars Lie died in 1829; in his last years he had lived as a widower with only his youngest daughter still at home. Born in 1815, this daughter, Edle, moved in with John, her eldest brother, following the death of her father. By that time John was already a practicing doctor in the capital. Edle would later live at the home of her brother Johan, and here, among other things, she would teach little Sophus. In her old age, Aunt Edle used to amuse herself saying that she had been the first to teach Sophus mathematics.

While he was a university student in Christiania, Johan Lie and his violin took part in the Music Society's overture concerts before various theatrical productions, and at the Society's own concerts in the theatre. In 1829 he did well in the public service examination in theology, receiving *cum laud* in both the theoretical and practical portions, before returning to Trondheim. Here he was employed by the hour to teach religion in the city's grammar school; he was also appointed temporary adjunct teacher at the cathedral school. And in order to pay off his debts as quickly as possible, he had an agreement with the vicar of the Church of Our Lady to conduct a number of holy services there. In the spring of 1832, having received good letters of recommendation from his various activities, Johan Lie was appointed to head the Molde Grammar School. But because the school in Molde[12] – the first of its kind between Bergen and Trondheim – had only just been established that year, and most of the funds used to establish it had come from subscriptions; because the school received little support from the state, Lie was appointed only temporary headmaster and senior teacher. Yet the position was sufficiently remunerative such that he could now marry Mette Maren Stabell. The salary was 500 speciedaler and included free accommodation in the school building. The newly married couple moved to Molde at the opening of the school term in the autumn of 1832. Mette Maren Stabell was twenty-five years old. Both her father, Major M. C. Stabell, and her mother, *née* Scharffenberg, had proud genealogies to look back upon. Right back to Councillor Stabell's time in the 1600s, generations of the Stabell family had belonged to Trondheim society and asserted themselves both in schooling and in commerce, while the branches of the Scharffenberg family tree bowed down heavily under the weight of its many gold-braided officers.

Johan Lie seems to have been a particularly brilliant and popular teacher at the Molde Grammar School. The number of pupils increased sharply under his direction, and the school came to employ three teachers. The teaching was directed at two types of pupil: "the studious" who had a view to going further with their schooling, and were taught in Greek and Latin, and "the not-studious", who were instructed in English and commercial subjects.

But when Johan Lie applied for a royal appointment as senior teacher and headmaster in Molde, it emerged that the Department wanted preferably to have a philologist as headmaster – an attempt by the state to rejuvenate the school system and make it more accessible to the general population. This rejection resulted in Johan Lie, for economic reasons, now beginning to apply for an ecclesiastical posting, even though he felt his "greatest calling was to continue as a school teacher".

[12] Molde is a coastal town on the outer Romsdal Fjord, 383 nautical miles northeast of Bergen, and 213 southwest of Trondheim.

Nevertheless, it was his merits as a school man that were given weight when, in 1835, he was appointed vicar of Nordfjordeid (locally known simply as Eid). Beyond this, it was a time when there was a surfeit of theologians. There had been seven applicants for the post at Eid.

In the course of the almost four years that the couple were in Molde, they had one son and two daughters. One of the daughters died in infancy. The two children who reached maturity were consequently Sophus' big brother, Fredrik Gill Lie, who was to become the father of the composer, Sigurd Lie, and the sister, Mathilde, who later became a doctor's wife in Tvedestrand, and for a period had frequent visits from her little brother, Sophus. Mathilde was also the mother of Johan H. L. Vogt, professor of mineralogy and geology, and Ragnar Vogt, professor of psychiatry.

The Priestly Family at Nordfjordeid

Johan Lie was appointed parish vicar in July 1835, but it seems it was not until May of the following year the ecclesiastical couple, together with the infant Mathilde and their three year-old son Fredrik, left Molde and moved to the vicarage at Nordfjordeid.

Eid was a large parish with rich traditions. After Selja, Nordfjordeid was the largest community in Nordfjord that had a church, and the Eid posting, or *calling*, as it was termed, was one of the largest in the bishopric at the time of the Reformation – the call at that time had six church parishes. By the time that the Lie family moved there, the posting had been halved, and in addition to the main church at Eid, included two subsidiary, annex churches – Starheim out along the fjord, and Hornindal located in the interior, on the long deep Lake Hornindal. The salary of the parish priest was 650 speciedaler per annum.

Nordfjordeid was a farming community, with hunting and fishing as important additional means of livelihood; freight transport along the fjord was also a source of daily bread. One feature above others that uniquely distinguished the community, and that was an important element in people's lives, was the Nordfjordeid drill grounds, most frequently called simply *The Square*. The history of this parade ground went right back to the 1600s, to the decades following King Christian IV and his son-in-law, Hannibal Sehested, organizing the peasants into a Norwegian parish army for war against Sweden. Whereas in most communities military exercises were practiced after Sunday service on the local church grounds, in Nordfjordeid an actual drill grounds was constructed, and this was one of the first in the country. The community and the Square gradually became a permanent site for military training for several battalion-level companies. The Square saw a particularly great amount of activity after it was decided in 1828 to abolish the training of recruits on church grounds in the local communities, and that military exercises should have their own permanent drill grounds. Thus it was that in the summertime, soldiers and officers streamed into Nordfjordeid from all over the Fjordane and Sunnmøre region. During this period, with parades of brilliant uniforms on foot and horseback, great "spectacles" were played out in The Square. The local people and the military got along well together. All the men, both officers and soldiers, were lodged in private houses in the community. Often people from the same communities came year after year to stay in the same quarters. The presence of the military also had economic significance, both for those who rented out rooms, and for local trade, as well as for the older women who arranged to wash clothes for the soldiers. The battalion music was also popular, performed as it was in the open air, and played all day long to the pleasure not only of haycutters but also of everyone else in the community.

For the local priest, the proximity of the military meant that not only did he have a greater audience and congregation for periods of the year, but also frequent social visits paid to the vicarage by the officers.

The Lie family seems rapidly to have found its feet in the community. With regard to practical relations around the vicarage and the priest's ecclesiastical work, one finds a certain amount of information in the church inspection records, particularly from W. F. Koren, Dean of Nordfjord, who lived at the vicarage in Selja. In other words, he would come to Nordfjordeid on annual inspection tours. As early as September 1836, Dean Koren paid his first official visit to the parish vicar, Johan Lie. The trustee and assistants of the main church at Eid, together with trustees and assistants from the subsidiary churches showed up together with the dean. The parish priest's official church records and protocols were examined and found to be "kept up with a clear hand" – and that requests, wishes and complaints from the various parties had been duly recorded. As well, the vicarage itself was gone over scrupulously, and it was noted that at the beginning of the vicar's tenure, a number of things were missing that the dean considered of such seriousness that he advised Lie to begin legal proceedings to get his predecessor to pay for "the requisite Reparations".

Johan Lie's predecessor at the vicarage was the Dane, Nikolai Nielsen, who had been one of the fathers of the Norwegian Constitution at Eidsvoll, and after fourteen years at Eid, with his large household, had received the far better post at Borgund in Sunnmøre County. But despite the fact that a number of the buildings at the vicarage were dilapidated and in bad condition, Vicar Lie did not demand any compensation from his predecessor. To the contrary, it seems that the Lies gradually began to stake their honour on working up the vicarage, improving it until it became a model farm. The later history of the Eid community states that "Neither before nor since had the vicarage been what it was in Lie's time." And it was said, the main honour was due to "the strong, tall vicar's wife who, from early morning, zealously and powerfully looked after all the farm matters."

The status of the main church at Eid upon the arrival of Pastor Lie was as follows: the church was in good conditions, but the Bible was missing, as well as the wine chalice and Eucharist paten bowl that were both used by the priest on visits to the sick. This had been pointed out on an earlier inspection tour, and the church verger, Captain Leganger, had promised to make good what was missing.

The church farm was in good order, although some stones had rolled down out of the stonewalls, and they ought to be "replaced at the earliest possible moment". Other than this, it was the church rituals and customs that the dean watched with an eagle eye. He noted that Holy Communion had not been neglected, and that the instruction of the youth had been advanced with a permanent school, and in addition, by employing eight ambulatory schoolteachers; it was also noted that at the end of 1835 the school fund was eighteen speciedaler and eighteen shillings. The

Above:
Eid of Nordfjordeid ca. 1880, seen from the vicarage. This photo was taken by parish vicar and founding member of the Norwegian Parliament, Nikolai Nielsen, Lie's predecessor in this parish posting.

Johan Herman Lie – Sophus' father, who was vicar at Eid from 1836 until 1851.

Top: *Sophus' eldest brother, Fredrik Gill Lie, head teacher, Kristiansand.*

Right: *Sophus' sister, Laura, director of the Eugenia Foundation of Christiania.*

Below: *Sophus' two-years older brother, John Herman, lieutenant-colonel in Bergen, together with his wife, Petra Thaulow Kloumann and their eight children.*

Left: *Cand. Real. Sophus Lie.*

Above right: *Taken at the home of Sophus' sister Mathilde and her husband, Dr. Fredrik Vogt at Tvedestrand, where Sophus was a frequent guest. The photo is from 1873 and shows, from the left: Tordis, Johan (later professor of metallurgy), Herdis, Eleonora, Mathilde (in the window) and Ragnar (later professor of psychiatry).*

Below: *The Lie Family moved to Moss in 1851. Photograph from ca. 1890.*

Photograph of the wedding of Sophus' sister, Dorothea, when on October 21st, 1866, she married Johan Vogt, who was the brother of her older sister (Mathilde)'s husband Fredrik. Fredrik and Johan were sons of the magistrate in Moss, David Vogt.

Back row, from the left:

Laura Lie, Mrs. Bredal (née Raven and married to Lawyer Bredal, later magistrate at Ringsaker), Fredrik Lie with his wife, Amalie (née Nielsen), Johan Vogt (bridegroom, and ship's captain by trade) with his Dorothea (called within the family, simply Thea), Fredrik Vogt, Amanda Lie (née Afzelius, and married to Sophus' cousin, Johannes), Lawyer Bredal.

Middle row, from the right:

John Lie (Sophus' father's brother) with his wife, Fredrikke (née Grønvold), Father Lie, Maria Magdalena Vogt (née Juul), Mariane Sofie Ross (née Stabell, the sister of Father Lie's deceased wife and Sophus' mother), Mathilde Wessel (née Ross, and daughter of the aforementioned Mrs. Ross).

Front row, from the left:

Sophus Lie, Magda Vogt, Albertine Vogt (née Bjerkstrøm, and mother of Magda), Jørgen Herman Vogt, Agnes Vogt (née Smith), Mathilde Vogt (Sophus' sister), Eleonora (daughter of the latter), and David Vogt (student).

dean delivered eighteen copies of an ABC book published by the School Teachers Association in Christiania, to be apportioned out among the children of the poor who attended the permanent school, as well as among those within the ambit of the ambulatory schools of the district. In addition, the amount of poor relief funds was noted, as was the carrying out of the annual smallpox vaccination. According to the reports, the two ancilliary churches were in good condition, but the church verger did not feel he should be responsible to supply either a special chair to the soloist of the choir at Starheim church, a special Bible, nor his collection of sermons for use at Hornindal.

In addition to these affairs of the church, it emerged that the local populace was angry at a merchant in the community, a certain Claus Wiese, who dealt in trade commodities and large fish, both in his shop and warehouse on the shore. What particularly upset people was the fact that Wiese hired labour from neighbouring parishes, usually married people with flocks of children, and people feared that these foreigners, when later there was no work for them, would remain in the community and become a burden on the funds of the local poor relief. There was also the complaint that the publican did not remain within the regulations regarding the sale of spirits. This was a recurring complaint and the dean and vicar were in agreement that it must now be sent to the civil authorities, such that the punishment of the law could be "visited upon the guilty".

With regard to moral conduct in Nordfjordeid, it was reported that inebriation continued to manifest itself "in a shocking Form". In response to the general state of affairs in the parish, a protocol on its condition at the time of arrival of Johan Lie, was drawn up, and signed by Dean Koren, Parish Vicar Lie, Captain Leganger, and six assistants who "had help holding the Pen".

When, nearly two years later, the dean paid another visit to inspect the parish, it was recorded that all the buildings had been expanded, the churchyard was in order, the church records and protocols were in perfect order, the Bible was in place, but the chalice and the Eucharist paten were still missing. The congregation was praised as unassuming and hardworking, and the former "so greatly prevailing Disposition toward Drink" had decreased. Merchant Wiese had promised to contribute to this orderliness. The poor relief was said to be in good condition, although beggars showed up from neighbouring communities from time to time.

Parish Vicar Lie had now, as well, demonstrated his zeal for the schooling of the young. At this point in time there was instruction given in *two* permanent schools and *seven* ambulatory schools. Furthermore, he seems to have satisfied the majority in his role as a civil claims mediator.

During that autumn of 1837 the ecclesiastical couple had their third child, their daughter, Laura, who in adult life would become the highly respected head of the Eugenia Institution[13], and the sibling with whom Sophus would have the most contact.

---

[13] Named after the Swedish-Norwegian Queen Eugénie Bernadine Désirée, the Eugenia Institution was founded in 1823. It took up the task of "raising" and educating the daughters of the poor of Christiania and the whole country, especially in subjects related to the field of home economics.

Otherwise, the year 1837 marked a milestone in the development of representative government in Norway. The law of municipal governance was finally sanctioned. Thus it was that municipal self-government saw the light of day in Eid, and Parish Vicar Johan Lie, elected the community's first mayor, began his tenure on January 1, 1838. Apart from a two-year period (1842 and 1843), Vicar Lie would continue to function as mayor for the rest of his time in Eid, almost twelve years in all. The affairs that he now presided over were not much more extensive that what he already had to deal with: church matters, schooling and poor relief. In the period that Vicar Lie was mayor, the municipality had no budget – bills were passed on for payment as they arose. The records of the municipal meetings report on matters that were currently to be dealt with: they had to do with how an appropriation for a general library could be obtained, and there was the question of who should have a concession to establish a "general merchandise shop" in the community, and if it was desirable to have a district doctor in Nordfjordeid, if indeed there was a need for several midwives, and several persons to deal with vaccinations. The executive committee of the council also strongly expressed its desire that each and everyone should reduce expenditure on weddings, christenings and funerals; it also encouraged more abstinence and moderation. In any case, one thing certainly had to come to an end, namely, the deplorable custom of noise and gaiety in the rooms where corpses lay prior to interment.

Those with adequate funds were the ones who most frequently obtained licences to engage in trade and commerce – as long as they did not sell spirits or come too near to the merchant, Claus Wiese – and the fees for such licences were most often put to use by the community's library. Apart from this, Merchant Claus Wiese was elected deputy mayor of the municipality, and he held this office for more than ten years. Judging by the records of the meetings, it seems that he was a great deal more active in the local council than was Mayor Lie. It also appears that Wiese, who had now opened an inn in the community, was a master at balancing on the fine line that lay between the various laws and regulations. There were complaints that he poured and sold spirits, and on several occasions it became necessary for him to promise to improve, and to mitigate the situation by making economic contributions for the common good.

A rather amusing story was later told about Lie's first period as mayor. When the district magistrate travelled around the communities to instruct the new municipal executives, he came to Eid, where it was said, he found the door to the assembly room standing open, and it was terribly cold. Mayor Lie's explanation in the subsequent meeting was that according to the law the proceedings had by law to be conducted with an open door policy.

Vicar Lie's greatest interest seems decidedly to have been in the field of schooling and teaching. As well as founding permanent and ambulatory schools he encouraged the most talented young people to obtain further education. One of these, Martinus Monsson Bjørlo, went on to the island of Stord, south of Bergen, to the teaching college, the *Seminarium*, and returned to establish evening school for "forward-looking boys", such as those who considered becoming teachers themselves. Martinus Bjørlo became parish clerk, the first graduate of the seminary examination, as well as deputy-mayor and mediator of the civil court, and it was he who took over the post of mayor when Vicar Lie left the community. Vicar Lie had also promoted another person, a young man called Hans Jönsson, to be sent to the Stord Seminarium. To be sure, this Hans himself was eager to go, but his parents refused: they wanted him to be a farmer and to take over his father's farm. And that is what happened. However, this Hans, later called simply called Old Hans Utigard, was an unusually skilled carpenter and he took on a number of boys as apprentices, some of whom went on to vocational school at Holmøyane in Hornindal. Right at the end of his term as mayor, Vicar Lie raised the idea of establishing in Eid a secondary school for the local people, an idea that would later be realized after great local efforts, and would make the community an important centre of schooling.

A daughter, Dorothea, came into the world at the vicarage at New Year's 1839, and the following year the pastor and his wife had another son, John Herman. Then, on December 17, 1842, Sophus Lie came into the world. The fact that he was christened two months later by his father, at home, might indicate that his health had suddenly deteriorated, although home christenings were certainly not uncommon. When the boy was five months old, he was also officially christened, and his name, Marius Sophus, was duly recorded in the church registry; his godparents were Dean Koren, the wife of Police Sergeant Steen, District Doctor J. W. Cammermeyer and the doctor's assistant.

It also seems that this period of 1842–43 was a turbulent time for Father Lie, and he may even have had a desire to leave Nordfjordeid. He did not stand for reelection as the community's mayor – he allowed a peasant farmer to take over, and he applied for the position of resident chaplain at the Church of Our Lady in Trondheim. But Jacob Neumann, Father Lie's bishop in Bergen seems to have found Lie's desire unacceptable and he wrote a sharp comment on the application, about "our daring supplicant", whereupon Lie withdrew the document. After New Year's 1844, Lie once more again took on the office of mayor, but the following year he undertook a tour abroad for three months, together with another parson, and the owner of an estate; then once again, in 1846, Lie applied to Trondheim, this time with the bishop's letter affirming him "as one of this Diocese's most valuable Priests". He did not get the chaplain post in Trondheim, and during these years Father Lie applied for several positions, including Christiania, without any success, and in one of his applications he expressed the desire for "a Transfer in order to be closer to the University and its Associations".

In recounting his experiences of foreign travel – with visits to Copenhagen, Hamburg, Berlin, Leipzig, Dresden, Vienna, Munich, Cologne, Paris and London – he described in great detail all the music he had heard. And on the vicarage at Nordfjordeid, particularly during the long near-polar nights, he played the violin, together with his eldest son while the eldest daughter accompanied them on the piano. It was said that the other siblings sat around the single candle on the piano and read their lessons during these musical seances. For little Sophus this would probably have been one or another form of play, and maybe it was at the table during one of these seances that he fell asleep with the half-eaten pancake in his mouth.

In addition to these musical entertainments, Father Lie also taught his children – and here German played a big role – both history and geography, together with light reading in German. For a period he was assisted by a tutor from Hornindal. In 1847 the eldest of the Lie sons, Fredrik, was sent to Bergen to begin studies at the Latin school. It was also during this period that Aunt Edle came to the vicarage to help out with the household and the teaching, something that led to her maintaining in later life that it had been she who first taught mathematics to Sophus Lie. But Aunt Edle did something else as well while she was at the Nordfjordeid vicarage. She met the assistant to District Doctor Cammermeyer, a Doctor Hirsch, and a short time later they were married and left the community.

Apart from this, the four- or five-year-old Sophus must have experienced a new fervour in the work going on around him at the vicarage – much of the praise and prosperity that was later ascribed to the running of the vicarage during Vicar Lie's time stemmed from these years. Order and prosperity seemed to have provided the melody both in the house and on the farm.

A new cattle shed and hay barn had long been planned for the vicarage. As early as the dean's pastoral visit in 1840 it had been noted that the cattle barn required new planking. Two years later the dean felt that conditions occasioned the building of a new cattle barn, since the old one was so very rotten. And when two more years had gone by, it was recorded that a new barn ought to be built for the livestock. A year later a building commission, composed of Vicar Lie and two farmers, was appointed. The work was not begun before 1847, but from that point, it progressed quickly. Much of the reason for this seems to have been the fact that during the spring of 1847, Father Lie got a new hired hand, Batolf Nilsen Gausemel. He was later referred to as Priest Batolf and seemed in practice to have functioned as the farm manager. From that point on, Mrs. Lie and Batolf led the work of transforming the property into the model farm that everyone later talked about. Batolf himself later told his son about these years at the vicarage. Maurseth subsequently wrote down what he had been told – he had often heard his father say, "I have Mrs. Lie to thank for this," and "It was Mrs. Lie who taught me that."

Batolf told the priest and his wife what he thought should be done on the farm. Lie was said to have declared, "I myself know very little about running a farm, and everything connected with it, and yet I can see that there are a great many things to be set right, and in particular, I am not pleased with the horses!" The vicar had need of a swift horse for getting around his far-flung parish, but during his first spring there, Batolf used the old horses and got the harness and vehicle – which had been in a miserable condition – into a good state of repair.

When the spring farmwork was over, the one horse was sold to a purchaser, and Batolf went to a neighbouring community where he bought a bay stallion. And this pleased Vicar Lie no end – such that later he could not do without this horse on his parish visits to Starheim, and to Hornindal which in winter was three Norwegian miles (34 km.) over the ice. On those occasions it was Batolf who drove, something he loved to do as it allowed him to visit with friends and relatives up the valley. Lie and Batolf were very good friends, and on their long journeys on the wagon they held lengthy conversations, no doubt many that had to do with religion and construction. But when they were journeying up over Hornindal Lake and they passed Spjøt Point where, according to rumour, Vicar Spjøt had perished, this event was a recurring topic of conversation. There was a story, almost a legend, that went back to the time of the Black Death.

When the devastating plague came to the communities of Nordfjord, it was compared to an old couple – the man had a broom and the wife, a rake – and wherever the man swept, everyone died, and wherever the wife raked, some survived. Others described the plague as a man who came riding over the hills and through the valleys on a white horse with a huge scythe in his hands, and wherever he rode, the plague followed behind. In any case, the ravages of the plague were everywhere; all the same, it was worse in Hornindal and Stryn – first it took the small children, then the adults, and finally, the old people. Priest Spjøt was said to have been terribly busy going around to attend to the dying, and persuade them to give their farms to the church and priest as a form of penitence. And in this endeavour he prospered very well: most of the farms of Hornindal became feathers in the priest's nest. But then one day Father Spjøt had come down to this peninsula, that later would be known as Spjøt Point. He travelled by boat in order to have an overview of the beautiful valley. He suddenly burst out, "Oh, Hornindal most fair! You used to belong to a long line of eldest sons, but now you belong to me!" He stood up, stretched out his arms, lost his balance, fell out of the boat and drowned.

In homilies on God's providence and the salvation of the soul, Father Lie would also have recounted, and interpreted, the narrative of what had happened to a certain man from Hornindal. The man had rowed down the lake to engage in some trading and to meet people he knew in Nordfjordeid. Then, several days later when he boarded his boat and began to row homeward, it was a hot and perfectly still day. He became drowsy in the baking hot sun and after having rowed a goodly distance, lay back in the bottom of the boat to rest. He fell asleep. The current from the in-flowing river, swollen as it was by the seasonal snowmelt, was extra-strong and began to pull the boat back down the lake. People ashore had seen the boat, with no one aboard, moving rapidly toward the terrible waterfall, Kviafossen, at the end of the lake. At the last possible second, the man awoke, saw the hopelessness of his situation, but nonetheless, pulled resolutely at the oars, sat back in the boat had did his best to steer as he was born out over the foaming waterfall, a fall that no one had thought it possible to survive. But strange as it seemed, both the man and the boat came ashore on a sharp bend further downstream. The man had managed to hold the boat clear, through the roiling mass of water, from the rocks and boulders that would otherwise have smashed it like an eggshell. And the moral of Vicar Lie's story was that despite the man's composure and presence of mind in those critical seconds, he did not pride himself in his ability, but placed his life in the hands of God.

There were six hired hands at the vicarage, and six housemaids, two horses, twenty-five cattle and sixty sheep and goats. They got firewood from the mountainsides and the salmon that belonged to the vicarage were so plentiful that the serving staff demanded, as a condition of employment, that it not be served more often than two or three times a week. Some of the hired hands took part in the annual cod fishery in January, and returned home with a large amount of salted fish. One winter when it was so cold that thick ice covered the fjord, they took a good supply of particularly rich herring. On that occasion the peasants from Eid and Hornindal drove sledges out over the ice toward the open sea, where they hauled in great quantities of herring and other species.

There was great joy at the vicarage over these rich gifts from the Lord, and Batolf was the obvious choice to drive out over the ice shelf and fetch the fish. Batolf later told his son that when he was standing there ready for the long drive out to the open sea in the biting cold, the vicar's wife had come out to see that everything was in order. "This is certainly going to be a cold journey, Batolf," she said. "Oh, it'll be all right," he had replied. "But wait a moment, Batolf," she said, and went into her husband's office, and returned with the vicar's travelling garments – everything from head to toe. And when she had dressed him in this armour of furs, she said, "Now, indeed, we can wish you a good journey!" Batolf thanked her politely for her goodhearted care and concern; the horse stood pawing with his foreleg; then they departed from the farm. On the way out over the ice he passed many drivers who lifted their hats to him, for they thought it was the priest himself that they had met.

That year the vicarage got much more fish than it needed, and Mrs. Lie gave a large part of it to the poor and needy in the community. But otherwise, there were no privileges associated with the herring fishery: anyone had the right to put his own effort into it, and a stimulus in this direction was the pride of being "the author of one's own fortune".

The shoemaker came to the vicarage at certain times and everyone on the property would be fitted with new footwear. Apart from this, everyone was clad in homespun cloth, and the furniture had homespun upholstery. The vicarage had a good economy and was prosperous, the house was well-supplied with everything. When guests came to visit, Mrs. Lie informed the housekeeper and servants how they should care for the visitors, and in this way, undisturbed by household duties, she could engage socially with her guests. It also happened that the vicar's family might accompany guests on to another visit in the district. In the summertime there was life and excitement on the parade grounds, and officers were frequent guests at the vicarage. There was often dancing on such occasions, and when the atmosphere was at its best, Mrs. Lie not infrequently summoned Batolf, who was light on his feet and supple, and urged him to dance for the cream of society.

From the reports, protocols and such, it appears that schooling and the poor relief functioned satisfactorily during these years. When there was evidence that schooling was neglected, it was quickly recommended that recalcitrant parents and guardians be punished. As in all other regions, undesirable beggars wandered around the community. The general library regularly received more books; in 1848, it had about 400 titles and was visited with diligence. There was a recurring desire to have a definite, permanent annual contribution made for the purchase of books but it proved difficult to get the support for this from the municipal executive. Vaccinations were carried out annually; there was a report one year that three vaccinators were in full swing. Moral conduct in the community was a constant subject in the dean's reports. Every single year it was reported that few children were born out of wedlock in the community, and the common folk earned praise for their industriousness and diligence, but with respect to drunkenness, the commentaries varied. One year it would seem that the propensity to drunkenness was "not on the increase", and another year it was "declining" – with the exception of Hornindal – while in a third year "Boozing" was still not an unknown vice, and several illegal, and unpunished, sales of spirits were reported. In 1844 it was reported that beer had surpassed the consumption of spirits, but "it seems to function in an equally destructive manner", and the following year, drunkenness was "on the increase" and threatened "to corrupt the up-and-coming Generation". Toward the end of Vicar Lie's time at Eid, they had managed to prevent the peddling of spirits, and drunkenness was again "on the decrease", but as for the sale of beer that took place near the churches "one could not still not consider this to be anything other than an act of perversity".

But this was also the time when homemade fruit wines became popular in the district, and in many places this surpassed beer and spirits as the drink consumed at Christmas and at other social gatherings, and there seems not to have been objections to wine-making, either from the Church or the politicians. Quite the reverse: the art of wine-making acquired great prestige, and vicars' wives often were involved in this art, which they had learned abroad, and for which they gave out recipes. Perhaps the praise that the village people later gave Mrs. Lie for all her work on the vicarage farm was also due to her talent as a maker of wine? And perhaps Mrs. Lie also taught the local people to have fragrant and highly visible potted plants in their windows? In any case, this was a trend in interior decorating that began to prevail at this time; indeed, it was a trend in which the vicarages were in the forefront. Be this as it may, one of Mrs. Lie's daughters, Laura, would later become an empassioned lover of flowers.

Sophus Lie grew up in a village community marked by orderly conduct and modest prosperity on all sides. Peasant farmers and fishers lived side by side, and the military arrived as a festive diversion every summer. And consequently, everything on the vicarage seems to have been exemplary. In these years, in addition to the extensive building of barns and stalls, Sophus experienced the building of a new church in the village – the biggest in the whole of Nordfjord – with room for 900 in its pews, big enough to contain the great explosion of population that the community experienced in the summertime. Vicar Lie most certainly stood front and centre in this particular work, that in most respects had been a long and laborious process.

With the completion of this construction, Vicar Lie may have felt that his work at Eid had been brought to an end, or perhaps it was his concern for his children's future schooling, that now prompted him once again to apply for a new posting. Indeed, his eldest son had already gone to school in Bergen, and soon the boys John Herman and Sophus would be old enough for further schooling. At the end of 1850, Father Lie received a posting as parish vicar in Moss. Six months later the family left Eid.

The vicarage was particularly well-equipped and well-run during the Lies' time there, and this long remained on the lips of the village people, as did the fact that Vicar Lie was good at teaching and particularly talented conducting the catechism on the floor of the church, and these were frequently brought forward as exemplary features. Dean Koren wrote about Vicar Lie: "When, with Life and Warmth, he practices his devotional Ecclesiastical Duties, he always undertakes them on the Church-Floor with the Youth in such a Manner, that one must acknowledge that he, as a Christian Cathecist occupies a remarkable Place."

Vicar Lie had improved the church choir in the community, and the people of the congregation now took up the idea of obtaining an organ for the main church. Later the church was restored and redecorated, and it stands there even today, with its commemorative plate with the year of its construction and the name of the architect, Claus Wiese.

The old vicarage burned down in 1902, but the livestock barn from that period is still standing on the farm today.

Father's Home in Moss

When he was barely nine years of age, Sophus Lie left Nordfjordeid with his parents and five brothers and sisters (one more son was born to the Lies two years after the birth of Sophus). They were rowed out the fjord to Gloppen, from where they travelled by horse and cart down to Vadheim on Sognefjord. They were then transported by rowboat for several days up Sognefjord to Lærdal, then they took the old highway over the mountains of Jotunheimen to the valley of Hallingdal, and on down to Drammen and Holmestrand on the west side of Oslofjord. Here they took the steamboat across the fjord to Moss.

This was quite certainly Sophus Lie's first great travel experience. In light of his later wanderlust, and the great delight he took in nature and the landscape, it might suggest that as far as little Sophus was concerned, the journey was highly enjoyable, that he loved the constantly changing scenery. At one place or another during this journey they experienced the great eclipse of the sun on July 28, 1851, which was written about and discussed long after the event had passed. The duration of the eclipse was quite long – it was reported that the hens flew up to their perches, the horses bolted in many places, and the strange, ceremonious atmosphere created by the darkness was broken in a peculiar manner when the light returned and the birds began to twitter.

For other members of the family the journey away from Nordfjordeid was perhaps a heart-rending and unsavory one, and a difficult break with a familiar and cherished everyday life. For the first period in Moss, Mrs. Lie fell into passivity and found things difficult. Their eldest daughter later recalled how the mother and the daughters quickly felt themselves like fish out of water. They arrived in Moss in homemade sun bonnets called "crows", a form of headgear sewn from thin, starched, floral calico with pleats and ruffles. A seventy-year-old lady in the city took pity on them, and Mrs. Lie was happy to accept the offer of the old lady's hat, as a loan. Even though she had to go through the streets of Moss to the milliner's in an old, well-used hat, this was decidedly better than having to wear her "crow".

Father Lie arrived in Moss with the best credentials and letters of recommendation. Dean Koren of Nordfjord had written that, among other things, Parish Vicar Lie distinguished himself by holding sermons with a rare diligence, such that in them one "retrieves the Richness of Thought, the Sacred Order, and the Essence of Christendom, in which one witnesses a fundamental Study and Facility, through which he demonstrates the order of his Thinking." Bishop Neumann in Bergen said that Lie stood out as one of the most competent clergymen in the diocese, who was an outstanding speaker, an excellent cathecist, and a devout and zealous spiritual advisor who with great thoughtfulness attended to his calling. Neumann's successor, Bishop Kjerschow, described Father Lie as an upright, accomplished and orderly man who applied himself with vigor and sober-mindedness to his calling. With his appointment by the Church Department it was pointed out that conditions in Moss, as pertaining to church affairs, made it highly desirable to have as an incumbent to the calling, such an uncommonly able and powerful man, and the Department pointed out that out of the nine applicants, Vicar Lie was "considered in all Respects the one who in ecclesiastical Activities was a remarkably competent and zealous Public Servant and a notably gifted Speaker."

Moss was a more remunerative post than Nordfjordeid; the position's income was estimated at 1,050 speciedaler, eighty of which covered the cost of the free lodging. During the past ten to twenty years, the town of Moss had seen great growth and prosperity, and the prosperity rested on three pillars: the ironworks, the lumber trade, and distilling. Distilling in particular had created an upswing in the recent period, and precisely in 1851, the year's production culminated at two and a half million litres of spirits.

Vicar Lie seems to have quickly come to take delight in the conditions in the town, and was particularly happy with the possibilities to be found in Moss for schooling – presumably this was as well a major reason for his having sought to be transferred here. Father Lie wanted his sons to attend university, and he had already sent his eldest boy to the Latin school in Bergen. The choice in Nordfjordeid had been between sending his next eldest sons, John Herman and Sophus, to such a Latin school – and there were seven of these in Norway at the time – or teaching them at home, either with himself as teacher, or paying for a competent tutor whom it could be difficult enough to find. But apart from the Latin schools, in the seats of the four dioceses and in three of the other largest cities as well, there had now been established in some towns and cities what were called *realskoler*, which were more science-oriented grammar schools that prepared pupils for the university entrance exam, the *examen artium*. It was exactly this type of grammar school that Vicar Lie had led and conducted during his time at Molde; moreover, the same year that the Molde school had begun to function, in 1832, a similar school was established in Moss; the latter had developed out of an existing school of commerce. Father Lie knew the system and as parish vicar, was now on the board of governors of the Moss school, together with the magistrate and three elected municipal representatives. Presumably it was also Vicar Lie's interest in and knowledge of teaching that was of added relevance to determining why he obtained the position of parish priest in Moss. His predecessor, Peder Monrad – father of the man who would later become the prominent professor of philosophy, Marcus Jacob Monrad – died at his post in Moss, but the fact that the town had been without its parish vicar for over a year was not fully explicable by the Ministry's desire to have a particularly competent public servant filling the post. Perhaps the real explanation was to be found on the schooling front: in Moss this was a period marked by a great deal of discontent about schooling. Teachers were quitting, temporary replacements came and went, parents removed their children or had them privately instructed, and there were disagreements about the teaching methods and the classification of school classes. While in his last term as the head of the board of governors of the grammar school, Peder Monrad had written "It is evident to me that people will go on organizing, reorganizing and reforming the School until they have totally disorganized it. The Question is to know what and where one should; and often when one speaks of the Practical and Impractical, the Useful and Useless, etc., one does not oneself know what one understands thereby, or what thereby ought to be understood."

Only a few months after the Lie family had arrived in Moss, in the autumn of 1851, their sons John Herman and Sophus were placed in the city-run Moss Realskole, which at that time had eighty-one pupils, the highest number in the school's history. The school had returned to an orderly condition and was earning respect and confidence, something due not least to the school's director, Nicolai Moursund Harboe, an experienced teacher who arrived in Moss in 1849. It seems that for Vicar Lie, his work and contact with the school were easy to deal with.

The school in which Sophus was now enrolled was divided into three classes, each of two years' duration, and the fees varied from six to eighteen speciedaler per annum. The pupils in the first class, which was usually the largest, ranged in age between eight and ten. Every six months the school put out an assessment report on the pupils, who were ranked according to their performance. The nine year-old Sophus seems to have been well prepared. In the first exams in December 1851, Sophus received the highest mark of *excellent* in Bible history and history; otherwise he received *very good* in most other subjects. Only in writing and penmanship were his marks poorer, and if one is able to judge from the recorded marks, then writing, penmanship and Norwegian style were subjects that young Sophus right away had little use for. Nevertheless, he applied himself, and gradually came up to *very good* or *almost very good* in these subjects too. After one year at the school he was advanced to the second class, where he was ranked fifth, but he climbed upwards and was ranked first by the time he entered the third class. In terms of orderliness and behaviour, Sophus seems to have been an exemplary pupil through his whole time at the school in Moss. In mathematics, which was divided into "domestic and commercial calculation", geometry and arithmetic, he was always the top pupil, and it was in the subject of mathematics that his classmates later said they had considered Sophus Lie to be a kind of "oracle". But the teachers attributed Sophus with even more constant top marks in Bible history, secular history, and natural history, such that his school days seem to have been marked by great versatility.

In addition to Rector Harboe and various chaplains, there were a number of seminarists who taught. When, in 1855, the school was looking for a successor to a seminarist, the appointment notice called for "Proficiency in the *English* Language will be an excellent Quality in the Selection for this Post." English was now a permanent language of instruction, but in Moss English did not replace French, as it had in other places, "since at the Moment, the City's Lumber Export is directed more toward France than England, and since most of the School's Disciples will be employed partly in Commerce and partly in Seamanship." Headmaster Harboe worked out new extensive plans for the school, pertaining not only to the curriculum and the allotment of time, but also the division into classes and the teaching methods. Physics and chemistry, for instance, were proposed to be incorporated at the expense of "the dead Languages", Latin and Greek. In this reorganization of the grammar in Moss, Headmaster Harboe sought out contact with and advice from the leading school reformer in the capital, Hartvig Nissen, a man who later would play a central role in Sophus Lie's schooling. Nissen commented upon Harboe's plans, and on all major points, he praised the reorganization in Moss. What is particularly interesting is perhaps Nissen's reference to the significance of instruction in drawing and sketching, with regard to the general acquisition of culture. Drawing ought to be awarded a greater proportion of time, and ought to be expanded in both "Free-hand Drawing and Linear Drawing", and as such, Nissen commented, should be related to instruction in mathematics.

Not all the features of the reorganization saw the light of day while Sophus was a pupil, but during the last year he was at the Moss school, there were nonetheless no longer examinations in Latin and Greek – and perhaps the natural sciences had been given accreditation. He received an *almost excellent* in the new subject of English, a mark that he received in the other "living Languages", French and German.

But it was life at home, rather than school and schoolwork that provided the greatest challenge not only to Sophus, but also to his brothers and sisters. At school everything seemed to go according to plans and expectations, and everything was safe and secure, overseen as it was by Father Lie and Headmaster Harboe. But for Mrs. Lie and her daughters, the move from the countryside to the city was different, and more unpleasant. They gradually became up-to-date with the new fashions in Moss, and the family seems to have glided into the social life of the town, in any case, they rather quickly developed a friendly acquaintance with the magistrate's family. Magistrate David Vogt was a prominent figure in the town who had represented Moss in Parliament for fourteen years. Certainly the fact that the pastoral Lie family and the magisterial Vogt family soon became acquaintances, had a lot to do with social rank and position – the vicar and the stipendary magistrate were both part of the public servant class and the same social rank in the town, and in their professional lives they often handled the same matters. But the two families also had a common background in the community of Nordfjordeid. Vogt's father had been the district magistrate of Nordfjord in the early years of the century. When his father died in office in 1809, David's eldest brother, Johan Herman Vogt, who later became a cabinet minister, applied for, and received the posting. But when Johan Herman Vogt became the district magistrate in Nordfjord, David Vogt moved there and worked as his brother's assistant for three years. He had lived right across from the vicarage, at Skaarhoug, and could recount rich good years in the community.

Memories of Nordfjordeid seem to have been particularly strong for Mrs. Lie, who never did feel at home in Moss, and never managed to build a position or status around herself in the city that was anything like that which she had enjoyed in the village. Be that as it may, the eldest daughter painted a picture of this state of affairs when, a year after they moved to Moss, catastrophe had struck the family. In April, 1852, less than a year after their arrival in the town, Mrs. Lie died. She passed away following a short illness, and suddenly was buried and gone from the family. The daughter's explicit explanation was that their mother had not been able to tolerate the shift from the healthy active life on the land, to the urban atmosphere of the town.

The prevailing tone of life in the home changed radically following the mother's death, and the well- being of the house seems to have been maintained and directed by Vicar Lie's strength of will. For the ten year-old Sophus it must have seemed that the doors to the riches of childhood had slammed shut once and for all. It was said that his sister Laura, five years his senior, was so downcast by the death that she wanted to follow her mother to the grave.

The loss of his wife was also very difficult for Vicar Lie to bear, and descriptions of his later life present contradictory views. His daughters explained that at home he became more forceful and vigilant, and in his ecclesiastical work many things point to the fact that he lost some of his power and energy. When he had been in Moss some years, Vicar Lie was considered a popular and outspoken man, who quickly won the congregation's respect and affection, however his sermons were said to be dry, and even more boring than was customary. Perhaps he was a "zealous Public Servant", but he certainly by no stretch of the imagination lived up to his characterization as "a gifted Speaker". In any case, Vicar Lie had to take some of the responsibility for the falling off of church-going in Moss, indeed, a decline so severe that the bishop of the diocese, Jens Lauritz Arup, expressed his alarm.

But during this period Father Lie also supported a series of initiatives. The same year that his wife died, the Mission Association was set up in Moss. The Association held its meetings twice a month on Sunday afternoons on the premises of the local schools. During the other Sunday afternoons each month, the same locales were used for meetings of the Temperance Union. Some years later, Vicar Lie also started his "Bible Readings", as a rule, every second Sunday at four o'clock in the afternoon. In these "Bible Readings" he was gradually replaced by the leaders of the grammar and the theological candidates employed by the school. But four or five years later when Lie allowed laymen to hold Bible readings, complaints arose from the public that these laymen did not always confine themselves to the text, and Vicar Lie had to rein them in.

Apart from this, Vicar Lie worked on the town's orphanage, for the poor, and other benevolent matters. He conducted catechism sessions on the floor of the church with the same zeal as previously – and at the advent of the bishop's visit of assessment, he instructed the previous two years' confirmants to meet in the church on Sunday for an examination. On the Monday, he took the bishop to the local school where all those of school age were assembled with the teacher.

Vicar Lie also eagerly supported the matter of the choir, and tried to improve church singing by supplementing the existing ladies' choir with twelve gowned boys from the local school. He held Holy Communion more frequently – as a rule, every second Sunday throughout the year, and every first Friday of the month, similarly with the day of prayer, Maundy Thursday, and Ascension Day. But despite these activities, it was said that he never inspired people to the religious life.

During the first year following the death of Mrs. Lie, the family had a servant girl working in the house, but since there were three daughters – aged fourteen, sixteen and seventeen – they, for the most part, each had "her week" being responsible for the house. Later, the two eldest daughters, Mathilde and Laura recalled one or two things about this period, and it might seem that they felt that the conditions were severe. There was bread and milk for breakfast, and generally porridge with milk in the evenings, except for Sundays. When the eldest daughter was having a new neckline fashioned for her dress, she and the seamstress had to go into the vicar's office so that Father Lie himself could oversee that the least possible amount of fabric be removed from around the neck. And it had been strictly forbidden for the daughters to promenade in the streets with any gentlemen. And with invitations to Christmas balls, the daughters had to show restraint by "refusing" the invitation to at least one ball.

Scarcely two years after losing their mother, the family lost its youngest son, Ludvig Adler, at the age of nine. Sophus was now the youngster of the house, and lived here together with his brother John Herman, and three sisters. Fredrik, the oldest brother, was now, in 1854, attending university in Christiania, and had begun the study of the natural sciences.

It was a strict rule in the house of Father Lie in Moss to go to bed at ten o'clock, and immediately blow out the candle. Even after the daughters had grown up and come home with their own children during holidays, they still had to obey their father's command about going to bed at ten o'clock – even during the long, bright summer nights. And even the eldest daughter, who recalled this much later, and who had opposed this, had her father's command so deeply rooted in her, that she too in her own home at Tvedestrand enforced the command that all children must be in bed by ten o'clock, be it winter or summer.

Father Lie ran a parsimonious house. The daughters had to calculate to the penny what they needed by way of clothing and accessories, and they received money from their father in accord with these precise calculations. The eldest daugher happily explained later that Father Lie remained so close-fisted in these calculations that when it came time for the dispersal of his last will and testament, he had calculated that she, for example, was to be awarded a credit in relation to these "pocket monies".

The expression "You are lying!" never occurred in the Lie household. Sister Laura recounted the occasion when she heard Sophus say something that was not exactly right, she had burst out, "Now once again, you're standing there and being mistaken!"

Barely four years after the Lie family moved to Moss, the eldest daughter, Mathilde, aged twenty, married the magistrate's son, Fredrik Vogt. He was seven years her senior, and had trained as a doctor. He had already made a heroic effort regarding a threatened cholera epidemic in Christiania, and a malarial fever in the islands of Hvaler on the northeast shore of the Skagerrakk. Some months before they married, Fredrik Vogt had obtained a position as district doctor in Tvedestrand, and there they lived for over thirty years.

Ten years after this wedding, the youngest Lie daughter, Dorothea, also married a Vogt son, Johan Vogt who became the master pilot of Moss. Down through the years the ecclesiastical Family Lie and the magisterial Family Vogt seem to have been on the best of terms. The two family heads had a common outlook, both in relation to school and church affairs, and certainly with respect to the question of temperance. Even though spirit-production had particular significance for the town, the question of temperance was what distinguished Magistrate Vogt, both in his public office in Moss, and as a Parliamentarian, and this did not always make him popular in his hometown. Alcohol was never served at the magistrate's home, and not even champagne when his sons got married. Father Lie was certainly not as fanatical about temperance, but he had seen enough of the obviously negative effects of alcohol, and after many years of work he finally had a breakthrough for his point of view when it was determined that there would be no sale of beer or wine in Moss on Sunday mornings between ten and twelve o'clock.

Father Lie was otherwise described as a strong man with a hardy nature, and they said he spent his nights on the sofa in his office. He often took long tours on foot, and he had the habit of walking all the way down the west shore of Oslofjord to Tvedestrand to visit his eldest daughter and son-in-law, and the growing number of grandchildren. In everyday life he wore an old gray wig, but on Sundays he donned a brand new *perruque* of black hair, which made him look much younger.

However, fewer and fewer attended church on Sundays in Moss. Both the merchants and the working people were in agreement that Vicar Lie was wretchedly boring to listen to. Church attendance fell off in Moss, but this was not entirely due to Lie's sermons. Some prosperous families stopped going to church as the holding of permanent pews began to go out of fashion, but the greatest reason for the fall in church attendance was the religious revival that in Moss, as all over the country, had begun to split the Christian congregation in the 1850s. Both the bishop and Vicar Lie were apprehensive that the revival would lead to formal withdrawals from the State Church, but this had not happened in Moss. Different sects held meetings, partly on local school premises, and partly in their own halls. Flocks of "the awakened" and "the salvation-seeking", and groups which neither agreed with the dogmatic teachings nor with the parish vicar, gradually created a climate of discord and an uncharitable spirit, and Vicar Lie knew it was his ecclesiastical duty to warn against cheap and superficial Christianity, which led so easily to spiritual pride and censoriousness. Vicar Lie warned against the exaggerated propensity toward "Edification" and religious lectures. Father Lie's motto was "big things happen quietly", and he expressed an evident understanding of this conflict. One day in the pulpit he said that "one can gorge oneself on the Word of God", and this led to heated debate in the city on whether one could "go too far in piety to the Lord", if there really were limits to piety and if "the stupefying Effects" occurred beyond these limits. "This Religious Fanaticism, which seems to be getting out of Control" became an unsavory part of everyday life in Vicar Lie's last years. However by then, only his daughter Laura was still living at home in her father's house. The other brothers and sisters came home only for holidays and short visits.

Sophus graduated from the grammar school at Moss as the top pupil in the senior class, in July 1856. His brother John Herman, who was two years older, had now also finished at the grammar school – a school that despite its attempts to organize the teaching as preparatory to further schooling, still sent most of its disciples directly out into working life. The Lie brothers were among the six or seven percent of the school's pupils who continued on to higher education. John Herman wanted to go on to the Military Academy in Christiania, and this required further knowledge of mathematics. Sophus seems to have decided to take the *examen artium*, and thereby obtain entrance to the University, but he lacked the necessary extensive knowledge of Latin and Greek. Sophus could now either be privately tutored at home, or be sent to a Latin school where he could be prepared for the *examen artium*, an examination that was still arranged by the University. Both options were equally common at the time, and often they were combined – first, some years of private tutoring, followed by attendance at a Latin school. It seems for Sophus, the combination solution was adopted – in any case, he was not sent on for further schooling in Christiania until another year had passed. What he did, and who taught him during that year, is rather difficult to know. In all probability it was Father Lie himself who taught Sophus the classical languages. In addition, his brother John Herman seems to have been tutored at home during the autumn of 1856 – the autumn when otherwise the measles were doing the rounds of the children in the Lie family, who at this time, excluding the father himself, was composed of Sophus, John Herman and two of their sisters, Laura and Thea.

In January, 1857, John Herman was accepted as a cadet at the Military College in Christiania. As well, at the beginning of the school year in August 1857, Sophus journeyed to the capital, to Nissen's Latin and Grammar School, to learn what he needed in order to take the *examen artium*. This would take a further two years. Perhaps Father Lie also had economic reasons for keeping Sophus home for a year – at Nissen's School one had to pay three speciedaler per month in school fees, in addition to outlays for food and lodging, and the payment of school fees had to be punctual. If payment were two months in arears, the pupil could be refused admittance to his class.

Back in Moss, Father Lie did his best to follow the further education of Sophus and his other sons. He also wanted to secure a good future for his daughters, and he certainly wanted to be a reliable shepherd to his congregation. Stories spread about Vicar Lie. Many subsequently maintained that the pastor's sermons were insufferably boring, but that he was nonetheless a sober man of many interests, who represented a mild and moderate form of faith in progress. He was interested in the natural sciences, and in time, he arranged open public lectures on natural science topics, designed particularly for the working people of the city. The city's teachers gave the lectures, which were held in the local schools, and the Moss workers' association supported them. Vicar Lie hoped that increased awareness of the laws of nature would reveal the creative hand of God. He supported, with reservations, the natural sciences "so far as these serve to reveal the Creator and benefit the Condition of Life."

PART III

School and Education

Nissens Latin- og Realskole (Nissen's Latin and Grammar School), an economic and pedagogical success that would form the pattern for all secondary education in Norway

Nissen's School. Examen Artium

Beginning in the autumn of 1859, Sophus now lived in a bed-sitter in Christiania with his brother John Herman. The two parsons' sons had found lodgings with a Miss Meyer, who seems to have lived somewhere in Pilestredet, one of the city's safe and good neighbourhoods. She was probably a somewhat well-off lady; she was in any case a great lover of flowers and had so many that her houseplants even overflowed into the boys' room, surely also in the windowsill, easily visible from the outside. In a letter home to his sister Laura in Moss, John Herman wrote that he and Sophus had nothing against all Miss Meyer's plants, which in a way reminded them of their sister's love of flowers. How were sister Laura's twenty-four different sorts of acacia coming along anyway? The two boys had a standing rose, two ivy plants and a geranium in their room, and they wondered if such a love of flowers was something particularly common to womenfolk.

Their big brother Fredrik, who was nine years older than Sophus, also lived in Christiania. He had become a university student in 1852, had passed his preparatory examinations two years later, and was now studying science. This course of study, oriented to the public service, was introduced in 1851 and its aim was to educate good science teachers for the secondary schools.

Their father's brother, the physician, John Lie, also lived in the capital, and the Lie boys often paid Sunday visits to their uncle's home at Øvre Vold Gate [Street] No.16, where he had lived with his wife and two children since 1840. Uncle John Lie was a well-known man-about-town and was deemed a great character, due both to his medical practice as well as his way of life. He had always provided free medical care to the less fortunate. He had an extensive practice among the city's craftsmen, and he counted many of the city's influential men among his circle of friends. He kept a hospitable and sociable home in Øvre Vold Gate. In summer he undertook various "Excursions", and in winter "very serious Outings by Ski and by Skate", according to contemporary reports. It was also said that "there was much of the old Norseman's wild energy in John Lie. He took pride that nobody could pull the wool over his eyes, and nothing could shake him once he had made a decision – however unreasonable it might be." He had been Henrik Wergeland's friend and physician – it had been John Lie who cared for and provided morphine to the poet on his deathbed in 1845. [14]

[14] Henrik Wergeland (1808–45) was one of Norway's greatest lyric poets. He wrote on the passions of love, war, freedom and oppression, history and Nordic mythology. Wergeland captured much of the fervour of the French Revolution, the Romantic Movement, and

The contact between the Lie family in Moss and the Lie family in Øvre Vold Gate seems to have been the very best for several years with regular annual summer visits and letter-writing. In their home in Øvre Vold Gate, John Lie and his wife, Frederikke Cathrine Grønvold, had valuable furniture, a quantity of etchings, lithographs and portraits of famous men, and a book collection consisting of several thousand volumes. Uncle Lie translated medical textbooks into Norwegian, and his collection of medical books was said to be rare and exquisite; otherwise he most certainly owned many books of historical content.

When Sophus Lie arrived at Nissen's School in the autumn of 1857, he was placed in the fourth Latin class – meaning the next-to-final form, a fact which speaks eloquently to his good previous grounding. He had ten or eleven classmates, all of whom had been at the school for a period of two to seven years – most of them had come up through the forms from the first Latin class. They were all approximately the same age, born in 1842 or a little earlier. About half of them lived in lodgings, the other half at home with their parents. The enrollment at Nissen's School was around 500, distributed between twenty-three classes – general classes, mathematics and science classes, and Latin classes. The fact that the school combined mathematics and sciences with Latin, and that these were preceded by general classes, was one of the innovative aspects of the school and something that had led to an immediate increase in the demand for admittance.

The school was located in the lower part of Pilestredet – the address later became Rosenkrantzgate 7. Nissen's School was one of the city's most handsome and stately buildings. The impressive and splendid school building was constructed in 1847 to accommodate the increasing influx of students and also to signal the new spirit of the times. The building was designed by the architect, J. H. Nebelong, who had also designed King Oscar's Hall at Bygdøy[15], and who had participated in the interior furbishing of the Royal Palace. Furthermore, Nebelong had initiated the restoration of various stave churches[16]. Architecturally, as well as pedagogically and organizationally, Nissen's School represented something new. The school was a sterling success. The reform ideas on which the school was based were soon copied by the publicly financed school system. The school, in which Sophus Lie was now enrolled literally shaped the school system of the country, thanks in large part to its founder, Hartvig Nissen, who became clerical officer and later Deputy Secretary in the Ministry of Church and Education.

In 1843, Hartvig Nissen and his friend, the mathematician Ole Jacob Broch, founded Nissen's School, and in the years that followed, Nissen became an increasingly important figure in the general campaign to reform the Norwegian school system. When in the years from 1829 to 1832, Sophus Lie's father, as a neophyte

---

liberal humanitarianism. His collected works, whose publication was completed in 1940, extended to 23 volumes.

[15] Bygdøy, an island in front of Oslo, is today joined to the mainland by landfill. It is home to the royal farm, Viking ships and other museums, beaches, residences and embassies.

[16] Stave churches are wooden churches in Norway dating back to the Middle Ages. Despite the ravages of nature and vandalism, a few remain extant in various parts of the countryside.

theology graduate, was a secondary school teacher at the Cathedral School in Trondheim, Ole Hartvig Nissen, who was twelve years younger than Sophus Lie's father, had been a pupil at that school. Nissen became a university student in 1835 and began to study philology – a course of studies particularly well suited and tailored to a future position within the school and educational system.

It was exactly such a graduate that the Ministry wished to have as director of the new secondary school of science in Molde – the position for which Vicar Lie had applied, and which he had coveted very much. His rejection at that time, in 1835, had led Lie to utilise his theology degree and join the clergy, a service that was also better paid. Although in theory graduating with a degree in language and literature was comparable to the other degrees – those of theology, law and medicine – salaries of philology graduates were not comparable. The low salary contributed to the fact that there was little demand for the degree. For every score of theologians, on average, scarcely one student graduated in philology, and at the beginning of 1840 the course of studies was changed. In practice the study of philology functioned as a line of defence for the old school system, which held that all real education was derived from the teaching and study of Latin and Greek. But with the advent of new technological innovations in many areas, a demand for more thorough knowledge of the natural science subjects was emphasised and advocated.

When it was founded in 1843, Nissen's School was an answer to these demands and expectations – Hartvig Nissen also started a secondary school for girls in 1849 in Christiania, Nissen's School for Girls. And it was a new triumph for the natural sciences when, in 1851, Ole Jacob Broch managed to establish a course of graduate studies in science and mathematics, resulting in the new public service exam that led to the science and mathematics teacher's certificate.

In his early days, Hartvig Nissen had chosen to study philology, largely inspired by Fredrik Moltke Bugge, headmaster of his school in Trondheim, who for a long time was the leading force in education within the so-called new humanities. This was a school of thought that vigorously stressed the kind of intellectual training that was acquired through learning the classical languages. In comparison, modern languages were considered too grammatically simple to constitute the foundation for stringent thinking. But new ways of thinking about education were forging ahead and in addition there was a demand to understand the basis for the technological developments that everywhere were changing most people's everyday lives. From the beginning of the 1830s, these divergent views on school and education – the classical languages versus the natural sciences and modern languages – had given rise to vehement debates. Everyone had an opinion about the dispute between the new humanists and the so-called "realists", or scientists, wherein the central question had always been whether or not the ancient languages – Greek and Latin – should cede their place to modern languages and the sciences. The dispute was in the balance, and there were advocates for both sides, but little by little more and more people came to support the natural sciences. The ancient languages began to be talked about as "the Old Age Course of Studies", and there was talk about cemeteries where vigorous new cypresses grew on the ghostly bones of Latin – a metaphor for education no longer arising from Latin grammar. On the other side, the new age and its realism was said to represent materialism, utilitarianism, encyclopedic knowledge and a science-oriented worldview.

This was the dispute in which Hartvig Nissen had engaged. With shrewdness and great diplomatic skill, he was to dissolve the frontiers between utilitarian considerations and formal ideals of education. Nissen combined his background in classical scholarship with a clear sense of the new educational knowledge, while at the same time he understood and defended people's economic and material demands and expectations. Hartvig Nissen's authority and credibility were originally based on the success of this Latin and Sciences school – "A School System constructed on Tenets that through the pedagogical Upheaval of the Age, have worked their way to general Recognition," as was stated in the invitation to the school's inauguration in 1843.

The school had been an immediate educational as well as economic success. Enrolment skyrocketed – after only three years of operation Nissen's Latin and Grammar School had 120 students, twice as many as the venerable Christiania Cathedral School. The average age of the teachers was a mere twenty-five years, many of them were friends from student days, some were scholarship recipients at the University, and many would later become professors, parliamentarians and cabinet ministers. Teaching at Nissen's School became a desirable occupation – P. Chr. Asbjørnsen and Jørgen Moe, Eilert Sundt and Ole Vig[17] were all teachers at the school. The school emphasised science and mathematics subjects more heavily than was customary in the established Cathedral schools and Latin schools, but the science and mathematics classes at Nissen's School were still no more than a sort of "higher citizens' school" for the urban areas – a three-year course that for many became a preparation for entering the Military Academy. Only the Latin line qualified a student for graduation with the *examen artium* and there was still a stringent requirement for the knowledge of Latin and Greek. But Nissen and most of the young teachers on his staff now expressed convincing doubt about the curriculum's ability to prepare pupils for higher studies. Perhaps, on the whole, it was not possible to train memory, intelligence and the power of discernment? Learning Latin words by rote could in any case not be said to train the memory in general, but rather, trained only the memory of words. The old theory of a formal education through classical languages alone was rejected on the basis of a more modern concept of what constitute human mental capacities. The advantages of ancient languages as an educational tool no longer could be maintained with the same conviction – for human intelligence performed the same operations whether one studied Latin or Norwegian, religion or mathematics. General education now had a larger framework and a broader content. Just as the Latin School, with its

[17] Asbjørnsen (1812–85) and Moe (1813–82), like the Brothers Grimm in Germany, and H. C. Andersen in Denmark, collected and published national folklore. Eilert Sundt (1817–75) documented cultural mores, social structure and social practices around the country; he is known as the father of Norwegian sociology, and Ole Vig (1824–57) was a lyric poet, promoter of folklore and folk literature, and worked to establish popular libraries and public education around the country.

concentration on classical languages and cultures, was one-sided, so too, were the old primary schools – both the itinerant and permanent schools – in their concentration on religion and reading. But a new time, with its technological developments and emerging industrialism, needed people with new knowledge at all layers of society – the entire school system needed to be recast. Hartvig Nissen managed, in his own way, to bring together these divergent ideals, the realistic, the national and the democratic. And he proposed arrangements that got both state and local authorities cooperating on the reforms. The municipalities received financial support from the state once local school conditions had been brought up to a certain level. And the state recognized its responsibility not only to educate good civil servants – meaning the Latin schools and the University – but also to produce good citizens, who could become leaders and work in local government around the country.

Young Sophus Lie landed in the midst of this school reform process. The very autumn when he became a pupil at Nissen's Latin and Grammar School, an educational feud was underway, a battle was later regarded as one of the most vehement in the history of Norwegian schooling. A decades-old tug of war came to a head this autumn of 1857 in an intense debate about the royal proposal to eliminate the Latin composition requirement in the *examen artium*. English was suggested as a substitute to the obligatory Latin. The newspapers printed a raft of polemic letters to the editor, often anonymously submitted in order that the writers could avoid being personally attacked and rendered suspect. The Student Society called a meeting to save the Latin composition, if possible, and following a fierce fight, the vote was 170 for, and 170 against. When the proposal to eliminate Latin composition was finally approved in Parliament on October 12, 1857, it was an event that was to have great significance for further developments. The elimination of Latin composition was a triumph for the "realists' " view of school and education; it was also in concert with Nissen's ideas of the equal standing of classroom subjects and of realistic and national ideals of education. For Sophus Lie, as for all Latin school disciples, the approval had immediate consequences in terms of their instruction.

Writing Latin was no longer required, but the requirement of understanding and translating Latin and Greek was still strong. Those students preparing for the *examen artium* still had to master the classical languages. For Sophus Lie and his classmates the two ancient languages consumed thirteen hours, eight for Latin and five for Greek, out of the total of thirty classroom hours every week.

Martin Kirkegaard Holfeldt was Sophus Lie's Latin teacher and form master. Twenty-six years of age, he had received the best marks in his *examen artium*, and in 1853 was awarded his public service degree in philology. He was an active member of the Student Association where he had been a member of the board as well as of different committees and panels, and he was a member of the executive of The Norwegian Theater. It has said of his Latin teaching he twice a week gave his disciples written assignments – one as homework, a translation of a Latin text, and one classroom assignment, which included not only translations, but also some of the old composition exercises. Virgil's *Æneid* was on the curriculum, as were the works by Livius and Horace, and they used the grammar by Scandinavia's leading man of arts and letters, the Dane, Johan Nicolai Madvig, with whom Hartvig Nissen had studied for a time in Copenhagen. The Latin curriculum also included textbooks in the history of literature and the "Antiquities".

Hartvig Nissen himself taught Greek. Homer's *Iliad* was on the curriculum with works by Plato and Plutarch. For Greek grammar and syntax they also used one of Madvig's textbooks. Nissen had long claimed that in the teaching of languages one should follow the golden educational rule of going from the easy to the more difficult, and he ensured that there was a good foundation before they started Greek. He stressed that the living word, the mother tongue, and history were important sources of knowledge about existence – both human and natural.

Professor Marcus Jacob Monrad was the teacher of Norwegian to the upper forms of the Latin classes. In addition to his leading role at the University, Monrad had been teaching these Norwegian classes steadily for twelve years at Nissen's School: two hours a week in Fourth Form Latin and three hours a week in Fifth Form Latin. At this point in time, Monrad was considered the country's most learned theologian and philosopher. He was a prominent figure in all aspects of public life. In school politics he had become a standard bearer for the conservative school of thought and on many issues was diametrically opposed to Hartvig Nissen. When the dispute raged most intensely over Latin composition, Monrad had been one of the leaders in the struggle to maintain it in the curriculum, and he had held a separate series of lectures on the importance of classical studies. The fact that he stayed on at Nissen's School probably speaks mostly to Nissen's willingness to compromise but also to Monrad's position, and his will to influence young minds. Nissen and Monrad had respectively been president and vice-president of the Association for the Advancement of Public Information at its inception in 1852. They agreed that reforms were needed, that public refinement at its most profound must be built up to serve the needs of the people, and that what was really at issue was the nation's culture. But Monrad was of the opinion that it was wrong if, by a higher general public refinement, they meant that the populace required the same level of reflective thinking "and a Multiplicity of Knowledge, which greatly departs from their Conditions of Life and their real Needs." The advice of the public must be sought, any actual progress had to adhere thoroughly to what already existed and not be a mechanical addition, and above all, there must not be too wide a gap between generations. The sovereignty of the national language must naturally be acknowledged in line with the political sovereignty of the country, but development had to come about slowly. "The Norwegian language" ought to be, in the first instance, such as "it really appears in the Script and the Speech of refined Norwegians", thereupon would come a gradual approach to the common language. With these attitudes he defended a staged language development for Norway. And he convincingly maintained that the dramatic arts must be a factor in the moral upbringing of the people. Art was necessary to the process of a nation's self-development. As a kind of prophetic "Prescience", art could give a glorified image of what Norwegians as a people were, are and will be. Yet, although slower than art, science was nevertheless more important. Science, Monrad proclaimed, represented the highest form of knowledge. But for his disciples at Nissen's School the great professor mostly taught lyrical and didactic poetry and in his sessions with his disciples he presented an "Overview of Rhetoric", the principal aspects of which the disciples had to repeat in written form. The school's annual reports also state that Monrad drilled his disciples in the delivery of oral lectures, and that every two or three weeks he required a written "Compilation of reasoned Content".

Half a year after Sophus Lie began at the school, he had the twenty-five year-old Oluf Rygh as his history teacher. A few years earlier, Rygh had taken the public service examination in philology and had been awarded special mention – a deed that naturally enough was accorded renown. In spite of his solid classical background, Rygh sided with Hartvig Nissen in his view on the role of modern languages and sciences in the school. Rygh taught history to all the Latin forms, each form had two hours a week for both the history of the world and the history of the nation. For the latter, P. A. Munch's *Extracts from Norwegian History* was the central text. In the senior class he evoked history from the French Revolution onward. While at the school, Rygh became a university fellow, and at age of thirty-three, he was appointed professor of history.

Religion was taught by Joachim Fredrik Buchholm. He had graduated nine years earlier from Nissen's School, had taken the public service examination in theology, and later became one of the school's directors. For two hours every week he went through the history of the Bible and the Church. The disciples had to go through the Gospel According to St John and the first five chapters of the Book of Romans, in the original.

In the modern languages – German, French and English – the upper forms had three teachers. German was taught by Bachelor of Divinity G. Hansen, complete with sight-unseen exercises in both written and oral forms, and every two or three weeks an essay to be prepared as homework. Schiller seems to have been the author of choice for German literature; they read all of *Wallenstein* and *Wilhelm Tell*. French was taught by the Bachelor of Arts, M. Schnitler, and English by the Bachelor of Divinity, Hans Ross – two hours a week were accorded both subjects and the curriculum consisted of both compulsory and elective readings from various textbooks. Selections from Charles Dickens' novels were read in English but no marks were yet given in this subject.

In mathematics, three and four hours a week were taught in the two upper forms. During the school's first few years, Ole Jacob Broch, obviously led all the teaching of mathematics. When Broch left the school in 1848, Hartvig Nissen's brother, J. M. Nissen, took over the teaching, then in the last five years before Sophus Lie came to the school, Sjur Sexe and Hartvig Caspar Christie were responsible for Mathematics. Both later became professors: Sexe, of mining and physical geography, and Christie, of physics. But it was still in many ways Ole Jacob Broch who ruled the teaching of mathematics. As professor of mathematics Broch not only taught future teachers: H. C. Christie was, for example, the first science candidate in 1855, that is, the first time this science teacher examination was conducted. Broch also determined the requirements for the *examen artium*. Broch was known to be such a strict examiner and censor that many could not pass his requirements –

both Ibsen and Bjørnson flunked when they went up before Professor Broch for the *examen artium* in 1850 and 1852.

During these years while Sophus Lie was a disciple, a change occurred in the mathematics textbooks at Nissen's School. Earlier it had been B.M.Holmboe's books from the 1820s – in a newer edition by J.Odén and C.A.Bjerknes – that were used in arithmetic and geometry, but the school was now in the process of changing over to Broch's textbooks. Broch's geometry textbook of 1855 had already been introduced in the lower forms and it continued to be used as the disciples progressed into the higher forms. Sophus Lie's mathematics teacher during his first year at the school was Ludvig Sylow, a man who was later to play a definite role in Lie's life, both as teacher and co-worker. Sylow himself would publish mathematical work of an epoch-making nature. Of the first meeting between these two at Nissen's School, Sylow in later years stated that although Sophus Lie was the most competent student in mathematics, he could not see "a future mathematician" in young Sophus Lie.

One teacher who seemed to have made his mark on the school was Danish-born Hans Smith Hiorth. He was fifty years old, had travelled widely in Europe, and had taught geography to all the science and Latin forms for ten years. Early in his teaching career, Hiorth had publicly critized the established Latin schools – in line with Nissen's program he had stated that "the School's Activities included the whole Human Being" and not just "intellectual, moral and religious Formation . . . where Body and Soul, where Comprehension, Emotion and Will, are to be developed in harmony with one another". In addition, he considered that the refinement of physical and aesthetic ideals were also essential.

Nissen's School, the handsome three-story building located in a large garden, had a tall, spacious gymnasium within its rear building. Gymnastics had been an important subject since the school's inception and many people regretted that circumstances did not permit more than two weekly hours per class – ideally it should have been one hour every other day. The newly founded Christiania Gymnastics Association had also used the school's gymnasium for a couple of years, until in the current year, 1857, when they built the country's first gymnasium on the neighbouring property. Sophus Lie seems immediately to have enjoyed himself in the gymnastics hall and he gradually became an enthusiastic gymnast. In summer there was bathing and swimming – the school did not have its own baths but had a rental agreement which provided admission to the military baths for the school disciples.

Lieutenant-Colonel Ulrik Sinding Rosenberg was the school's gymnastics teacher, and who also led its rather extensive drill practice. The youths' preparation for "the defense of the Fatherland" was made explicit in the school's program. The school had taken over 200 used rifles from the Army Department and the older students were individually trained to load and fire. There were chain formation exercises and salvo firing along the same lines as those of the infantry. Six or seven days a year were allocated solely to such practice, which terminated in something the disciples regarded as a celebration with "forest manoeuvres" led by the gymnastics teacher, who was in turn assisted by junior officers and two buglers. The benefit of these exercises, as made explicit by the school, was not only military preparedness, but also the *esprit de corps* created among the disciples, as well as the installation of precise obedience and mutual dependence.

The school's motto and slogan was "Maintain the Chain!" and its symbolic meaning was interpreted as follows: "If during a Working Day at School one does not do one's best, one will hinder the Others; he will disturb their Progress and will deprive his Comrades of the smooth and uninterrupted Joy of Progression. Therefore – *Maintain the Chain!*" And should the good spirit and tone "cease to inspire One Person or Another, then he was betraying his Teachers', his Comrades' Trust: *he was not maintaining the Chain.*" During Sophus Lie's years at the school the disciples also began to use a black school cap with a white cord around it and a white silk button attached – the upper forms were also allowed a special kind of bow with the inscription, Maintain the Chain!

Among Sophus Lie's classmates there was one in particular – Ernst Motzfeldt – with whom Lie bound himself with ties of friendship that would last for the rest of his life. Ernst Motzfeldt would become an invaluable supporter in the years to come, and Sophus Lie would often frequent the Motzfeldt home. Sophus Lie became almost like a member of the Motzfeldt family, and he would later be a lodger in their house for several years. Ernst and Sophus were the same age, but at the time when Sophus started at Nissen's School, Ernst Motzfeldt had already been a pupil there for three years. He was the son of Supreme Court Judge Ulrik Anton Motzfeldt and Pauline Birch.

U. A. Motzfeldt had long been a renowned figure in the world of Norwegian officialdom. At this time, in 1857, he was President of Parliament and sat on a series of commissions and committees. In the 1830s he had been part of the editorial board of the journal *Vidar* and he had edited *Den Constitutionelle* [The Constitutional Journal], the organ of the intelligentsia and the newspaper of the government and the conservatives. Motzfeldt belonged to conservative circles, and in his home – with five children from a previous marriage – a conservative spirit reigned. In the intense debate raging about the elimination of Latin composition, Motzfeldt senior, as a parliamentary representative, had objected to the issue being raised in the first place. He probably did not completely share Hartvig Nissen's views on school and education; nonetheless he wanted his son to receive this education, which in deepest conservative circles was regarded as the most progressive. Other men in leadership positions also preferred that their sons receive this more modern education. The son of Cabinet Minister Hans Riddervold of the Church Ministry was another of Sophus Lie's classmates, and several others were the sons of society's prosperous and leading men.

Christiania – the country's capital with the National Assembly, the University, the Supreme Court and other institutions that otherwise characterise a sovereign state – at this time had a population of scarcely 35,000 inhabitants. These were prosperous times and there was much activity in the city. New and bigger buildings shot up, trade and commerce blossomed. The railway had made its inroads with construction work and regulations, and a new meeting place was created around the railroad station. The university buildings along the main street (which that same year had been named Karl Johans Gate) were finished in 1852. Much prestige and authority congregated around the University, the seat of learning for public servants. But this was also a time when the class of high officialdom and that of higher finance sought out each other's company and created alliances that became a driving force in the country's social and economic life. As an example, The Norwegian Credit Bank [Den Norske Creditbank] was founded in 1857 at the instigation of Professor of Mathematics Ole Jacob Broch. Broch was also counted as the founder of the insurance company that later took the name Gjensidige. Professor Broch was equally engaged in both the construction of the railroad network and the first telegraph installations; he also had a finger in countless commissions and boards concerning statistics and measuring systems, finance and insurance – both private- and government-operated.

At about the time of Sophus Lie's arrival in Christiania in August 1857, a summary appeared in a weekly magazine, *Illustreret Nyhedsblad* [News Illustrated], of an account of how a travelling Frenchman, Louis Enault, regarded the milieu in the Norwegian capital. The following appeared in a travel description in Paris: "The University in Christiania is the Seat of zealous Studies. One might perhaps accuse it of pursuing a far too one-sided Course toward the Utilitarian, and of too quickly reaching for the practical Results of Science, an Accusation one would never make against the Universities at Göttingen, Heidelberg, Bonn or Jena, where they seem prepared to abandon the Earth and lose themselves in the Clouds of Ideology. One must reprehend both Excesses. This Goal, pursued so obviously, brings the Norwegian University somewhat far Afield from the more noble Road, such as the one embarked upon by the Schools of France and England. In Norway one finds the Preparations for a Refinement that seeks to be finished all at once, and unintentionally apart from that educational Preparation that can only be acquired with several Years of patient Effort. One overlooks the classical studies [les humanités anciennes] far too much, those that reveal the Man of Education. During my Stay in Christiania I visited many Students. I have seen cold Men of Reason with good Qualities of Comprehension and otherwise, nice people; very few Poets and far too many Mathematicians."

This piece was translated and presented by Ludvig Daae, later a professor and power figure with whom Sophus Lie also had to associate. Ludvig Daae had read Latin from the age of eight and later was said to be the last Norwegian to speak and write Latin with ease and facility; hence, in connection with the elimination of the Latin composition, it is understandable that he called his opponents "flat-bottomed reasoners" and "anglicised snobs". This item also appeared in *Illustrert Nyhedsblad*. Professor Monrad, on the other hand, represented "both the Study of ancient as well as modern Civilizations in general". Nonetheless, it was Voltaire's words about Archimedes being more imaginative than Homer that ruled the day in the Norwegian capital.

In terms of the cultural arena, it seems these years were an in-between time, an intermission and an apparent stagnation, but in reality it was a time of growth. In future years, the 1850s were regarded as an interregnum of minor artists before the big "modern breakthrough" of Henrik Ibsen, Bjørnstjerne Bjørnson and the golden age of literature, accompanied by painters and composers of top European rank. Camilla Collett excited attention when her novel *Amtmandens Døtre* [The District Officer's Daughters] came out in two parts, in 1854 and 1855, and it was a signal of new times when Bjørnson's peasant tale *Synnøve Solbakken* came on the scene in the fall of 1857, and Ibsen's *Fru Inger til Østeraad* [Mrs. Inger of Østeraad] was staged at the Christiania Theatre the same year. Otherwise the 1850s were golden years for student theatre and student comedies; the big theatrical event of this year 1857 – when Sophus Lie came to the capital – was the hullabaloo surrounding the satirical piece *Gildet paa Mærrahaugen* [The Feast of Mærrahaugen, with its subtitle, *The Charming Cucumber*] [18], under the signature of Jokum Pjurre, in which the concrete, the physical and the down-to-earth were presented as the jolly alternative to late romanticism's use of the supernatural with its ghosts, wood nymphs and sorcery. Behind the pseudonym Jokum Pjurre, stood Olaf Skavlan, a philologist, and someone Sophus Lie would come to know very well.

Thus it was that Sophus Lie stayed at Nissen's School for two years before he took his *examen artium* together with nine of his classmates. There were also disciples from the Cathedral School in Christiania who took the examination. From the entire country a total of 109 boys sat the exams, and 98 of them succeeded. Approximately half of the candidates were privately tutored. The *examen artium* that year began on August 3rd, and consisted first of three written tests – two compositions in the "mother tongue" and one translation of a Latin text. Later there were oral examinations in Latin, Greek, German, French, Religion, History, Geography, Arithmetic and Geometry. On the first day of examinations the students could choose between two subject questions: "Description of a Summer Day in the Country" or "The State of Europe at the Time of Martin Luther's Emergence as a Reformer." Sophus Lie chose the first subject. He became ill and could not sit for the next two days of examinations. The grading of the written examinations was to take place on August 11th – Sophus Lie had passed the first exam and was allowed to take the two he had missed. In the first of his postponed exams he answered the theme "A Comparison between Lycurgus and Solon as Lawmakers" and the Latin text he had to translate consisted of approximately 270 words. Lie now passed these exams as well, but in the school's printed schedule for the days of the oral examinations his name is merely inserted by pen, probably as a result of his illness and the postponed tests.

The oral examinations took place over a period of sixteen days and the candidates were distributed among sixteen groups of six to seven each. These were

[18] "Feast at Mærrahaugen" (Mærrahaugen being the hill in Christiania (today, St. Hans-haugen) where horses were stabled) was an ironic reworking of Ibsen's play, "Feast at Solhaug", which was presented for the first time a year earlier.

tough and extensive exams and caused great stress to many of the aspirants. The day before the oral exams were to commence, one of the Nissen's School candidates died. The deceased was Sophus Lie's classmate Johan Fredrik Johannessen, whose father had been a merchant in Christiania. The school issued the following statement: "In him the School lost one of its most promising Disciples ever to stand for Admission to University and his Mother, the poor Widow, lost the best, the dearest, her only Son, her bright Hope for the Future."

Sophus Lie succeeded with his *examen artium* and received the top mark, that is, Number 1 in all oral subjects except Greek, in which he received a 2. Only in his "Preparation in the Mother Tongue" did the mark go down to a 3, but of his classmates only Riddervold received a better mark for his Norwegian compositions. In the Latin translation, Lie received a 2. The marks were totalled up to make up a summary mark, which for Sophus Lie was 15. His best friend Ernst Motzfeldt received a 14, while Riddervold had a 16. These were the three best in their class. Ernst Motzfeldt later intimated that Sophus Lie was disappointed in his mark in Greek and that it was a contributing factor to his not choosing to study philology. No discernible enthusiasm for mathematics had yet emerged into the light of day from within Lie, even though Sylow, his teacher, regarded him as the best student in the class in this subject. Lie knew what he needed to know and received the best mark in Arithmetic and Geometry – with Professor Broch as internal examiner and Senior Teacher Sylow as external examiner. Lie's mathematics education during his last year at Nissen's School seems to have been somewhat uncertain. When Ludvig Sylow at the beginning of the school year in the fall of 1858 was appointed Headmaster at Fredrikshald[19], a whole year passed before Nissen's School once again had a permanent mathematics teacher for the upper forms. Then the newly graduated science teacher, Cato Maximillian Guldberg, arrived – someone who perhaps had also substituted at the school during Sophus Lie's last year as a disciple. Ten years later C. M. Guldberg became a professor of applied mathematics – and three years after that Sophus Lie also held a professorate in Christiania.

[19] Fredrikshald, a garrison town on the southern boundary with Sweden is today known as Halden.

Student Life

A short time after he had become a university student, in the fall of 1859, Sophus Lie enrolled at the Royal Fredrik's University of Christiania. We know little about his reflections concerning the choice of studies, except that according to comments by his best friend Ernst Motzfeldt, Lie might well have been thinking of studying philology. The two friends and neophyte university students, Ernst and Sophus, had mounted a little hiking tour after the *examen artium* and before the start of lectures at the University. They headed west from Christiania, through Asker to Drammen, on to Kongsberg, and returned across Ringerike, north of the capital.

A definitive choice of studies could in any case be put off for the time being. Before the young scholars were eligible to embark upon their public service curricula, they had to take a set of rather comprehensive preparatory exams, the so-called secondary examination, which was divided into three sections – the *examen artium* was considered the first examination. The secondary examination could be concluded after one year of study, but many – Sophus Lie among them – took a year and a half to complete the programme.

Exactly during that autumn of 1859, two variations of the secondary examination became available. The fierce struggle for hegemony – classical languages versus the sciences – had made its imprint here as well. The fact that Sophus Lie chose the natural sciences may indicate that he had already decided to study the sciences, but on the other hand his choice was still in line with the University's Universal Act of 1845. The freedom to choose that became a reality in the autumn of 1859, provided more room for the humanities and classical studies; moreover it was considered to constitute a confession that things had gone too far, too fast, in the direction of the natural sciences in the years since 1845, when the new university regulations had been voted into effect. In the interim, an examination for science teachers had been instituted as a separate course of studies – a big triumph for the sciences, although these studies still came under the Faculty of Philosophy. The two variations in the course of studies toward the secondary examination could also be supplemented by additional examinations and did not involve any definitive choice of future public service studies.

The background to this broad-based secondary examination and its significance for general education was a subject Sophus Lie would later be interested in, and would come to take a position on, both as a professor and in letters to newspaper editors.

Having to go through two sets of academic examinations – the *examen artium* and the secondary exam – before one could apply to study in public service fields – had been a long tradition at the Danish-Norwegian universities. The regulations went back to 1675, to Imperial Chancellor Peder Griffenfeldt, who was of the opinion that the scientific level in the twin realms must be increased and that the foundations of knowledge among aspiring professionals had to be strengthened. This was probably in counterpoint to Sweden, where the quality of teaching had already been made more effective in the Latin schools and universities.

From the moment of the first University Act of 1824, it was determined that a supplement to the Latin school was necessary. The students must be tested in their knowledge of subjects useful to everyone, such as philosophy and mathematics, astronomy and natural history, classical languages and history. The old Danish-Norwegian specialisation that required only theology students must take the test in Hebrew was continued in Norway. But when a new University Act came into effect in 1845 – primarily due to Professor A.M. Schweigaard's efforts at the University and in Parliament – Latin and Greek were eliminated from the curriculum in favour of the natural sciences. The secondary examination now consisted of philosophy and five natural science subjects: Mathematics, Astronomy, Physics, Chemistry and Natural History. The examinations took place in three sections. In the first, the subjects of mathematics and natural history would be read, in the second section, zoology, botanical classification and astronomy, and in the third, physics and chemistry. Philosophy held a special place: lectures in philosophy were held in the course of three semesters and the students were required to take philosophy for at least two of these. During the first semester at University no one was allowed to take more than the first section and no one could sit the examinations in the two other sections without first having finished and passed the first section (mathematics and natural history). The regulation of 1845 marked a definitive triumph for the importance and position in general education of the natural sciences. At this time the secondary examination was the highest degree one could acquire in the natural sciences, apart from a degree in mining. And Latin had long since disappeared as a terminal degree – a Doctorate in Norwegian was first defended in 1847 by Ole Jacob Broch – the University curriculum catalogue was the only place where the text was printed first in Latin, and then in Norwegian, a practice that persisted until the fall semester of 1894.

But the secondary examination regulation of 1845 was criticised by theologians, philologists and lawyers, who claimed that they had little use for what they learned in mathematics, astronomy and the other science subjects. The Faculty of Theology in particular was of the opinion that they should rather emphasise their real calling – the examination of the soul, and preaching. The times when parsons taught people natural history were irretrievably gone. And without saying anything demeaning about the natural sciences, the Faculty of Theology commonly claimed that in principle the times did not require a broadened education – that "General Education must not be advanced at the Expense of Thoroughness" – and good sense and concentration were underlined as the basis of all scientific development. This was the ongoing discourse that, as of the fall of 1859, led to a choice of subjects for the secondary examination. Now those who took three of the four subjects – History, Old Norse, Latin or Greek – would be exempt from Chemistry, and in addition, would have their Natural History curriculum reduced to a general introduction to this science.

The principle that all students should follow the same course of studies was abandoned. One year later, when the Faculty of Philosophy was divided into the Faculty of History and Philosophy and the Faculty of Mathematics and Natural Sciences, a new division in the old educational foundation became a fact. The "unity of education" would be further destroyed through the school and university disputes of the 1860s, that among other things resulted in the old Latin school being divided into a Classics line and a Science line, which respectively graduated Latin *artium* and science *artium* students.

In a newspaper article Lie later wrote about the regulation concerning choice of subject for the secondary examination: "The Purpose of this Freedom of Choice was supposedly to provide a certain Latitude for the Interests and Desires of those with different Gifts. In fact the whole Thing largely amounted to the future Science Teachers and Medics choosing the Sciences while Philologists and Theologians rather preferred the Humanities. In any case, the newly-instituted Freedom to choose, which did not exactly coincide with the original Perception of the secondary Examination, *had to lead to Thinking that this Examination might not only be regarded as a Termination of General Education, but that it also might serve as an Introduction to the various Public Service Courses of Study.*"

And when later, in 1883, this secondary examination was the subject of intense debate, Professor Sophus Lie was one of those from the Faculty of Mathematics and Natural Sciences who proposed that "one ought to return to the Original Intent of the Secondary Examination, that this Examination alone ought to fill the Holes in the General Education of Latin and Science Examinees." His proposal, which involved the secondary examination having compulsory subjects, found support in the Faculty of Medicine, while the other three faculties wanted to maintain the freedom to choose, to the extent that the secondary examination would serve as an introduction to the public service degree in these faculties.

Probably no other exam in Norway has been so much discussed as was this secondary examination. Literary memoirs are full of anecdotes about legendary professors and outlandish questions and answers both in the auditorium and around the examination writing table. One of the reasons for this was that the professor or the lecturer who did the teaching also presided over the examinations, and since most of the teachers wanted very much to be well-liked, they told their eager students what they should answer in order to come through the examination gracefully. This in turn led to the students swatting and cramming – often only shortly before the exam – what they felt certain they would be asked about. Holberg's[20] old expression "half-educated bandits" was a much-used characterisation

[20] Ludvig Holberg (1684–1754), born in Bergen, was poet, philosopher and historian, as well as dramatist. Considered the Molière of Denmark-Norway. His plays, often satirical, burst upon the public in 1722.

and some professors were so butter-fingered with regard to the marking that the University's executive had to step in and demand they be more prudent.

For a student with a good *examen artium*, the subjects of the secondary examination were considered easy, and the subjects were laid out such that to some degree they could be mastered by individual study. There continued to be many who combined the secondary examination with one or another private tutorship in the country.

But many others also put much work into the secondary examination and felt that it paid off in their later studies. Indeed, Sophus Lie seems to have been of this opinion. He would apply himself with great industry to his studies, even though in the autumn of 1859 he started off with a superficial attitude and was absent from classes.

When Professor O. J. Broch began his lectures in mathematics on September 9th that year, and continued with three hours each week, Sophus Lie had left Christiania and was back home with his family in Moss. It seems he had been feeling ill for a period – what he called "a terrible Stomach Pain", in a letter to Ernst Motzfeldt.

Over the years the correspondence between these two friends would become extensive, and fifty-eight of Lie's letters to Motzfeldt are still extant, although for his part, Sophus Lie did not take care of the letters he received from Motzfeldt. The very earliest letters in this collection date from the beginning of September 1859, while Lie was home in Moss even though the lectures for the secondary examination had begun in Christiania. He reported that the terrible stomach pains continued and left him quite inactive. For several days he had "no Mind for anything other than to read Novels, something that in and for itself is not such a wrong Occupation, when one, as I, on this Occasion, have good (that will say enjoyable) Novels." Lie complained that he had very few friends in Moss, and those there were, "were so busy with the Business Life that they have no Time for anything else." When the stomach pains began to diminish he took a three-hour walk to the neighbouring parish of Råde, a distance of about 20 km. from Moss, in order to visit one of his classmates from Nissen's School, Lauritz Bassøe. And Lie reported that he grew both disappointed and angry when Bassøe was not home: "But this would be to the discredit of Bassøe when I got back to Town; the Blockhead invited me himself, and then he went out; now indeed, the Time, the Grief !" About the forthcoming studies he wrote: "Naturally enough, I am living in a full Ignorance concerning the Lectures and such, [...] you would do me a great Service if you would give me a Touch of Enlightenment thereon, together with how you have found the Lectures, and in any case, whether or not you have found them to be indispensable." Motzfeldt's answer to this would be decisive for it would determine when Sophus returned to Christiania. Motzfeldt was determined to study law, but at the beginning of the secondary examination course, it was in any case a common course for all.

In addition to Professor Broch's lectures in mathematics, Professor Halvor Rasch's lectures in natural history were now a topic of conversation – on average there was an audience of about fifty. Rasch lectured six hours a week ("on the Anatomy of Organs in the Plant and Animal Kingdoms, Physiology and common Systematics"), and in addition to the lectures at the new university buildings on

Sophus Lie – in the middle of the back row – had planned, while still a university student, on a military career, but poor eyesight seems to have put a stop to this plan.

In 1856 the students got their own gymnasium, the "gymnastics hall", at the corner of Universitetsgaten and Kristian IVs gate. The photo shows the room as it is today.

Top left:

Ole Hartvig Nissen, in 1843 founded Nissen's School, and later became the country's leading reformer of school policy.

Top right:

Carl Anton Bjerknes, Professor of Mathematics, famous for his experiments and theories concerning hydrodynamics.

Ole Jacob Broch, Professor of Mathematics, Parliamentarian and Cabinet Minister, and from 1879, leader and director of the international office of "weights and measures" at Sèvre in Paris.

Lie's friends and hiking companions: Ludvig Sylow (top left), *Kristofer Janson* (top right), *Axel Blytt* (bottom left) *and Gerhard Armauer Hansen* (bottom right).

The Green Room, a circle of students and academics who, following dinner at the Student Society, gathered in this room for coffee, tobacco and discussions on questions of the day.

Karl Johans Gate, he would also take the students out for demonstration tours, to see living plants at the Botanical Gardens located in the district of Tøyen.

After his absence during the first half of September, Sophus was probably quite soon back and attending lectures in the capital. When the examination came in December he had mastered the course of studies so well that he had obtained the mark of 1 in both subjects. Moreover, Professor Rasch was one of those who never gave a mark lower than a 3; that is, a *laud* or *second class* – it was said that Rasch was, moreover, one of those who would give top marks to a candidate who did nothing greater than distinguish between a rose and a lily, and he would certainly have given good marks to anyone who explained that the blood's dark colouring was caused by a substance that resided in the veins and resembled soot inside a pipe. But as for Professor Broch, his severity at the examination table never relaxed, and the mathematics course included "Stereometry, Trigonometry (the plane and the sphere), Equations and the first Fundamentals of Algebra, including Knowledge of common and quadratic Equation Solving". Presumably this teaching was in accord with the textbooks in arithmetic and algebra that Broch published in 1860. In comparison to Holmboe's old textbooks, Broch's starting point was closer to the realities of life: the accounts and the material were broader and richer, but the books were heavy-going, and dependent upon a teacher. It is perhaps interesting to note that Broch erroneously maintained that Euclid's parallel postulate could be demonstrated using the other four axioms alone.

In January 1860, Sophus Lie went on to the second part of the secondary examination course, and during this spring semester he read astronomy and the systematics of natural history. The natural history lectures were given by Professor Rasch, but astronomy was conducted by the lecturer, Carl Fredrik Fearnley. The astronomy section included "an Overview of the most important Theses of spherical and theoretical Astronomy, together with a Description of the most common Furnishings and Use of Astronomical Instruments." The latter, the instruments' furnishings and use, were explained at the Observatory, the fine University building on the Solli estate. C. F. Fearnley was said to have been a silent, absent-minded and withdrawn person who wished everyone well and had no ill-will toward anyone. But for Sophus Lie, who after completing the science teacher examination, thought to pursue a career for himself in astronomy, Fearnley became the one who stuck a spanner in the works and brought such plans to a halt.

In the exam in June – in astronomy and systematic zoology – Sophus Lie again received *præ cæteris*, the mark of 1 in both subjects. To complete the secondary examination there was now only the third section left, that is, chemistry, physics and philosophy; physics and philosophy continued over two semesters, and during this spring semester Sophus had listened to his old teacher of Norwegian, Professor Monrad, lecture for four hours a week in psychology. Young Sophus seems not to have enjoyed this at all.

He was home in Moss during the summer holidays. He had actually planned a journey to Trondheim this summer together with his father – for the great and ceremonious crowning of King Karl XV, at Norway's medieval Nidaros Cathedral.

But it seems that Father Lie could not be away from Moss for such a lengthy period due to the many religious sects and itinerant preachers who at that time were impressing their stamp upon the Moss congregation. Thus Sophus had to console himself that his father had only postponed the Trondheim journey until the following year. Then, according to his father, they would, together with Sophus' eldest brother, Fredrik, who would be finished with his science teacher's examination, visit Trondheim and the regions where Father Lie had spent his childhood – and at that time brother John Herman would be finished at the Military Academy. That summer of 1860, Father Lie thought that Sophus "with the greatest of Unction" should, "out of brotherly Piety" visit his sister Mathilde in Tvedestrand. It was more than three years since he had last seen her and her husband, the district doctor, Fredrik Vogt. And Sophus had never seen the three children who had been born during that time. Sophus confessed to his friend Motzfeldt that he had no desire to go to Tvedestrand, but that he could not announce his lack of desire without being called "a Scoundrel of a Brother" – and: "When anyone begins to talk to me about Piety, it is almost as bad as when Monrad goes on about his Morality. I do not know whether to go forwards or backwards."

Sophus visited his sister that summer, and stayed at the doctor's estate in Tvedestrand – with the sun and the summer over the fjord and the town giving life a heightened taste. During the years that followed, Sophus would make several summer visits to Tvedestrand, something from which both he and the Vogt family would take great pleasure.

The autumn semester at the University began again in September. In physics, which also included meteorology, the lecturer, H. C. Christie, who in the spring term had lectured on electrotechnology and thermology, now continued with mechanics and acoustics. Christie was known for laying out the course clearly and simply for the students, and frequently he had an audience of seventy, and he inspired enthusiasm with the lively and humorous elements he included in his lectures. Associate Professor Hans Henrik Hvoslef taught chemistry. In philosophy it was Monrad who held forth on ethics and the history of classical philosophy. In addition, Johan Sebastian Welhaven, the poet and professor, contributed to the philosophy course by lecturing on "philosophical Propadeutics [the nature of preliminary instruction in philosophy]". The two professors alternated their lecturing in the various fields of philosophy, and they based their teaching on some small pamphlets that Monrad had allowed to be published. Everyone said that Welhaven's lectures were brilliant in their eloquence, no one had a voice of such surpassing quality, nor such a command over the art of oratory as Professor Welhaven. He mounted the podium, always at the precise moment, in the largest auditorium – No. 6 in *Domus academica* – cast a sharp glance at the gathering, fitted a small piece of paper to the lecturn with a pin, and then captivated the students for three-quarters of an hour. He then removed the pin, put the slip of paper in his pocket, and silently departed. Most reported later that they could not recall much of what Welhaven had said, but they had a memory of an extraordinary master at work, bending and joining his words. Welhaven could be extremely demanding as an examiner, and derisive of the less knowledgeable candidates. Sophus Lie was hardly in their ranks, but philosophy was nevertheless the only subject from which he emerged without a top mark – in philosophy he was left with a 2. Apart from this, he completed the examination in December 1860, with the mark of 1 in both physics and chemistry. As his overall standing in the secondary examination Sophus Lie stood *præ ceteris*, and with this he was one of the two best in his cohort.

After this came what was termed "a well-deserved Christmas holiday" at home in his father's house at Moss. Then Sophus Lie began the public service course in science in January 1861. He was nineteen years of age, and was thus beginning the same course that his brother Fredrik was now finishing. The total number studying at the University at this time was 560, and only ten or twelve of them were science students.

The public service course in science had been founded by Professor Broch ten years earlier. It took as its primary task the education of competent science teachers; yet the plan was also to give "the Future of technological Activity a powerful Support". The first candidates graduated in 1855, and of the fifty or so who took the examination over the course of the next twenty years, all of them, so to say, became teachers, most adjunct teachers and head masters in the higher schools, nine were linked to the University, and only a very few had anything to do with practical economic life.

The course of studies was divided into three sections, each with its specialised centre of gravity. The first section was the heaviest, and it included mathematics, geometry, mechanics, mechanical engineering and technical drawing, and it was this part of the curriculum that was most time-consuming. Sophus Lie would use three years on this part of the curriculum – and only one year for each of the other two sections.

It was Professor Broch who led the subject, and he was the predominant teacher in this first section. He gave lectures over a series of broad themes – "as good as all the branches of mathematics", according to the subsequent science student and associate professor, Elling Holst. The University's annual reports show that in this particular year, Broch lectured on integration, analytic plane geometry and statics (knowledge of the equilibrium between energies and bodies), and that in these lectures he had from six to eight students. In addition to Professor Broch, Carl Anton Bjerknes also lectured. During these years he had advanced from a research fellow to a lectureship, and in 1866 became professor of applied mathematics. Bjerknes lectured on complex functions, statics, mechanical engineering theory, analytical stereometry, and went through Cauchy's residue theorem. This was composed of a series of lectures stretching over several semesters of the science students' first section. These students also heard Lector Christie go through the material that dealt with the earth's magnetism.

Sophus Lie seems to have thrived during these first years as a student. Apart from all the lectures and studies, he was often together with friends – in festive situations, in academic gatherings, out on tours and inside the gymnastics hall.

Getting involved in gymnastics was a very popular activity of the time. Nissen's School had had a large gymnasium – The Christiania Gymnastics Association had been founded in 1855, mainly by immigrant Germans, and went directly to work building a gymnastics hall. Thus, in the autumn of 1856, the students had been able to put to use a gymnastics premises that had been built on the corner of the University property, facing Kristian IV Gate, and this was the fulfilment of a decade-long demand by the students. It is highly likely that it was here that Sophus Lie and his friends met and practiced gymnastic exercises. Apart from Ernst Motzfeldt, Gerhard Armauer Hansen was one of Lie's closest friends at that time. Armauer Hansen had become a student the same year as Motzfeldt and Lie, and following the secondary examination, he had begun to study medicine; he took the public service exam in medicine the same year that Sophus Lie completed his own public service studies. At this time it seems that it was these three – Armauer Hansen, Motzfeldt and Lie – who were the core of a group of comrades, who were all avid students, gymnasts and mountaineers. In his *Memoirs of a Life* (1910), Armauer Hansen wrote about how they often amused themselves swinging from the gymnastic rings and doing backward somersaults. It is in these memoirs that Armauer Hansen recalled the episode in which Sophus "sailed through the air a great distance and fell to the floor on his posterior" and cursed them up and down for laughing when "he had nearly done himself in". And it was now, in the middle of this cursing that Sophus was said to have abruptly stopped and muttered, "It's strange though, because I even calculated exactly when I should let go." Armauer Hansen commented, "It never occurred to him that perhaps he let go (of the gymnastic rings) later than he calculated he should do so."

Also in relation to the gymnastics hall, Armauer Hansen recounted that Sophus Lie was very methodical when he vaulted the wooden horse: "He would spit on the floor at the point from which he would spring, then he would place his left foot on the spit mark and take seven paces backwards; he would now run and vault."

The venerable Student Society was certainly the great meeting place. Here the students gathered for discussions, parties and theatrical performances. But during the 1850s an active life of societies and associations had blossomed, with various fields setting up associations – and much of the professional and scientific discussion took place there. "Literaten" focused on literature, while the Language Association focused on languages; theology had its own student association, as did medicine. And in 1859 the science students formed their own association, called "Realisternes og Mineralogenes Forening" [The Scientists and Minerologists Association]. As a rule there was considerable interaction between these associations and the Student Society, and most of the associations held their meetings on the premises of the Student Society on Universitetsgaten [University Street]. Here they held lectures in their respective disciplines, and here they comforted themselves with a glass of beer, and among other things, a smoked herring – seldom did they resort to the punch bowl – and when the spirits were particularly high, they danced "the Round Dance and the Chain of Friendship" and strolled in the park of the royal palace afterwards. It also occurred that topics that had been taken up in these associations were brought into wider discussion in the Student Society.

For example, this happened when the theologians studied Søren Kirkegaard, and took the discussion to the Student Society under the question "Is is defensible to become a Priest in the State Church?"

Sophus Lie was never a very zealous "societies man", but he was one of the most diligent about the meetings of Realistforeningen, the Science Association. He probably shared his friend Armauer Hansen's assessment: "Otherwise there was not much excitement in the Student Society during those years" – on Saturday evenings one simply sat around a long table and drank "barely endurable punch and made boring speeches; one had to be sophomoric and witty." And thus it was that all these jurists did their best to outshine one another in sophistry. After the founding of the Medical Association, Armauer Hansen almost never attended the meetings of the Student Society, except for the larger celebrations. Sophus Lie seems to have followed a similar pattern; he went instead to Realistforeningen.

The Science Students Association started its activities independently of the Student Society – in rented premises at an organ builder's establishment, a place in Akersgaten – but the relationship with the Student Society created disagreements about what sort of an association it should be, about what it should focus upon, and what its social and scientific activities should be. In January 1861 – that is precisely when Sophus Lie became a student of the sciences and a member – the Science Students Association held its meeting for the first time on the Student Society premises, and this was seen as a defeat for those who were the most interested in science.

In these gatherings – at the outset, every Thursday evening – there were normally ten to twelve in attendance, and Sophus Lie was among those who attended most meetings in these first years. The meetings generally consisted of a scholarly or scientific lecture by one of the members, but over half the meetings had no lecture, and as such they scarcely had high scientific ambitions. In a description written thirty years later, Lie wrote about the science students of the 1860s: "We got together at Realistforeningen for Lectures and Discussions on theoretical, and occasionally also, pedagogical Questions. The very fact that we had any Lectures whatsover by means of this Method, although seldom of any higher Sense of scientific Value, I am still convinced most were of great Benefit by virtue of this *independent* Activity."

In any case, the list of themes taken up by Realistforeningen's members seems quite impressive: lectures were held on glaciations, floral and faunal remains in bone cavities, on the propagation phenomena among plants, on languages and the origins of species, on the summation of six-sided stacks of cannonballs, on Gauss' development of the magnetic axes, on Euler's formula series for numbers, on the history of statics, on the origin of Arabic numbers, on the history of mathematics – and in addition, the concept *work*, over a period of time became a source of vehement discussion. Sophus' eldest brother Fredrik had also been active in the Association, and during this first year that Sophus was a member, Fredrik gave a lecture on the Keplerian laws – earlier he had spoken on functions in the development of series, on Taylor's theorem, on some geometric theorems, and on conic sections. Two years would pass before Sophus himself would hold any lectures in the Association, but then he also gradually became one of the most avid. One of the Association's most active members during these years was Thorvald Ingolf Broch, an all-rounded sciences candidate who seems to have been an inspiring force in the milieu. Thorvald Broch's contributions had both professional content and at the same time took up and discussed the regulations governing the whole of the science teacher examination, and in addition, he was a zealous reader of all kinds of *belles-lettres*. During the spring of 1861 at Realistforeningen, Thorvald Broch began a series of lectures on the geometric significance of differential equations – the material was difficult and after six lectures, with everyone's consent, the series was discontinued. The most thought-provoking of Thorvald Broch's enterprises was probably that he posed far-reaching mathematical questions that did not have secure answers, and it might seem as though Sophus Lie later took over a little of Broch's active and querying role in the Association.

Whether Sophus Lie went to Trondheim together with his brothers and his father during the summer of 1861, as his father had promised, is unknown. His two brothers, Fredrik and John Herman, in any case were not finished with their education, as the previous year's plan had assumed they would be. John Herman was discharged from the Military Academy in September and began his service in Bergen a short time later, while his big brother Fredrik was not finished with his science teacher examinations until December – and Fredrik began employment as a teacher at Nissen's School almost immediately. Perhaps little brother Sophus had to confine himself to being home at Moss that summer – in any case he hardly had the money in his own pocket for a journey to Trondheim – or perhaps he visited his sister and her family again at Tvedestrand. There are no definite clues as to where in the landscape he was to be found that summer. But all the same, he was back in the capital in September when the fall semester began, and at every Thursday meeting of Realistforeningen, through both fall and spring semesters, he was in attendance. Professor Ole Jacob Broch continued his lectures on "almost all mathematical branches", and the lecturers Bjerknes and Christie punctually conducted their planned lessons.

There is every reason to believe that Sophus Lie was performing gymnastics as avidly as previously, and undoubtedly he took part in overnight outings in the forests of Nordmarka north of the city, and out along the fjord to a cabin on one of the islands near Sandvika on the west side of Christiania. There they played cards late into the night, and were underway again by four o'clock in the morning to go shooting the ducks that made their way out to the fjord. Armauer Hansen would later recall, "I do not recall whether we took any life; we would shoot throughout the morning at half or wholly full bottles that were slung out to sea; whether we hit any of them I do not recall. But carefree and healthy it was, and then at the end of the day we rode back to the city again."

During the summer of 1862 Sophus Lie's first mountaineering tour came into being – a tour he mounted together with Ernst Motzfeldt and theology student Kristofer Janson, who had also become a student at the same time as the other two. The three student friends left Christiania on a beautiful afternoon at the end of June in the direction of what had long been considered one of the country's foremost sights: Telemark and Tinn, on to Rjukan Waterfall and the valley of Vestfjord and the towering peak of Gaustatoppen. The three students hiked to Rjukan, viewed the impressive waterfall [21], and climbed Gaustatoppen. In his memoirs written fifty years later, Kristofer Janson wrote about how, after Gaustatoppen, they arrived at Hjartdal, dropping from fatigue, in order to overnight there in the post house, as was the custom. Due to a christening on a neighbouring farm, with lots of people and lots of drink, there was only one serving girl on duty, even though it was late in the evening. In the middle of the night the people of the house came home, and the wife announced that she really wanted to go back to the christening party at the farm, and she insisted at the top of her lungs such that the husband, in order to stop her, locked his lady into the granary. The husband beat and tormented her so vehemently that the serving girl ran off to alert people at the farm. Then even deeper into the night the party-clad celebrants from the neighbouring farm arrived with shouts and yells, pounded on the door of the granary, freed the captive wife and bore everybody back to the christening party. And now it was the husband's turn to be cursed. Kristofer Janson commented in his memoirs that he later explained how ungodly it was to hit one's spouse, but the peasant had only responded, "You got to beat the fleece when it's full of fleas."

Kristofer Janson had more to tell as well about this tour, and there is a letter from Sophus Lie that recounts their subsequent journey over the mountains. When they reached Byrtevatnet and Mo, they were hospitably invited to the vicarage, where they stayed a week. That is to say "there was a whole measure of lively, young, cheerful daughters there, and we were – students," commented Janson and recounted how both the vicar's wife and the boys of the farm joined in when it came to Kiss in the Ring, which was played out in the farmyard. Apart from this, the three students experienced "the old Vicarage's Amusements" with "Hare in the Oven", "Billy Goat on the Bedpost" (The billy goat did not behave in an appropriate manner and was dismissed), spruce sprigs under the sheets, and so on. The vicar's daughters – six in number – were, according to Janson, particularly interested to make closer acquaintance with Sophus Lie. Motzfeldt had earlier on the journey tried to clip Sophus Lie's hair, but as a prank had done it in such a way that there was "a deep valley across his skull". Sophus' head had consequently been christened Vestfjorddalen (the valley) and Gaustatoppen (the peak), and it was certainly a continual delight for the vicar's daughters to try to level off the Gaustatoppen with their scissors. Whether Sophus Lie had found this intimacy enjoyable is not known – his travel companion wrote simply: "The good-natured Sophus kept his patience, but he was still fortunate, for when he left, his Vestfjorddal was scarcely visible."

From Mo, Lie and Janson hiked further west, that is, first they had to make a water crossing by rowboat in the direction of Vinje, a main town in Telemark, and then further westward. The plan was to reach Bergen where Janson's parents

[21] During the occupation of Norway during World War II, heavy water was produced here – an element in what British and American intelligence saw as part of the development of nuclear capability, and hence part of the Axis war effort. With assistance from England, the Rjukan installation was sabotaged by the Norwegia resistance movement.

lived, and where Sophus' brother John Herman had moved. For his part, Motzfeldt continued on over the mountains to Bykle, and down the long north-south valley, Setesdalen, to his fiancée in Kristiansand, and then back to Christiania. It was to him that Sophus later reported in a letter on how the continuing westward tour had gone. Motzfeldt had left Mo the day before the other two, and Sophus began by describing how he and Janson had spent the evening prior to their departure from the hospitable vicarage – how they had been astounded by all the ladies urging the vicar's wife to set herself up out in the garden, whereupon with great solemnity, she had sung a serenade composed for the occasion, in honour of the two students. The vicar's wife, Louise Marie, née Poppe, was the cousin of the collector of folk melodies, Ludvig Mathias Lindeman, and the year before, during the summer of 1861, Lindeman had also been there at the vicarage, where he had written down seven melodies that Madame Louise had sung for him, including what later became the well-known "Ola Glomstulen".

Other than this, the leave-taking from the vicarage had been a lengthy affair. After the touching farewell serenade in the garden, Janson had to make a speech – Lie commented that he had certainly thought it was "one of his Copenhagen Speeches; he really got cooking again," and here he was alluding to the fact that earlier in the summer Janson and Motzfeldt had participated in a student meeting in Copenhagen. Moreover, according to Lie it was precisely *Janson*, with whom one of the vicarage girls had fallen in love, and she had informed Janson in confidence that "for her these Days would be unforgettable". When the two students had gone down to the beach to be rowed across the lake, Janson gave a new speech and the two then "chimed in with a rousing, ninefold Hurra for Mo, whereupon the Priest [Elling H. Friedrichsen] responded ." Lie concluded his description, saying "Whilst the Boat glided across the Water, we sang a Song, and kept it up as long as we thought we could be heard; whereupon we swung our Caps over our Heads, the Ladies waved their Handkerchiefs, and so it was over."

When Lie next wrote to his friend Motzfeldt ten days later, he had reached Bergen. The trip over the mountains had gone well; they had "walked well, eaten well (that is, much) slept well, and had Rain the whole Time." They had been taken in at the vicarage at Vinje and got to hear that two other students they knew from Christiania – Otto Lund and Georg Kent – had been there four or five days earlier, and had headed further west towards Haukeli. Lie and Janson caught up with the other two students the following day but they had the misfortune that their flask of spirits fell out of the rucksack and smashed – "Ever since then Janson kept thinking back with Wistfulness to that Event," commented Lie. He went on to describe the westward journey: "As a Rule, the Route was easy to find. At two Places we had to wade across Riverbeds. This was extremely unpleasant. We met a number of horse traders in the mountains, plus we encountered several Sætres (summer farm camps in alpine pastures), so we suffered no Want." They reached Røldal, and the next day, Odda: "At the Beginning, the Way was very arduous, one Hill was particular wild and seemed never to end; for the last Two Miles we had the Highway, right the way down." They stayed in a hotel in Odda – "like almost everywhere around Hardanger and Voss, one finds Hotels where one can have an excellent Time, although the cost was stiff. We had to break into the Reserves."

From Odda they took the steamship down the fjord; the other two students got off at Ullensvang, while Lie and Janson continued on across Hardangerfjord to Granvin – from where "we took a Tour over to Ulvik, the loveliest Place in Hardanger, but we were extremely sleepy," and their clothes were wet and never really dried out again. When they reached Voss, tired and almost penniless they walked past the hotel because they hoped they might be taken in at the home of a student friend, Jens Stub Irgens, who they knew lived at home with his parents on the sheriff's estate. Lie wrote: "We asked after the Student, who also came out, but he excused himself, that hollow Trumpeter, saying he could not invite us in, that the Family had just sat down to Dinner. To us this seemed a greater Reason to invite us in, but when we could not bring this about, we shamefacedly turned around and wandered back to the Hotel. Otherwise, Voss was "a large pretty Village" according to Lie, and he then described the onward journey – out along Osterfjord by boat they had met such bad weather conditions that they had to lay to at Dale and proceed on to Bergen the next day, where: "In the Middle of the Square we parted politely", and even though the exterior conditions during the last part had been unfavorable, they were "extremely well pleased with the Tour."

Sophus Lie now spent a couple of weeks at his brother's home in Bergen, during which time he met a number of students he knew, attended a student ball in full evening dress (which he borrowed), and began the return journey to Christiania at the beginning of August, presumably on foot. To Motzfeldt he simply wrote: "I must do the Home Journey rather all in one Go, otherwise I fear the complete collapse of my Exchequer, and now I no longer have Reserves I can resort to."

In any case, Sophus Lie was back in Christiania well before the beginning of the autumn semester in early September. He was now midway through the demanding first section of the curriculum for the sciences. Professor Broch had lectured on "as good as all the branches of mathematics". But more important in the mathematical sense, was the fact that Sophus Lie now got Ludvig Sylow as teacher. In the spring semester of 1862 and the following spring, Ludvig Sylow substituted for Professor Broch, who had been elected a Member of Parliament.

Sylow taught function theory, differential and integral calculus, the theory of the rotation of fixed bodies, hydrostatics and hydrodynamics – and most importantly, Sylow lectured as well on Broch's request, on the theory of algebraic equations, and he was thereby among the very first in Europe who taught Galois' group theory. In these latter lecturers he had an audience of only two or three, and Sophus Lie was one of them. Sylow's manuscript for these lectures is still extant, and it shows that he went through both Abel's and Galois' works on the solution of algebraic equations. Sylow seems to have been up-to-date in everything to do with contemporary research – later he himself would also make important contributions to algebra, and "Sylow subgroups" is a standard term in mathematical theory.

In the course of the spring of 1863, Sophus Lie gave his first lecture at the Science Students Association. The minutes of a meeting in April show that Sophus Lie proposed "that to study the Evolutes of Curves was a good Exercise in Differential Equations." (An evolute is a curve that is the locus of the centres of curvature of another curve {its involute}). And a week later he held "this aforementioned Lecture on these Curves, in which Evolutes of the nth Degree are the same as the Curves themselves." Notes and sketches from this lecture are still in existence. Evolutes certainly are a feature in mathematical history, and probably had been mentioned in one of Professor Broch's lectures, even though it did not appear as a lecture theme in Broch's varied lecture series. The well-known *cycloid* is a curve whose evolute is identical with the curve itself – the problem Lie posed seems thus to have been an attempt at generalisation, and to evince in any case a fundamental and accurate analysis of the relationship between geometric forms and differential equations.

When the summer holidays approached, Sophus planned his first foot tour in Jotunheimen, the range of high mountains in the southern half of Norway that at that moment was becoming the most popular of all for hiking expeditions. It was on this tour via Lillehammer to Vestre Gausdal that they were invited to join a wedding party on a farm, and where they amused themselves "excellently", and when Sophus impressed the village lads by lifting himself on the rings and bonking his head against the ceiling – three or four times more frequently than anyone else – and it was further along the way that they met the three beautiful, quick-witted alpine milkmaids, who, in Lie's words, were "free of every Type of superfluous Reticence". However, the distance they penetrated into Jotunheimen that summer seems uncertain.

In the first meeting of Realforeningen in the autumn of 1863 Lie presented the following – and from present-day perspectives, rather opaque – challenge: "Can one, through Differentiation, reach the Series Development of a Function of Sinus to Multiples of the variable Cosinus Series, and vice versa?" Then in October he began to borrow books from the University Library – the first book was *Élements de calcul infinitesimal*, by J. M. C. Duhamel, one of the leading mathematicians in Paris. Otherwise this was his examination semester for the comprehensive first division – Professor Broch lectured on aerostatics and aerodynamics, but most of the work was repetition of the required materials for the course. Lie's goal was to be the best of those who sat the examinations, but quite certainly he also took time to attend the Student Society's great celebration that autumn. With great panache the Society's fiftieth anniversary was celebrated with speeches by cabinet ministers, professors and students – Ernst Motzfeldt spoke on behalf of the students "at the fraternal Universities" (in Sweden and Denmark) – and two songs which had been especially composed for the occasion were sung – one written by Henrik Ibsen and the other by Jokum Pjurre, the *alias* of Olaf Skavlan. Ibsen urged the students: "Onwards! That's the demand./ Onwards Day and Night! / Go on, wheresoever Times demand,/ In the Youth of Norway's Fight!" Jokum Pjurre generated the golden glow of the Student Society (= the hero) now, in comparison with its early, thin years – so skint, so skint: but now it was as though the hero (= The Student

Society) "raised itself up like a Beggar with a Chalice/ and sat as regal as a King inside a Palace."

Sophus Lie sat the examination in December and received the top mark – *laudabilis præ ceteris*. He knew to perfection what was required. Yet despite this, Ludvig Sylow maintained his view that in meetings with Lie at this time, he in no way glimpsed signs of an emerging mathematician in him.

In a newspaper article written more than thirty years later, Lie explained these first years at university. He praised "Broch's strong Personality" that stamped itself on the course of studies and made up for the shortcomings of a subject that had only one professor. Lie maintained "that Broch, by means of his skilful, lively and spirited Lectures managed to impart to his Audience an Enthusiasm for Mathematics", but Lie also pointed out that "Broch's whole Personality was pulled along by stronger Powers of the practical Life than of Science [. . .] that his scientific Deeds were to a greater Degree those of a Physicist than a Mathematician." But despite this, Broch had made mathematics a major subject of in the programme of studies, and according to Lie, the basis for this was that "*this practical Man* knew that a good School had greater Use for Mathematics Teachers than for Teachers in all the Natural Sciences."

Sophus Lie finished this major part of the course of studies during the autumn of 1863. He took the two remaining parts of the science teacher exam course in the following two years – in hindsight it seemed to him that these two last parts were almost a subsidiary subject. He wrote: "Physics, Chemistry, Astronomy, Mineralogy, Geology, Botany, Zoology and physical Geography were regarded by many at that time as a Subsidiary Subject. In any case there were *certainly few who managed to get a solid Grounding and Knowledge, and real insight, in these Subjects*." As a result, he had little to say about the lecturing in these "subsidiary Subjects" and maintained firmly that "*Broch's, and later Bjerknes' and Sylow's Lectures on Mathematics in my Student Days, were those that from the whole Field of Science best answered the Demands of the Science Teacher Examination*." Lie must have been thinking about others than himself when he maintained that often those who were bright in mathematics did not come to the fore in the natural sciences, and vice versa. For his part, he himself wanted passionately to master everything and be the best in every subject. In any case, his ambitious goal was to get *præ ceteris* in all three sections, and therefore take the public service examination with distinction. He almost succeeded.

The spring and summer of 1864 were a turbulent time. The minutes of Realistforeningen for March show that "Lie began a Lecture on the Elements of Impact Theory" (the physical study of colliding bodies was popular at the time), but a month later the Association disbanded from lack of interest and poor attendance. At that moment great political questions had seized the students' attention. (Yet, when four years later Realistforeningen rose from the ashes, this was due to the efforts of Sophus Lie.)

The German-Danish War created great consternation among both students and Members of Parliament. The beautiful words of student-Scandinavianism about fraternity and solidarity, which had marked Nordic student meetings for years, was powerfully put to the test in February 1864, when a German military force crossed the border into the Danish county of Slesvig [Schleswig], and the Danes had to retreat. The Student Society – where among others, Ernst Motzfeldt was at this time in the leadership – became the arena of passionate discussions. A whole-hearted commitment to the fatherland was most frequently presented as the greatest of all human duties. At the end of March an extraordinary session of Parliament had passed a motion that allowed the king to use "line troops" to help Denmark, but the conditions were formulated such that everyone understood that any Norwegian war effort was in reality out of the question. It was understandable that the peasant farmers among the Members of Parliament were unwilling – the peasant population still made up the major portion of the Norwegian army – but what about "our educated, those familiar with Ideas"? It was said to be "pure Humbug" that only peasant boys were fit to be soldiers. On April 2nd, the young theologian, Christopher Bruun held a flaming speech in the Student Society – Bruun had reported to the army as a volunteer for the war effort on the side of Denmark, and challenged others to do the same thing. Still, when Bruun left for the war, he left the country all by himself. Bruun would come to make a great heroic contribution to the war, but it was a war that everyone knew Denmark was doomed to lose. Roughly at the same time as Bruun's departure, a disgruntled Henrik Ibsen also left the homeland, and would remain abroad for twenty-seven years. Ibsen was a zealous spokesman for Scandinavianism, and as early as December 1863, had written the poem "A Brother in Need" in order to awaken "Strong and courageous, a sleeping Folk, to Deed!"

Other great issues of that spring included the celebration of the fiftieth anniversary of Norway's Constitution. But should the commemoration be May 17th – when the constitution was adopted – or November 4th, when, in 1814, the revised constitution was adopted, after the union with Sweden had become a reality? The various points of view concerning Norway's links to Denmark and Sweden, both in the past and at present, all continued to smoulder. A Scandinavia Society was formed in Christiania, but the movement was quickly split into a three-states- and a two-states-Scandinavianism, depending upon whether one spoke up most vehemently for the cause of Denmark, or suggested that an integration – an amalgamation between Sweden and Norway – would be best. Following the paradigm of the Swedish "Fourth of November Stipendium", King Karl XV established an annual scholarship to be awarded to a Norwegian student to visit one of the other Scandinavian universities. The first award of this travel scholarship went to Ernst Motzfeldt for a year of study at Uppsala.

Meanwhile reports were coming in about the war situation in Denmark. Many members, clad in soldiers' uniforms, met at the Student Society and established "A School of Instruction for up to 200 Reserve Officer Aspirants". Sophus Lie, along with other student volunteers enrolled in this school during the summer of 1864. A ceasefire had been declared in the war, and the Danes had been forced to retreat on all fronts. Nevertheless, there was talk that the war could flare up again. In August, a provisional treaty was entered into, and with the final treaty, a couple of months later, Denmark lost both Schleswig and Holstein.

It is uncertain how far Sophus Lie's military plans and aspirations stretched. One finds on the military rolls that Science Student Sophus Lie was "appointed by the Armée Department" on June 18th, 1864. The course for aspiring reserve officers was planned to last eight months, but in reality it continued only three or four weeks, and in the course of this time it seems that the student Sophus Lie introduced himself to the idea of a career in the military. It was said subsequently that his oblique corneas and associated weak eyesight were the real reason that nothing came of Lie's military career. He completed his normal period of service, "the encumbent Conscription", that in peacetime lasted five years. During this time he would do service for some months as a reserve lieutenant with the Trondheim Infantry Brigade, and he spent one year assisting with examinations in mathematics at the Military Academy. Following this period, he was transferred by normal procedure to the territorial militia.

The second section of the science teacher's examination course, which was composed of physics and chemistry, included the following pursuits: land surveying, mathematical geography, physical geography, physics and chemistry. In these subjects Sophus encountered once again many of the lecturers he had had in the secondary examination. Lector Fearnley was responsible for land surveying and mathematical geography. Lector Christie taught physics, and his lectures seem to have been spread over several semesters. Lie had probably already heard these lectures when he prepared for the first section. Sjur Sexe gave the lectures in physical geography, and Lector Peter Waage was responsible for chemistry, with Hvoslef as his laboratory assistant. Sophus Lie took the examination in these subjects in December 1864, and a couple of months later he wrote to Ernst Motzfeldt in Uppsala:

> During the Whole of last Semester, as You can well imagine, I lived almost exclusively for the Second Section. Cramming organic chemistry and doing the laboratory work, this was my most important and extremely pleasant Pursuit. Now, thanks to Our Lord God, Everything has indeed come to an End, and Everything has even come to a ridiculously happy End.

He had come through the analysis of chemistry "tolerably unscathed", and in organic chemistry "I quite drove Waage around in Circles with some of the Calculations that he was careless enough to have included." And in physics and land-surveying, as he described it himself, he lived up to his "Repute from the old Days" – an expression that probably referred to his reputation arising from the second examination.

> But, alas – in physical Geography and Meteorology – Sexe's Subjects – I came up quite short. I was asked Things that I never in my Life had heard spoken of – certainly quite simple Things – but I lack sufficient Audacity to pretend that I am familiar with Things that, to me, are absolutely unknown. Nevertheless, Christie and Waage were reasonable enough to declare that I should unconditionally have Præceteris. 'Physical Geography was a much too insignificant Subject to be taken into any real Consideration, and so much more so since no Textbook exists thereof'. But Sexe maintained that he had given Lectures about this, and that I had not been there more than 3–4 Times. The Result was thus, finally, after a ½ hour's

violent Dispute, that on the Protocol I got *Laudabilis* with an Addition that this Laud should not exclude the Possibility of a Distinction (actually 3 *Præs* are a sufficient Requirement), an Outcome that I was particularly pleased with.

During the autumn of 1864, he had participated in military drills under the leadership of First Lieutenant Halvdan C. E. Wang, who three years later was employed as the University's teacher of gymnastics. Lie reported to Motzfeldt that for the time being he was involved with fencing, and what particularly interested him was fencing with foils, but at the same time he commented that the switch between parry and thrust was "especially difficult for me, who am so terribly raw when it comes to distinguishing the Elements." Lie admitted to Motzfeldt that he, with good reason, felt inferior when it came to such activities as fencing and shooting (as Motzfeldt for his part had for many years been one of the city's best shots and had taken first prize in the competitions of the Christiania Association for Weapon Training, where they specialized in bayonet work and fencing with foils). In the wintertime up to seventy men participated in these exercises, held in the gymnastics hall of Nissen's School. The Association was also involved in skiing, and with its training sessions during the winter of 1861–62, it was presumably the very first association in Norway to have ski activities as part of its programme.

This letter from Lie to Motzfeldt was written at the beginning of February 1865. As for other aspects of his life, Sophus recounted that he enjoyed playing billiards, but he was not nearly as zealous about it as his friend Theodor Blehr. Blehr was to be found, "according to my certainly not absolutely reliable Observations", every day at "The Billiard", from two until midnight. Beyond this, ever since New Year's Lie had been substitute teaching at Nissen's School, replacing a teacher who had fractured his leg. Lie continued – "I am not particularly busy. I go to visit, or to put it more correctly, should be going to visit, the zoological Collections; but You cannot imagine how exceedingly pleasant it is under the present Temperature Conditions."

The third section of the study programme, that of natural history, and which Sophus Lie had now begun, included mineralogy and botany in addition to zoology. Professor Jens Esmark was responsible for zoology, and already during the winter he had begun to take the students on field excursions out into the environs of the city. A zoologist with an international reputation was also lecturing at the university at this time. This was Professor Michael Sars, who despite his renown, had difficulties getting an audience for his specialised lectures. It is not very probable that Sophus Lie was one of the four or five who came to hear Sars lecture on molluscs. Indeed, there is a notebook among Lie's papers that contains lecture notes under the heading *Molluscs, Crustacea, Radiata*; however, these were topics that Professor Esmark talked about as well, although at a more elementary level, and student Lie was not one who always had to get to the bottom of a subject. About lecturer Theodor Kjerulf, who had responsibility for geology and mineralogy, Sophus Lie commented as he looked back thirty years later, that "Kjerulf's most certainly excellent Lectures" were "geared towards the Students of Geology and therefore included more Details than necessary for the Science Teacher's examination". Be that as it may, the lending records of the University Library show that during the spring of 1865, Sophus Lie borrowed Kjerulf's book, in German, on the geology of southern Norway – and he also borrowed other books about Norway's flora and Scandinavia's fauna that were relevant to his course of studies. In relation to lecturer Frederik Christian Schübeler's botanical lectures we find that among Lie's papers there is also a nicely kept little notebook of botanical notes and drawings. Lie wrote again to Motzfeldt at the beginning of June 1865:

> During the recent Months I have regularly been out 2–3 Times a week on all kinds of excursions – botanical, zoological, geological. In this Sense it is quite a lovely Thing to be studying for the third Section. I do not agree with the majority of Science Students who complain so bitterly that at the End, after having for such a long Time been concerned with abstract Things, that they should now be forced to bother themselves with Things, which are appropriate for Children and for the Cramming Years. If one studies the Subjects as they ought to be studied: out in Nature, above all else, they are interesting, and besides, I believe that they fill in a great Hole in the Education of abstract Mathematicians.

Sophus Lie was very concerned with the subject of science and its position at a broader level of "acculturation". In the letters to Motzfeldt he expresses optimism about the subject's place in and significance to society. Hartvig Nissen, exactly at that moment, had been appointed Deputy Secretary of the Ministry of Church and Education, which had just established a "School Commission". Sophus Lie felt that Minister Riddervold, who headed this department, had been somewhat influenced in this matter by his Scientist Sons-in-Law, and alluded to Professors Peter Waage and Cato M. Guldberg, who had each married one of the minister's daughters. Hartvig Nissen had written a sort of manifesto in *Morgenbladet*, that from Lie's point of view was good news, and that inspired him to give the following views on the future: "I therefore have the Hope to be, before I am 30, a Head Teacher, and in any case, there is always Something." Apart from this, Lie reported that he had a desire to engage in private teaching, for those who were preparing for the secondary examination, and later, to coach those who wanted to enter the Military Academy. It was the science subjects which he was thinking of, and he wrote that even though what was at issue was certainly only teaching at an elementary level, they were interesting subjects, that could "be useful to coach, for therein one became even more well-grounded in them." Otherwise, confiding in Motzfeldt, Lie apologized for not having paid back the 11 spd. The reason was that he had gotten some of the money back so very late from those to whom he, in turn, had lent it.

With summer approaching, Lie was planning another mountain expedition – this time together with Axel Blytt, the son of the professor of botany, M. N. Blytt. Axel Blytt was one year younger than Sophus, but had already become assistant professor and conservator of the university botanical collection. The book *Flora*, by old M. N. Blytt and young A. Blytt was something that all science candidates were obliged to study. That summer of 1865, young Blytt was to go on a study tour to the Sogn area, and Sophus wanted to accompany him, at least for the period of one month "to educate myself as a Botanist under his Auspices", and otherwise study the natural sciences in a practical way. Axel Blytt later was appointed Professor of Botany, and the year after this summer tour to Sogn, he journeyed to London, where among other things, he encountered the vehement debate around Charles Darwin's Theory of Evolution.

After having accompanied Axel Blytt to Sogn, Lie had made an arrangement to meet his friend Blehr, who at that point had just finished military exercises in Bergen, and together the two of them – and perhaps another student friend as well – would "journey through Søndfjord and Nordfjord, over to Lom and the Høifjeldene [the high mountains] where I rather swiftly will begin to feel myself again."

Sophus Lie completed his university education during the autumn of 1865 – the final exams of the third section of the science teacher's examination programme included, as mentioned earlier, mineralogy, zoology and botany. He received *laudabilis*, that is, the mark of 2, such that any public service examination with distinction was now out of the question. Later, friend Motzfeldt explained that this had been a great disappointment to Lie, and that he was very hurt by not having succeeded in what he set out to do. For his part, Lie maintained that already after having completed the second part the year before, he had given up the hope of taking the examination with distinction, and that the reason for this was that he anticipated a military career. In the meantime however, he must have come to know that he suffered from oblique corneas and poor eyesight; moreover, in the letters to Motzfeldt during the spring of 1865 he expressed his desire to end up a head teacher of the sciences in one or another school. But now these views of the future no longer seem to have given him any peace of mind. After finishing his public service examination Sophus Lie was still perplexed and in great doubt about what would become of him. The cause of this change is uncertain. Nor is it clear whether the autumn's great topics of debate in the Student Society had made an impression on him. It was precisely the significance of the natural sciences for teaching that was discussed in a series of well-attended meetings. Strong voices maintained "that The Study of the Humanities had to stand higher than the Study of Nature, because the Spirit stands above Nature."

The Lack of a Calling

Sophus Lie had taken his examinations; he had finished his education; and it worried him that he did not know in all certainty what he would use his education and his life for. His last period as a student seems to have been difficult, and when he went home to Moss at Christmas time, 1865, he was depressed. His friend Motzfeldt later felt that this dejection was linked to Lie's disappointment over not having achieved the top mark in the final examination. No doubt there were also other conditions that played a role, and most certainly the depression was worse than his friend Motzfeldt had believed. After he had been home at his father's house in Moss for almost three months, Lie answered a letter from Motzfeldt, writing: "Thanks and again Thanks for your Letter. I regard this as a Sign that, at least You will not reject me, although You know that I am a forlorn Subject." He continued, saying that he "in truth" had been "aimless, thoughtless, superficial, bad", and despite all the time they had spent together in recent years, Lie wrote: "If You were to believe me, I should, when the Opportunity arises, tell You my Story." He insistently urged Motzfeldt to go on acting as his friend, for thus could he (Lie) "perhaps, although the Probability has almost disappeared", manage to struggle for "a Place in Society." Toward the end of this short letter to his friend in Christiania, as has already been mentioned above, Lie wrote:

> When I bid You Farewell before Christmas, I believed that it was for Now and all Eternity; for it was my Intention to become a Suicide. But I do not have the Strength for it. So in consequence I get another Chance to try to live.

What had happened? What did he mean when he said he had been "aimless, thoughtless and superficial"? What sort of demands had he made on himself, and failed to achieve, which would lead him to decide that the only or the best solution would be a wretched death?

The idea of a calling and a purpose in life, of taking upon oneself a task, and being responsible for a very personal and committed choice had come up for debate, especially in student circles. Set up against this ideal perhaps – and regarded as completely negative – was every kind of doubt and the lack of knowledge about what one wanted in life. Duty to the fatherland and faith in Scandinavianism had been powerfully put to the test in relation to the German-Danish War of 1864. Christopher Bruun's ringing call to put aside ideas and thought in favour of life and action – and his own heroic contribution to the war – was an illustrative sign of ideal and action. Bruun had left Norway all alone for the battlefields of Denmark, but many students, and among them, Sophus Lie, had thereupon signed up for voluntary military service within the homeland. The way people behaved, against the background of everything that had been said, was nonetheless in various ways considered cowardly indifference toward the ideal of the Scandinavian union. Some spoke of treachery; others used theoretical superstructures and set up an aesthetic attitude to life against ethical demands and fighting ideals. Henrik Ibsen, who had written the poem of struggle "A Brother in Need", and had earlier expressed his ideas about a life's calling – "I must, I must, deep down within my soul/A voice commands and I will do its bidding" – felt that what happened was a broken promise, and he abandoned his fatherland, disappointed about Scandinavianism and the people's romantic yearning, who had, by their actions, revealed themselves to be powerless and lacking stout-hearted will. Ibsen travelled to Italy, and in Rome wrote his famous verse play "Brand", and thereby gave a dazzling expression of a man with a god-given task in life, and with a drive to a vocation that ended with the demand for "everything or nothing". "Brand" was published in March 1866, and two months later, Ibsen received a poet's pension, awarded by the Norwegian Parliament – and this only a few years after it was said in Christiania that the poet Ibsen had almost perished for lack of a calling.

Everything points to Sophus Lie being ambitious, but at the same time not being sure *what* it was he intended to be the best in. Anguish and depression were perhaps an integral part of this, for he now suddenly began to wonder, if – with only second best examination results – he really did have a place set aside for him among the great? Did such a position even exist, that is, a position to which he aspired so very unclearly? Maybe it was a sort of identity crisis he was going through, in which death, compared to a mediocre life, seemed preferable? Be this as it may, he chose the mediocre life. The first task he now took up, in March 1866, after he had explained his earlier "Intention to become a Suicide", was to decide whether he would take up a position as teacher. He returned to Christiania where he hoped accordingly to ask Motzfeldt for advice. He had further support in his big brother Fredrik, who rented a house at Dronningensgate 22, where Sophus now stayed. For the past four years Fredrik had been teaching "Mathematics and Assignments in Geography" at Nissen's School, but had just been appointed adjunct teacher at the Drammen Learnèd and Science School, and it was the position that Fredrik was leaving at Nissen's that Sophus had the opportunity to take over.

Eighteen years later when Sophus Lie himself commented on this segment of his life – in a letter to his mathematical friend, Felix Klein – he used the words "melancholy" and "eccentric" to characterise his state of mind, and the reason being that he thought he had lost all mental ability and power ("geistige Fähigkeiten"), and felt that he lacked any professional interest and engagement. In retrospect, he said, this lasted about six to eight months, until he again had faith in his own abilities. It was at this time that Ernst Motzfeldt had taken special care of him, and they had gone on long hikes in the Nordmarka forest north of Christiania, with the aim of tiring out Sophus physically so that he could begin to regain his ability to sleep at night.

This close relationship with Ernst Motzfeldt also involved Sophus in the course of the year 1866 moving into his friend's family home. Ernst's mother, Anna Pauline (née Birch), who had become a widow a year earlier, would gradually become a person with whom Sophus Lie would have much to do. In the coming eight years, right up to his marriage, Lie would continue to rent a garret room in the Motzfeldt home at Grottebakken 1.

There are signs that indicate as well that Lie's melancholy continued throughout the spring and summer of 1866, perhaps throughout the autumn as well. Being together with his brother and staying in Christiania gradually convinced Sophus that tying himself down to permanent position at Nissen's School was not the way to go. In recent years he had had a certain income from giving private lessons to students who intended to go on to the second examination or for the Military Academy, and he thought he would continue this coaching activity, and expand it. Another reason for not taking up the position at Nissen's School was probably that he hoped to be able to obtain a position at the astronomical observatory, a post everyone knew would fall vacant in the course of the year. It was during this summer of 1866 that indeed the decision was taken to establish a meteorological institute, and this was to be led by the assistant-director of the observatory, Henrik Mohn. Sophus Lie's borrowings from the University Library indicate that he was orienting himself toward this vacant position at the observatory.

Sophus Lie spent that summer at his sister's in Tvedestrand, that is, at Doctor Vogt's home. And it was during the course of this summer that Uncle Sophus held his swimming school for his nephews and their friends, out in Tvedestrand Fjord – something that on one particular day almost ended in catastrophe. Sophus' sister's son, who would later become the distinguised Professor of Mineralogy and Geology, Johan Herman Lie Vogt, and would look back upon this summer when he was eight, and recount that he had been tossed overboard in a brisk breeze out in the middle of the fjord, in order to learn how to swim. According to those who heard Professor J. H. L. Vogt speak about this period, the starting point was to have been when he, little Johan, had irritated his uncle by entering the old discussion that began: "since animals can swim, so must humans be able to do so as well", or else it had been the opposite, that Uncle Sophus maintained this and would momentarily demonstrate his stand – in a desperate attempt not to drown, the boy would probably develop the spontaneous ability to swim. But whether he learned to swim from this treatment, the erstwhile honourable Professor Johan H. L. Vogt had nothing to say.

In Christiania that autumn Sophus Lie would continue to behave in a manner that engendered anecdotes and stories. The scene on this occasion was the observatory building. Lie kept firmly in pursuit of his goal of a future career in astronomy, and during the autumn semester he went to "give a Hand to Professor Fearnley"; he followed the professor's lectures and adapted himself to the work routine at the observatory. But for his part, Professor Fearnley, according to others, did not in the least want Lie as his assistant – they did not get along well together, they were so very different in temperament, it was said. In any case, there were many who could tell of the alarm that had gripped Fearnley when on cold days he observed Lie vaulting over the instruments and installations, enacting the gymnast over Fearnley's erstwhile wooden horses, in order to keep himself warm. And then one day – either in a state of distraction, or simply to dampen his assistant's fresh air inclinations, the professor shut Lie into a room on the second floor, and the young Lie resolutely opened the window and jumped out. This signalled that the trial period was definitely over. Fearnley would not have anything to do with such a raw, unrefined force of nature! Fearnley wrote to the Academic Collegium and requested that the appointment of an observator be postponed – not because there was not an applicant with remarkable abilities and examination results, but because the applicant's "Inclination, together with physical and intellectual Disposition for an astronomical Observator's Studies and Activities, would scarcely allow the Applicant, and still less, the Undersigned, be able to arrive at any basic Opinion without preliminary Testing." The Collegium denied the postponement of the announcement, but Sophus Lie understood the way the wind was blowing, and did not apply. Thus, in all likelihood, Professor Fearnley got things the way he wanted them, with the subsequent employment of the not-yet graduated Hans Geelmuyden.

It seems that in some ways Lie's suicidal condition continued, even though he had decided to go on living. He must have known that his behaviour at the Observatory did not serve to promote his career, and his defiant attitude toward the authorities can be seen as underlining this. That he had now been stopped and rejected, did not prevent him from continuing his studies in astronomy. He borrowed a series of astronomical books, and he wanted to develop his own insights further, not only through private tutoring, but also through lectures. During the following semester he applied to the university authorities, the Collegium, for permission to hold "popular Lectures" on university premises. It seems that the decisions regarding such requests were left to the faculties, and when the Mathematics and Sciences Faculty was asked to give their opinion, Lie's request was rejected – only two, Bjerknes and Waage – felt that Lie ought to be allowed to hold his "popular" lectures.

But even this rebuff did not stop Lie. He applied to the Student Society to hold "popular Lectures in Astronomy" in the small hall of the Society, and here he did hold free lectures – two or three times a week during the spring semester of 1867 – with an audience of about fifty, and he attracted attention with "the unusually lively, eccentric and almost grotesque means by which he illustrated his lectures," wrote Elling Holst, who was later his student and biographer. What was meant by "almost grotesque means" is not further clarified.

There is much to indicate that the group that was known by the name "The Green Room" meant a great deal to Lie at this time. This was the circle of students and academics who, every day after dinner at the Student Society, gathered for coffee and tobacco in the room beside the dining hall, because smoking was not allowed in the dining hall. And this side room – "The Green Room" because all its fabrics were green – saw fifteen to twenty people gather as a rule; it was open to all, but all were not equally welcome. The poet and journalist, A. O. Vinje, used to drop in from time to time, read from his articles and joined in the discussions. Other central participants were Nordahl Rolfsen, Olaf Skavlan, Otto Blehr, and Amund Helland,[22] among others, who once in a while called themselves "the Cream of Society". Lie was close to several of these men, and Amund Helland and his camp would in particular win adherents to new outlooks and theories in the field of geology, and this would be a discourse in which Lie would later engage himself. It does not seem, however, that Lie was an especially central participant or enterprising debater in "The Green Room" – his favorite place seems to have been in front of the blackboard – but the circle's social radicalism and outspokenness in the treatment of the questions of the time must have appealed to him. Moreover, the fact that Lie was an important person in this milieu is attested to by the fact that during that spring he was chosen "Director-General of Punch" in the Student Society, an honorary position that upon festive occasions required the incumbent to oversee the condition of the coveted punch bowl.

When the summer of 1867 approached, he had to do his military service, and he was inducted into the battalion at Trondheim. This was a direct result of the voluntary military service he and many other students had incited when, during the Danish-German War three years earlier, they had signed up. His desire for the military was perhaps not as great now; in any case, the head of the Inherred Battalion at Skogn wrote to Ernst Motzfeldt and asked for a clarification as to why Sophus Lie had not replied to his conscription call, which required him to report to the recruiting school on May 23rd: "Was he ill, or was there another Reason why he was not responding?" Be that as it may, Sophus Lie now appeared at the recruiting school at the stipulated time, and a few days later he wrote to Motzfeldt from the camp: "We are having a good Time, terribly much to do, lovely Weather, pleasant Relations." That aside, this was also the period, during his service that lasted a good two months, that Reserve Lieutenant Lie was said to have made an impression with his unusual strength and endurance. One day, "while in the Services" he walked from the camp at Skogn to Trondheim and back again – four Norwegian miles [48 km.] each way – and still managed to be up at six o'clock the next morning.

This period of service in Trøndelag County was over at the end of July. Sophus Lie returned to Christiania, and probably also to his father's home in Moss before the students convened again for the next semester in the capital. His main source of income continued to be giving private tutoring – coaching – and he seems to have been a sought-after teacher. One of those Lie taught that semester was the student,

---

[22] N. Rolfsen (1848–1928) was an educator, poet and writer of popular Scandinavian school books. He contributed significantly to Norwegian national consciousness and language development. O. Skavlan (1838–91), as already mentioned, was a poet of the Norwegian romantic movement, man of letters and scholar of Norwegian literary history. O. Blehr (1847–1927) was a jurist who brought about juridical reform, and a leading member of the Venstre (Liberal) Party; active in the cabinets of the Norway-Sweden Parliaments, and later the independent Norwegian Parliament.

Alexander Kielland[23]. And later a close collaborator of Lie's, Friedrich Engel, told about how Lie in this connection was also said to have pronounced that Kielland's mathematical proficiency was of such mediocre calibre that he could could not conceive of Kielland becoming a famous writer of novels! But Lie did not become rich from his teaching – records show that he gradually ran up debts, both to the restaurant at the Student Society, and to the tailor and book-dealer, as well as to Mrs. Motzfeldt.

In addition, during the fall of 1867, Lie continued his free lectures in astronomy, and again on the Student Society premises, but this time in a smaller hall and less frequently than in the spring semester. Later in the semester he was to hold lectures in mathematics as well – in trigonometry and stereometry – for "a Majority of the new Students, who in their Artium, obtained the best Examination Results in Proficiency in Mathematics." Such was his characterisation when, at the end of the year he returned again to the Academic Collegium with a request to use the auditoria of the University of Christiania "for the Holding of Lectures on Astronomy, up to 3 Times a Week, with the Income going to the Student Society Building Fund". Presumably, to strengthen his application he wrote that he would soon give out "a Textbook in Astronomy", and that 100 students were "desirous of hearing my Lectures on Astronomy", and he appended a list of those he had signed up in the Student Society. Lie concluded his application to the Collegium: "I believe this allows one to find Proof that the Students understand my Activities, which cost me a great deal of Time and Work without bringing me any Income, which even fundamentally do Injury to my Manuduction's praxis, but which is based upon the proper Love of Science."

But once again Lie received a rejection. Probably his lecturing activities were seen as unnecessary and in competition with the astronomical teaching of Professor Fearnley. It might appear strange that Lie, with the ambitious nature he seemed to possess, would continue his activity in a field for which there were such clear signs that he could make no advance. Could it be that his desire to challenge the established order was as strong as his engagement in the subject?

The plans for writing a textbook in astronomy in any case came to nothing. In his surviving papers one finds some lecture manuscripts which perhaps could also have been worked up for such a book. But instead, Lie made another little venture into publishing at this time, a little volume presumably written as a sort of compendium in relation to the above-mentioned mathematical teaching "for a Majority of the new Students". He had this booklet printed in the cheapest possible manner, what was called contact printing, and this little trigonometry textbook, with a conclusion on *The Trigonometry of Spherical Triangles*, must in this context count as Lie's first mathematical work. In later times it was said that Lie wrote this trigonometry pamphlet to convince friends that he was not in such bad shape after all – according to Elling Holst there were many who still "feared for his sanity."

[23] Alexander Kielland (1849–1906) was a controversial writer who had taken his law degree in 1871, ran a brickworks for ten years, was imbued with radical thinking, particularly of Darwin, Heine, Stuart Mill, Brandes and Kierkegaard, and became a prolific writer of strong, socially critical novels and stories, particularly about class and religious hypocrisy.

On the surface of things it thus seems that those around him continued to regard Lie as an unbalanced person. He himself felt that he had gotten over this depressive period a year earlier. When in the middle of November 1867, he applied for a large scholarship – which was called the adjunct stipend – given to talented men of science who had completed the public service examinations, Lie summed up the past two years in the following words: "Since the final Examination I have – with the Exception of the first Half Year, when my Condition of Health prevented me – continued my mathematical Studies, even though I have not employed myself with more Teaching than had been Necessary for a modest Outcome." This application was submitted to the Academic Collegium. But the Collegium rejected this applications from Cand. Real (science candidate) Sophus Lie as well. The University leadership found that "for the Time being" it could not recommend such a stipend as he had "explained too little about his Studies". Six months later the Collegium also refused an application for a travel grant "from the Hjemstjerne-Rosenkrons Legacy", even though the Faculty of Natural Sciences this time had recommended Lie's application be granted.

Perhaps an ideal that permeates an epoch has certain consequences for the individual experiencing the era, an ideal that in retrospect cannot really be grasped by anyone. Under such conditions, when a human life is seen and lived as a persistent and on-going struggle between constructive and destructive forces – an arena in which the decisive victory of good should triumph – one's very awareness of one's destiny and vocation on earth was etched into one's very strength, autonomy and identity. And this preoccupation with having a vocation in life perhaps sets things in relief in more than one way – a calling also implies having a sort of pact, and a duty to carry out in the deepest and most thorough-going way what one was destined to be. A gifted personality was a Divine image one felt compelled to realise. When Sophus Lie was unsure about what he should choose, at the same time that he in some ways sawed off the branch upon which he sat, this perhaps could have been an expression of the wish that the pact itself ought to be more evident – that an even stronger rejection might get him on track again – once all that which was irrelevant had been rejected and peeled away.

During the autumn of 1867 Christiania was visited by one of the time's leading contemporary thinkers, the Danish Rasmus Nielsen. He had been invited by the Student Society, and in the course of sixteen well-attended lectures, under the title "Hindrances and Conditions for the spiritual Life in the Nordic Lands", something that also represented a discourse on the relation between religion and the natural sciences – Professor Nielsen set the agenda for a philosophical discussion in the academic milieu. Sophus Lie's close friend, Armauer Hansen, wrote in his *Life's Memoirs* that "the spiritual movements out in Europe" for the most part were discussed "in private gatherings", but when "the well-known Danish Professor Rasmus Nielsen came to Christiania", he (Armauer Hansen) too went to the Student Society, but his expectations had not been fulfilled: "I listened to these lectures with great interest and became very disappointed." Rasmus Nielsen frequently used mathematics and mathematical illustrations in his lectures, and some years earlier this had resulted in a vehement struggle in Denmark, between Professor Nielsen and Professor of Mathematics Adolf Steen. Steen felt that Nielsen used mathematics in an erroneous manner. This discourse came to be known in Armauer Hansen's "private gatherings", whether indeed Sophus Lie also heard of this, is unknown, but it was probably Nielsen's lack of knowledge of the natural scientific challenges to which Armauer Hansen was reacting negatively. Others however were very impressed, and before Rasmus Nielsen left Christiania, a great party was held in his honour, at which the poet, Andreas Munch, hailed him in song as "Thought's Hero from Denmark's Valleys".

Sophus Lie's activity and energy at this time gradually brought him into a prominent position in the Student Society. During the spring semester he had been chosen Director-General of Punch, and at the great carnival-related List of Honours of His Majesty The Pig, now, in December 1867, he was appointed an honorary squire; that is, he entered the ranks of the noble dignitaries who led the stately and pompous Festival of the Pig. This was the third time the Student Society arranged the Pig Festival, and this year it was considerably expanded compared to the earlier ones; indeed, this year's came to act as the model for later Pig Festivals. The point of origin of these festivities had been the Christmas Tree Festival of 1857, when a money pig – that is, a piggybank – was introduced "to receive the Student Society Members' priestly Offerings of Small Change". This was an innovation made by the Building Committee whose members were working to erect the Society's own building in Universitetsgaten. To commemorate the completion of concrete plans and drawings for this building, the Committee arranged a celebration, which took place on April 9, 1859, and inaugurated The Royal Order of the Pig. This was subsequently looked upon as the first Pig-Festival , the second was two years later when the Society building was finished and inaugurated. All the building expenses were of course not paid for, and it was to this "Student Society Building Fund" that Sophus Lie too would contribute with his free lectures in astronomy.

The old piggybank that had been the Student Society's symbol and logo, was accidentally broken in 1865. But some of the Pig's faithful knights had provided a new pig – carved in wood at the Debtors' Prison – and this was now, in 1867, ceremoniously bid welcome with cantatas and hymns, and an intricate ceremony. Eight people were made Knights of the Pig – two of the most faithful were promoted to Commander and to the Great Cross – and a ceremonial toastmaster also belonged to the lofty court, along with two chamberlains and a squire. And this squire was Sophus Lie himself. Verse was written about all the members of this court, and ceremoniously sung. About Sophus Lie they wrote sympathetically and expressed faith in his future possibilities, with expressions that referred not only to his enthusiastic lectures, but also his zeal at getting people to sign petitions, and getting his mathematical work out in the form of contact printing. The years that followed would be decisive for Sophus Lie; the year 1868 must be characterised has his breakthrough year. He seems finally to have achieved clarity about what his vocation in life was. He became clear, as he himself later put it, "that there was a Mathematician" in him.

Right enough, the year began in the same old track. He continued his lectures in astronomy at the Student Society – now in the "Room at the Side of the small Hall", and indeed with a rather modest number of listeners – and still for a time he kept to his plan to write a textbook in astronomy. Although it is impossible to have an overview of who his private pupils were, he most certainly continued his coaching activities as before. But according to the lending records of the University Library he had, as early as January, borrowed Euclid's major work, *The Elements* – all fifteen volumes had been collected into one magnificent edition dating from 1546 – and in his working notebooks from this period one finds his ponderings about the place of number theory in geometry. He had read some of Euclid before – and earlier he had also borrowed Descartes' *Géometrie* – but it was now, during the spring of 1868, that his study of the fundamentals of geometry became goal-oriented. At the same time he also took up another theme, namely the study of elliptic functions, one of the most topical subjects of research at the time. Both of Lie's earlier teachers, Professor O. J. Broch and Senior Teacher L. Sylow, had worked with elliptic functions, and this was also one of the fields in which Niels Henrik Abel had opened up in such an innovative manner. In the course of that spring, Sophus Lie borrowed a series of books – most by French mathematicians and most on elliptic functions – in May, he borrowed Abel's collected works, as had been edited and given out by B. M. Holmboe in 1839.

The reason for Lie's interest in elliptic functions perhaps lay in the desire to be well-prepared for the summer's great event, a comprehensive Scandinavian meeting of the natural sciences, that this year would be organised in Christiania. Lie probably wanted to update himself in relation to these meetings – perhaps he aimed to give a paper, and at least, he wanted to be able to participate in debates. This Scandinavian meeting of the natural sciences seems to have been a decisive factor in Lie's mathematical development. In the course of that month of July 1868, Lie listened to a series of mathematical papers and lectures, and he became familiar with Scandinavian mathematicians – above all, it seems the friendship he developed with the two Danish professors, Adolf Steen and Hieronimus Zeuthen, inspired him to pursue new directions. Adolf Steen held lectures on "The Integration of Linear Differential Equations with Definite Integrals", and Zeuthen was an enthusiastic promoter of the newer geometry. Zeuthen had defended a thesis on conic sections and was involved with fundamental problems of geometry, and he had studied in Paris under the renowned geometer, Michel Chasles.

The two previous meetings of the Scandinavian natural sciences had been held respectively in Copenhagen (1860) and Stockholm (1860). The meeting in Christiania was the tenth in the series of such gatherings of Scandinavian researchers, and as well the first following the devastating Danish-German War. A total of 368 participants met in Christiania that summer, and the best and newest of Scandinavian research was presented. The Swedish botanist Areschoug, for example, gave a lecture in which he spoke about how Charles Darwin's new theories intruded into the posing of the problem "Of the Origins of European Vegetation" – although Darwin's theory of evolution was still at this point in time not considered the arch-fiend and threat to Creation and Faith in God. Otherwise, the natural science meetings were divided into nine sections, and those who wanted to give lectures, made themselves known to the organisers on the opening day, and as it was reported in the minutes, "Most of the Foreign Natural Science Researchers were Guests in the Homes of the Inhabitants." Thus, in the physics and mathematics section, the Danes Steen and Zeuthen gave their lectures, and Ludvig Sylow presented his paper "On the Hallmark of the Resolution by Radicals of an irreducible algebraic Equation of Prime Degree." Of the other Norwegian mathematicians, the research fellow, C. M. Guldberg, and Dr. A. S. Guldberg, both gave lectures, and Professor of Medicine Christian Boeck demonstrated his own component thermometer. But what captured Lie's interest was geometer Zeuthen's lecture, "On a new Coordinate System of Space". Zeuthen referred to and discussed the work of Chasles and the German mathematicians, A. F. Möbius and Michel Plücker. During the approaching autumn, which would be a constantly intense work period for Lie, he would read and study precisely the mathematicians Zeuthen had referred to.

The meeting between Sylow and Lie at these sessions of the natural sciences meetings seem to have been the first since Sylow had worked as a substitute teacher at the university five years earlier. Later, following Lie's death, Sylow told about how Lie had complained during these meetings that he had lost his notes on Sylow's lectures about Galois' group theory, and he had thus asked to borrow Sylow's manuscript. In relation to this, Lie was said to have made a remark that, in light of Lie's later work, stuck in Sylow's memory; namely, "I believe that Group Theory will become very Important." Sylow thereby thought to signal that Lie had already begun to use group theory in geometric transformations, and already understood that group theory's theorems and conceptual framework would be of significance to his own research.

Sixteen years later in a letter to Felix Klein, Lie himself wrote that it had been in the autumn of 1868 that he had seriously begun to study geometry, and that at the beginning he had found the study to be very heavy-going. He wrote that it was only with very great effort that he got through Chasles' works and his attempts to reconstitute Euclid's lost books. When he looked back on the years before 1868, Lie felt this had been a period wherein he never imagined that he would ever conduct independent, original research. At the most it had been mathematical pedagogy that he had hoped to prepare for, and his work had been concentrated on showing the shortcomings of textbooks, in terms of presentation of the inner consistency of the fundamental concepts of mathematics: "The Road to Mathematics was long and heavy", Lie was also said to have remarked. In his memorial speech in honour of Lie in 1899, Elling Holst pointed to the transition from a youthful period of all-sided versatility, to the concentration on only *"one single Thought"*. "The Transition between these two Conditions", Holst maintained, was "a huge Crisis, that can be most closely compared to the religious Awakening of a great Prophet."

PART IV

In Tune with the Times

In Berlin, 1867

Into Mathematical History

It was modern geometry that Sophus Lie now encountered with such fascination and zeal. This was an area within mathematics that stemmed mainly from Euclid's two-thousand-year-old books, and a field that now, during the 1800s underwent a furious development. Outlooks and methods, both old and new were being combined together into integrated theories, and these theories developed and diverged more and more radically from what earlier had been understood as geometry – first and foremost perhaps as a liberation from a direct perception of the perceivable world. This modern geometry – called at the beginning astral or imaginary geometry – gradually came to be called non-Euclidean geometry. In many places during the course of the nineteenth century, it would come to engender oppositional reactions just as vehement as those that greeted Darwin's Theory of Evolution. In England particularly, non-Euclidean geometry, Darwin, and ethical relativism were lumped together and said to represent those forces that were absolutely inimical to God. On the other hand, the theory of evolution and non-Euclidean geometry were manifestations of what, at the time, were radical new methods of thinking. Whereas established faith and methods of thinking tried to pack away every single phenomenon within one eternal framework – to define thought according to a single necessary formula, and explain human variation as a result of the existence of different types and stocks – the modern theories advanced conceptions about how scarcely anything could be seen, known or described in real, absolute terms: instead, everything must of necessity be seen and described according to known conditions, and therefore, in the final analysis, be relative. With that, the pivotal question arose about what it is that knowledge is really composed of, whether it be God-given or man-made.

In France and Germany the link between mathematics and philosophy was not as integrated in the general cultural and intellectual milieu as it was in Britain. In Britain the discourse on non-Euclidean geometry was found not only in learned journals, but also in more popular publications, and here it was asserted that, with the same certainty that God exists, the sum of the angles of a triangle is 180 degrees, and vice versa. The old Euclidean geometry had been hammered into generations of young heads as the prototype for clarity of thought and logical reasoning, and it stood as a shining example of the type of knowledge that was absolute and true – absolutely true because it was descriptive of reality and completely in conformity with human comprehension of the world as encountered and experienced. This new, non-Euclidean geometry, which presented the fact that the sum of the angles of a triangle, such as found, for example, in a spherical surface, is not 180 degrees, represented a threat not only to ethics and morals but also to the hope of finding true knowledge in science. This non-Euclidean geometry showed that mathematical theory could give rise to a series of dissimilar points of view, and this consequently became an important part of the outlook that threatened the authoritative approach to knowledge. In addition, the looming philosophical figure of the day, Immanuel Kant, considered the old Euclidean geometry as something firmly established in the human brain, and a fundamental quality of the human means of observing and experiencing the external world.

Classical geometry had long occupied a position atop the consolidated hierarchy of knowledge. The form of certainty that one found in geometry, had also to be found in physics, biology, ethics and religion. All knowledge, so to speak, found itself located in the same sphere, and consequently this new non-Euclidean geometry was seen as something that undermined perfect knowledge. Geometry was no longer something that could be given *a priori*, no longer something that from its inception was dictated by the world we live in. More and more, geometry was projected as something man-made. The old, assembled "treasury of knowledge" was about to fall. The oppositions this inspired were strong, and consequently a split and a division came into being during the nineteenth century. Despite new technological developments with their associated material optimism about the future, there arose as well a more and more bounded view of what science was, what it was capable of accomplishing. A number of independent sciences were established, while the study of ethics and psychology were separated from science without anyone clearly wanting to pursue this restricted point of view, or saying anything about what, for example, ethics would now legitimately be in a position to accomplish. The knowledge of different cultures – each with its own legitimacy and integrity – also transformed notions of absolutes in all forms into almost meaningless superstitions. Change was natural, while the constant and the orderly were the exceptions.

It was these fundamentally new scientific developments in which Sophus Lie would now come to participate. There is little however to indicate that the vicar's son, Lie, would from time to time link modern geometry with his own personal relationship to God. When all is said and done, it seems that he kept a kind of childhood faith throughout his life. He was certainly "little religious", and he did not have "the appropriate Candour in Prayer" as he wrote his wife – in 1890 when he was a patient at the psychiatric clinic at Ilten – but nevertheless, he bade her and the children "to pray to God as before to look after [him]."

Sophus Lie's encounter with his "life's work", modern geometry, became – as he himself would subsequently come to express it – "long and heavy", but when he first discovered "that there was a Mathematician hidden inside him", things happened quickly. Less than a year after the Scandinavian natural sciences research meeting, Lie had finished his first scientific paper, and with economic assistance from Ernst Motzfeldt had had the work published. This publishing event unleashed a remarkable degree of goodwill, even though the country's two professors of mathematics, O.J. Broch and C.A. Bjerknes, did not understand the work, and at first reacted negatively when Lie sent the paper to the Christiania Academy of Sciences as a submission to the Academy's proceedings. Once again it was Ernst Motzfeldt who stepped forward to help. Motzfeldt went to the general secretary of the Academy of Sciences, Professor M.J. Monrad, who rapidly expressed interest in Lie's work. Monrad created necessary support beyond the mathematical milieu, even among persons who knew still less than Bjerknes and Broch. The result was that gradually the Academy of Sciences in Christiania supported the publication of an expanded version of Lie's "Imaginary Theory". Lie was now encouraged by Broch to translate the work into German, and the travel grant that he now received was given to him on the basis of this work. During the year 1869, the treatise entitled "Über eine Darstellung des Imaginären in der Geometrie" was published in the well-known German periodical, *Crelle's Journal*. This was the same periodical in which Niels Henrik Abel had published the greater part of his works, and it was due to Abel's treatises that the periodical had become world famous. With his paper published in *Crelle's Journal* in the autumn of 1869, the Norwegian Sophus Lie quickly became a name of which many central mathematicians out in Europe began to take notice.

The new geometry had its foundations in the classical "geometry" of land-surveying, and had been something that people from time immemorial had been dependent upon. Euclid, who lived and worked in Alexandria about the year 300 BC, was not the first geometer and perhaps not even the greatest of his own time, but he was the great organiser and collector of geometric knowledge, and he had given a clear and logical form to the subject in his books. The books were collected together under the title, *The Elements* – which in addition to geometry contained all the mathematical knowledge that had been mastered at that point in history. Euclid gave definitions and proofs systematically and with precision – and built up geometric knowledge in such a systematic manner that ever since it has been called Euclidean geometry. And for more than two thousand years this geometry formed the basis for mathematical instruction in schools and universities across Europe.

Before it was possible to speak rationally about different figures and plans, the system defined concepts such as "point", "line", "straight line", "triangle", "circle", and so on. A point was defined as "that which cannot be divided", and a line as "a length without width", and about the straight line, it said "this lies straight between its two points", and so on. As well as a series of such *definitions*, the foundation pillars of Euclidean geometry include *postulates* and *universal concepts*. The postulates are a series of assertions that one was forced to accept due to the following: the universal concepts or axioms are assertions employed as self-evident, and therefore taken as universally known. From these postulates and axioms there follows a series of propositions that demonstrate geometric truths and how geometric constructions are to be executed.

Euclid formulated five postulates – the first four were easy to agree upon; they are short and easily understood, but there was always doubt about the fifth, and in the end this doubt was to lead to a completely new geometry. The first three postulates insist that the following propositions be accepted: that it is possible to draw one and only one single straight line between two given points, that any straight line can be extended without limit, at both ends, that one can inscribe a circle with whatever point one desires as its centre, and with a given distance as its radius. The fourth postulate confirmed quite simply the similarity of all right angles; that is, all right angles are of equal size. But the fifth postulate had a more complex formulation, and down through the epochs there have been many from the scientific and scholarly world who have felt that this was a blemish in an otherwise perfect and elegant work. The famous and notorious fifth postulate says when a straight line intersects two straight lines, and the sum of the interior angles on the same side is less than two right angles, then the two lines, when expanded indefinitely, will meet, and they will meet on the side upon which the sum of the two angles is less than two right angles.

This is Euclid's theory of parallel lines, that is, it has to do with lines that will not meet. The theory can also be formulated in other ways, as for example, through a given point beyond a straight line there can be drawn one, and only one, other straight line that is parallel to the given line – and again this is equivalent to the fact that the sum of angles in a triangle is always 180 degrees, or two right angles.

A series of translations and commentaries of Euclid's books have come out through the centuries – and these, as a rule, accommodate different levels and goals of teaching, but also include some scattered attempts to challenge Euclid's fifth postulate.

Gradually as mathematics and science turned more and more to application and gave solutions to technical and practical tasks, it consequently came to be felt that some of Euclid's presentations were unnecessarily theoretical and sophistic for normal use. Already by the 1600s the practical and utilitarian were being set up against the theoretical and speculative – and the utilitarian application of mathematical knowledge gradually became an aspect of the field that was just as conspicuous as the mathematical ability to develop formal refinement of thought and the ability to think with logical order and vigour. Committing Euclid's definitions to memory, and understanding his subsequent conclusions – the classical legacy with its theoretical and formal style of presentation – began to lose its dominant position in the teaching.

A variant of this debate broke out with great intensity in Norway as late as the 1830s, between two of the professors of the day at the University of Christiania – B.M. Holmboe and Christopher Hansteen, who had both been friends of Niels Henrik Abel, and in different ways, both had stood in intimate relationship to the country's scintillating mathematical star. Beginning in the pages of the textbooks each of these professors had respectively published on plane geometry, a harrowing public debate broke out. The practically-inclined Hansteen had reacted to Holmboe's theoretical approach and lack of practical utilitarian considerations. He attacked Holmboe and the whole Euclidean school, claiming that they had made rigorous form and adherence to theory the most essential features of geometry. Hansteen wrote: "The Euclidean Definition of parallel straight Lines, as is accepted by almost all Geometers, according to my Assessment, has got it as Wrong in terms of Logic, as a Definition can." Hansteen wanted to improve Euclid, and in more than one theoretical exercise into logical thinking, he called for a geometry built on "direct Perception", and to equip students to be able to solve practical tasks. Hansteen's motto was: "You do not need Proof that the Sun is up when it is shining in your Eyes."

Hansteen expanded his conception of parallelism to include circular lines, but this mathematical dispute in Norway – conducted by the country's first generation of university teachers – the forebears of Sophus Lie's professors, O. J. Broch and C. A. Bjerknes, was in no way of the same type as was now occurring abroad in Europe. Euclid's definition of parallel lines was certainly under attack, and moreover represented a veritable parting of the ways, between old and new geometry, but in a fundamental way that was located far from the positions taken in the Norwegian milieu. The new geometry had detached itself considerably from questions of natural science, and in every way was more theoretical than the old, and much time would go by before it led to any practical application. Pure mathematics – and a form of investigation which did not have a main objective linked to a field of application – began in the early 1800s, and not least for geometry, this century would become a golden age. In the beginning, what prepared the new ground was the field called projective properties geometry. The central names in this field were, among others, Monge, Poncelet and Chasles, and their German rivals, Möbius and Plücker. But the two men who led the breakthrough to a coherent non-Euclidean geometry were the Russian mathematician, Nicholai Ivanovich Lobachevsky and the Hungarian, János Bolyai. Quite independently of one another, they each developed their theories in the last years of the 1820s, and at the same time the great German mathematician, Carl Friedrich Gauss came out with pronouncements that he had for a long time been tracking down a non-Euclidean geometry, but that due to the commotion he anticipated, primarily from Kant and his followers, had refrained from publishing. To be sure, Kant maintained that it would be very difficult to know things that lay beyond the realm of the filters and lenses of the human senses. Now here was a geometry that gave entirely new concepts about the space around us.

Commentaries and reactions to such non-Euclidean geometry certainly presupposed a receptive mathematical climate, and such was not in existence in Norway – and out in the rest of Europe such a milieu did not exist seriously before the 1860s and 1870s.

The first book that Sophus Lie borrowed in the autumn of 1868 was, according to the loan records at the University Library, the first volume of Jean Victor Poncelet's *Application d'Analyse et de Géometrie*. However, according to Elling Holst – who became Lie's student three years later – Lie also had access to Poncelet's first major work, published in 1822, and entitled *Traité des propriétés projectives*. Poncelet, born in 1788, was one of the main architects of projective properties geometry, one of the veritable building blocks of modern geometry.

After having had the first volume of Poncelet's *Application d'Analyse et de Géometrie* for a month, Lie, borrowed the second volume as well, in October 1868.

Lie must quite certainly now have read with great interest the forewords to these two volumes, and probably took courage from the passages in these prefaces where Poncelet described his own long and dramatic course towards geometric research. Following his exam at l'École polytechnique, Poncelet became a lieutenant in 1812, and he swiftly drew attention as an outstanding officer-engineer. Later, his lift bridges would stand as technical paradigms, and above all, Poncelet's water wheel was copied and used all over Europe; that is to say, Poncelet constructed dipper blades in a manner that allowed the energy of the water to be utilized much more efficiently. He doubled the degree of efficiency compared to that of earlier water wheels. Later in life Poncelet was also to receive a series of high and responsible positions in the practical military and teaching fields. But in 1812, when Napoleon directed half a million soldiers against Russia, the twenty-four year-old lieutenant had to join "the great army". Here he found himself in an engineer battalion and most certainly contributed good technical solutions to a series of bloody altercations, but he was also enmeshed in Napoleon's army of unfed and starving troops which, in November 1812, were forced to retreat from Moscow. Poncelet was left for dead on the battlefield at Krasnaya, and stricken from the rolls of the French army. But a Russian patrol noticed his officer's uniform, discovered that the man was still breathing, brought him in, and took him to the leadership. After examination, Poncelet was sent eastward as a prisoner, to Saratov on the Volga, a march that lasted four months and covered 1,200 kilometres. And, as he himself wrote in these forewords to his mathematical work, he had come through the experience "alive by the special Grace of God", and thanks to "the physical and moral strength and energy with which nature had fortunately equipped him", he travelled "clad in the threads of a French uniform and provisioned by the black bread of the Russian peasants", over "the desolate and frozen steppes, where the cold in that fatal and unusually fierce winter of 1812 was such that even the quicksilver in the thermometer was often frozen." And thus it was at that time and place – as a prisoner for almost two years, with no recourse to books or facilities, in Saratov on the Volga – that he worked out new geometric concepts and approaches.

Which properties and relations of a figure remained unchanged when a figure was projected on, or to, another surface? Such unchanging properties were called projective, and became basic elements in the new approach. During his imprisonment, Poncelet worked with such paradigms (projections), and he had the idea of giving certain geometric elements fixed infinite extensions. For example, a straight line was assigned, in addition to all its finite points, one and only one point of infinite extension – and in addition, a surface received one and only one infinitely extended straight line. Thus, Poncelet was able to contend, among other things, that as a universal statement, two straight lines in a plane always have a point of intersection – or where the lines are parallel, they meet at a point on the line at infinity.

Projective mathematics does not study all the properties of a figure, but only those that are preserved – that are invariant – in a given perspective of the figure. Such invariant properties are brought together in a class of their own, and are called projective properties. Thus in the projective plane there is no possibility of distinguishing, for example, between the classic figures of the circle, ellipse, parabola and hyperbola. However, this perspective provided a useful tool for obtaining an overview of changes in such depictions – in, among other things, the form of studying the relation between given points on a line, and what was termed the cross ratio between four randomly chosen points. Poncelet advanced the view – by means of his imaginary line – that this double relation is invariant in projections. This was one of several ideas that now led to a firmer grip on what happens in geometric spaces, and what these geometric spaces really were.

Moreover, one of Poncelet's fruitful ideas was what came to be called the continuity principle; namely, for Poncelet it was intuitively clear that a figure's adequate, universal and recognised relations must also apply to all other figures which, by means of continuous changes, can be derived from the original figure. He drew challenging conclusions from this principle about how actual points of intersection have, as analogues, points where imaginary straight lines meet the line at infinity. Poncelet did not feel any need to demonstrate this continuity principle, but other mathematicians found it impossible to find peace of mind before they had proven the assertion, and managed to assign the imaginary points numerical value in accordance with a system of equations. This difference of approach and method of working has made it common to speak about two types of geometers: the synthetic and the analytic. Poncelet was a synthesist who used pure geometric methods in their traditional form to arrive at his geometric truths – in opposition to the analyst who occupied himself with analytic and algebraic techniques from other fields of mathematics. Gradually there developed a greater demand for the formal character of mathematics as a common language of communication, and thus the analytic side became dominant within the field. Despite this trend, however, Sophus Lie would, in his mathematical pursuits, work in the tradition of Poncelet, as a synthesist.

Poncelet died in 1867 at the age of seventy-nine. Eleven years later Sophus Lie was to be publisher of Elling Holst's paper, "On the Significance of Poncelet for Geometry. A Contribution to the Developmental History of the Ideas of Modern Geometry." The discussion about who had been first, who had priority and who was the greatest, were important contemporary issues to which great attention was devoted. This paper by Holst, Lie's student, in reference to Poncelet, maintained that "no mind among the pure Geometricians of the Century surpassed him in terms of fundamental Significance" – and it was a strong refutation of what the French mathematician, Chasles, had done, namely to place Poncelet on the same footing with other of his contemporaries, and definitely to set him behind mathematicians like Gaspard Monge and the man of revolution, Lazare Nicolas Marguerite Carnot. To Lie and Holst, Poncelet represented a new approach to geometry, a universalising of geometry in the same way as analysis and algebra. In their evaluation of the Poncelet's position among "the Century's pure Geometricians" it seems however that they did not know the works of Bernhard Riemann.

The mathematics of this period can roughly be viewed as falling into three main areas, and these were analysis, algebra and geometry. As early as the end of the

1600s, Isaac Newton and Gottfried Wilhelm Leibniz had introduced the most important tool of analysis, namely infinitesmal calculus. With the use of this concept, functions could be handled by means of analytic methods, and to obtain an understanding of space, a completely new type of equations was introduced – differential equations. From now on, the laws of physics could be expressed in functions, and these functions could be studied by means of equations. The main problem of mathematical analysis was the solution of differential equations, and mathematical analysis was a major field within the natural sciences. But only a minority of the differential equations that arose could be solved, and nobody had an overview about which properties were the hindrances to solving these equations.

Functions and equations were the central focus of work in analysis and algebra. In the 1820s, Niels Henrik Abel had instituted a new epoch in the study of algebraic equations. His proof for concluding that equations of powers of a higher degree than four, in general could not be solved by means of the usual computation operations, opened completely new fields of investigation. But if such equations could not in general be solved, there were certainly many special equations that had solutions. The questions that followed were: which equations *could* be solved? What criteria allowed for the possibility of a solution? Abel had not found a complete answer to this by the time of his death at the age of twenty-six in 1829, but a short time later, a Frenchman, Évariste Galois was able to describe and classify these solutions from certain properties of symmetry in the equations. Galois' grasp of this problematic would revolutionise mathematics. Specifically, Galois introduced the concept of *group*, and this mathematical tool would prove itself to be tremendously fruitful. This occurred around 1830 – Galois was laid to rest in Mother Earth at the age of twenty-one, killed in a duel over a woman in 1832. In the decades that followed, groups and group theory were applied almost entirely to algebraic equations – but new fields lay waiting.

Sophus Lie admired Abel, and he would soon be urged to participate in the editing of the works of his world-renowned countryman. Abel's life and fate would turn up inspirationally both in public and private discussions around Lie. But Lie did not share Abel's interest in pure algebra. Nonetheless, he, like Abel and Galois, would have his name immortally linked to this algebraic concept, *group*. From the conceptions of Abel and Galois about the solution of algebraic equations, Lie consequently got an idea about how differential equations could be solved. And with inspiration from geometry, Lie created his own group theory – about what was called continuous groups, a mathematical tool that would prove itself particularly well-suited to express symmetry in geometry and analysis.

Lie's ideas were the starting point for a new and important mathematical discipline, which is now called simply *Lie theory*. Concepts such as *Lie algebra, Lie group,* and *Lie symmetry* first came into use about 1930, and were then used in quantum mechanics. For present-day mathematicians and physicists the world over, Lie theory stands central, and Lie theory continually finds new areas of utility in modern natural science. One of the foremost experts in mathematics and mathematical history, the French scholar Jean Dieudonné, wrote in 1980, "Lie theory is in the process of becoming the most important part of modern mathematics. Little by little it became obvious that the most unexpected theories, from arithmetic to quantum physics, came to encircle this Lie field like a gigantic axis."

While he was still alive Lie received a great recognition, when, in 1897, he was the very first to receive the Lobachevsky Prize, a prize that had been instituted two years earlier to honour geometrical achievements, particularly in the field of non-Euclidean geometry.

There was a Mathematician in Him

It seems that for Lie as well, the autumn of 1868 had been a constantly inspiring period of work. Apart from Poncelet, what brought him the first wave of inspiration, there were the books of Plücker – in any case, it was among the works of Poncelet and Plücker that he first found a conceptual apparatus, a mathematical language that he could use to express his own ideas. But besides these two mathematicians, he briefed himself on the works of Carnot, Hamilton, Cremona, Möbius, Hunyadi, Townsend, Grassmann, Salmon and others, and he borrowed various volumes of leading mathematical journals: *Crelle's Journal* from Berlin, *Liouville's Journal* and *Comptes Rendus* from Paris, and *Philosophical Transactions* published in London.

And as he wanted very much to discuss the ideas that were now streaming out of him, he was eager to create a forum for mathematical discourse in Christiania. To this end he himself tried to re-establish Realistforeningen, the old Science Students Association, which had not functioned since 1864. Together with ten or twelve others – the foremost of whom being Carl Berner,[24] Lie's former classmate from Nissen's School and who had been active in the Association – he decided to move *Realistforeningen*'s activities from the Student Society on Universitetsgaten to a rented premises at Nedre Voldgate 7. Later there were rumours about how Sophus Lie had carried the Association's massive blackboard through the city streets on his back, quite probably on October 25th, 1868 – in any case, the following day there was a celebration to mark the Association's inauguration. Three years later, to commemorate the founding day, a festive song was composed:

> Sofus *Lie took the board upon his back,*
> *Like Samson did with Gaza's gate,*
> *Bore it up and over "Svartfejerbakk";*
> *That was the source, at any rate.*
> [. . .]
> *Thereafter he, who show'd best tenacity*
> *Was as always Sofus Lie;*
> *For himself he built a dignity,*
> *In* moderne géometrie.

The fact that Lie was the most active member of the Association is recorded in the minutes. Three days after the move, Lie gave his lecture "Concerning the Newer Ge-

[24] Carl C. Berner (1841–1918) was, like Sophus Lie, a graduate of the sciences. He was an active Liberal Party politician, technical school director in Bergen, Minister of Church and Education, and active as a parlamentarian who worked toward the ending the Norway-Sweden Union in 1905.

ometry", a lecture he continued some weeks later. Lie allowed himself to be chosen chairman of the Association, and its seems that when others failed to contribute he was able at short notice to take over the Association's weekly Wednesday lecture, and even the minutes of lectures given by others included additional commentary by Lie. For example, when Carl Berner lectured on "The Objects of Geometry" – in which he defined addition and multiplication in terms of geometry, and among other things, showed that "the Algebraic Numbers, positive and negative, could be viewed as a special Case of the Geometric Picture" – Lie added some historical remarks "concerning the Development during our Century of the Concept of Number Systems through Cauchy's Imaginary Forms and Hamilton's Quaternaries." Apart from Lie's frequent lectures during that first year of the reconstituted Realistforeningen, there were lectures on the Jostedal Glacier, on the planet Venus, on the nebula theories of Laplace, and on "the Age of the Human Family".

Among the papers that Lie left behind is a draft of his first lecture in the Association that fall. It begins thus:

> Gentlemen!
> I have given myself the Task in this *Realforening* Lecture to arouse in the University's young Mathematicians' Interest in the newer Geometry.
>
> Unlike ordinary Geometry, the newer Geometry is not characterised by its Method. In Reality there exist different distinct Methods for treating the same Things within the new Geometry. One has only to name Poncelet, Möbius, Chasles, as Representatives of such distinct Methods."

He maintained that the old geometry referred to figures in a right-angle system of coordinates, and in this way defined figures with the help of equations between the coordinates, while the new geometry applied a "Diversity" of coordinates: the old system had been inherited from Descartes and analytical geometry, and the new was represented by Möbius, Plücker and the others. This new relationship was something that Lie hoped to go into in more detail in a later lecture. He then gave a short summary of the methods of Möbius, Poncelet and Chasles, and referred to their published works by name and journal volume. He stressed that through Chasles' works one found "the newer Geometry set into a surprising Combination with the Old-fashioned", and Lie regarded this as evidence that Euclid had known "The Principles of our Century's Geometric Future". The reason for this was Chasles' reconstruction of Euclid's three lost books, *The Porisms*, where the subject addressed is the ancient Greeks' knowledge of the mathematically important curves that go under the collective term, conic sections. The conic sections are the curves one gets when planes cut a cone – to give us curves such as circles, ellipses, parabolas and hyperbolas. Through his work, Chasles had shown that Euclid and his contemporaries probably had an understanding that these curves were different projections of one another – that in reality all conic sections were perspective projections of a circle – in all likelihood the Greeks also could be demonstrated to have had a considerable understanding of the strong reciprocal relationships between these curves. In any case, Lie concluded his scattered fundamental speculations that autumn of 1868 by contending that the difference between the old and the new geometry was more than by the *method*, characterised as "the very Nature of the Things studied". Whereupon he switched over to his main theme, "namely, Poncelet's Method". He gave a definition of the ellipse as an orthogonal projection of the circle; he defined conjugated diameters in the ellipse and showed that the sectors between two such have the same area. The draft for this lecture at Realistforeningen in October 1868, thus ended with an emphasis that one – instead of studying a given figure, can make the investigation easier by studying a projection of it – "One can say that one thus transforms the simpler Figure into the given." Lie stressed projection as a fruitful tool, and cited various mathematicians' knowledge and treatment of the invariant relationship between points on a line, known as the cross ratio.

In addition to his frequent lectures before an audience of about ten to fifteen at the Science Students Association, Lie also gave private tutoring that fall, which was still his major source of income. But above all, he also began his own original research at this time. According to later descriptions he formulated his "Imaginary Theory" in December 1868. Whether Sophus took time off to celebrate Christmas that year seems uncertain. But his already-mentioned nephew, Johan, now wrote from Tvedestrand and invited "dear Uncle" there for the Christmas holidays – it had been so long – wrote the ten-year-old Johan – since they had seen one another.

We find that in February 1869, Lie presented his comrades at the Science Students Association with "The Principles of his new Imaginary Theory" and "by way of Illustration he applied this by means of a Number of Examples" – and a couple of months later in another set of minutes we find the following description: "S. Lie communicated, in Part, the Results he reached with the Help of his Imaginary Theory, principally regarding Properties of the one-sheeted Hyperboloid" – that is, a hyperbola that rotates on its "imaginary" axis.

It was this "Imaginary Theory" that came to be Lie's first independent work, and which was subsequently published in *Crelle's Journal* and quickly awoke interest and found success.

Lobachevsky had called his work an imaginary geometry, and now an astral geometry was being talked about – this was certainly to stress that this work turned on something other than what one, without much ado, could see, and about which one could conceptualise. In mathematical history there is a tradition of dealing with the not obviously visible, that which is not clearly comprehensible for the imagination.

Yet, in a completely different setting, the so-called imaginary numbers had been an acceptable part of the mathematical conceptual apparatus for quite a long time – for at least two or three centuries before they became comprehensible in a more natural manner. As a result of this historical trajectory, the designation "imaginary numbers" has remained, and these imaginary numbers, along with real numbers, make up "complex" numbers. By analogy with Cartesian geometry, geometry based upon complex numbers, became known as imaginary geometry. To give a lucid account or representation of what were called imaginary points and curves of the complex plane, was consequently a task that not only Poncelet, but also Hamilton and Grassmann tried unsuccessfully to handle. It was the challenge posed by this geometry to which Sophus Lie now began to apply himself, and which he treated by means of his "Imaginary Theory".

Just as contour lines on a map replace the missing third dimension on a particular plane, so now Sophus Lie assigned a *real weight* to a point in ordinary 3-space, as a substitute for the missing fourth dimension in our *perceivable* space, and he regarded the result as the "imaginary" complex plane. Lie showed that the points of a complex or "imaginary" line, with weight zero, form an ordinary line in space. He was able to make this line, called the zero line, represent the desired, imaginary straight line in the complex plane. He was thus able to demonstrate a relationship that is not visible with "the naked eye", and to do so in a manner that was simple enough to be recorded; that is, "imaginary" straight lines in the complex plane could be devised from the actual straight lines that are "visible" in ordinary space. In other words, Lie had succeeded in transferring geometric relations and information from the world of the real to the world of the imaginary – and it was precisely this principle of transfer, this ability to relate properties from one sphere to another, that was a matter of great interest. The fact that relations and properties from one area allowed themselves to be represented by relations and properties from another area had far-reaching implications. For Lie, this treatment represented a transfer principle whereby any proposition from plane geometry could be translated into a proposition in space geometry, and he was able to point out and calculate geometric correlations between figures depicted in the real plane and those from the complex.

Sophus Lie certainly wanted to visit the mathematical centres abroad, and in February 1869, he applied to the Academic Collegium for a travel scholarship. One month later the application was supplemented with his newly published paper – and the application was also supported by recommendations from the faculty and by the news that Professor Broch very much wanted to see Lie's work translated and published in German. When, at the beginning of April, the Collegium reviewed the applications for the travel stipend, Sophus Lie was placed securely in ninth place, out of the thirteen who had applied – as for the rest, the list included Gerhard Armauer Hansen, and Candidate Henrik Ibsen. For Lie, the money – 400 speciedaler – would be sufficient for half a year in Paris, and a similar period of time in Germany.

It seems that that spring of 1869 was otherwise devoted to diverse editorial projects in relation to publishing the various versions of the "Imaginary Theory". The first was the eight pages published with the financial support of Ernst Motzfeldt, then two versions for the Proceedings of the Christiania Academy of the Sciences – the second being an elaboration of the first – and then there came the task of getting the paper translated into German, which also led to publication in *Crelle's Journal*, and thus to making contact with an international public. However, before the work became accessible in the German periodical, Lie sent the first version to the mathematicians he knew. He sent his work to the Danish professor, Adolf Steen, whom he had met the previous summer at the Nordic Meeting of the Natural Sciences; in the accompanying letter he said, among other things, "If Herr

Professor would be so kind as to look through the enclosed Work, which makes a Claim to the *fruitfulness* of the Theories here included, it would please me exceedingly to hear your Views on the same. Up here in my little Backwater there is Nobody interested in Geometry."

The importance of signalling he was the first out in print was abundantly clear to Lie. In a letter to the Academy of Sciences in relation to the aforementioned publication of the work, and with a resumé of the ideas enclosed therein, he wrote, "It is my Wish that by this Means possibly to secure my Priority over Ideas, that by my Assessment are fruitful, and as I have Grounds to consider, constitute an Innovation."

In addition to his lectures, his editorial work and private tutoring, Lie also found time that spring to work as a substitute teacher at Nissen's School. His income, however, was too small. He was in debt both to Mrs. Motzfeldt – as he continued to lodge at Grottebakken 1, and he remained in debt to Ernst Motzfeldt, who had financed the printing of Lie's work. But both Lie and Motzfeldt were convinced that this was only the beginning of a great career, that good fortune would continue, and since Lie already had high hopes that in the course of the year he would also be awarded the university's large Adjunct Stipend.

That autumn, when Professor Broch resigned from the university to go into the government, the students held a party for him – the chairman of the Student Society gave a speech, and Sophus Lie, on behalf of the scientists and mineralogists, expressed deep gratitude and devotion to their respected teacher. A little later Lie wrote to his friend Motzfeldt that he hoped Sylow would apply for the professorship vacated by Broch. Lie seemed to know that Bjerknes would be in favour of such an appointment, as undoubtedly was Broch, "who certainly indirectly (as minister) would have decisive Influence over the Filling of the Post."

During the summer of 1869 Sophus rambled through Telemark with his sister Laura on their way to spend summer holidays with his elder sister and her family at the home provided for the doctor at Tvedestrand. He reached Tvedestrand at the end of July and remained in that town for a month. From here he wrote to Motzfeldt and reported that parts of the anticipated travel stipend would have to be used to pay off debts – the tour of Telemark had proven more costly than he had planned – otherwise, he reported that he spent his days revising the paper for the Academy of Sciences, and mentioned that the paper was "becoming more longwinded" and that it had "required more work than I had thought."

Immediately upon his return to Christiania at the end of August he borrowed six volumes of *Crelle's Journal*.

The First Tour Abroad

One day in September, Sophus Lie left Christiania and set course for the great outside world. He had planned to travel by way of Copenhagen, but when the ship berthed at Fredrikshavn in Denmark, he disembarked and instead, journeyed down through the Jutland Peninsula to Germany and Berlin. The reason he had wanted to visit Copenhagen was that he felt he could obtain good advice there about how he could best spend his time abroad. At least this is what he expressed some while later in a letter to Professor Zeuthen in Copenhagen, where he simultaneously explained why he had not come through that city. Lie wrote, "Unfortunately I am such a wretched Sailor that when we landed in Fredrikshavn, I am ashamed to say, Matter overruled Mind, and I preferred Jutland to Copenhagen." Were he to have continued on to Copenhagen, he would have had to board another ship. Instead, he proceeded onward by land.

Lie had been in Berlin for a couple of months when he wrote to Zeuthen. He had found lodgings at Kronenstrasse 52, and had already met and sought advice from German mathematicians about how his time abroad could best be utilized. "But as You know," he continued to Zeuthen in November 1869, "the Germans and the Frenchmen do not love one another, neither in Politics nor in Science." And about his plan to spend the winter semester in Berlin and the spring semester in Paris – "in both places to study particularly *Geometry*" – he had come to hear, from the German point of view, that he "would not find Anything in Paris, that the French Mathematicians of any Significance are decrepit old Methusalahs." Lie did not completely agree that this was so, and pointed out in his letter to Zeuthen that a stay in Paris would be profitable "due to the Fact that my whole scientific Development is French. Only after I had begun to engage myself in Geometry (a Year and a Half ago) did I acquaint myself with the German and English Literature."

Lie would remain abroad for about fifteen months, and the time would be fairly evenly divided between Germany and France; in the course of things, plans also developed for a visit to England. In letters to friends and acquaintances, Lie recorded much of what he encountered and experienced, and a number of these letters are extant – not the least of which are the more than twenty letters he wrote to his friend Ernst Motzfeldt, and which contained many illuminations of his life and work at this period. In the first, from Berlin on Wednesday, October 6th, Lie wrote, "I am living here in a Deluge of Scandinavians; this can be a rather good turn of events; one receives Visits from the One and the Other, however, People cannot understand that Others may have more to engage them than They themselves do, and therefore on occasion they can become intrusive." In positive terms, he described meeting a certain Dr. Svendsen from Lund, Sweden, a certain Meyer, Cand. Mag., from Bergen, and Peter O. Schjøtt, who would later become professor of Greek and a member of Sverdrup's government in Christiania (1888–89). Lie looked forward to the beginning of the semester, for then the young Germans would arrive in the city, and "maybe then I can make some good scientific Acquaintances." Otherwise, regarding his everyday life, he reported, "I have seen a Bit here: the Museum is splendid, even though I myself am able to discern that it has many weak Points (for Example, the Collection of Old Norse Artifacts). In the Mineralogy Collections I often had the Gratification of finding Norwegian Names describing the Locations of Acquisition. The Aquarium is interesting even though the good Berliners have some of the general Human Weakness of lauding their Own." He also commented on the famous Orpheum Ballroom; he thought it was beautiful, and that in particular it "displayed a splendid Taste in the Employment of Gas-Lighting with coloured Lamps in the ballroom and the Garden."

In the course of these fifteen months, the twenty-seven year-old Lie would build personal relations with men of science in both Germany and France, and in many ways, this stay abroad laid the groundwork for his future. The missing amiability "in Politics and Science" between these two great powers of Europe would become obvious in the acts of war that were to break out a year later, in July 1870 – and in which Sophus Lie would find himself embroiled. In later life, through his position and his field of activity, Sophus Lie would become an important bridge-builder between men of science in Germany and those in France.

When Lie came to Berlin that autumn of 1869, there were two centres of mathematics in Germany – Berlin and Göttingen – which represented strict, rivalrous trends in mathematical research. Three great men ruled over what was called the Berlin School: Ernst E. Kummer, 59 years of age; Karl Weierstrass, 54, and Leopold Kronecker, 46. A special characteristic of this milieu was that mathematical topics and ideas were treated, above all, according to the pure reasoning of abstraction. Properties of mathematical objects were revealed with the critical power of logic. There was an infallibility intrinsic to mathematical material, and mathematical relations reflected something that one might consider analogous to divinely inspired truths. Nevertheless, the emphasis on strict logic as the road to insight hindered neither Kronecker nor Weierstrass from imbibing in the vocabulary of Romanticism: "Dichter sind wir" (Poets are we) said Kronecker, and ascribed the work of mathematicians to a position close to that of artists. Kronecker maintained that the starting point of a mathematician – the rational numbers – came to us fundamentally by intuition. And Weierstrass, for his part, maintained, "Ein Mathematiker, der nicht zugleich ein Stück von einem Poeten ist, wird niemals ein vollkommener Mathematiker." (A Mathematician who does not at the same time have some of the Poet in him, will never be an adequate Mathematician.) The great field of endeavour that Weierstrass followed was function theory, and in this field he advanced the work of Niels Henrik Abel considerably. For his part, Kummer had long been interested in the field of number theory – in particular, he had worked with "ideal prime factors" in complex numbers – but around 1860 he quite suddenly began to study geometry, and a while later he discovered the wellknown surface that bears his name, the Kummer surface. But for Kummer too, all the developments in geometry were also dependent upon the strictest argumentation and presentation of evidence.

On the other hand, in Göttingen, geometric perceptions and methods were to a much greater degree the basis for mathematical interpretation, something that inspired complaints from Berlin about a lack of stringency. The leading figure in the Göttingen School was Alfred Clebsch, who was also described as an inspired and gifted teacher. Clebsch was a great connoisseur of the relationship between algebra and geometry, and, to use a modern expression, used *geometric intuition* in his demonstration of relationships between Abelian integrals and algebraic curve theory.

In the course of his studies at home in Christiania, Sophus Lie had become acquainted with the stringency of the Berlin School's programme, but in his own work he had taken what he needed – ideas and methods – from wherever they were to be found. Lie was an eclectic. He had sent his first treatise, on "the Imaginaries", to Clebsch in Göttingen, and best of all, Lie knew the geometric outlook of the French syntheticists, Poncelet and Monge. During his stay in Berlin, Lie now attained still better knowledge of the mathematical literature, and he had come abreast of the time's current research, and noticed the correspondences between his own approaches and the theories of others. His closest companion at this time was Felix Klein. One of the first to whom Klein directed Lie's attention at this time was Theodor Reye, at the technical college in Zurich, who some years earlier had treated the "line complex" that Lie had dealt with "by means of the Imaginaries". Lie mentioned this in a letter to Zeuthen, with the following admission, "Of course what I have published to date contains nothing really new. But at any rate, in my Opinion, my Method demands Interest."

Felix Klein was only twenty, but he had been Plücker's pupil, and at the invitation of Clebsch had already edited Plücker's last treatises – papers that Lie had read at home in Christiania. Thus it was that Lie and Klein had a common interest in Plücker's line geometry. Klein seems quickly to have been impressed by the audacity of Lie's ideas, and for his part, Lie seems to have valued Klein's breadth of reading knowledge and scientific orientation. From now on there developed a personal friendship and scientific cooperation between Lie and Klein that would support them in both life and work.

Lie and Klein met for the first time at the Berlin Mathematical Association, and that must have been on October 24th or 25th. In any case, both wrote letters dated October 31st and dated their meeting to "a week earlier". Lie wrote to his friend Motzfeldt in Christiania, and Klein wrote to his mother in Düsseldorf. Klein wrote, "Among the younger mathematicians I have become acquainted with, one impresses me strongly. This is Lie, a Norwegian, whose name I was already familiar with from an article published in Christiania. In a particular manner we have both been concerned with the same things, so that there is no lack of topics to discuss.

But we are not only united by the same love, but also a bit in terms of our critique of the way mathematicians here [in Berlin] express their importance at the cost of the work done by others, and particularly by foreigners."

In Lie's letter to Motzfeldt, he introduced Klein in connection to explaining that he had received a reply to his own letter to Professor Clebsch. That is to say, Lie, in addition to his first paper, had also sent Clebsch "two elegant Theorems to which my Theories have led me", and in reply he got to know that Clebsch was not really familiar with the theorems. But Clebsch, from Göttingen, had urged Lie to make contact with Felix Klein, who for the time being was also in Berlin and lived at Carlsstrasse No. 11. Lie commented to his friend in Christiania: "Quite by chance Klein and I had already a Week ago made one another's Acquaintance, and are now already good friends. We were from Beforehand, mutually familiar with one another's Work, which has several Points in Common." And besides, Lie had otherwise made a couple of other "good scientific Acquaintances" in Berlin, and he stressed that it was a great help to have "written for *Crelle's Journal*." On the whole, Lie seems to have been very satisfied, and he added, "In recent days (as before) my scientific Work has met with good Luck and I continue to see my *scientific* Future in bright Colours."

The fact that Motzfeldt back in Christiania always maintained such strong faith in his friend's mathematical future was important to Lie. Consequently, in the letters to his friend Lie emphasized his faith in his own powers, his state of being, and the praise he gradually received from the mathematicians he met. While Lie was abroad, while he was growing as a mathematician, both in his own studies, and through his meetings with other mathematicians, he remained concerned with conditions in Norway, and how he could secure himself a future position in his homeland. And back in this domestic arena he thus had Ernst Motzfeldt, a faithful support player who was constantly being requested to go on errands for Lie, and speak on his behalf in relation to applications for positions and research grants. For some years Motzfeldt had been working as a high court lawyer, and in 1869 had been advanced to the position of supreme court attorney. He had married Else Gram the year before, and the young couple lived elegantly on Skippergaden in Christiania. Motzfeldt was also authorized to remove money from Lie's travel scholarship fund and send it to him abroad, and in appropriate payments to pay Lie's creditors in Christiania. These were book dealers, restauranteurs and tailors.

Sophus Lie's state stipend amounted to four hundred speciedaler, and in reserve he had a promise from his father to get a further one hundred daler during the spring of 1870. This should last him through his stay abroad, but after that, with his homecoming, everything was uncertain. With letters of recommendation from Nissen's School, together with attestations from some candidates he had tutored in higher mathematics, Lie had applied for a position at the Institute of Sea Cadets in the town of Horten, on the west side of Oslofjord, where an elementary technical school had been established by Parliamentary resolution in 1854. In his application, Lie had written that he would be available for employment from the summer of 1870, and in a letter from Berlin he urged Motzfeldt to enquire into the matter: would it be possible to postpone the remainder of his travel scholarship until later if it happened that he must begin work at Horten at an earlier date: was it a permanent position, and did Motzfeldt know who the other applicants were? When, a short time later Lie got to know that a Lieutenant Jens Koch had been awarded the position as teacher at the Sea Cadet Institute in Horten, he took this "with greater Serenity than perhaps I ought," he reported to Motzfeldt, and added, "Moreover, if the Truth be known, I have very little Desire to be buried at Horten."

During this first period in Berlin, Lie was also eagerly awaiting the first proofs for correction of the work that the Christiania Academy of Sciences was to publish. And, "Since it is in many Respects a Question of Life and Death to get my Work published as quickly as possible," he urged Motzfeldt to go to the printers, Brøgger & Christie, to have the work dispatched to him. He also urged Motzfeldt to send him as expeditiously and cheaply as possible the forty free copies he had requested. Actually, Professor Monrad had promised Lie more than fifty free copies – "I want to have 100," Lie wrote, and wondered if, at some point in the future he could make an arrangement such that the Academy of Sciences would put funds at his disposal so that he could get the works printed abroad: "This would have been my particular preference (much quicker and cheaper)." Lie mentioned Professor Monrad in several letters, and the assistance he had given, and felt "the Courtesy he [Monrad] now as before shows toward me" might give grounds for hope about such printing abroad in the future. To Motzfeldt he confided that this "Haste I have over the Publication of my Works" mainly arose from the fact that he was constantly finding in foreign periodicals such "Results as I already have obtained or Further derived from my Theorems." Fortunately, through his discussions with Klein, he was able to ascertain "that my Method is completely in all Essentials new to Science, as it were, and that a Series of interesting Propositions of mine have not been published by Others to this Date."

But: "What should I do with Respect to the Research Fellowship?" he wrote to his friend in Christiania, and urged Motzfeldt to find out what had happened to persons who had sought and received this stipend earlier while they were abroad: had they immediately returned home? Worries and remarks about this "adjunct stipend", which, thus, according to the statutes, was given to young deserving scientists, came up over and over in Lie's letters during that autumn. (The Adjunct Stipend was, in reality, a research stipend, which it later came to be called.) Lie urged Motzfeldt to use the same letters of recommendation that he had used in the application at Horten, for the stipend application, and he asked his friend to retrieve the papers from the Marine Command Headquarters. Lie was afraid that as the stipend application was processed, he would be penalised by Professor Christie, his former physics teacher, for what Christie referred to as his "Arrogance", and therefore that Christie would use his great influence in the faculty to Lie's disadvantage. He tried to puncture his fear and anxiety with the following comments: "But what will be will be; these Gentlemen shall not be able to ignore me forever."

Lie wrote this in a letter dated Berlin, October 31st. He was beginning to attend lectures: "There is no trace of Language Difficulties," but conversations were more difficult to manage. He reported that there were no lectures in several of "the Things I am specially interested in", and that he therefore had "more Time for my own Work", and through direct discussions with Klein and "a couple of other good scientific Acquaintances" he had "great Benefit" from his stay.

The lectures that Lie attended were a "mathematics seminar" that was alternatively led by Professors Kummer and Weierstrass – and under the guidance of Kummer, Lie would reap considerable praise here.

As for other aspects of his life in Berlin, Lie had very little to report: "By and large, Things transpire as at home, with some necessary Changes. Occasionally in the Evenings I visit the Theatres. On a couple of Evenings I have attended scientific Student Societies. '*Bier*' [beer] plays a great Role everywhere here. I meet many Norwegians." And at this point he asked Motzfeldt to send him eighty speciedaler, the equivalence of which in Berlin was one hundred and twenty *thaler*, as there were some books to be procured. He wrote that in the libraries, it was precisely the good books that were out on loan. As well, there were university fees to pay in Berlin; whether or not this charge would be demanded in France, he did not know.

Once again, when Lie wrote to Motzfeldt in the middle of November, it was the adjunct stipend that most preoccupied him. He enclosed a laudatory letter of testimony from Clebsch, and asked Motzfeldt "in Recognition of how important it is for me to get the Stipend" to show this to Christie. Lie commented, "Although Christie is no Scientist, he does however have Respect for Science," and he counted on Motzfeldt – if he were able to use the letter in Lie's favour – hopefully to do so. Two weeks later Lie sent two new testimonial letters to use in Lie's "Favour with respect to the Adjunct Stipend". One was from Zeuthen in Copenhagen, and the other from Reye in Zurich. Zeuthen commented upon Lie's mathematical methods and praised Lie for the way he applied his methodology: "You show that this can be utilized, and surpasses Everything; You show that You understand how to use this and to draw rich and interesting Conclusions about geometric Transformations." For his part, Reye found Lie's treatment of the representation of "imaginaries" to be extremely interesting, and he was surprised to see how Lie, with a completely new and simple principle, had produced the same line complex that Reye himself had treated in the second part of his renowned *Geometrie der Lage*. Reye continued with the following laudatory description: "You, with your geometry of imaginaries have made a very happy discovery."

For a week at the end of November, Lie was together with one of his old student friends from Christiania, the medical doctor, Axel Lund, who had unexpectedly come to visit him in Berlin. Lund was on his way home because he, to use Lie's words, "could not tolerate studying" away from home, and since his stay abroad had been so short, he had "Money to the point of Overflowing, so I, to the best of my ability, have helped him lighten his Burden." About what the two friends might actually have used the money on, we have only Lie's words "around about Museums and the Theatre". Throughout his life Lie would maintain good and regular contact with Axel Lund, who later would become a respected city doctor in Christiania, and the doctor whom Lie sought out during his last, fatal illness.

Otherwise, about his life in Berlin during the autumn of 1869, Lie wrote that one Sunday he had attended a social evening at the home of Professor Kronecker:

Hermann von Helmholtz, Professor of Physiology and after the death of Gauss in 1855, the man who reigned as the most prominent presence in the sciences in Germany. Von Helmholtz made important discoveries in several fields – a treatise from 1868 on the foundation of geometry was an inspiration to many mathematicians, and his geometrical axioms were situated not far from Lie's works. Lie, with his new theories would gradually come to complement von Helmholtz. But then there was a reaction in Germany about the manner in which Lie demonstrated shortcomings in von Helmholtz's work. Taking the data of the human senses as his starting point, von Helmholtz (in a manner more radical than Freud would treat this question) maintained that most of what goes on in the head, is unconscious.

The three great mathematicians of the Berlin School whom Lie met in the autumn of 1869.

Right:
Karl Weierstrass.

Bottom left:
Leopold Kronecker.

Bottom right:
Ernst Eduard Kummer.

Paris' famous institution of learning, l'École Normale Supérieure – most frequently referred to simply as École Normale, the elite school where the most theoretically advanced mathematical teaching was conducted. Several of this school's best pupils were sent to Leipzig to sit in front of Lie's lectern.

In recognition that he was the person who had to the greatest extent developed Galois' ideas in new fruitful fields, Lie was invited on the occasion of l'École Normale Supérieure's 100th anniversary celebrations in 1895, to speak about Évariste Galois, the school's most famous pupil.

For almost his whole life, Lie had contact with the French mathematical milieu. In 1892 he was appointed corresponding member of l'Académie des Sciences, the highest recognition that France could bestow upon a foreign man of the sciences.

Some of France's leading mathematicians.

Top left: *Charles Hermite*. Top Right: *Camille Jordan*. Bottom Left: *Emile Picard*. Bottom right: *Gaston Darboux*.

Dagbladet, October 12, 1870: "Norwegian Man of Science Jailed as German Spy". (The event was also covered in the other newspapers that same day.)

Below: *For a period of time Lie made a series of sketches from his travels in the mountains. This is from the summer of 1887.*

En norsk Videnskabsmand fængslet som tysk Spion. (Fædrel.) I Frankrige er der for Tiden den største Skræk for tyske Spioner, hvilket er rimeligt nok, da Preusserne særdeles meget bruge slige Folk og ved Hjælp af dem har været vel underrettede om den franske Regjerings og Krigsførsels Hemmeligheder. Men denne Skræk gaar ud over mange Uskyldige, da Franskmændene nu er saa opskræmte, at de tro at se en Spion næsten i hver Udlænding, som træffes inden deres Enemærker. Flere Nordmænd have været udsatte for denne Mistanke og har havt adskillig Uleilighed deraf. Værst er det dog vel gaaet en af vore dygtigste yngre Videnskabsmænd, Realkandidat Sofus Lie (Søn af Sogneprest Lie paa Moss). Denne 28aarige unge Mand har med offentligt Stipendium i et Aars Tid været paa en videnskabelig Udenlandsreise og har herunder skrevet mathematiske Afhandlinger, som efter ansete tyske Mathematikeres Vidnesbyrd vække de allerbedste Forhaabninger. Siden Marts har han paa Studiernes Vegne opholdt sig i Paris og har der ogsaa været rosende omtalt i flere Blade. I Begyndelsen af August var han færdig og agtede sig nu til et italiensk Universitet. Men paa

tede sig nu til et italiensk Universitet. Men paa
Veien havde hans videnskabelige Vaner nær faaet en brat Ende. Han er en ivrig Fodgjænger og har gjennemvandret sit eget Land paa kryds og tvers*). Istedetfor nu som andre Reisende at tage med Jernbanen, besluttede han at gaa til Fods for at lære Frankrige bedre at kjende. Denne Fodvandring kom ham dyrt at staa. Han var ikke kommen mere end et Par Dagsvandringer fra Paris, før dette i Frankrige aldeles usædvanlige Syn, en udenlandsk Turist til Fods, vakte Mistanke. Han blev anholdt ved Staden Fontainebleau den 11te August for nærmere at undersøges. Uheldigvis var der flere Omstændigheder, som styrkede Mistanken. Han havde ikke faaet sit Pas paaskrevet af fransk Øvrighed, han havde for Reisens Skyld en større Pengesum hos sig (som man naturligvis tog for tysk Guld) — og hvad der var det værste, man fandt hos ham tyske Breve, ja endog Breve fra den preussiske Prins Fredrik Karls Hovedkvarter; han stod nemlig i videnskabelig Brevvexling med en tysk Ven, der nu som Værnepligtig havde gaaet i Krigen og under denne vedblev sin Brevskrivning. Dette var nok til at føre ham i Fængsel og sætte ham under Forhør. Han var naturligvis aldeles uskyldig. Heldigvis fik han da renset sig saavidt fra Mistanken, at man ikke strax gjorde kort Proces med ham (d. v. s. hængte ham). Efter nogle Dages Forløb sagde man ham, at man var blevet overbevist om hans Uskyldighed, men han maatte dog sidde i Fængsel, til den kommanderende General gav ham fri. Under al den Forvirring, som i Slutningen af August opstod i Frankrige, har man vel glemt ham, og i Begyndelsen af September fandt han det derfor raadeligst at skrive til sine Venner i Norge. Gjennem dem blev Diplomatiet sat i Bevægelse; men førend dettes Hjælp var naaet frem, var han heldigvis sluppet løs den 11te September. — Paa Kosten nær havde han forøvrigt havt det ganske godt i Fængslet, og han havde faaet god Leilighed til at arbeide paa en mathematisk Afhandling. Da han var sluppet ud, vilde han før Afreisen gjerne se sig lidt om i den berømte gamle Skov ved Fontainebleau. Men da Folk kjendte ham igjen fra hans forrige Anholdelse og begyndte at stimle om ham, fandt han det raadeligst øieblikkelig at gaa til Jernbanestationen, hvorfra han da paa hurtigste Maade i god Behold kom til Schweitz.

Kvindernes Stemmeret ved Spörg maalet om ny Salmebogs Indförelse. Anledning af Indførelsen af Landstads Salm skriver Bergp.: Det synes i vor Tid at v billigt Forlangende, at hele Menigheden eller benytte den nye Salmebog, fo ale sin Mening og ikke under tals Mening, som „H oruden den Ting, ltryk ha

"Ladies were in Attendance and the Time was applied somewhat as in Norway – Conversation, Parlour Games, etc."

A month later he reported: "It is with Wistfulness that I think about Snow and Ice in Norway. Here we have Fog, Rain and Mud, and we shall see nothing Other in the coming Months."

When Lie got to hear from Christiania that Axel Guldberg had also applied for the adjunct stipend, he decided to write immediately "a very precise letter to Bjerknes" in which he would remind Bjerknes, who was also a member of the Stipend Committee, of how he earlier had expressed his opinion; namely, that according to his view there could not "be any Discussion about the Comparison between A. Guldberg and me as Mathematicians." Motzfeldt also informed him about who the other applicants were for the adjunct stipend, and Lie commented that he himself could not offer any opinion on these, "but it is certainly an Absurdity to draw Parallels between me and some of them (Sars excepted)" – and he confided in Motzfeldt that he was now so eager to obtain this stipend because he hoped to be able to extend his time abroad by half a year, and would thereby in the spring journey to Milan, because: "For the Time being (curiously enough) the condition of science in Paris is not favourable."

In the letter to Professor Bjerknes, dated Berlin, December 3rd, Lie came out bluntly and urged Bjerknes "urgently" and "forcefully to bring to Attention this Information on A. Guldberg's Scientific Character, that You share with, for example, Sylow" – and Lie was surprised that the Natural Sciences Faculty in Christiania did not think that he deserved "a Particle of Support" more than the other applicants, and he mentioned the zoologists, Axel Boeck and Georg Ossian Sars, both of whom already had permanent, salaried scientific positions. Lie wrote that it was clear in his mind "that within the different Faculties there were normally intrigues waged by various more or less obscure Persons", but it had not occurred to him that he could be passed over after his last publication. Lie excused "the Passion" in which the letter was written, after having fired off the following:

> Should I be passed over once more, I can take consolation in the fact that in recent Times, among Men such as *Kummer*, Zeuthen, Klein, Reye (and partially also Clebsch) I have met Recognition that certainly has greater Value than, under the current Conditions, the doubtful Honour of being Adjunct-Stipendiat in Christiania.
>
> When some Years have passed, if not before, You will agree with me that my continually being ignoring might by then be justified, only if I have failed to show that my Science is nothing more than a Potted Plant.

To Motzfeldt, Lie also wrote, "The Honour of the University depends upon it not passing me over". In the middle of December he received the news that the adjunct stipend had gone to the brothers, Ernst Sars and Georg Ossian Sars, together with the philologist, Gustav Storm. Despite his disappointment, Lie was pleased to find it was "People like the Brothers Sars and Storm who had been preferred over me." However, through Bjerknes he got to know that Christie really had supported Lie over Axel Guldberg. But Bjerknes seems to have indicated that Lie's application in part "was arrogantly drawn up", and Lie observed, "I readily accept that [. . .] My *Application* was addressed to a *layman* Collegium. As in the Spring, I used Diligence to choose a Form that would force the Faculty to make a Pronouncement." And in the same letter: "When I write with style about myself and my Scientific Qualities, I do so *in Relation to the Yardstick that one is justified to apply in Christiania.*"

The reason that Lie now explained himself in this way seems to have been that Motzfeldt, together with their common friend, Carl Berner, had planned to write a critical article to the press on the whole "character of stipends" – and Lie asked that his name not be involved in the debate, and reminded his friend about "a Couple of actual Facts": *"Up to the present time I have expressed myself in such a Manner that my Works remained so lofty that People in Christiania could not understand them.* The Branch of Mathematics that I have been involved in for the past 1½ Years is not the most difficult in Science", and he maintained that this branch of the sciences was relatively new, and that therefore few mathematicians had occupied themselves with the subject and felt that for this reason, on one hand, he was very lucky: He had had "the Foresight to find Something in a Field that was little sought-after". But on the other hand, he was unlucky, for he, in the most most fortunate case was only able to "render Account to an extremely limited Public."

Having suffered the stipend rejection, Lie now wanted to ask the University of Christiania for two hundred speciedaler in addition to the travel stipend he had already received – when, as he wrote to Motzfeldt, he considered that many medical students received for one year, more than the 400 daler that he had received for his whole stay abroad, "was not this Matter extremely absurd?" Lie stressed that such an application must certainly go through the former professor and now cabinet minister, Ole Jacob Broch, and he formulated what amounted almost to a prayer to Motzfeldt to take up the matter. Lie wrote that he wanted very much to visit the mathematician Cremona at the technical college in Milan, and he concluded, "Now, during this Year, I should build Relations for Life."

Despite the rejection in Christiania, December was a happy time for Lie in Berlin. In a sense it was now that he made his breakthrough in mathematical circles, in any case, this was how he himself experienced it, and it gave him a powerful lift. It happened in the "mathematical Seminar" under the leadership of Professor Kummer. Lie related his "Triumph" to Motzfeldt: "Professor Kummer proposed that we test our Powers in the Discussion of all Line Congruencies to the 3rd Degree. Fortunately for me I had solved a Problem (about which nothing is published) a Couple of Months ago, that in certain Respects was a Special Case of the above-mentioned, but in other Respects was far more general. Kummer, who knew that I occupied myself with Line Congruencies, urged me to give a Lecture."

Lie prepared a lecture (a further elaboration of § 28 of his latest paper) and Klein presented a resumé of Lie's work at Kummer's seminar, and this resulted in "Kummer, in a *particularly flattering* Manner," thanking him for the lecture and urging Lie to "treat in greater detail a certain Congruence of the 3rd Degree, with which he had involved himself recently." In the course of his own work, Kummer had encountered difficulties "that he had not overcome", and the treatment of which he now thought, thanks to Lie's method and principles, would be simplified. "In a terribly long Lecture he [Kummer] presented characteristic Properties of

Certain Instances of Congruence" and formulated questions "whose answers he would be inordinately interested in." And it was now that Lie stepped forward and impressed both Kummer and the other participants:

> I was in the happy Position of making Kummer aware that his Problem fell within what could be treated as one of the Special Cases of what had already been treated. One of the propositions that he had referred to was wrong (in all likelihood due to a lapse of memory on Kummer's part, for this was something with which, he was occupied in only a preliminary sense) [. . .] In the end, I was in a Position to answer several of the Questions he particularly wanted solved.

Again Lie was given the task of preparing a lecture, and this time, with the assistance of Felix Klein. Lie did not hide from his friend Motzfeldt in Christiania that this was a dazzling personal triumph, but he also described the incident as "an extraordinary Bit of Luck", and stressed that it was normally very difficult "to get the floor" in scientific circles, particularly in Berlin:

> In Germany, Berlin Professors constitute an Aristocracy of the Sciences that considers itself to possess all of the World's Wisdom [. . .] Contrary to Conditions in several lesser German Universities, it is considered particularly difficult to get personally into Touch with the above-mentioned Gentlemen. Until now, I have been able to speak only on one single Occasion with each of Berlin's three Greatest (Kronecker, Kummer and Weierstrass). When such an Opportunity does arise, one feels like one is standing before a King of the Realm of the Sciences. Hopefully in the Future I will get more frequent Occasion to discuss particularly with Kummer.
>
> One can go on and on speaking badly about the Prussians. But they can certainly think; about that there is certainly no Doubt. If I were to say anything about their Defects it would have to be about their Manner of *giving Lectures*. In any case, from my own experience one cannot absolve them from the sin of undue Verbosity. Another Drawback of University Lectures is that they are not associated with definite Works. Here one is constantly bringing out new material.

Lie also commented on the difference between German and Norwegian conditions of study and research. There were, for example, no ordinary students in this mathematics seminar, but rather doctors and professors who had finished their university studies and now took part in "one or several Problems whose solution now queued up for solution in the Sciences." Also students worked in a completely different manner of organisation and concentration in Germany, a fact that often resulted in a person going "into his Speciality" at quite an early age – "and by the Age of 20–22" having achieved "Outstanding Competence". Another great difference between Germany and Norway was that while attending the University of Christiania "one got lectures on the state of the Sciences as they stood a Couple of Decades ago, while in Berlin one was treated to the scientific Questions of Today." The result was that in Norway after graduating "one did not have a very high level of scientific competence", and one still had to apply many years – "if not exactly the Time I have Used" – to be able to participate "In the Treatment of the Tasks with which the Times are engaged." In Germany it was almost a rule that each brilliant scientist had, while still a student, produced "Works of Value."

Finally, Lie concluded – regarding his activity in Kummer's seminar, and the nature of the latter's response thereto – "I am unable to discern whether Kummer meant what he said."

Some days later Lie commented to his friend in Christiania on both his former state of depression and others' views on his work projects: "One should certainly not admit it, as one should hardly judge from external appearances, but still the fact remains that I, for many Years (from 1864–68) have *undervalued* my own Intellectual Powers. *This* has been my Misfortune, not the Opposite."

It was getting close to Christmas. Motzfeldt received a message to greet his mother, who otherwise had received many special greetings from Sophus that autumn, and greetings also went out "to my Comrades, whom You on occasion will meet. You yourself and your Family, I wish a very happy Christmas." And Lie thanked his faithful friend for all his "Efforts for the Future of my Interests", and produced the following self insight: "As You have seen, my Letters deal only with me and my Science, but my world is not any broader than this."

In January 1870, he wrote again: "For the most part, I spent Christmas peacefully; more recently I have been to a Couple of Parties." From "a *Scandinavian* Arrangement with Ball and Comedy" during the week between Christmas and New Year's, Lie was able to report on "a little Scene". One of the players was to have ripped his trousers a bit, seemingly by accident in the course of the play, but this had been done in such a manner "that one complete Leg had been visible." Lie commented that "Fortunately the Norwegian Ladies who had accepted the Invitation had already found Cause to remove themselves."

Otherwise Lie reported that he was low on money and asked Motzfeldt to send him a bill of exchange for 100 daler to Kronenstrasse 52, Berlin. From his friend Amund Helland, Lie had learned that he could possibly receive the Nicolaysen's Stipendium, and perhaps also something from the Rosenkrons Legacy – this would at least allow him to travel to Milan during the autumn semester, otherwise he would have to turn homewards at the beginning of August, and in that case, hoped that upon returning to Christiania he would be able to earn some money immediately as an examiner in mathematics for the *examen artium*. He asked Motzfeldt – "If your Time allows it" – to find out if he could get this examination job, and suggested that Hartvig Nissen was probably the correct person to ask, but added, "I am well aware that *if I am not wanted as Examiner*, then it is certainly possible to find Objections; I shall not involve myself in countering them."

Lie had now found out that none of his papers would be published in the journals of the Academy of Sciences in Christiania while he was abroad and he asked Motzfeldt to inform Professor Monrad of this. A short time later Lie would, however, through Clebsch, get a little article published in *Göttinger Nachrichten*, and later works would also come to be accepted by Clebsch's journal. Meanwhile, together with Klein he worked "on an Enquiry, which we no doubt will publish in the above-mentioned Journal this Summer."

Lie's plans were now to remain in Berlin until the end of the month of February, and then to go to Göttingen "personally to make the Acquaintance of Clebsch". In all likelihood he would also spend "some Days along the Rhine in the Company of Klein", before he and Klein journeyed together to Paris.

Apart from this, now in the New Year, Lie gave a new lecture in the "mathematics Seminar" – on what he had been working over the recent months – but this time the reaction from Kummer had been somewhat disappointing. Kummer considered that "a rich Abundance of original Ideas had been invested" in Lie's lecture, "but that he had not quite been able to follow it." Lie commented, "I had certainly not intended to give more than an Idea of my Thought Process." But he had no doubt hoped that Kummer would have been able to follow the lecture: "I am convinced that he then, if he had been able to follow, would have uttered a Couple of Words, that I could savour."

In his last letter from Berlin to Motzfeldt in Christiania, Lie reported that he would follow his friend's advice; namely, he would ask the Collegium for a supplement to the travel stipend he had already received, but left it to "the learned Gentlemen" as to what the amount would be. However Lie did not believe that he would get anything, nor did he want to make "any extraordinary Efforts in this Regard". On many occasions he had been thinking to ask Kummer or Clebsch "for a Couple of Words of Support", but he had always been reluctant and did not pursue this. To Motzfeldt in Christiania he wrote, "During this long last Year it has pained me much to have had always to support myself with Letters of Testimony from Others. It has cost me a great deal in terms of Self-Conquest to ask Someone to put in a good Word for me, and I have only done so as Necessity dictates this." And he concluded by saying that despite the fact that it was very useful for him to remain longer abroad, he would rather renounce this in preference to asking someone else for a letter of attestation.

Before Lie left Berlin in February, he wrote another letter, this to Professor Bjerknes, and asked Bjerknes "to once again, if it be possible, to speak on my Behalf to the Collegium regarding my Petition for 150 Spd. from the Hjelmstjerne-Rosenkrons Legacy to be used to continue my Studies for 3–4 months either in Milan or at Cambridge." Both the English and Italian mathematical circles were considered in Berlin "with the greatest possible Recognition", and those mentioned as representatives of these circles included, according to Lie, Arthur Cayley, James Joseph Sylvester, George Salmon, William Rowan Hamilton, Luigi Cremona, François Brioschi and Guiseppe Battaglini. As an assignment from Clebsch, Klein had written a paper on Battaglini; in his seminar, Kummer had taken Hamilton's work from the 1840s as his point of departure; and Cayley's work was something with which Klein had long been preoccupied, especially his *Notes on Lobachevsky's Imaginary Geometry* (1865). Otherwise, about prevailing attitudes in Berlin, Lie added that "one readily, according to one's Ability, is engaged in demolishing contemporary French Mathematics." Kummer's ill-will toward the French no doubt stemmed from his traumatic childhood experience, from the period when Napoleon's army invaded his hometown of Soreau and infected the populace with typhus – including Kummer's father, who, as the town doctor, succumbed to the disease after some weeks. And it was rumoured that one of Kummer's first contributions to science – in order to repulse the enemy assault – was to calculate the trajectory of cannon-fire.

Lie however maintained his "old Love for French Mathematics", but would also gladly entertain a stay in Cambridge "to be inspired by modern English Mathematics (Quaterns and modern Algebra), which no doubt ought to be regarded as the most recent Decades' most outstanding Advance." All the same, perhaps he should go first to Milan, because, as he wrote, "It is easier for me to entertain the Possibility of some time in the Future visiting Cambridge than Milan."

Lie also sent Bjerknes the same citations from Zeuthen and Reye about his work, as he had sent to Motzfeldt. As well, Bjerknes got to know about the two lectures that Lie had given in "the Mathematics Seminar", and about the chance he had received from Kummer following the first lecture, to apply his methods "in a rigorous Discussion of a Class of Line Congruencies of the 3rd Degree". This work, which he had consequently taken up together with Klein, would however "at the earliest be finished in the Course of the Summer". Lie now wrote to Bjerknes about the second lecture that he had given in Berlin, and Kummer's somewhat disappointing comments were given in the words that he had used in his letter to Motzfeldt: Kummer had felt that the lecture contained "a rich Abundance of original Ideas" but admitted that he had not been able to follow Lie. Lie's own reflection on this was that it would probably require "considerable Time to work oneself into my Way of Thinking". However, it was with a certain pride that he was able to inform Bjerknes that it was precisely this lecture which "with the Goodwill of Clebsch" was now being published by the Academy of Sciences in Göttingen, and now, when he arrived in Göttingen in a few days' time, he would send Bjerknes an offprint of the paper.

Apart from this, Bjerknes was informed that before Lie and Klein left Berlin, Klein was to give a lecture on their "common Investigations", and Lie commented, "I regard it as an extraordinary Stroke of Luck that Klein, who is an outstanding (if still young) Pupil of Plücker and Clebsch has remained in Berlin this Semester. We are travelling together to Paris, and if I get the Stipend in question, also to Milan or Cambridge." Once again Lie recalled that at the moment when he received his 400 speciedaler, two medical students had received respectively 500 and 600 daler for a one-year stay abroad. Otherwise, he reported that he had been to two social gatherings at the home of Bjerknes' old acquaintance, Poggendorf: "This old Man is still quick and active", but Lie had had no opportunity to convey Bjerknes' greetings to Poggendorf, but now he would certainly bear Bjerknes' greetings to Professor Schering in Göttingen. Lie concluded, writing, "I shall stay 14 Days in Göttingen. My Course within Mathematics falls within this School, which in Germany is represented by Clebsch."

Lie left Berlin on February 28th, and on the way to Göttingen he passed through both Potsdam and Magdeburg. In Göttingen it was the end of the semester and Lie now experienced the students' "End of Term Binge", and he found himself in agreement with something that Motzfeldt had earlier said: "That the very fine Speeches on

Ideas are Something we [in the Nordic countries] rather have to ourselves." Despite all the "Loathsomeness of its considerable Hollowness, the Student Character at home," Lie had to admit, was rather better organised in Christiania than at most foreign universities – in Berlin of course there were some "Student-Corporations – scientific and drum-beating societies", but in Paris it seemed that there was no student society to be found.

The stay in Göttingen with Clebsch seems to have lived up to expectations. Lie later wrote to Zeuthen that "Clebsch, both as a Human Being and as a Mathematician has something strangely attractive about Him," and admitted that he could have wished to stay longer in Göttingen. One evening (the 10th of March, at eight o'clock) Lie was also invited to Clebsch's home "zu einer Tasse Thee" [for a Cup of Tea] together with a number of other friends of Clebsch, among whom was Professor Stern, who with great pleasure recalled that Lie's countryman, C. A. Bjerknes, had attended his lectures, and now asked Lie to give his greetings to Professor Bjerknes in Christiania. Felix Klein, on the other hand, was not in Göttingen, as he had not been able to leave Berlin before the end of the semester, and there was as well the lecture on Cayley's generalizing of the concept of distance that he had promised to give before leaving Berlin. (In Germany and France the winter semester at university continued until the middle of March; this was followed by one month's holiday before the summer semester began.)

In the course of the two weeks Lie spend in Göttingen he received four letters from Klein, who would be delighted to come as quickly as practicalities allowed. It would however be almost a month before the two friends met again.

Three or four days after the gathering at the home of Clebsch, Lie left Göttingen in the company of a Norwegian friend from Bergen, and he reported on the journey in a letter to Motzfeldt: They spent one day in Cologne – "the Dome is crushingly beautiful" – they inspected "Brussels superficially in the Course of an Afternoon, which in many Respects is particularly worth seeing. Many medieval Constructions." Then they arrived in Paris: "Right away on our first Evening we went up along the Boulevards (Montmartre . . .) You can imagine what a Bustle it is with the splendid boutiques and cafés and milling Crowds."

In Paris, the new semester had begun. Lie went to lectures a couple of times in the mornings, "partly in order to train the Ear to the Language", and naturally enough what he sought was mathematics: "Mathematical lectures are not difficult to follow in a foreign Language." On the other hand, what was difficult to understand was ordinary everyday speech, and when after a couple of weeks in the city he went to the theatre he would have understood nothing if the action had not been so simple and "the Performance so graphic and transparent".

Nonetheless, money ran through his fingers and he hoped intensely that he would soon receive the one hundred speciedaler that his father had promised him, and asked Motzfeldt to send him immediately ten speciedaler in francs to the address, *Rue de l'École de Médicine 32, Passage de commerce 30, Hotel Molinié* – where he had accommodations that he called "Student Digs". Now, since once again an adjunct stipend was being offered in Christiania, he calculated that Motzfeldt would pursue this on his behalf. Lie quoted from a letter he had received from Bjerknes, in which he (Lie) had been urged to apply for the stipend in the following manner: "Take care that you apply for the Adjunct Stipend; but do not have great Expectations, as in all Probability the Faculty that wins will be the one with the highest Unemployment." In response to such a view from Professor Bjerknes, Lie commented to Motzfeldt, "It is most tragic to have the Support of a Man whose Principle is that Truth in this dull World is determined by constantly losing."

Lie found no student society where he could go to find like-minded people. "People who know each other go to the same Taverns, the same Dancehalls; that seems to be All", he reported. In the letters he wrote now to Motzfeldt in Christiania there was much talk about an eventual Paris visit that Motzfeldt planned that summer, and Lie acted as his advisor: Paris absolutely deserved a visit. "Berlin was nice enough, but still Nothing compared to Paris." And certainly Motzfeldt ought to come "during the Springtime before it became so very warm." Nevertheless, the first weeks in Paris had been marked by "desperately bitter Weather", and Lie complained that "there were no regular tiled Stoves or Wood to be had at any reasonable Price. But the most important thing was the language: "Language is the Main Issue. Quite decidedly. But take careful note that it is not Necessary to be any sort of Master of the Language in order to *have the full* Pleasure of the Visit." Lie mentioned which book would be best for improving his friend's language skills, and suggested how long each day he ought to devote to language studies. In summary, he wrote "It is true enough that one never has sufficient Grounding in Languages – but there is Nothing to do about it." The greatest saving grace would certainly be the theatre, and for that, one could buy the manuscript of the plays at the bookshop, so long as one knew beforehand what one was going to see, and thus one found it "remarkable how much easier it was to understand what was said", Lie wrote, and added a postscript that the play was "transparent", and "the Mimicry and Gesticulations were so characteristic that You [Motzfeldt the accomplished marksman] who have such good Sight will have a tremendous Help therein."

In other respects, according to Lie, there were certain characteristics about Paris:

> One enjoys oneself here *fully as much* with the *Eyes* as much as the *Ears*. [...] Thus far I have (although my Eyes are so bad) restricted myself considerably in wandering around and looking at Everything. Here there is a Comedy played out in the Streets. The Spirit of Speculation lures the Shopper with every conceivable Art. When a Man is like me, who has no Tendency to buy the various beautiful Things, then one can have Fun at little cost.

Lie recounted in detail his expenses, at least partly to prepare Motzfeldt for what his stay in Paris would cost him. Apart from clothes and books, which were among his "more considerable Sources of Expense", he paid five speciedaler per month for his room, three speciedaler for a breakfast consisting of *café au lait*, bread and butter, and for a mid-day meal, or "Dejeuner [Lunch] as it is called here he had to expend another five speciedaler, and finally, for dinner, a further 4–4½ speciedaler, and he remarked, "Compared to Conditions in Christiania it is frightfully dear; but when one has a normal Appetite, one has difficulty doing it more cheaply." And he warned Motzfeldt that he, as a newcomer to the city, would quite likely spend more on food, the "Norwegian Traveller" gladly doing so. Lie continued by calculating what one had to shell out for a cup of coffee at a café, a visit to the theatre or a museum, or a coach tour in the city, although he had to admit he was not precisely familiar with the coach fares "as, so far, I have always used my own two Legs." "At le Café de la Régeance where I read Norwegian newspapers, one can get a Coffee for ten sous" – "drinking money included"; otherwise, fourteen sous was normal "on the Boulevards". Theatre prices were high. The two times that he had been to the theatre during his first month in the city he had paid three francs each time for a seat in the third balcony – the best seats cost sixteen to twenty francs. He added that the smaller theatres were somewhat cheaper. All in all, he was convinced that Motzfeldt could "enjoy himself" and live "quite decently" for between two and two speciedaler fifty per day, but if he wished to live in a "veritable Hotel" and eat "Table d'hôte, etc." then every day amazing Sums will go out "without any real Extravagance."

Lie thanked Motzfeldt for his letter and the money and for having, "to the best of his Ability", advanced his (Lie's) "Interests", and added, "Pity the Fact that Lady Fortune does not agree to a Reconciliation." That is to say, Lie had received the news that the adjunct stipend in Christiania had gone to the medical man, Jacob Worm-Müller, about whom he had no other complaints: "But I am growing afraid that this is my Fate, always having to stand back." A position had also become available at "the polytechnical College" in Trondheim. Lie wanted the post, but did not think that he would be considered, and concluded, "It would be with small sanguine Expectations that I could soon return." There was however a ray of sunshine in this letter from Paris in the middle of April: "I continue to be content with Respect to my scientific Work."

When scarcely a month later he reported from Paris, he summed up his first four weeks in the city in the following manner: "I studied terribly hard – constantly freezing. In the afternoons I went out in General for 3–4 Hours around Paris. I made few Acquaintances during this Time and learned a little French."

Great change came over Lie's life finally when Felix Klein arrived in Paris and stayed at the same hotel. Since they had parted they had exchanged letters, and according to the last letter Klein wrote before his arrival in Paris, they had an agreement that Lie would meet him on the morning of Tuesday, April 20th, at le Gard du Nord, the station to which Klein would come on the night train from Aachen, where the day before he had planned to meet Reye who had gained employment at the Aachen Technical College. And indeed, things went as planned. Lie and Klein met at the Gare du Nord that Tuesday, and over the next eight to ten weeks the two friends together would experience great mathematical progress in the French metropole. On May 2nd, Lie wrote to Zeuthen in Copenhagen: "It was for this that I went to Paris. One Main Reason was that by so doing I had the Company of Klein, a straightforward, amiable and bright young Man." And he explained that together with Klein, he had continued to study Reye complexity with "Congruencies, Complex Curves, etc.", and especially a "Family of Surfaces" that had some properties such that curves resulting from the cross section of two such surfaces were "remarkable insofar as they, with a Diversity of Transformations, can be transferred into themselves."

Lie wrote to Motzfeldt in Christiania on May 12: "After Klein's arrival we began to gather Information in French. We have made the Acquaintance of several of the foremost of the Young among the Mathematicians, have been to dinner at the homes of a Couple of Professors, etc."

The two French mathematicians with whom Lie and Klein would now spend the most time with were Gaston Darboux, who was the same age as Lie, and Camille Jordan, four years older. Klein had earlier that spring corresponded with Darboux, and had been asked to write an article on Plücker in the new French journal, *Bulletin des Sciences Mathematiques et Astronomiques*. Klein had replied that he would gladly give an account of Plücker's last works in relation to the very latest developments in the field. Darboux had a thorough knowledge of curves and smooth surfaces – what was called a geometry of differentiable functions, where the objects are given as functions that are differentiable, and where the starting point is studies of the properties of local neighbourhoods of points on a curve or a surface. The aim is to have an overview, to know and be familiar with what happens all along a curve, and over a whole surface. Darboux worked with a general theory of all surfaces, and succeeded in combining together differential geometry and differential equations in a way that allowed geometric questions to be dealt with analytically. New clear and verifiable relations were created between geometric objects and formulated differential equations. Darboux's work would become a source of inspiration for Sophus Lie.

Jordan had just published his great work on substitution and algebraic equations – *Traité des substitutions et des équations algébriques* – where for the first time the thinking of Galois was made accessible to wider circles of mathematicians. Ten years earlier Jordan had the task of editing an edition of the collected mathematical works of the recently deceased Cauchy, and during the hunt for unpublished materials he had found Galois' letters and been greatly astonished and interested. During the 1860s Jordan had published a series of articles in which he explained and elaborated Galois' ideas. But for the first time, in this year of 1870, he had pulled together and given form to his insights, and for this systematic account of group theory he was awarded the Prix Poncelet. Lie and Klein now became Jordan's eager audience, and Jordan considered both Klein and Lie to be two of his brightest pupils. In any case, it is certain that Klein and Lie, by being together with Jordan during that summer of 1870 quite rapidly focused upon the group concept as a tremendously valuable tool for the study of geometry and other fields of mathematics. Lie began to use the concept in geometric portrayals and transformations, and he got the idea of invariant properties in respect to such a transformation group. Together with Klein, he looked into curves and surfaces where infinitely many interchangeble projective transformations were possible – a study that resulted in two articles on what were called W-curves that Chasles presented before l'Académie des Sciences, and which were published in *Comptes Rendus*. The great advantage of this periodical was, in addition, that here, after a mere week, one could see one's articles in print.

When Lie reported to Motzfeldt on May 12th it had warmed up: "For the Time being, we have lovely Weather. The Boulevards with their magnificent Trees are indeed at this Time of Year at their most attractive Appearance." Otherwise, on precisely that evening he had immediately sat down at his desk upon coming "home from the Boulevards, where one had expected a Spectacle: yesterday, they said, a considerable number from the Mob were killed. But today, as often before, a good Rain Shower has saved Paris from a Bloodbath."[25]

Motzfeldt seems to have planned his month-long stay in Paris to begin in the middle of June, and it was Lie's opinion that it would be better if he arrived at the end of May or the beginning of June "because the fashionable *beau Monde* has still not at that Time left the City for the Countryside, [and] moreover there is much to see in the Theatres, the Bois-de-Boulogne, etc." – later the heat could be bothersome as well and "the Whole World abandons Paris in the course of the Summer".

He himself, together with Klein, would certainly be in Paris until the end of July: "Great Changes in the Time of our Departure are hardly imaginable. We have still not definitely made a choice between Milan and Cambridge." Klein had just written to Cayley in Cambridge and asked him about various affairs, and the response they received would determine their choice. In any case, Lie had begun to prepare himself by reading a little English.

Turning to other matters, he had seen in the Norwegian newspapers that a new adjunct stipend was available, and he wrote that he would apply again, but without any consideration of getting it. Later in the year however, he was extremely certain that he would get it, and he wrote to Motzfeldt:

> When one has been so – I might say, rash – or in any case, audacious as to throw himself into Science, then one gets to thinking about the ideal Prerequisites such that one does not fall into the Ranks of the Halt and the Lame.
>
> As an Adjunct Stipendiat I believe I could work in a beneficial Manner for the Development of scientific Realism in our Country. This is a vain Thought: but I cannot work with any Desire if I do not have an ideal Goal. In other Respects I have adequate Experience to think better of necessarily making a Sacrifice to the Material.

Lie had come to know that Professor and Cabinet Minister Ole Jacob Broch might be coming to Paris. Broch had been on sick-leave since May and was resting at the baths at Ems, near Koblenz on the Rhine, and Lie hoped to find the opportunity "to confer with Broch" about his future: "I believe I can say that I have such a Degree of

[25] This description is an indication of the social unrest which later, in March 1871, and exacerbated by the effects of war between Germany and France, would see the working population of Paris rise against the government, shoot 150 hostages, set fire to the Tuileries and form the Paris Commune which was subsequently revenged with the massacre of 20,000 communards and the deportation of a further 10,000 (E. S. Mason. 1967. *The Paris Commune.*)

Excellence relative to a Lot of the Science Candidates at home that I am too good to go in as a Private or something even lower."

Lie had also received news from home about friends and comrades who had married, and he commented, "Every Time one of my Comrades becomes engaged or marries, half-wishful Thoughts come over me, who for the longest Time has regarded himself a Bachelor. I have passed the Rubicon, and if I am rightly aware of my Obstinacy, I shall never turn back."

Ernst Motzfeldt and his friend, Axel Bruun arrived in Paris in the middle of June, and stayed for a month, until, according to Motzfeldt "immediately before the Outbreak of the French-German War." Apart from this, we find nothing else written down about the three Norwegian friends' time together in Paris, although it seems that Klein too developed good relations with Motzfeldt; in any case, later on Lie was often the middleman for greetings between the two.

Apart from Darboux and Jordan, Klein and Lie were also in contact with two other mathematicians in Paris: the sixty year-old Joseph Liouville and the seventy-seven year-old Michel Chasles who, according to Lie, "looked terribly decrepit", but nonetheless spry enough to meet "every Monday at L'Institut".

But it was the contact with Darboux and Jordan, a contact that seems to have been quite regular, both at public lectures and private gatherings, and in which new mathematical ideas took form – especially, perhaps, for Sophus Lie. Many years later Felix Klein recounted an episode that occurred one morning at the beginning of July 1870. He had gotten up early and was ready to go out when Lie, who was still in bed, called him into his room and began to tell him about the link he had found during the night, between the principal tangent curves of a surface (asymptotic curves) and the curvatures of another surface. Lie laid it out and explained it in a way that Klein was unable to follow – but it concerned a line-sphere transformation where instead of operating with spheres, he used straight-lined hyperboloids which cut a real, given conic section, and he maintained that the principal tangent curves in the Kummer surface had to be algebraic curves of the 16th order.

That night's insights were the beginning of the theory of what came to be called contact transformations. The presentation of evidence for the link between a surface's principal tangent curves and another surface's lines of curvature demonstrated that with the appropriate choice of constants, one could associate lines to spheres in space, and study the resulting transformation. Later this resulted in his line-sphere transformation, which gave rise to the generalisation of Plücker's line geometry and a demonstration that it was possible to establish a space geometry with spheres as basic elements. It was these remarkable transformations and correspondences from one space to another that, a year later, would form the basis for his doctoral dissertation "On a Class of Geometric Transformations".

A short time after this nocturnal breakthrough in Paris, Lie wrote to the Academy of Sciences in Christiania to give an account of his findings. Through much of the nineteenth century this was a not-uncommon means of securing for oneself priority for ideas, in the event that later there should be doubt and discussion on the subject, and Lie wanted to make sure that others did not steal his ideas. The letter to the Academy of Sciences was dated Paris, July 5th:

I allow myself to communicate to the Academy the following scientific Findings with the Aim of possibly securing my Priority.

(1) By means of my Imaginary Theory I have found a geometric transformation, that conveys a descriptive Theorem about straight lines in a corresponding Theorem about Spheres. Herewith there is a Correspondence between two straight Lines that intersect one another, two spheres that touch one another.

(2) Of this I have derived the fact that it is always possible by means of algebraic Operations to reverse the Determination of a Surface's Principal – Tg. – Curves to the Determination of another Surface's Curvature – Curves, and likewise, vice-versa.

(3) Kummer's Surface of the 4th Order and the 4th Class has algebraic Principal-Tangent-Curves of the 16th Order and the 16th Class. Herewith it is also certainly declared that the mentioned Curves are algebraic on a Wave Surface, the Plückerian Complex – Surface, etc.

The letter contained a further four points in which he also explained his results about minimal surfaces and logarithmic transformations.

Two days after sending this letter to the Academy of Sciences in Christiania, Lie and Klein together wrote a long letter to the mathematical section of the University of Berlin. This was a briefing on the state of mathematics in France, the mathematical public, the mathematical journals and the mathematical production of recent years. They tried to be objective in their descriptions but were aware of how difficult it was to evaluate the merits of mathematical works that did not fall within their own fields of investigation. Above all, however, they firmly maintained – and this was not simply their own personal feeling, but commonly known in all mathematical circles – that the study of mathematics in France was not at the level it had reached fifty years earlier. And the reason could be both that the prospects for a mathematical career were lacking, and that the teaching situation was so centralized. But still more at fault, according to Klein and Lie, was the fact that for far too long the mathematical milieu had basked in the glory of earlier enterprise and that the aspiration to find new knowledge was lying fallow. Self-satisfaction will always lead to deterioration – this was also acknowledged in Paris itself, and efforts were being made to turn things around: priority was given to the knowledge of new mathematics, the German curriculum of studies was being examined with interest and a mathematician had been appointed to direct the ministry responsible for education. As well, the new periodical, *Bulletin des Sciences Mathematiques et Astronomiques* gave account of mathematical works and was part of this readjustment – but the difficult task for such a journal was obviously to have good collaborators from a wide number of fields so as to be capable of awakening interest among a broader public. Fortunately Darboux was a major player in this journal, and his great professional knowledge and clear exposition augured well for the future.

In this letter report on French mathematical conditions, Klein and Lie also emphasised the differences they felt existed between German and French methods of editing a mathematical paper. While to a great degree mathematicians in France stressed clarity and simplicity such that the greatest number should understand, in Germany there was, perhaps, a tendency toward telescoping together accounts in such a manner that they were almost unreadable for anyone other than those who were actually in that field. Klein and Lie had no doubt that the French method was much to be preferred: "The Intention of a mathematical Work is reasonably to be understood, and not simply to engender admiration for the writer."

Klein and Lie maintained that the clarity and geometric acuity found in the mathematical output of France in recent years lay in the tradition of Monge. And, apart from the writings of Darboux and Jordan, they mentioned the works of Serret and Bertrand. From Darboux's works they cited his finding that at each point on a surface it was possible to calculate two curvatures. They mentioned this work not only because it was near their own studies, but also because, according to Klein and Lie, Darboux's works were of a particular importance. Jordan's great *Traité des substitutions* was also mentioned, and specifically that there, working in the tradition of Galois, he had given criteria for the solution of algebraic equations.

By this time, at the beginning of July, Klein and Lie were making plans to travel to England. Before they went there, however, Lie had planned a little tour to Switzerland, presumably simply to see the country and go hiking in the mountains. But then war broke out between France and Germany. On July 19th the Empire of France declared war against the Kingdom of Prussia. Some days before, probably on July 16th, Lie had written in "flying Haste as Klein will probably be leaving in a Couple of Hours". "My dear Ernst! People are furious. Since yesterday War between France and Prussia is considered to be inevitable. By the Time these Lines have reached you, you will know what has happened."

The political situation meant that precipitately Klein had to leave Paris and report for service at the front, on the other side. Lie, for his part, with a neutral passport, decided to remain in Paris to the end of July – mainly in order to submit "a new Note to the Paris Academy" – and then journey to Switzerland, Italy and back through Germany and a homecoming planned for December. Due to the uncertainty of the times, Lie asked Motzfeldt to send him whatever money he had to the good, immediately, to "this Address until further notice: Mont Parnasse no. 35", and he wrote:

> The Parisians are a curious Race. Recently they would not give two Shillings for the Honour of the King, the Government, or especially the Senate. *Au sénat!* was a Curse Word. Now all of them are enthusiastic: united in Hatred against Prussia. In the Evenings along the great Boulevards there is a terrible Bedlam. So Black with People that one must flow with the stream of Humanity. However, Everything is functioning in a calm and orderly Manner almost without Exception. The only Altercation was provoked when a little Crowd came along shouting *Vive la paix!* One can still laugh off such Demonstrations. Otherwise, when one of these odd Bands passes by, the Watchword is *Au Berlin, Vive la guerre!*

But, he reported, he saw little sign of troop movements "in part because they take place during the Night." In the next letter to Motzfeldt (dated Paris, Sunday, July 31st) he complained about "an appalling Temperature over the last 14 Days." Such abnormal conditions day after day and night after night "almost made one sick."

He was overjoyed to be leaving Paris. He would send one suitcase of books and clothing, home to Moss, and perhaps Motzfeldt too could receive a parcel of copies of his article that was ready to be printed in *Comptes Rendus*. Over the past fourteen days he had also heard a couple of times from Klein, who in the first round had been declared unfit for active service and therefore would probably spend the war fulfilling a sort of civil-militia position. The correspondence between Klein and Lie focused on mathematics. In one letter at the end of July, Klein corrected his own calculations of the singularities of the principal tangential curve of the Kummer surface, and he praised Lie for his works. The final letter Lie received from Klein in Paris was dated Düsseldorf, August 8th.

Some days later Sophus Lie left Paris on foot in a southerly direction. He had taken with him only what he could carry in a sack on his back. His plan was to go on foot through the French countryside and into the mountains of Switzerland, and then on to Milan to visit the mathematician Cremona. But Lie did not get beyond Fontainebleau – little more than fifty kilometres from Paris – before he was stopped, and mistaken for a German spy. Lie did not manage to prove his innocence, and came to sit in prison for a month at Fontainebleau. There were several stories about how, what and why he was mistaken for a spy, detained and put in prison – much had been spun from Lie's own accounts, although some undoubtedly were cooked up to create a sensation. In any case, the headline in Norwegian newspapers – NORWEGIAN MAN OF SCIENCE JAILED AS GERMAN SPY – made Sophus Lie a famous personality in his homeland.

What seems certain is that Lie had no stamps in his Norwegian passport from the Norwegian legation in Paris. His passport was invalid, and notwithstanding, it was certainly of no use to insist to the French police that he was Norwegian. Above all, the reason he had been stopped and investigated was said to have been because he talked aloud to himself, and anyone could hear that his speech was unmistakably foreign. And the situation certainly would not have been improved when the police found German letters among his papers, letters stamped "Metz" – where the German army was billetted – and full of strange cryptic signs and formulations. There might also have been some landscape drawings in Lie's notebooks, that showed French fortifications. Thus, when Lie tried to defend himself by saying he was a mathematician and these were not codes but rather mathematical signs, this was taken as an old dodge – secret codes and cyphers were something with which mathematicians had long involved themselves. But Lie ought to have the opportunity to prove his mathematical profession. Knowing full well that the fewest possible number of contemporary mathematicians would profitably read his works, Lie was supposed then to have burst out, saying, "You will never, in all Eternity, be able to understand it!" But when he realized what danger he was in, he was said nevertheless to have made an effort, and he began thus: "Now then, Gentlemen, I want You to think of three axes, perpendicular to each other, the x-axis, the y-axis and the z-axis . . . " and while he drew figures for them in the air with his finger, they broke into laughter and needed no further proof.

When Lie asked a guard what was done with prisoners like himself, the reply was as follows: "We usually shoot them at six in the morning."

Now, for the next four weeks Lie sat in prison at Fontainebleau, he worked on mathematics – particularly with his spherical geometry and the theory of contact transformations. According to his own account he also spent many dark depressing hours in prison reading Sir Walter Scott novels that had been translated into French. But he also managed to send some letters – at least Darboux in Paris came to hear about Lie's situation. And it was due to Darboux's political contacts – and probably with the support of Gambetta himself, the interior minister – that Lie's prison sojourn came to an end on September 10th. In fact, Darboux went to Fontainebleau and convinced the authorities there that the misleading notations were indeed mathematical formulas and figures, and that the German names belonged to known mathematicians.

The day after his release, safely established on the train, and well beyond Neuchâtel in Switzerland on his way to Geneva, Lie wrote to Motzfeldt: "If the Truth be told, the Sun has never seemed to me to have shone so clearly. The Trees have never been so green as those I saw yesterday as a free Man on my way to Fontainebleau's station." Late than night he had got himself by train to Switzerland – at the railway station, and at the many subsequent stops along the way "there were many Scenes to see: young Men who openly, for the first Time, with Rifle in Hand, were being drawn toward Paris, and Women and Children, toward Switzerland." About his time in prison he also had this to say: "The Misfortune is not great *now that* it is over, but it could have become an atrocious Affair in these tumultuous Times." And apart from the beginning when he thought it would be over in a couple of days, he had "taken Things in a veritably philosophical Manner", and came out with the following conjecture about the experience: "I think that a Mathematician is comparatively well suited to be in Prison." He felt that in one way he had himself to thank for the misfortune, but the Norwegian legation in Paris, whom he had expressedly asked for advice, certainly ought as well to have told him "that visas were necessary". Later Lie was encouraged to protest that the legation had not done its duty, for he could then, as far it went, consider the possibility of doing so if he thought he could "thereby benefit my Countrymen", and he asked Motzfeldt, if he had an opportunity, to look into the matter. Now he wanted to "wander through" Switzerland, and the next address at which he could be reached was *poste restante*, Milan.

Sophus Lie arrived in Milan on October 3, 1870, and wrote about his journey in a letter to Motzfeldt: The walking tour through Switzerland had been fine, the weather had been propititious most of the way:

> Switzerland is a lovely Country with magnificent Nature – Beauties are concentrated in the most striking Contrasts. When I gather together what I have seen piecemeal in Norway, so indeed is it comparable, but in Switzerland, everything is gathered together. One Thing we still have that is better than in Switzerland, and that is our marvellous Fjords.

Of the places that had revealed themselves "in their full Glory", he mentioned "Chamouni with Mt. Blanc and Mer de Glâce, Ravyl (Gemmi), Berner-Oberland with Faulhorn, Grindelwald etc. etc." – "The outlook up to the Interior of Snow and Ice Massifs is perhaps the most singularly magnificent." It was only at Mount Rigi that he had been unfortunate: "The Mountains along the Horizon showed themselves clearly, but what lay below my Feet – Vier Waldstädter-See, Zuger-See etc. was hidden by Fog, which lay deeply packed." He had not met many other travellers – "the War has destroyed the Tourist Sector as well" – but everywhere one encountered a number of English, both men and women, who, whatever else they might have been, were, to be sure, "accomplished Hikers". Otherwise it was unbelievably "strange up on the Mountaintop – a veritable Peak, not a Plateau as in Norway – to see 4–5 storey Hotels". As for demanding alpine hiking tours, he had only encountered five or six, and then only those that the guidebook declared: "Führer nicht unbedingt nöthig" [Guide not necessary]. He wanted very much to climb Mont Blanc, but it would have cost "quite an enormous sum, 300–400 frc." and that very day he had climbed Argentière, directly opposite Mont Blanc – and if he had had his binoculars he could have followed "the Search Party which was just setting out after the Remains of 11 Persons who had perished."

About the mountains, "so steep in a rather different way that in Norway", he wrote, "It is most peculiar that in the Mountains one follows a broad cut Track, from which one, in the true Meaning of the Word, could in one leap, come down into the Valley, Hundreds of Feet below Oneself." The transition in nature, in vegetation and in human types was really remarkable as one passed through St. Gotthard – from where one saw "only thick black Hair", and the Italian women were outstandingly more beautiful than I have seen elsewhere in my Travels, especially the Italian peasant girls. Here in Milan they have far too many of the Parisian Type."

Milan was not a large city and in the course of two days he could walk through it and see "a Number of its Peculiarities". Milan "has a Triumphal Arch that is more handsome than any in Paris, and the Cathedral Domo seems to me more handsome that the Dome in Cologne, even though not as large." Lie also wanted to see Rome, and thought of asking Motzfeldt to send him more money – the stay in prison had been an opportunity to save money, but he had had to pay for his sustenance, and part of the expenses, "God knows why!" His plan was now to stay in Milan for three weeks, followed by three weeks in Göttingen, where he was to meet Klein again, and then arrive home at the end of November. He had already heard that the mathematician Chasles had just died, and Lie remarked, "With him a whole Generation has gone extinct."

Motzfeldt had reported from Christiania that Lie was now quite certain to get the adjunct stipend and finally become a research fellow. He replied, in writing, that if he had been able to bring himself to believe this, he might have been in a mood to travel on to Rome, and he reported that he was working on a treatise that he hoped to submit for a doctoral degree in the new year, but he asked Motzfeldt to say nothing about this in Christiania. On the other hand, he wanted very much to hear if the position at "the Polytechnical Institute of Trondheim" had been filled. The College in Trondheim had been established just that year, 1870, and was caught up in a constantly recurring debate about what the terms should be for a higher technical education in Norway. Many of the university's teachers were engaged in the question despite the fact that the university, as an institution, was not willing to make way for such education.

Lie remained in Milan for twenty days, and seems to have spent most of his time working on his forthcoming doctoral dissertation. There is no available information on his meeting with the illustrious Cremona, and later letters suggest that perhaps Lie did not meet the Italian mathematician, for Cremona was probably away from the city during this period.

Lie bought a 14-day "Tour Ticket at a reduced Price" and left Milan on October 24th. He then stopped for a day each in Turin, Genoa and Florence, and spend a week in Rome – then returned via Bologna and Venice, arriving in Munich on November 9th.

During this tour by train he wrote a couple of letters to Motzfeldt and one letter to Bjerknes, and here he described his experiences:

> In Italy they have an infinite Wealth of Monuments, Churches, Castles etc. There is much I have no Opinion on; but there is also much that makes a strong Impression on me. [. . .] Italy is of a quite special Interest in that it offers the Aroma of History, which permeates Everything.

The two-week tour was "naturally exceedingly Little, but for me, who has no Sense of Painting and essentially seeks only Impressions, it was adequate."

The letters to Motzfeldt were full of requests for support and the dispatching of money, and Motzfeldt again received the task of attending to the application for the adjunct stipend, arrange it with "a Note to the Academy of Sciences" and expedite "an Application for the Post in Trondheim". To strengthen the application for the research stipend, Lie wanted as well to send a statement by Clebsch, if Clebsch now replied favourably to "a scientific Epistle" that Lie was about to send to him. Nevertheless, Lie calculated that Cabinet Minister Broch would put in a good word for him this time, particularly since Broch had no doubt heard from the professors in Berlin about the praise for Lie. In any case, Lie had heard from Klein that three of "Berlin's Greatest" had described him as "an intelligent Mathematician". Lie felt that the only snag with asking Broch for a recommendation was that, as a cabinet minister, he would perhaps not want to involve himself in university matters, and Lie added, in that case, he could get "Words of Attestation" on the "State of his Science" from his good friend Amund Helland, "or another of my Comrades in the science courses, but the Matter requires a Drop of Diplomacy. In fact, You must be careful, and do things in style, on my Behalf." Lie asked him to send sixty speciedaler to "*poste restante*, München". As well, Lie urged Motzfeldt as well to ask his mother if he (Lie) might continue to rent a room in her house in the event that he would be staying in Christiania, and he concluded: "I assault You as usual with a Mass of Bother. And I can only give You my Thanks for all your inestimable Services." Then, on October 30th, Lie wrote to Professor Bjerknes from Rome:

> When after some Time I return from my Tour abroad, it would be my Desire to enter such a Position that would allow me to continue my scientific Studies under not too unfavourable Conditions, and on the other Hand, to have a working Environment that is no worse than that of any Student.

Consequently, I would like either to receive an adjunct Stipend, or be employed as a Teacher of Mathematics at Trondheim's technical Teaching Institution.

He hoped that Bjerknes would speak to his case in relation to the adjunct stipend, and with regard to the position at Trondheim he did not know "which Principles" would be relevant: "In Germany where these Teaching Institutions have a grander Character, they always try to get Men of Science as Teachers." Otherwise in his letter to Bjerknes he mentioned that he had continued to work on the material described in the note he had sent via Motzfeldt to the Academy of Sciences, and that it was a work which he hoped by New Year's to be able to submit for a doctorate.

When Lie wrote to Motzfeldt from Munich on November 9th, he had learned that Hans Geelmuyden's father, the fifty-four year-old ship's captain, Christian T.H. Geelmuyden, had been awarded the post in Trondheim. Lie felt that his own application had reached Trondheim too late, and he added, "Thank goodness I have never thought that I had the Possibility of getting the Post in Question."

The sixty speciedaler that he had asked Motzfeldt to send him, *poste restante*, in Munich, had not arrived. Lie was more or less penniless, but wanted to wait on in the city for some days in the hope that the money would arrive.

I find myself in miserable Circumstances. München is already gripped by full Winter with Snow and Storm, and I have only thin summer Clothes. As soon as I get Money I shall obtain outer Garments.

München is otherwise an interesting City, splendid Galleries of Art, Monuments, etc., were it only a little warmer.

Lie's plan was now to go on to meet Klein, who had not gone to Göttingen, but was at the home of his parents in Düsseldorf. But Lie was penniless and could travel no further. If the money did not arrive from Christiania he would have to ask Klein for money for the fare. Lie waited for over a week in Munich. He requested that his mail be forwarded from *poste restante* Munich to *poste restante* Düsseldorf. And to Motzfeldt he wrote that if the money had already been sent to Munich, he should send a telegraph to *poste restante* Düsseldorf, or preferably to Klein's address, Bahnstrasse 15 in Düsseldorf.

Since Lie and Klein had parted company in the middle of July in Paris they had remained in constant contact through letters, and Lie took great care of Klein's letters. Since being exempted from military service, Klein had placed himself at the disposal of the emergency aid organization that had been organised in Bonn at the beginning of August 1870. Klein had been overjoyed to be busy among these young people who would thus go out to the battlefield to give the wounded food and refreshment, meet their last wishes, write letters, etc. He wore a cap and white band on his left arm, with a red cross. During the period he was waiting to be sent out in the field, he stayed at the home of his parents in Düsseldorf. Here he worked on mathematics and wrote to Lie that if it so happened that he did not return from the battlefield, all his papers would be sent to Lie, at Motzfeldt's address in Christiania. Among these papers Lie would find some peculiarities about a system of fifteen permutable point operations. On a happier note, Klein hoped that he would quickly and often receive letters from Lie – one way or another, he would always find time to reply: "One can think well under the most peculiar Conditions." Over a month later, after having worked for almost four weeks as an assistant among the wounded on the battlefield, Klein wrote to Lie again, from Bouillon where after over-exertion he was convalescing for a while, and therefore had time to relate his impressions of the war. He had never been under front-line fire, but both at Metz and Sedan he had worked transporting the wounded and evacuating them from field hospitals. Perhaps, he thought, his own contribution had been of no use, but the fact that during this period he had been so completely torn away from any mathematical or philosophical speculation, had had a special significance. In the first place – before he had over-exerted himself – he had felt much happier and lighter of spirit than he had felt for a long time, and he hoped that this would have propitious effects in the future. Klein felt that Lie was in a condition to understand this, for Lie had certainly seen how mentally exhausted and sick Klein had actually been that summer – how time after time he had had to force himself to work on mathematics, the subject that now after all was his profession. But Klein hoped and believed that this difficult period was over and that he could soon again find happiness and satisfaction in mathematics. As to his plans, Klein reported that the moment that a truce was declared, he would journey home to Düsseldorf, and from there, as soon as possible, resume scientific activities in Göttingen. And how were things going with their joint work on the Reye line complex? As a convalescent in Bouillon, Klein had nothing to reply except: "Wisdom comes with Time" – for his part, he would do everything within his power.

Klein had been on the way back to the front when he became ill again at the beginning of October, with the diagnosis "gastric Fever" (typhus). And thus was he still at his parental home in Düsseldorf when Lie was on his way north in November. Since the imprisonment at Fontainbleau, Lie had written at least two letters to Klein with mathematical contents, letters that Klein, due to illness, had had to lay aside "with a kind of feeling of sadness" [mit einer Art von wehmüthigem Gefühl]. But he was steadily getting better, and when Lie now arrived in Düsseldorf in the middle of November he was on his feet and could eat almost normally. Both were overjoyed to be reunited, and when all was said and done, the meeting met their expectations. Lie seems to have laid out his contact transformations, and they collaborated again on the work they had taken up in Berlin and which would be published there in *Monatsberichte*.

Klein gladly wanted Lie to stay as long as possible in Düsseldorf, and Lie remained in the parental Klein household for ten days. Eventually, Lie had left Düsseldorf at the time when a letter arrived there from Kummer. Dated Berlin, November 26th, the letter was addressed both to Lie and Klein. Kummer was aware both of Klein's heroic contribution to the fatherland, and of Lie's misfortunes as a result of being taken as a German in France. Nor could Kummer restrain himself from expressing his deep-rooted hatred of everything French: French conduct toward Germans was a "blott of shame" that wounded the French nation: "In terms of decency and morality, France as a nation had sunk to the depths, and its fall would only continue." But what Kummer really wanted to discuss was Lie's and

Klein's treatment of surfaces of the fourth degree with sixteen nodal points. He was eager to publish the work in the Berlin Academy's next issue of *Monatsberichte*. But since he, like all mathematicians in Berlin, considered the development of geometry to be a question of having sufficiently thorough evidence, he wanted to urge Klein and Lie to substantiate and give the grounds for their results about this fourth degree surface in a slightly more elaborated manner.

Lie received a copy of this letter via Klein later in the month of December, when he had returned to Christiania. And in an accompanying letter, Klein stated that their "desires and expectations" were more than fulfilled by Kummer's reaction, and that he was already involved in editing the work – "Über die Haupttangenten-Curven der Kummer'schen Fläche" – and as soon as possible he would send this to Lie for approval. Klein was sorry that Lie had had to leave so soon. Since Lie's departure he, Klein, had done nothing great – the only thing was some observations on W-curves, that he would continue to work on – and he reported as well that Lie had left his "Bädeker" (his guidebook) at the Klein home in Düsseldorf.

Lie sent his corrections and comments on the paper, together with a photo of Motzfeldt that Klein had requested. And thus, on December 15th in Berlin, Kummer was able to send off for printing the paper on principal tangent/asymptotic curves of Kummer surfaces.

Parliamentary Professor

When Sophus Lie returned to Christiania in December 1870, he was immediately a great topic of conversation: the Norwegian scholar and scientist who had spent time in a French prison, mistaken for a German spy! And Lie himself was more than forthcoming about his experiences, at least at the beginning. An aura of celebrity and sensation enveloped him – the mighty conqueror of the mountain tracks and the suspected spy who was also involved in a mathematics that was incomprehensible to most people. From this period many anecdotal stories attached themselves to Sophus Lie's name, something that he himself came to fear, would stand in the way of his significance as a mathematician in people's eyes.

One of those who met Lie for the first time at this period was the science student, Elling Holst. Holst, who rapidly became Lie's student, and later, associate professor of geometry, was also a collector of traditional nursery rhymes and children's verse. Holst was the first in Norway to publish such traditional folksongs as "Row, row to the fishing reef", "Shoe, shoe the horse", "Hop, said the goose", "The old crone with the broomstick", and many more. Holst would eventually author several biographical articles about Sophus Lie, and on the subject of his first meeting with Lie, he said: "Right from the first moment he made a deep impression on me. Tall, erect and slim, yet of an athletic build, blond, with an open, friendly face, obliging and amiable, and who, on closer examination, appeared to be definitely manly of will, warm and engaged – something that was revealed as well in his scientific enthusiasms or in his personal sympathy."

On the mathematical homefront things did not go exactly as Lie had hoped while he was abroad: the professorship in pure mathematics – after Broch's departure and his entry into the government of Frederik Stang – had not been awarded to Ludvig Sylow, whom Lie and many others felt was the person most qualified for the position. What had happened instead was in effect a new round of discussions on what the centre of gravity should be with regard to qualifications and skills in the nation of Norway: once again the theoretical-scientific was set up against the technical-practical. The university firmly supported its system of having teachers in *both* pure and applied mathematics – while Broch had been professor of pure mathematics, Bjerknes was professor of applied math. But actually in recent years Broch had given lectures in what was nearest to his heart, among which was mechanics and other more practically accented subjects. For his part, Bjerknes had kept mostly to the theoretical, and the only purely practical field he was responsible for was mechanical engineering, and in this field others more competent were employed as well. It also seems that Bjerknes himself realized that he had not managed to energize the practical-technical applications of mathematics, such as were outlined in what today would be called his job description. In any case, according to rumour, Bjerknes had allowed himself to be persuaded, with little pressure, to take over the professorship in *pure* mathematics after Broch. Thus it was that the vacant position which came to be advertised, was in *applied* mathematics. And since Sylow's mathematical field was equation theory, where he was among the foremost in Europe, he was completely out of consideration for the position of professor of applied mathematics. Bjerknes was painfully aware of Sylow's qualifications in the field of pure mathematics, and campaigned at length "that the Distinction between pure and applied mathematics be done away with", something that would be of keenly *à propos* for Sylow and his future prospects. However, the intention that the University "facilitate a fruitful effect" upon "Engineering in general and Mechanical Engineering in particular" was given great weight. In an apologetic letter to Sylow, Bjerknes expressed his regrets that: "People have so little possible use for the likes of You and I. Even if there were an Abel lodged inside You, it would make no difference." And in another letter he wrote: "People have as little Use as possible for 'pure' Persons. They ought as far as possible to be so 'applied' as to be pure Machines; and then they can be used for almost Anything and at any Time or Place."

Thus it was that Cato Maximilian Guldberg had been appointed Professor of Applied Mathematics. Cato Guldberg was forty years of age at the time and, following his graduation from the science teacher's examination, had gone to Germany and France to study mathematics, mechanical engineering, and what was the latest development in mechanics – heat theory. Since 1863 he had been editor of the journal *Polyteknisk Tidsskrift*, he sat on many different commissions and directorates, and had received a university research fellowship, and became a research fellow in 1867, at the same time that he was nominated to the Academy of Sciences. While Broch was serving as a parliamentarian in 1868, Guldberg substituted for him at the University of Christiania. Guldberg busied himself with practical and technical problems; he was involved in chemical-physical research, and together with his brother-in-law, Professor of Chemistry Peter Waage, Cato Guldberg was behind the discovery of what later became the famous "law of mass action" which revealed the laws of chemical reactions.

Throughout this period the university held the position that it was not the appropriate place for pure technical education; nevertheless, it was the practical and applicable functions of science that attracted attention and focus during those years. In the Botanical Gardens Professor Schübeler mounted displays of useful plants, and he founded the Garden Society. In the field of zoology Professor Rasch was most concerned with bee-keeping, fisheries and appropriate convex formations for growing oysters. Research of a protracted nature and of uncertain utility was of less interest.

While Sophus Lie had been abroad, two of the great men of the fatherland had passed away: Sars and Schweigaard. When the marine zoologist Michael Sars died and was buried during the fall of 1869 everything had been done quietly, and was marked by poverty. French and English men of science initiated a large-scale international collection of money to support the family Sars left behind. There had been a complete sea change by the time of Professor A. M. Schweigaard's passing in February 1870. He was buried at the expense of the City of Christiania, with the funeral held at Trinity Church, outside which stood ten metre-high spruce trees, and inside, potted cypresses along the wall. The church choir was bedecked in black mourning crêpe, under a baldachin canopy with a mass of silver cords hanging from it – a choir performed, composed of one hundred men selected from the Students' and Merchants' Choral Societies, and all that the capital city could muster in the way of men of authority and representation filled the pews.

New times were in the making, particularly in student circles. Once again, student life had begun to set trends in terms of the debate about society, the tone and way of life were in ferment, and discussion was underway about altering the character of society – toward something freer and more radical. Much of the honour for this change was ascribed to Bjørnstjerne Bjørnson, who led the Student Society from January 1870. Ernst Motzfeldt was also elected to Bjørnson's executive, as one of four representatives. At that time one did not have to be a student to be elected to the Student Society executive; it was sufficient to *have been* a student. Motzfeldt was not one of Bjørnson's partisans, but nor was he in the camp of the declared opposition. Rather, Motzfeldt garnered praise for his polished comportment during the many vehement and entrenched battles that were to come. One of the first acts by Bjørnson's executive was to open the Society's meetings and parties to everyone who was interested. The exceptional participation in Schweigaard's funeral – and also in a number of cities outside Christiania where condolatory commemorations were arranged – degenerated into an open battle about what the national cultural values should be. An assembly of students had also taken part in the great interment ceremony around Schweigaard's bier. Bjørnson had written a cantata that was sung in the church, and he had stood at the forefront of the students' own evocative and ceremonial gathering that was held afterwards at the Society's premises. But some days later when the conservative *Morgenbladet* initiated a large-scale campaign to have a monument of Schweigaard built – and who was also written about and described as the country's eternal providence and a man of unique integrity had left – this caused Bjørnson to react in a manner that led to a clash of interests. In an obituary published in *Norsk Folkeblad* Bjørnson set Schweigaard up against the poet, Henrik Wergeland, and argued that Wergeland must get his monument *first*. Schweigaard had not given an *initiative* to the young life of the nation; rather, he had been the "supreme logical regulator" of his time. It had been Schweigaard's *regulatory* activity, in the form of distinguised juridical interpretations – which the era so highly prized – that was the nature of the man's greatness, argued Bjørnson. And by setting up Wergeland as a brilliant example of the type of initiative-rich men the nation wanted to live with for all time, Bjørnson caused a storm of vehement reaction, but also a more open and freer discussion on national priorities.

When Sophus Lie returned to Norway shortly before Christmas 1870, he saw a bright future if front of him, despite his own small practical research experience.

The rumours that he would receive an adjunct stipend proved true. He became a university research fellow in mathematics from January 1871, and right away he began his career as a university lecturer. Elling Holst characterized Lie's teaching and lectures as "exceedingly lively and inspiring", and described how Lie would have few or no notes with him when he was teaching. The lectures, therefore, were "a series of inspired improvisations, full of spontaneous ingenuity. He had a masterful understanding of what constituted a new, transcending idea."

In order to spin out his income and pay down his debts, Lie also became an hourly teacher at Nissen's School from New Year 1871. He also allowed himself to stand again as Chairman of the Science Students Association where, at the beginning of February he held a lecture on Euclid's *Porismata*, and Chasles' interpretation of these, and he lectured on line complexes. He generally ate his dinner in the Student Society dining room, and in the room next door – in "The Green Room" among his circle of friends – he took his coffee, but he never seems to have been a smoker. Moreover, as before, he rented a room from Mrs. Motzfeldt's at Grottebakken No. 1.

The university's two ordinary professorships in mathematics were accordingly filled by Bjerknes and Cato Guldberg, but unforseen circumstances would lead to there suddenly being four professors of mathematics in Christiania in the course of the next year and a half – and Sophus Lie would be one of them.

During this winter and spring of 1871, Lie was first and foremost occupied with the writing of his doctoral dissertation, *Concerning a Class of Geometric Transformations*. Lie had written most of it in German, and now translated it into Norwegian since, according to the rules of the university, Norwegian and Latin were the only two languages in which one could defend a dissertation. Following Ole Jacob Broch's first doctoral defense in Norwegian in 1847, twenty years went by before another doctoral defense was granted in Christiania, and during these years the "institution" was almost abolished. In the period between the founding of the Fredrik's University of Christiania, in 1811, and 1867, only three doctoral defenses were held in Norway.

Part of the reason was no doubt the fact that the candidate was given only a few hours of preparatory time before giving his test lecture on a given theme. Now in 1871, the preparatory time had been extended to twenty-four hours, but a doctoral defense was still a great rarity. In the decade between 1867 and 1877, only eleven doctorates were conferred in Norway.

Lie led into his thesis with some general comments:

> It is recognized that the rapid Development of Geometry in our Century, rests upon its intimate Dependent Relationship to philosophical Observations about the Character of Cartesian Geometry, which in its most common Form has been advanced by *Plücker* in his earliest works.
>
> For those who have pressed on into the Soul of Plücker's works, there is Nothing substantially New in the particular Idea, that in terms of Elements of Spatial Geometry, one can treat any Kind of Curve so long as it is dependent upon three Parameters. The Reason that Nobody yet, as far as I know, has given Reality to this Thought, must of course be found in the fact that one has not seen any Advantage, that might result therefrom.

I have been the first to undertake a general Studium of the above-mentioned Theory, thereupon finding that by a particularly curious Depiction, one can convert Principal Tangent Curve Theory into the Theory of Curvature.

Lie's doctorate was a fundamental advance of the line-sphere transformation he had discovered in Paris the previous year. His starting point was contact transformation, which with the appropriate choice of constants had the property of transforming straight lines in space into spheres. In a fundamentally new way, the geometry of projective properties was placed in a relationship with a metrical geometry, and the reasoning was contained in a radically new mathematical language. It became an extremely difficult task for the three who were appointed to evaluate Lie's work: the two professors of mathematics, Bjerknes and Guldberg, supplemented by Professor of Metallurgy E. B. Münster.

During the whole oral examination, Münster did not say a word, Guldberg commented no doubt only on inconsequential aspects and could not get into Lie's theme at all, and the classical mathematics of Bjerknes seems to have been located at such a distance from the modern mathematics of Lie, that, according to witnesses, no point of contact arose in the discussion. Elling Holst, who was one of the many who followed the disputation with anticipation, and later wrote about this scene of June 1871, reported that Lie was awarded the doctorate with distinction, but: "Not one Mother's Son had grasped a single Word of it." Subsequently, at the celebratory doctoral party that evening, a ballad to Lie had been written and performed to mark the occasion:

Messrs Bjerknes and Guldberg, of high'st reach,
Were wrapped in Beauty's splendid Fold;
So delicious, the flitting of your Silver Speech!
While Münster's Silence, was to all, pure Gold.

* * *

A Circle is round, and a line is straight,
and the Stars shine down from the sky,
You are as a Star, indeed, first-rate,
So brightly do You circle the crowd on high.

Long live Filosofus Sofus Lie!
Who's captured the Degree so sweet,
Thanks to his Friends' support, you see,
Here to rejoice with beer and meat.

* * *

The ballad had nine stanzas in all, and had been written by his friend, Olaf Skavlan, who had received an adjunct stipend and become a University Research Fellow at the same time as Lie. Skavlan's field was literary history, and he himself had taken his doctorate only three months earlier with a thesis entitled *Ludvig Holberg as a Comedy Writer*. After the success of his abovementioned drama, *The Feast at Mærrahaug or The Charming Cucumber*, he had gone on to write a new comedy, *The Tower of Babel*, which had played to great acclaim the previous year under the direction of the Student Society at one of the leading theatres. And then six years after receiving his doctorate, Skavlan became Norway's first professor of European literature.

Lie's new status, marked by the doctoral mortarboard, was alluded to as well when the Realistforeningen celebrated its third anniversary in October 1871, when great thanks were given to Lie's contribution to the re-establishment of the Association three years earlier:

He, who showed the best tenacity
Was always Sofus Lie:
He construed himself a dignity
In moderne géometrie.
Frightful imaginary functions,
That seduced the brain to snap,
And oiled the supremest unctions,
When Lie received the doctoral cap.

Lie seems to have gone back home to his father in Moss for part of the summer holiday – the doctoral dissertation had to be made ready for publication by the Christiania Academy of Sciences, while an expanded version in German (112 pages in *Mathematische Annalen*) was ready in the course of the autumn. But a summer without long and thorough rambles in the mountains was inconceivable – thus Sophus made a trip to the peaks of Jotunheimen as well that summer. In any case, he wrote a letter from the mountain town of Fagernes, in the Jotunheimen region at the end of July, to Professor Bjerknes, asking for a list of the professor's scientific output – that is, Lie had undertaken to reference Norwegian mathematical literature for a German periodical, *Jahrbuch über die Fortschritte der Mathematik*, which had just begun to come out in Berlin.

Lie's doctoral dissertation resonated across Europe. The trend-setting Darboux in Paris characterized Lie's doctoral work as "one of the most handsome discoveries of modern geometry". And it seemed that Lie would rapidly be offered greater challenges than to continue as a research fellow in Christiania. When a position was announced at Lund in the course of the autumn 1871, Lie applied for the professorship of mathematics, which had been held by the Swedish mathematician, Carl Johan Danielsson Hill. Hill had dominated the mathematical milieu of Sweden for over forty years. Right from the first issues of *Crelle's Journal* in 1826, where Abel published his great works on elliptic functions, one also finds articles by Hill, who most frequently wrote in Latin. Hill's post was like a reigning institution in Sweden. The fact that Sophus Lie was one of the five applicants for this post in the "brother land" of Sweden was greeted with dismay by radical circles in Norway. It was considered a defeat, in which the country's greatest mathematical talent was being forced to abandon his fatherland and move to Sweden. From many quarters there arose protests, not the least from Lie's friends in "The Green Room" and the Realistforening – and as well, statements of support came for Lie from Clebsch in Göttingen and Cremona in Milan. Now Ole Jacob Broch became a key figure in the events – he, as cabinet minister and former professor of mathematics had contacts in all camps. As a matter of course all Norwegian cabinet ministers in turn spend time in the Cabinet Department in Stockholm – indeed, for some in Norway it came to be considered almost as a banishment. Long after the deferring of his stay, and after numerous requests for further deferment and reduction of time, it happened that Broch's obligatory year in Stockholm was halved. After serving six months he returned to Christiania, in February 1872. Broch had consequently been in Stockholm when the Lund professorship was advertised, and he wanted rather decisively to make sure that the Norwegian Sophus Lie would not be drawn into the vicious struggle over Hill's successor in the brother country. In any case, Broch now did all he could to secure for Lie the best working conditions in Christiania: Broch had just received a letter from Professor Clebsch, a letter that contained the most laudatory words about Lie's work, the power of his work, and its originality – and when Broch sent this letter on to the Academic Collegium in Christiania, he quite certainly did so to emphasise the possibility of establishing a new professorship.

The Realistforening also sent a letter asking for the establishment of an extraordinary professorship for Dr. Lie, "to the Profit of our University and the Honour of our Land". This also pointed to possible changes in the examination regulations that would make it necessary to have a larger teaching force, and they had all known the learned Dr. Lie to be "a skilled and gifted Teacher". The letter was signed by thirty members of the Realistforening, among them Elling Holst, Carl Berner and Amund Helland – the latter having taken the initiative – and the letter was addressed to Cabinet Minister Broch, who they knew had a heart beating for "the Success of the Sciences in our Land".

Without doubt, the greatest action was that initiated by "The Green Room". This radical group of friends clearly advanced the view that the Collegium would not take action to keep Lie, and they had to build broad public opinion for the creation of an extra professorship, thus they involved the press. The thirty-three year-old jurist, Theodor Blehr, had for a couple of years been a permanent correspondent covering the capital city for the daily paper, *Bergensposten*. He now wrote articles tinged with emotion and sensation about the issue. He began by way of a letter to the editor, and after having declared flatly that it was "a rather well-known Fact" that "the greatest Minds" had difficulty being recognized and rewarded in small countries, and he pointed out that this Norwegian, who had been "so blessed as to win, at a relatively young Age, Recognition from Abroad". Then the story of Sophus Lie was recounted with all the ingredients one could expect to be used to awaken interest, and it was done in pure fairytale style: once upon a time there was a man who received the best marks in all his examinations, but nonetheless became known as "an eccentric Tourist" – how his fabulously long and rapid marches on foot became the stuff of folklore, and the verses that were sung about him, such as: "Fifty Miles in one shake / With Rucksack and its Ache / Is to Him a Piece of Cake."

And when he was out in the mountains he was always clad in the same four garments plus hat. When he first went up into the mountains he was properly well-clad, and he rolled up his sleeves, rolled up his trousers and down his socks. However, with such a clothing style it was not difficult to mistake him for a tramp, or something worse. It was told how he had been suspected of being a murderer in Trøndelag County and a thief in Halden, and special emphasis was given to the the story of how in Fontainebleau he had been taken for a German spy.

But Lie's scientific triumphs abroad were also described in the newspaper: the letter from Professor Clebsch in Göttingen – which the readers learned was addressed to Cabinet Minister Broch, and who had forwarded it to the Collegium – was cited in a Norwegian translation: "Your excellent Lie constantly delivers the most handsome Works, and I likewise greatly admire his Labour Power for the Originality and Independence of Thought in his approach. It will perhaps not be easy for him to press onwards, when his Work is too deep and good to be grasped by the general Public, but Time will have its way, and I have no doubt that the Future will have much to say about this great Norwegian Mathematician." Of all the words of praise and recognition from "our Era's most celebrated Mathematicians" – and Theodor Blehr mentioned that at Lie's request statements of support had come from Berlin, Göttingen, Copenhagen, Milan and Paris – it was explained that there was room only to cite one more in addition to the letter from Clebsch, that of Cremona in Milan. Cremona admired Lie's "Ideas, which are true Discoveries, disguised, so to say, under a too modest and concise Form." Furthermore, in Lie's treatises he had found "original and acute Thinking associated with the Solution of very comprehensive and difficult Questions". The greatest impression with which the newspaper's readers were left, was certainly made when the august Cremona made the following description: "I would have gladly renounced all my own Works, if I had only been so fortunate as to discover what Mr. Lie has discovered, and on many Occasions have I said to my Friends, that the Fatherland of Abel has already another young Talent, to whom the Future of Geometry will come to owe an extraordinary amount." The only thing that Cremona could wish that Lie had paid more attention to was "the limited Comprehension of the Public". Lie's ideas were presented in a far too concentrated form, Cremona felt, and he concluded with the following request and expectation: "Were You to dedicate Yourself to the eminent Clarity of Abel, as I believe You have inherited his creative Genius, Your Renown and the Dissemination of your Works would know no bounds."

In closing, *Bergensposten*'s correspondent, Theodor Blehr maintained that such commendations would certainly make a strong impression at Lund. He thereupon made the following incendiary assertion: "But Mr. Editor! our Country cannot be so poor that it cannot offer such an outstanding Man of Science a decent Position in his Homeland."

This was published on February 14th, and when Bergensposten reached Christiania some days later, the local press – *Dagbladet, Aftenbladet, Den Norske Rigstidende,* and *Morgenbladet* – all ran the greater part of Blehr's contribution – and the phrase "our Country cannot be so poor that ... ", etc., was underlined in both *Aftenbladet* and *Morgenbladet.*

A breadth of public opinion had been created on the affair, and the next step was to have a proposal made in Parliament that an extraordinary professorship be offered to Lie. And this was rapidly carried out. Only ten days after the article in *Bergensposten* the proposal was formally presented in Parliament. The reason for this rapid action on the issue seems to have lain with a small group of parliamentarians, the so-called *collegium politicum* – seven men who for a number of years had been political power brokers with clearly liberal attitudes combined with populism and an expressed distrust of excessive bureaucracy. In any case, it was this circle which formulated and signed the proposal. Among the signatories was the central politician Johan Sverdrup, together with the jurist and later cabinet minister, Ludvig Daae, and Head Doctor Daniel C. Danielsen, who perhaps was the prime mover of the proposal and who had much the best contacts in *Bergensposten*. Doctor Danielsen, head of St. Jørgen's Leprosy Foundation in Bergen was highly distinguished internationally as a man of medical science, and had published a series of works on leprosy and other diseases.

The nominator of the motion, Ludvig Daae, wrote in his diary about the events of February 24th in Parliament: "At the very Moment that the Proposal was being presented, Mr. Lie, trying to force himself into the gallery in order to hear the proceedings, thereupon molested Others, and was grabbed by the Neck by a Constable and was cast out. We shall see whether this was a good or a bad Omen."

It was not in any way a bad "omen"; with Sverdrup's signature there was no doubt as to the outcome. The issue of an extraordinary professorship for Sophus Lie was carried, following two or three hours' debate in Parliament, on March 26th.

Normally of course it was Government which appointed professors, and there was dispassionate and formal disagreement over Parliament's right to take its own initiative in such matters. The motion concerning a professorship for Lie was tossed into this debate and it became a card in a great political game during the parliamentary session of 1872. The structure of power relations between Parliament and Government was on the agenda, and the great issue, where the nature of this power structure really found expression was on the motion about the Cabinet's use of Parliament, and its participation in procedures there. In his whole cabinet career, Cabinet Minister Broch had eagerly supported a softening of the relationship between the government's exalted isolation, and the elected representatives' discussions, while the government chief, Fredrik Stang, had a different view. Broch, who had come back from Stockholm in February, had already caused Stang a stinging defeat in relation to the Conscription Law, which from the ideals of uniformity and equality, now imposed military service on all. Stang had wanted to preserve the old social privilege of rights stemming from social position. Many hoped that Broch would also win out in the matter of Cabinet's access to Parliament. In order to secure a thorough handling of this and all of the other matters, the elected members had unanimously requested that the parliamentary session be expanded beyond its customary three months. The government refused. And it was precisely Cabinet Minister Broch who was entrusted with informing Parliament of this very negative decision. That this was to be done the very day that Sophus Lie's professorship was to be ratified, meant that it was clear to all that this matter too was really part of the underlying theme of the structure of power between Parliament and Government. To support the motion on an extraordinary professorship for Lie, was in reality, to support Cabinet Minister Broch and his stand on the Cabinet access to Parliament. Fifteen representatives took part in the debate on Lie's professorship, only two spoke against – one, O. Løvenskiold, felt that in principle it was the government's responsibility and duty to appoint professors; the other was the leader of the peasants and farmers, Søren Jaabæk, who in a familiar style, maintained that the professorship was a "needless" expense. But apart from this, there were only great and laudatory words spoken in the parliamentary chambers in relation to Lie's position: one representative wanted to preserve for Norway "this Man who reminds us so much of Abel"; another recalled that one must not "allow this remarkable Mathematician, this brilliant Cultivator of Science suffer the same Fate as Abel"; a third wanted to secure "The University the Fund of Endowments and Scientific Talents that are here spoken of"; a fourth said it was a "National Honour to support this outstanding Genius" – and the flood of such expressions as "extraordinary Qualities", "Talents which justify great Expectations for the Future", "uncommon Gifts", that nothing in the world should allow "to be lost to our Land", etc. In the end the vote was 85 in favour, and 16 opposed, to allocating 800 speciedaler in annual salary for an extraordinary professorship in mathematics. Sophus Lie had become a professor at the age of twenty-nine.

But in the big issue of cabinet access to Parliament, it was Parliament that suffered and Cabinet Minister Broch who was defeated. Broch, who usually went around with the description of "moderate Man of Progress", saw himself in particular in an intermediary role, as a man of the centre in the new dawning political landscape where the Conservative Party was more and more being challenged by the maturing Liberal Party. At an early stage in this development there were many who considered Broch would be a unifying figure and a possible government leader – a vote of non-confidence in the Stang Cabinet was openly contemplated in many quarters. The conflict between Broch and Fredrik Stang was the beginning of a twelve-year constitutional struggle between Government and Parliament. But the first round was won by Stang, and when in 1872, this cabinet affair was denied assent, Broch sought to resign from Cabinet. Broch's resignation was consented to on May 28th, when simultaneously he was appointed an extraordinary professor of mathematics. In the autumn of that same year, Broch was elected chairman of the Collegium Academicum at the University (what would later be the rector position). Otherwise, what would occupy Broch the most was the establishment of the metric system and the struggle for Scandinavian monetary union and the switch over to the krone (crown) system of currency based on the gold standard. He was designated "The Nordic countries' foremost Writer and Authority on Monetary Affairs", and gradually he assumed responsibility for a number of international commissions. As a national politician he was pushed into the background.

Discussion still continued in Sweden as to who should be the new professor of mathematics at Lund, and the position was not filled until the autumn of 1873, by Carl Fabian Emanuel Bjørling, after such a vehement and agonizing debate that the Swedish government concluded that henceforth such employment issues should always begin with a working group of three experts who would give their views on the applicants' respective merits before a final decision was taken by the government. One of the most active participants in the on-going debate in Sweden was Bjørling himself, and in relation to the matter he was also in contact with Professor Bjerknes in Christiania. Bjerknes had unequivocably maintained that Sophus Lie, of all the five applicants, absolutely had the greatest gifts. The applicants, in addition to Bjørling and Lie, were the Swedish mathematicians Göran Dillner, Edvard von Zeipel and Albert Viktor Bäcklund – the last of these was someone with whom Lie would later be much in contact.

In Norway, for several years still, Parliament would take the initiative to establish extraordinary professorships, and in conservative circles the concept "Parliamentary Professors" had a negative connotation.

Eleven years after Sophus Lie had been made a parliamentary professor, he commented upon this appointment in a letter to his friend Felix Klein. Lie stated that the notion had been spread in Norway that he had been awarded this position due to his political sympathies: Parliament's benevolence toward a specific individual could be due to nothing else, or so it was said. Christiania's conservative intelligentsia considered it doubtful that he was awarded his post as a result of his contributions to science and scholarship. "It always annoys me to hear the absurd story that I never would have become professor if Parliament had not been aware of my political sympathies." And Lie went on to assure Klein that he had not said a word to any member of the opposition in Parliament before his post had been proposed: "This is the truth."

During the summer of 1872 the newly-appointed Professor Lie found time for two alpine foot tours. In July, via Kongsberg, he walked through Telemark on his way to his sister and brother-in-law at Tvedestrand. Along the way he wrote to Motzfeldt and asked him to send money – "It is going to be much better now that I get a higher Salary", he said by way of comment, and complained about "the most appalling Sunshine" on the tour through Telemark having wounded his eyes. Lie had planned to make a longer journey on foot with Motzfeldt and his wife Else later in August. The plan had been to go to Lillehammer, northeast of Christiania by way of Gausdal and take a boat across Espedal Lake, then carry on to Sikkilsdalen and into the high mountains of Jotunheimen. But while in Tvedestrand Lie reported that perhaps he was not sure that he was up to such a demanding foot-tour at the present time:

> In spite of the fact that I feel I am capable of enduring Exertions as strenuous as in my best Days, I still consider it best to avoid any such Activities. On the Whole, for Good or Bad, first and foremost I have to take Care of my Eyes.

Lie himself felt that the pressure of work and the general strain of life had sapped his strength, and in a letter that he wrote later to Adolf Mayer in Leipzig, he admitted that in that summer of 1872 he had "become a little nervous", but added, "Luckily my appointment as professor made it possible for me to live a little more sensibly."

His stay for some weeks in Tvedestrand seems to have been good for him; in any case, he decided to accompany Motzfeldt "to the high Peaks" when, about July 20th, he took the steamer back to Christiania. The main reason for this early return journey was quite certainly an unexpected letter Lie had received from his French mathematical friend, Camille Jordan. Together with a group of friends, Jordan was touring in the Nordic countries, and in a letter written from Trondheim, asked if Lie would meet him in Christiania on July 22nd – Jordan's friends were in a hurry and wanted to continue their tour the following day. Lie managed to get a message to Sylow (across the fjord at Halden) about this rare meeting, and in Christiania, Lie brought these two mathematicians together. According to Elling Holst, who perhaps had also been invited, as they walked down from Frognerseteren, near Holmenkollen in the hills to the north of Christiania, Sylow had explained to Jordan his famous theorem on the existence of maximal subgroups with the order of a prime power. Sylow's work, "Théorèmes sur les groupes du substitution", was to be published later that year in *Mathematische Annalen*; and thus Sylow groups would become, and continue to be, a well-known term and important mathematical working tool. To be sure, there and then in Christiania, the great Jordan from Paris had difficulty grasping Sylow's surprising pronouncements. But he was soon convinced, and only a few days later – from a steamer on the Swedish canal system – Sylow received a letter from Jordan, saying that he had already found application for Sylow's theorem, and back in Paris he accelerated the publication of Sylow's work.

During 1872, the Christiania Academy of Sciences published four of Lie's articles – on integration methods of first order differential equations and on invariant theory of contact transformations.

Lie continued to maintain a lively mathematical correspondence with Felix Klein. Since the two friends parted in Düsseldorf in November 1870, they had exchanged ideas, interpretations and plans for new works. The summer of 1871 saw their third joint project come into being – on curves which could, by linear transformations, be mapped into themselves – with its publication in *Mathematische Annalen*. Klein, who in the meantime had been made docent, or associate professor, in Göttingen also had plans to visit Norway that summer so that he and Lie could collaborate further, a journey that however, for economic reasons, never came to pass. From Göttingen, Klein reported on his work and lectures in themes both mathematical and physical. He depicted the milieu of the young men of science working in a series of different subjects as extremely inspiring, and he explained that he himself had attracted notice in a public debate (with O. Stolz) on non-Euclidean geometry. With a commendation from Clebsch, Klein too was made professor of mathematics in 1872, at Erlangen, and at the age of only twenty-three.

Lie began his first semester as professor by taking two months' leave of absence. At his own expense, he travelled to Germany to meet with Klein. His itinerary took him via Copenhagen where he met Zeuthen and "chatted 2–3 Hours", and on to Göttingen where he wrote to Motzfeldt: "My Journey went well, with the Exception that in Fredericia [in Denmark] I did not notice that there were two trains to choose between, and I unfortunately took the wrong one." Thus he had arrived in Hamburg later than planned and similarly in Göttingen, where he stayed with Klein, and otherwise reported that he was "remarkable well", and how he would probably stay there for three weeks before travelling together with Klein to Erlangen. On the homeward journey he would probably stop off for a couple of days in Berlin, he wrote. Otherwise, the contents of the letter to Motzfeldt was the customary request for the forwarding of money: from his salary of sixty-seven speciedaler he wanted fifty sent on, and the remainder Motzfeldt should pay his mother for the rental of his lodgings – and apart from this, it was 'quite conspicuous' how much 'more expensive things had become since 1869.'

After not seeing one another for almost two years, Lie and Klein now met again in Göttingen at the beginning of September 1872. In a letter written a couple of weeks before the meeting, Klein had written: "You will find me to be an almost completely different person, even with sides that are visible in ways different from what they were earlier, above all, in Paris. For the most part I have become a very jolly being, who, of all things, sticks to mathematics, indeed, who with great seriousness, sticks to mathematics, however not with that gloomy sepulchral seriousness I had before." In a letter a couple of months earlier, Klein had complained that he lacked intensity and perseverance in his work – and he maintained that in terms of his science and scholarship, Lie's visit was crucial to him.

In addition to the happiness of meeting Klein again in Göttingen, Lie was delighted to meet the mathematician Adolf Mayer. Mayer was a professor in Leipzig and he worked with partial differential equations of the first order, and via analysis, he had developed integration methods which corresponded well with Lie's geometric methods. Mayer, who was an economically affluent man, and newlywed, invited Lie to his home in Leipzig to continue their discussion on various integration methods. Lie remained at Mayer's for three weeks, and a lively exchange of letters followed. When fourteen years later, Lie too became a professor at Leipzig, the friendship deepened. But now, in the autumn of 1872, it was Klein's journey to Erlangen that was foremost on the agenda. On the way there – in both Frankfurt and Heidelberg – Lie spent the whole time meeting mathematicians, among others, Max Noether and Jacob Lüroth, and he wrote to Motzfeldt:

> I have, to a tolerable extent, extended Relations among the more prominent young German Mathematicians. I have less Interest to know the older ones, since Production Capacity in our Field, almost without Exception, decreases to an alarming Degree with Age.

In relation to his professorship in Erlangen, Klein according to the custom, had to hold an inaugural lecture. This *Antrittsrede* was the point of departure for what would become a treatise called *The Erlangen Programme* and was accorded very great importance, not least because the theme was the nature of geometry and its general principles. Lie and Klein were together in Erlangen for two or three weeks from the beginning of October. They discussed mathematical questions and, above all, the forthcoming lecture Klein had to give. The title was "Vergleichende Betrachtungen über neuere geometrische Forschungen" [Comparative Considerations of newer Geometric Research], and the aim was, as Klein wrote, to present methods and results he and Lie had worked with. *The Erlangen Programme* contained the formulations that were common in different geometric methods, and which lines of connection could be drawn between the geometry of projective properties, Plücker's line geometry, and Lie's sphere geometry. The first goal was to specify the geometric object, and thereafter define what could be done with this object. A geometric object must be seen to be a thing in a space, wherein this thing's external form can be assigned co-ordinates, and this figure can thus be shifted, reflected and twisted over and along lines and surfaces by a number of means, and these means can be determined as different groups of transformations. The task was thus to study the properties of this figure that remained unchanged – that is, those which were invariant in different groups of transformations. This comprised a classification of geometric points of view that was new, and which would – perhaps more via Lie than Klein – spread in the mathematical milieu. The order and unity that was thus created, became the basis for a distinctive structural way of thinking – and in the work of locating and identifying structures in what is variable and changing, mathematicians became forerunners to research in other fields of science.

Lie probably would have liked to stay longer in Erlangen that fall, in any case Motzfeldt received a letter requesting that he try to have Lie's leave of absence extended to November 15th. Lie wrote that he would certainly be back in Christiania in the first days of November, but: "I would very much like to have free Hands. It is possible that it will be of *Importance* for me to stay some Time in Leipzig and Berlin."

Apart from this, about the time he spent that October with Klein, he reported, in addition to the "pleasant Autumn Weather", that "Here there is a Flood of many good Things, especially the famous Erlanger Beer which however, in my Eyes, is less attractive than the Grapes, which just now are being harvested." The two friends also spoke about the future in the form of possible marriage and home life – if we are to judge from later correspondence: they were of one opinion that it was a good thing "to find a safe Haven in a good Home", but neither of them reported any concrete plans, at any rate not during the very real and actual walk they took from Nürnberg to Fuerth, and which they both later recalled with fondness.

It does not seem that Lie returned home via Mayer in Leipzig, and his stay in Berlin was probably only a day or two. From all accounts, he was back in Christiania at the beginning of November.

Professor Clebsch died suddenly in Göttingen on November 7th, and from now on, Klein was considered the new leader of geometric research in Germany – many of Clebsch's students went to Klein in Erlangen for help and supervision.

Lie had special memories of the time around the death of Clebsch. During this period, Mrs. Motzfeldt, from whom he continued to rent lodgings, had her niece, Anna Birch, visiting from Risør (in Sørlandet, on the northwest shore of the Skagerrak, in the same region as Tvedestrand). Anna was only eighteen years of age, and Lie had met her earlier when she had come to visit her aunt in the capital – in fact, Lie had had a photograph of her for some time, and he now invited her to the theatre.

During the remainder of November and up to Christmas, Lie seems to have lectured according to the planned postponement, and in December the newspapers reported that he had been chosen a corresponding member of the Academy of Sciences of Göttingen. He had thought to spend Christmas with his sister and brother-in-law in Tvedestrand, but before he left, he wrote a letter of proposal to Anna Birch. He wrote that on his "Christmas Journey to Tvedestrand" he would travel by way of Risør, and thereupon he, on "Christmas Eve between 3 and 4 o'clock in the Afternoon, would try to "meet [her] to have my Fate decided."

PART V

Professor in Christiania

Studenterlunden and the University, Karl Johans gate, Christiania

"My Inner Life Has Been Most Mighty"

Anna Birch gave Sophus Lie her yes on Christmas Eve 1872, when he visited her parents' home in Risør. Her only condition was that for a period of time their engagement must be secret – by way of justification she claimed that she thought she was too young to commit herself.

Their period of engagement would come to be twenty months, much too long from Sophus' point of view. They met several times, both in Risør, in Christiania and at his home in Moss, but most of this period was marked by an avid correspondence between the two. Of this correspondence, the majority of the letters that have survived were his letters to her – almost ninety – he, for his part, seems to have kept none of hers from this period.

He formulated the letter that contained his proposal, and to which he received his response on Christmas Eve, as follows:

Miss Anna Birch !

Since the first Time I made Your Acquaintance, it has been my greatest Wish to win Your Love.

I am well aware of the Fact that I lack many Qualities, that You must hold dear; but when now, in spite of everything, I dare ask You to become my Wife some Day, the reason thus is I am sincerely convinced You shall never meet any other Man who will be as fond of You as I am.

I cannot offer You an outwardly brilliant Station in Life; but Fortune is smiling on me in a peculiar Manner in my Work, and I can place my Hope on a creditable Future.

Should my Letter reach You as somewhat unexpected; should You consider it impossible to fulfil my Desire, then I implore You to wait to speak to me before You formulate a final Conclusion.

It is my Thought to make my Route via Risør on my Christmas Journey to Tvedestrand. On Christmas Eve between 3 and 4 in the Afternoon I will seek a Discussion with You to have my Fate decided.

Sophus Lie

After receiving Anna's assent in Risør, he went on to Tvedestrand to celebrate Christmas. At this time he wrote letters to Mayer in Leipzig, and here he described for the very first time what later would come to be called Lie Algebra. Lie gave an answer to what occurred to certain rotations – which he earlier had described with his groups, and which later would be called Lie Groups – that examined what happens when such rotations occur consecutively, or in different orders.

After celebrating Christmas in Tvedestrand, Sophus was back in Risør for New Year's Eve. He most likely walked the approximately thirty-to-forty-five kilometres each way; in any case, he walked away from Risør on New Year's Eve, even though Anna's father had advised him to hire a carriage and coachman. Some days later in a letter written from Tvedestrand he commented upon this first stay at her parents' house: "My own dear Anna! I cannot comprehend, how it happens, that I who certainly have never been a timid Man, have, in the last few Days, turned into a poor Wretch. Yet, despite the Fact that I have no known Reason to embarrass myself before Your Parents, though I seem to have done so, when I decided to walk on New Year's Eve, that I forgot to remind You about Your old Photograph." Sophus had gotten a new picture of Anna, but he also wanted very much to have the old one back, because, "I had thus had so many good times gazing at it in the course of a Year." However both photographs gave "an extraordinarily incomplete Idea" of her "sweet Face". Photography, as an artistic craft, was coming into its own precisely at this time.

Already with these first visits – on Christmas Eve and New Year's Eve – there was discussion about how the engagement could be kept secret. The advice of Anna's father that Sophus not walk, but rather drive from Risør in a carriage, was certainly an attempt to hide him from the townspeople. He explained himself to Anna, saying, "When I left Risør on New Year's Eve, I somehow thought that I would not meet anyone I knew. In any case, You can rest assured that my tour on Foot to Risør did not surprise Anyone, since People know I find Pleasure in eccentric Tours. It would have been a lot more conspicuous had I followed Your Father's Advice and driven."

Other than reporting on the celebration of New Year's at Tvedestrand, Sophus mentioned the abominable weather, but had amused himself indoors as well as he could with his sister, brother-in-law, and children: "They have invited us to Tvedestrand for Easter. I know that You would enjoy it here; for I do not know more amiable People than my Brother-in-law Vogt and my Sister, and the Children are also good, all of Them." After a number of earlier sojourns at Tvedestrand he had many other friends in that town as well, and in particular he mentioned to Anna his stay in the summer of 1866, when "for a Period of 6 Weeks" he was "the Source of Amusement for all the Youth in Tvedestrand": "Everyday we had Outings to sail, to row, as well as ashore." Now, together with one hundred others, he was invited to a New Year's dinner "and afterwards a Ball" at the home of Tvedestrand's man of stature, Fritz Smith. Sophus hoped that Anna would enjoy herself at the ball in Risør on New Year's Day, and that she also had "many other Pleasures during these Days". Regarding his own relationship to such social activity, he declared:

> When I was a young Student, I did not know which I valued most, Foot tours in the Mountains or Balls in the winter at Moss. In recent years I cannot deny that my Joy of Balls has, in my Eyes, faded somewhat; but the Matter will present itself differently now that I have such a charming Sweetheart, who, even on the long Term, shall now have the Desire to dance.

Otherwise there were three points about which Sophus was concerned. He was eager that they should exchange rings, he wanted to discuss how their correspondence should be kept secret, and he wanted to meet her again on the way back to Christiania.

With regard to the rings, he wrote that it would be "inexcusable to go to any of the Goldsmiths here in the Small Towns where there are certainly Pockets of Gossip," and he asked if it would not be better if he got rings when he arrived in Christiania. He would then consult with Mrs. Motzfeldt, and could insure that in this matter Anna need have no fear about "the Conservation of our Secret". In addition to his sister and brother-in-law in Tvedestrand, and Mrs. Motzfeldt and her daughter Ida in Christiania, those who knew of the betrothal were only his father and sister Laura in Moss. As for their future correspondence, he thought that she could alternatively send her letters "under separate cover" to Mrs. Motzfeldt, and to his sister Laura in Moss, who would send the letters on to him. And so that the postal authorities in Risør should not get the least inkling that it was she who had written these letters, he could give her a stack of envelopes on which Brother-in-law Vogt had written Sister Laura's name and address. For his part, he could alternatively have Mrs. Motzfeldt, Ernst and Ida write on the envelopes, alternating between those addressed to her, her mother, and her father. And he wrote:

> Right away from the Beginning, let us write to one another often. Indeed You still know so little of me, and Letters are the only Way to increase Familiarity, that for the time being will allow us to reveal ourselves. As for myself, I quite believe that I have long since come to know You thoroughly. Your pure charming Soul opened itself to me from the first Moment.

Sophus had planned to journey back to Christiania by the steamship that followed a coastal route from Kristiansand, on Norway's main southwestern peninsula, to Christiania, and which passed Risør on Tuesday, January 7th about three o'clock. He very much wanted to arrive in Risør on foot the day before, and about four o'clock on Monday afternoon he would love to pay her a visit "that should not last long", and if she were to be going out to a ball or a party that evening, then he would pay his visit on Tuesday morning: "Remember, Dear Anna, that much Time shall come to pass before we see one another again. We ought not to pursue our Caution too far."

Sophus closed his letter to Anna, writing that indeed he did not dare "touch the Word Marriage" this time, but when they met again at Easter, "we shall get to build our Plans for the Future."

But things were not to go exactly as Sophus had hoped and planned. Anna must have responded very promptly to his letter and explained that in order to keep their betrothal secret it was not advisable that he return yet once again. Already in Risør there were those who had noticed his New Year's Eve visit. Sophus did not take this news with equanimity. He told his sister and brother-in-law in Tvedestrand that he would walk to Arendal and board the steamer there. He also sent his baggage to Arendal. But on Sunday morning, January 5th, he took the highway in completely the opposite direction. With his normal forced march speed he calculated that he could reach Horten before the steamer did, and board it there for the journey across the fjord to Moss.

He actually did reach Horten before the steamship, and while he waited, that Wednesday morning, he began a letter:

Dearest Anna!
You are Right that when our Betrothal is secret, we must arrange things in such a Manner that the Manner itself does not become public. To say the least, Your Letter was doubly dear to me as the first, even though it dealt me a bitter Disappointment and grieved me greatly, Anna my dear, something which certainly will not surprise You. Indeed, I have counted the Days, the Hours until once again I am able to look into Your fine clear Eyes.

About his journey along the highway, he had this to recount:

I have used my Legs in a decent manner during these Days; yet the Weather was so good and in the Evenings I had the most lovely Moonlight. People think that I have very strange Tastes; but this strengthens the Muscles and Nerves, and besides, I find myself with such good feelings, as I roam the Countryside as a Hobo. By habit, I am always deep in my own Thoughts, and building Castles in the Air for the Future: and You quite understand who it is that plays the Main Role . . .

The only Change from Routine occurred when, in the Neighbourhood of Brevik and late in the Evening, I encountered a Man who was dead-drunk, sleeping in the Snow, at a Temperature of a Couple of Degrees of Frost, in the Middle of the Woods. When I surmised that he had all the Signs of freezing to Death, I hauled him along a quarter Mile (and him weighing almost a ton). Further on, I was almost assaulted by a Swede in the Vicinity of Tønsberg, but if it had come to serious Fisticuffs, I am afraid it would have been the Swede who would have come off worse.

The steamship reached Horten and Sophus took it over to Moss, visited his father, and his sister, Laura, and "Upon my Arrival here, I found a Couple of Letters waiting from Abroad. I have such good Fortune with my Work. I told You that shortly before Christmas I had been chosen corresponding Member of the Academy of Sciences of Göttingen, which for a long Time has been a Centre of Mathematics. It was a great Honour, especially since it came to me so very much more unexpectedly, since Professor Clebsch in Göttingen, to whom I am grateful for so much, died a Couple of Months ago."

The two letters from abroad were from Klein and Darboux. During the Christmas holidays, Klein had been in Göttingen and worked on the biography of Clebsch and the *Nachlass*, the papers Clebsch had left behind. He wished his friend luck in his work with the partial differential equations, and offered all the editorial help Lie might want or need. Darboux too had only good things to say about Lie's works, which he still had not found time to study with the thoroughness they deserved. Otherwise, he corrected a small mistake in a differential equation he himself had earlier mentioned, and reminded Lie that there was great interest in France to have published a new edition of the works of Abel, Lie's "famous compatriot". The first edition of Abel's works had been edited by B. M. Holmboe and came out in 1839, but indeed, it had not been complete, and moreover, had been sold out. Therefore Lie, together with Ludvig Sylow had been asked to oversee a new edition. Darboux, writing from Paris, reported that *Gergonne's Annals* had located a paper by Abel which was not in the first edition of the collected works. Chasles too had made Lie aware of some of the works that were not included in Holmboe's old Abel edition – and Weierstrass in Berlin sent a letter with his comments on the forthcoming edition. Lie now reported further to Sylow on these matters and about all those

Lie's portrait in oil, by Erik Werenskiold in 1902, said to be based on an earlier drawing.

Top left:
Motzfeldt, high court judge, and for a period, cabinet minister.

Top right:
Axel Lund, doctor of the City of Kristiania, and Lie's private doctor in the final period of his life.

Professor Georg Ossian Sars, famed for his extensive zoological studies.

Geologist, writer and professor, Amund Helland, painted by Erik Werenskiold.

Bottom left:
Reader in mathematics, and writer, Elling Holst.

Bottom right:
Professor Axel Thue, who started off the strong Norwegian tradition in number theory.

Top left:
Bjørnstjerne Bjørnson, one of Norway's national poets who in the 1890s worked actively to get Lie back to Norway.

Top right:
Edvard Grieg, who met Lie several times in Leipzig.

Fridtjof Nansen. Nansen and Lie met on several occasions, and had common plans for, among other things, the Abel Jubilee in 1902.

who looked forward to a new and complete edition of the works of Abel: "We can be particularly satisfied, it seems to me, from *my* Point of View, it would have been correct to have had Something said about Abel's Position in Science."

Eight years of much contact and many letters would pass before Lie and Sylow reached the end of this task, and at that point, in a letter to Sylow (Dec. 20, 1881) Lie claimed that the idea for the project had come from Clebsch; it was "Clebsch's Approach to me that occasioned the Publication."

Now, in January 1873, Lie told Sylow that he was "right in the middle of partial Differential Equations", something that Mayer too was preoccupied by, and according to Lie, for the time being, they were both working "keenly", and it was "advancing surprisingly well": "I have in any case taught Mayer to respect the Manner of Thinking that, by great Error, is called modern *Geometry*." As for his own position, Lie pointed out that he himself was working in a "Segment of Mathematics that in any case belongs among the most important", even though he was "inclined to consider Invariant Theory in an expanded Sense, as the Basic Character and Nucleus of Mathematics". Lie also explained to Sylow that Darboux, according to Klein, used certain of Lie's things "in an incorrect Manner", and that therefore Klein should now have "found the right way to take Darboux". Lie commented: "I hope that this does not disturb my Relationship with Darboux, which I value exceedingly." But nonetheless, Lie had his reservations about Darboux "who, like most Frenchmen, is not particularly honorable in terms of Science, even though he remains both a charming and honest Man."

Sophus wrote to Anna that he would later tell her more about his foreign friends: "There are not exactly so many with whom I exchange Letters out there; but the Correspondence with each one is that much livelier. Besides, my Friends have complained bitterly that since my last Journey Abroad I have become remiss in my Letter-Writing. You certainly know who bears the Blame." When he was back in Christiania he would write her a long letter. Although, in his own words, he was "wretched at using the Pen", he wanted the correspondence to give her "a full-blown, and as far as possible, a complete Knowledge of my Personality." Therefore he wanted to tell her in his next letters about what he had experienced, and added:

> At any rate, my Life has been poor in external Incidents for long Periods; but my internal Life has been mighty; frequently my Plans, my Expectations, have been bitterly let down; frequently they have remained overcast for long Periods.

Sophus was eager to take in hand Anna's reading matter, and he asked her if there were "any Reading Materials" in particular that she valued. "You must not believe that I intend to make some Blue-stocking out of You. But as your Mind is rich and gifted I see that You, in later Years, might regret it if You have not utilised well this Time now." He asked if she had done "great Readings in the Works of the Natural Sciences", and said that, for example, he could get out from the University Library the two popular books by the French author, Professor Louis Figuier – *The World Before the Flood and The World After the Flood*. But he remembered that Mrs. Motzfeldt had told him that Anna read not too little, but too much, and he added, "But of course You must certainly look after Your Health. Nothing would distress me more, when next time we meet, to find that You have lost any of the rosy Glow in Your Cheeks, or Your youthful Vigour in general." Because he himself had always had strong good health – "My Health has always been as firm as the Mountains" – earlier he had had nothing but contempt for health rules, but recently however he had "tried to live rationally, with Strolls in the Outdoors, etc. Nonetheless the best Advice: seek Happiness and Satisfaction in being outside: go on Boat Tours in Summer, into the Mountains, and in Winter, diligent Use of the Skates. A propos the Skates. Shall I send You some good Skates?" In any case, did she not know the Horn family in Risør so well that she could tell them about their engagement, so that alternatively he could send letters to her there? And for his part in Christiania, he could tell one of his good friends, so that she could occasionally send her letters there: "If I take care to provide You Envelopes with the address written in the Hand of a Stranger, so that there would be no Danger, no matter how closely Your Risør Friends spy on Your Affairs and Your Correspondence."

With regard to the "Risør Populace" who bothered her – and in particular a student by the name of Opsahl was mentioned – Lie commented, "I can well understand that Risør's young Gentlemen with their Argus Eyes will defend Risør's best Pearl," and he concluded:

I hope that You can tell me that You have enjoyed Yourself over Christmas, and that You have not been bothered terribly much by Your Friends with their Banter over

Your
Sophus Lie

But Sophus Lie had no sooner arrived back in Christiania than he "Head over Heels" had to return to Moss. His father, Vicar Lie, died suddenly on January 18th, at the age of seventy. Two days later, Sophus wrote:

Dearest Anna!

Sadness and Joy mix in this Life. I did not have any inkling that today it would be my Lot to inform You that suddenly my dear Father, following two pain-free Days of Illness, calmly and peacefully departed into Death, Saturday, the 18th of January, at 5½ o'clock in the Afternoon.

Like Most of my Relatives, my Father was always a more than commonly powerful Man. His physical Strength maintained itself to the End. As Laura said, even during the last 10–15 Years only once was he absent from his Clerical Duties due to Illness. His mental Abilities had not maintained themselves quite as perfectly, but the Changes were so slow and so little that he himself had no Idea thereof.

Sophus and his two sisters in Moss had only through "a Couple of small very innocent Traits" noticed their "Father's Failing with Age". Half a year earlier, that is, their father had decided to exchange his old wig, which he had worn for over twenty years, for an elegant new one, which decidedly "better suited the Head of a Youth than that of an Old Man." However this would not have been so bad if, at the same time, he had not hit upon the idea that when his hair had become so youthful, his beard should follow suit. Whereupon he had dyed the "white Beard that he had worn with Honour for so many Years." Otherwise the only symptom of his father's illness had been his loss of appetite, and an increasing drowsiness had caused him to take to his bed for a couple of days, and Sophus remarked, "His Death came so imperceptibly, and in all Ways was as easy and good as Anyone could desire." Despite his sorrow he was happy that his old father "had departed so lightly and so well from Life, which in the Future would most certainly have caused Him more Disappointment and Sorrow than Happiness."

Sophus well knew his father's difficulties in his final years. Indeed, right from the pietistic awakening and the appearance of lay preachers into the 1850s, Vicar Lie had at times been attacked for standing for a liberal Christian outlook. From the pulpit he had warned against easy and superficial religiosity, and pointed out that Christianity in its rather romantic form easily led to pride and censoriousness. But popular, personally-professed Christianity more and more became a part of the country's state church religion, something that to no small part was due to Gisle Johnson, the professor of theology, who had often performed together with lay preachers, and zealously treated them as his peers. The low-church trends were spreading. And despite his hard-up position, Vicar Lie at the age of seventy, had decided either to resign or hire a young chaplain, whom he himself would pay – a solution that in other respects was quite common. And since he had felt in good health, he decided on the latter course. It was common knowledge among the congregation – Vicar Lie would celebrate his seventieth birthday in March that year – and some were afraid that he would now hire a chaplain who was too liberal in his basic theological views. Precisely the week before he died, he had been visited by three men who purportedly were said to have doubts about the priest's spiritual activity. And when Vicar Lie died only a few days later, rumours began to circulate that he had been deeply offended by the unseemly visit – that the three men were said to have come with orders and almost threats if Lie did not follow their advice – and that this was probably the cause of his demise. On the day that Vicar Lie died, the newspaper *Moss Tilskuer* had come out with an article in which the three men who had sought out Lie, were mentioned by name and strongly criticised – one of the three even taught at the secondary school and ought therefore to be let go for opposing his superior.

On January 22nd, the *Moss Tilskuer* published an obituary, entitled "Our Parish Vicar". After a short sketch of Vicar Lie's *vita*, it reported: "Consequently, for close to 22 Years the deceased, an Elderly Man of Honour, lived and worked in this Parish. He was surely not known as one of the most gifted of Clerics, as no one would be more willing to admit than he, but he possessed Divine Confidence, with which he carried out his tasks, imbuing them with Justice – fulfilling his Calling with Integrity, Faith and Good Will." More than "any spiritual discourse", during his whole life, he had been that sort of priest who was mild, peaceable and affectionate. He always cheerfully presented the Gospel to his congregation – and in his conduct he was calm, honest, humble and human toward all – "Therefore the Memory of Pastor Lie, the beloved Father to his Family, conscientious Public Servant, as a Man of Honour and a Christian who deserved universal Respect and Goodwill . . . will be blessed and cherished for a long time in the Parish of Moss."

The funeral and interment took place two days later. Sophus had sent letters and announcements to relatives and acquaintances, and was otherwise occupied with all the practical details surrounding the funeral. *Moss Tilskuer* reported a large funeral cortege – "more numerous than has ever been seen here in the City for any Officials or Elders" with the the city council, the labour council and all three deans taking the lead. According to the account of the funeral, the eldest dean, Dean Emeritus Gude, addressed the mourners and dwelt upon the deceased as a family man and father, and wished him "Farewell from those nearest". A male choir then began the first notes of a seven stanza song written especially for the occasion. A coffin, beautifully decorated – according to Sophus' letter to Anna – "with the between 200 and 300 lovely Wreaths" was borne into the nave of the church, conducted in by a choir composed of 50 to 60 girls from the previous year's confirmation class. There was singing, Dean Bassøe spoke, and then there was more singing again. Then the casket was carried out. The city's leading men and political representatives took turns with members of the labour organization and city council in bearing the casket to the churchyard. Presiding at the graveside was the third dean, Dean Hartmann, who for many years had been a teacher at the city's secondary school and now held the position of vicar in the town of Horten, across Oslofjord. The city's shops remained closed for the duration of the funeral proceedings. The newspaper would publish the seven-stanzas of the song sung during the service.

Five days later, the *Moss Tilskuer* printed a reply from the three men who had been blamed for exercising undue pressure on Vicar Lie in relation to the forthcoming fulfilment of the chaplain position. The three could assure the readers that the discussion had taken place in the greatest amiability, that Pastor Lie had been in agreement with their expressed wish that it be the cabinet minister in the Department of Church Affairs who ought to "face up to the choice between applicants" – and indeed, it had been Vicar Lie in the end who had said that the discussion had been both pleasant and desirable, and he would be extremely pleased if the congregation consulted him more often "when it is done in such a friendly manner". Under this article there was a clarification from the district doctor, who had been treating Vicar Lie, and who said that neither Lie himself, nor those nearest to him, had in any way put forward anything to suggest that the meeting between the three men and the pastor had contained anything untoward that had served as "the Antechamber" to Lie's fatal illness. Sophus expressed the same view in a letter to Anna, but he added, "Another Thing is that Father, like most, and particularly, the elderly Priests, had in recent Years been badgered by Pietists, who did not find the Work he carried out to be satisfactory."

A stone was erected over Vicar Lie's grave with the following inscription:

Peace glistened from your Eye,
Faith in Christ from your every Act.
Thus the Palm must bend and sigh
Over Life, so tenuously tracked.

Sophus stayed in Moss for a week – as he himself described it, "overburdened with Matters in Relation to Father's Death and Funeral". He wrote three letters to his brother John Herman in Bergen, and received a reply that the brother was quite the most interested in the bed linen and silverware, as well as the cash from disposing of the dwelling – nobody would be interested in the furniture, the brother felt –

and he otherwise was of the opinion that if possible the household effects should not be sold, but rather, be divided up.

During these hectic days in Moss Sophus received three letters from Anna. He answered these letters and thanked her for permission to write her so often: "You can well believe that I shall utilise the Permission", and he urged her in future to call him Sophus. He got to hear that she was suffering from a strong bout of influenza, and he swiftly presented two pieces of advice: "The first is that one ought not to pass up any Day of Sunshine without taking a good long Walk; the second being that one restrict oneself to the Rule that one guards against sudden Chills."

Otherwise, they busied themselves with the task of obtaining rings and they discussed the time at which it would be appropriate to make public the knowledge of their engagement. He had received from her "The Measurement for the Ring", but he replied that his sister Thea had made known that it was extremely difficult to take a precise measurement: "I am therefore thinking about finding a better Method. I have exactly at this Moment an Idea." And in the next letter: "I have strained my Abilities to find a passable Means by which to take the Measurement. In partial reply I am sending You a little Stick of Wood, on which there is a black line on each of the two Ends marking the Location that has the same Thickness as my right Ring Finger at its Thickest. In addition, there follows a small steel Thread-Ring, whose Precision is presumably the correct Size."

Both Sophus and Anna, from their respective circles, had gotten reactions to the secrecy of their betrothal. Most seemed to think that such scrupulous principles were an absurdity, most particularly because under such circumstances another man could come to propose to the secretly betrothed, and so receive an embarass-ing refusal – without knowing the real reason. It seems that Sophus' view was that at Easter, at the beginning of April, they should assemble their families at Tvedestrand, and there make public the engagement. He wrote that his whole family were overjoyed with the relationship: "They began to fear that I was becoming an old Bachelor Fuddy-Duddy" – and also his deceased father had "talked with such Happiness, in his last Days, about our Betrothal". Anna for her part had made it known that she felt it would be "ghastly" if the engagement were to be made public, but Sophus himself was convinced that her "Apprehension about this Thing will change little by little". He wrote that for him it would be "painful in the highest Degree" if they did not meet at Easter, but, he added, "All the same, I speak with full serious Intent, when I say that on this Point, You should not take any undue Consideration of my Desires."

Otherwise, in Christiania, Sophus' relationship to Mrs. Motzfeldt, his hostess and confidant, and Anna's aunt, became problematic. Without doubt Mrs. Motzfeldt took a great interest in their engagement, and she had also explained that Anna's father was the one "of her Brothers of whom she was most fond", but her humour had been bad and unstable all winter long – and Sophus wrote: "I do not rightly understand Mrs. Motzfeldt . . . When before Christmas my spirits too were extremely irritable, for Reasons You well know, the fact could not be avoided that we had some small Battles." After Christmas, and after she had come to know of the engagement there had also been a couple of such conflicts. But since Sophus had made "exceedingly respectable Endeavours to make her happy," he absolved himself of all guilt – he had kept away from Mrs. Motzfeldt and although he was grateful for all that she had done, he urged Anna not to misunderstand now when he nonetheless gave her "a little Warning":

Mrs. Motzfeldt has this Weakness, that she seeks to penetrate all human Confidences, and that afterwards she misuses this Trust, such that one must watch her. Perhaps You think these are strong Words. But You have to believe that she has told me of many Betrothals that she has intrigued either to bring about, or to frustrate. About giving the Boot to unsuccessful Suitors, etc.

Sophus felt that Mrs. Motzfeldt would also prevent the relationship between himself and Anna, and explained that when he came back to Christiania after the Christmas holidays, he had been outraged and indignant about the way she had received him: "Right away she would scarcely congratulate me, and she said that it had been Your Duty to report the Engagement to her, and so on." The conversation had continued "with Tears and meaningless Outbursts". Mrs. Motzfeldt had spoken about the great age difference between Sophus and Anna – even though she herself had been fifteen years younger than her husband – and she had held up in front of Sophus all his "bad Habits", and came out with "a Multitude of imprudent Observations." In his own words, he had kept himself "amazingly calm", but confided to Anna: "That I have not been able to drive away the Bitterness I feel inside, ought to be forgivable." At the same time he was trying to put everything back into its old order again, but this "has its Difficulties and I still make blunders from time to time." Sophus was receptive to the fact that Mrs. Motzfeldt perhaps in some ways meant well, but she had "the Weakness of not being able to allow a Thing calmly to go its own Way" – and in any case she had no right to behave as she did. He wrote to Anna: "I know that You are in agreement with me, that no one ought to stand between two who are betrothed", and if there was anything that disturbed her about this, she should consult him, "and in any case never Mrs. Motzfeldt", her aunt. Both now and later it seems that Sophus was blind to the possibility that Mrs. Motzfeldt would have been very interested to have her own daughter, Ida, joined with Sophus, who of course was a professor with brilliant prospects.

Anna – christened Anna Sophie – Birch, was on both her father's and mother's sides connected to distinguished relatives. Her father, Gottfried Jørgen Stenersen Birch, studied law. He was Mrs. Motzfeldt's older brother by two years. Their eldest brother, who had been named after their paternal grandfather, was the nationally known Christian Birch-Reichenwald who, during the period between 1858 and 1861, had been Norwegian government leader. But in subsequent years Birch-Reichenwald became more and more of a tragic figure. His friendship and intimacy with King Karl XV of Sweden-Norway had brought him to power, but efforts to utilise this royal friendship to obtain political results had been a complete failure, and he himself had been humiliated. By his faith in, and reliance upon royal power he had managed to create a space for his own civil service, while at the same time he had given the opposition the belief in the possibility of a change of political system. Birch-Reichenwald was consequently left standing as a sort of organiser of a fundamental change of system that would come into being, but presumably without consciously having desired this, and in any case, without the political camps having any confidence in him. Birch-Reichenwald became a bitter man, and left public life, becoming a magistrate just outside the capital, in the County of Aker in 1869, all the while his manic-depressive features became increasingly obvious.

His brother, Anna's father, also suffered from this illness. Already four years before Sophus Lie became engaged to Anna, her father, due to mental illness, had been pensioned off from his position as senior customs officer at Risør. Three years earlier, in 1866, he had arrived in Risør from Vadsø in northern Norway where he had been customs paymaster for six years. Before that, he had worked in Christiania in various office positions within the ministry and the military. It had been during this time that he had married Marie Elisabeth Simonsen, and in 1854 their daughter Anna was born. But three years later Anna's mother died, at the age of only twenty-nine. Anna's father then married Ellen Marie Johansen, who seems previously to have been his housekeeper, and who consequently became Anna's step-mother, and whom Anna herself always referred to simply as "my Mother".

Anna's actual mother, Marie Elisabeth Simonsen, born in 1827, was the daughter of the bailiff, Daniel Barth Simonsen, who in turn was the son of shipowner and trader, Niels Henrik Saxild Simonsen of Risør, who was also the maternal grandfather of Niels Henrik Abel. Thus it was that Anna's maternal grandfather was Niels Henrik Abel's uncle – his mother's brother.

Already during Sophus Lie's first meeting with Anna's family in Risør there was talk about certain records in the city that pertained to Abel's genealogical connections. Sophus wanted very much to see these papers, and a few days after the funeral in Moss he wrote to Anna:

> You yourself have no Conception about what an odd Coincidence, it seems to me, is the Fact that my Fiancée is a close Relative of Abel, my great, though unattainable, Ideal.

And there were other close relations between the families of the two who were secretly engaged: Christian Birch-Reichenwald had married his cousin, Jacobine Ida Sophie Motzfeldt, who again was a cousin of Ernst Motzfeldt. And as well, for one reason or another, the Birch-Reichenwalds and the Lie family had frequent relations: at certain periods Mrs. Birch-Reichenwald took care of Sophus' sisters, and Birch-Reichenwald had attended Father Lie's funeral.

Due to the death of Sophus' father, most now felt that it would be best to wait until Easter before making public the engagement of Sophus and Anna. But Birch-Reichenwald and his wife however felt that it would be sufficient to wait one month after the funeral: "He [Birch-Reichenwald] in particular recalled frightful Examples, in which secret Betrothals had brought about Misfortune; they customarily had to do with the Lady in question receiving new Offers in the Meantime." And in a letter to Anna, Sophus expressed his agreement: "The Man who has been shown the Door by a secretly engaged Lady has some Basis to deplore his Fate." In addition, the Birch-Reichenwalds were worried that Anna's father, if he grew ill again, would "speak of the Matter among the People". Mrs. Motzfeldt certainly felt there was little danger of that occurring, and, Sophus wrote, "According to the Impression I have of Your Father, it seems to me that such a Thing would be completely unthinkable. You, dear Anna, who know him, would understand if there were any Danger in that Respect." As early as 1864, Anna's father had been admitted to the Gausdal Asylum in Christiania for a period of two months, and upon obtaining leave of absence from his job in Risør, had, from the autumn of 1867, become a patient at Gausdal Asylum for ten months – Cabinet Minister Birch-Reichenwald then paid sixty-eight shillings a day for "better Board" for his brother.

Anna went along with the decision to have the engagement made public at Easter. Sophus answered immediately that he hoped Anna would accompany him to Moss after Easter, and become better acquainted with his sisters Laura and Thea. Otherwise, he was very busy. Work and correspondence had piled up whilst he had been in Moss, busy with his father's funeral arrangements, but, he commented: "For the Time Being, the only Thing that I really *fear* is to lose Your Love. If I can hold on to that, then I am as sure as it is given for a Human Being to be, that the Future shall be bright and good to us both."

He was also able to report that his economic situation had become significantly better following his father's death:

> Now there is no economic Hindrance to our Wedding. It is up to You and Your Family to decide when it shall take Place. Please realize, my dearest Anna, that as great as my Longing is to be able to call you my Wife, that I would never attempt to exercise any untoward Influence over Your Will on this Point.

During the month of January, Sophus had sent Anna several books – most seem to have come from the book collection at home in Moss – but he was not sure that they had "measured up to the appropriate Criteria" that Anna had explained must be found "for deciding whether a Book is enjoyable or not". And he urged her simply to lay the books aside if she felt "they are boring". In addition he had just then seen an illustrated English edition of "Dickens' *Collected Works*, of which 19 or 20 Volumes have come out", and he would send these unless she wrote back immediately to say that she would prefer to have something else: "I myself am still a Wretch when it comes to English (I can only read the scientific Papers, or an easy Newspaper Article); but I have also thought to perfect my Command of this Language: perhaps You at some Point could become my Teaching Master?" And if she did not have a good English dictionary and a grammar, he could also send such. At the end of January, Sophus also wanted to send the ring to her – if the goldsmith kept his promise – and if it did not fit, she was to send it back immediately with instructions about how it should be changed.

Sophus was very concerned with supplying books to Anna – not out of any "Obligation, that I am thinking to settle upon You, but only an Opportunity that I want to obtain for You." Moreover, he felt that she was far too modest in her wants: "You forget to tell me which English Books You want. In the Meanwhile, I am sending You therefore only a classical Work by Goldsmith which You perhaps are already familiar with." He asked again if she would not like to study English, and added:

It seems to me there is an ongoing Error in our social Relations, that our young Ladies, upon leaving School, have become crammed with much more Learning even than desired, and thereafter lay aside all serious Dealings with Books. Do You not agree with me on this? I do not mean to say that our Ladies ought to study the way we do, but only that they, throughout Life, ought to nurse along a Particle of Interest in Books. And if You agree with this, which is quite certainly the case, then the best Beginning is Engagement in one of our Languages of Culture. It is a marvellous Thing to be able to read German, French, English, or some of these Languages with the same Ease as Norwegian. Indeed, the large Nations have such an infinitely longer and more interesting Literature than does our poor little Norway.

At the same time I am sending You a Work by Hartvig, which from a cursory Survey, seems to me to be rather interesting. I myself have only a basic Understanding of Botany, Zoology and Geology; but I have busied myself so much with these Things that I have acquired a lively Interest in Nature and its Wonders. This Spring and Summer, if You will permit me, when we are out walking, I can tell and teach You the little that I know.

In Risør, Anna often visited the home of Bailiff Zwilgmeyer. Here there were four daughters between the ages of ten and twenty-two, and the second eldest of these was Dikken Zwilgmeyer, who later became a striking and well-known writer in Norway. Dikken was a year older than Anna, and the Zwilgmeyer daughters – Dikken and her eldest sister Louise – were Anna's very best friends in the town. They were of the same social background, took part in the same social life, and went out together promenading and flirting with the opposite sex. Bailiff Zwilgmeyer allowed his civil service daughters be educated in domestic pursuits such that they might become "prized goods" on the marriage market, but he also took care that they also attained a certain level of cultural knowledge – apart from preparation of meals, sewing and embroidery, they read books by known writers, they read newspapers, and played the piano. Anna was part of this milieu – her "distinctive criteria of an enjoyable book" certainly came from this circle. Also the discussion about whether a woman's fate and destiny really being "to marry" seems to have been a lively part of this milieu – in any case, Dikken would come to be preoccupied by this topic in her writing. Dikken herself, and two of her sisters, would remain unmarried, and the fourth sister married her cousin, but did so late in life.

Anna had told Louise about her secret engagement, but made her promise not to say anything about it to Dikken. As early as the end of February, however, Anna received a letter from Dikken, congratulating her and wishing her happiness "your whole Life through at the Side of your beloved Professor". But, Dikken commented, "It was mean of you" not to tell immediately about the engagement, and reminded her of how they had discussed together who they would marry – Anna to a priest, and Dikken to a magistrate. Now Anna was to become the wife of a professor instead of a priest, Dikken pointed out, saying that she could hardly grasp the fact that Anna really was engaged. She was pleased to see Anna together with Sophus Lie, and wrote that she would have given anything to have been present, sitting in a corner, while Sophus Lie proposed.

During his Christmas and New Year's stay at Tvedestrand, Sophus had also ordered skis and a sled from the ironworks at Nes, just outside Tvedestrand. And when these items arrived in Christiania by steamboat at the end of January, he had to go out right away and try the sled, about which he reported to Anna: "It seems absolutely perfect and I hope that You will come out sledding with me many Times." He reported on "the Cold and the remarkable Sled Runs" in the capital, and told about an earlier sledding tour with his comrades Ernst Motzfeldt and Axel Bruun, and "a Number of Gentlemen and Ladies":

Otherwise, we Comrades used to go out a Couple of Times a Winter on Saturday Afternoons, a Couple of Norwegian Miles [ca. 24 km.] from the City, and return on Sunday. We went sledding on the Way out and the Way back, and on Saturday Evening we drank Toddy. Ernst was the Soul of these Enterprises, whether they were Sledding Tours, Shooting Parties across the Fjord after Seabirds, or on Hiking Tours, etc.

Apart from this, Sophus Lie's sled from Nes Ironworks would "acquire a certain Fame" in the capital – the finely crafted woodwork and the runners forged from choice iron made the sled both fast and easy to manoeuvre with the steering bar. Sophus explained that he "as a Beginner in the Art of Steering a Sled" was out on several sled outings far from the city, together with comrades and acquaintances – and that his comrades, when it came to impressive "Sledding with Ladies", would come to borrow his sled.

This aside, he had begun to make plans for the summer with Anna, and he thought there were three things they could do: spend a period of time in Moss, and then a period in Ringerike at her Aunt Burchhardt's, and then together with his sister Laura, take a lengthy hiking trip in the interior of Norway. He, as "an experienced Hiker", knew where the finest tours were: up through Valdres, "with little Detours up among Jotunheimen's Mountains, around the lakes, Bygdin and Tyin, and then on to Sogn, and possibly a little more of the Bergen Diocese. A more lovely Tour cannot be taken in Norway." Both in Moss and Ringerike she could, if she so desired, "learn many things, and much" and perfect herself in "the Deeds of the Housewife".

Sophus always sent greetings from his sister Laura in Moss, who still could not write to Anna due to an injury to one arm, something Sophus felt had been a consequence "of a neural Trembling brought on by Father's Death". But he wanted to include in his letter to Anna a portrait of his sister, with the comment: "You see, she is no beautiful Lady", but adding: "You can have full Confidence in Laura, as though she were Your own Mother." When he had sent the letter he discovered that he had forgotten to include the photograph in the envelope, and he explained in the next letter: "I do not think that You really could call me distraught, but I do have too much in my Head. This time I am quite certain that I shall send You her Picture."

He reported on his Christiania relatives, his uncle – his father's only living brother – Dr. John Lie and his wife: "Both of them are kind but a Touch eccentric." And then there was their son, Anatomy-Demonstrator Johannes Lie, "who himself is a quite ordinary Person, but his Wife, a Swedish Lady, Amanda, is an uncommonly engaging Lady, whom You will like immensely. She is about 33 years of age and has 6 Children, but still, she is a remarkably beautiful and handsome Lady."

Sophus had hoped that Anna would come for a visit to Christiania before Easter, and when she reported that it was not convenient for the Birch-Reichenwalds that she stay with them, he tried to arrange things differently, and reassured her: "You can be certain that I shall exert all my Powers of Invention to arrange it" – and he went on to explain the reason for this ardour:

When I so much wanted to have You come here, this was of course partly because I long so intensely to have an Opportunity to speak properly with You; but also because I want exceedingly to be able to show You a pleasant Time. I feel a half Pity for you , because whilst You have become engaged, it is still to a Man whom Many would not consider a young Man. I want so very much for You to have great youthful Joy and Pleasure in the coming Time leading up to our becoming married.

The "Pleasures" that Sophus was thinking of were a "Ball at the Association" and theatre performances whenever she had the desire. As a professor, he had been invited to the formal opening of Parliament, and he confided to Anna that it truly would be "of more than commonly interesting, when certainly the new King would address serious Words to Parliament during these Days of Discord." This was King Oscar II, who now following the death of Karl XV in September 1872, began his reign over the Kingdom of Sweden-Norway. There was particular reason to mention to Anna one of the issues that this session of Parliament would take up:

A Motion will be placed before the forthcoming Parliament to the Effect that Mathematician Abel's Works be published anew by Headmaster Sylow of Fredrikshald [Halden] and I, through the Academy. This being the case, I would get a small Travel Stipend. If this comes about, as I have Reason to assume it shall, then I have a Plan to flesh out, which possibly at first Glance, You will think is a Touch wild, but in any case, You must never speak about it to a Soul. I would like to wait with the Trip Abroad until after we are married, and if You have such a Desire, to take You with me to Paris for a Couple of Weeks. I have always had a Lust to travel, and I have the intense Desire that You, whilst You are still young and have the Interests of Youth, should get to travel and see many Things. The idea is a little audacious; but trust me, I would never have mentioned it to You if I did not consider that in Truth, my Plan could be realized.

About King Oscar's first stay in the Norwegian capital, Sophus was able to report that the King did his utmost "to win over the whole World", and that he had "had no small Luck in this Respect" – every single day one heard that the King had been to one place or another to become familiar with the people, with conditions, and institutions, and one evening he even attended the Student Society Theatre, when Sophus had also been in attendance. Furthermore, he would be together with the king at a meeting of the Academy of the Sciences, but: "I cannot however justly say that I feel the Need to sun myself under the Courtly Gaze."

But it was still early February, and two months until Easter. In Christiania, Sophus held his lectures, and worked away at his scientific work:

My sole University lecturing before the Examinations is composed of 3 afternoon Lectures; apart from this, I teach one Hour a Day at Nissen's School. All the Rest of the Time I can dispose of according to my own Pleasure, and although I never devote especially many hours a day to my Scientific Work, I have, as a Rule, enough Time.

In relation to the traditional, annual market days at the beginning of February, there were some extra holidays, and Sophus reported to Anna that he, then, had been out with some friends "from the Student Society (where we eat together)" on "a terribly long Sled Outing, half a Mile [ca. 6 km] from the City," and in addition: "Now, during the Day we also have the Skating Rink." He hoped that there were equally good winter weather in Risør, and that that being the case, she too was "out on some such Kind of Pleasure" , and asked, "Am I boring when I urge You when finally it comes to the Use of Time, when the Weather is good, to get outdoors and breathe fresh Air?"

The market days were also a time for parties, and Sophus told Anna about "a convivial Evening" at the Motzfeldt home: Ida had been visited by five of her girlfriends, and Sophus named them all – Julle and Alette Wedel, Caroline Wedel, Miss Ihlen and Miss Platou. And about Julle, with whom he liked to joke, he wrote that she, in scarcely a year, had changed from a slight little girl until now she could "compete with any Peasant Girl in Terms of great rosy Cheeks and general Well-being." Nevertheless, he could not understand why a girl as sensible as Alette would use her Time learning domestic science, and he added:

> Is there any one who can explain to me why it is that a young Girl, who is not engaged, and who is considering the Possibility of going through Life unmarried, would not be getting herself a Vocation, be it great or small, so as to provide Life with some Content. Nothing is more awful than to hear such half-grown Girls talking about the Old Maids' Home as a Future. It is no better than being buried alive – I become quite overheated when I think about these Things.

Yet Sophus was not thriving as well at the Motzfeldt house now as he had earlier. Mrs. Motzfeldt was still in a "miserable and wretchedly variable Humour", something that extended out over her nearest surroundings as well as her relatives. To Anna he directly admitted that he considered changing his lodgings, even though he knew that this would anger Mrs. Motzfeldt, and he added: "In the old Days we had exceedingly affable Times during the Evenings within these Walls, but now this seldom happens." Mrs. Motzfeldt certainly noticed that he went "out of his Way for her on every Occasion", and to mollify her he had once invited Ida out for a sledding tour.

In order to size up conditions at home after the death of his father, Sophus had planned to make a trip there. Since he was the only son who lived in the vicinity of Moss, he had taken it upon himself to look after everything that had to do with the dwelling, and he thereupon engaged in "a terribly long-winded Correspondence" with his brothers, brothers-in-law and sisters, even though almost everything was in proper order: "Father was a Man of Order." Sister Laura had still not overcome the "nervous Quaking", and the new, interim vicar was living at Laura's.

> I shall certainly come to use the Legs on the Trip to Moss, or else on the Way back. It is really a boring and monotonous Track to walk, but in part there is one Thing about it, a whole Day to be in Motion, it is an important Pleasure to me; in part because I feel so good thereafter. You see, the Lesson I got last Summer has put Fear into my Blood. [. . .] I never feel as well as when I have hiked 3–4 Miles [36–48 km]; it is then that I become an 18 or 19-Year-Old again. This is something that Ladies seldom have the Opportunity to feel, the Sense of Well-being that real and proper Motion brings.

At the end of February, Sophus reported on the "tiresome Roads deep in Snow, that melts almost as it comes to Earth." He had heard from Anna in Risør that she had been enjoying herself "on Skates and at Balls", and Sophus replied, "Just take care that You do not go out in the Cold when You are too warm. Nothing is more dangerous. Otherwise, I suppose the same Thing has happened to the skating ice in Risør as here, where the Snow has come down on top of it."

Sophus was eager to tell Anna how Klein and Mayer – "both remarkable Friends of mine" – had written to him on the occasion of the engagement, and he translated parts of their letters for Anna. Mayer wrote: "My best and most heartfelt good Wishes to Your Engagement. You shall see that a quite new Life begins when one no longer stands alone; but Another merges with our Life, and fills the Heart just as much, and even more, than the Intellect fills the Study. Then one gets the right Sense of Life, the right Joy of Living." Sophus added to Anna: "I do not need to tell You that I have already for a long time admitted the Truth of Mayer's Words, and that I long for the Time when I even have the Opportunity to experience them."

Klein had also sent his "heartiest best Wishes", and wrote: "It seems to me really unbelievable, that You should overtake me. Do You recall our Discussion on the Hiking Tour from Nürnberg to Fuerth? On that Occasion we were both indeed in agreement that it was a beautiful Thing to find a Harbour in a good Home. And now that You shall do this first, amuses me so much the more, since I, for my Part constantly used to have similar Thoughts going through my Head, and I would immediately say to myself: No, sir, in such a Condition You can no longer work, and Lie would never permit that! Whether I shall soon follow Your Example, I cannot rightly say. That one Day I shall do so however, can easily be taken as a Certainty." Sophus commented for Anna's benefit:

> He [Klein] is a quite handsome and quite uncommonly charming Man. His own Countrywomen are strongly intrigued by him. I have certainly not had much Opportunity to make such Observations, but as You shall see from his Letter, I did not allow him, whenever we were together, to think of Dancing and the like. It seemed to me then that we could spend the Time better. He has thought of coming up to Norway during the Summer. That should be enormously enjoyable for you as well; so, Klein is one of the most charming Persons one could meet.

Sophus also wanted to prepare Anna for the planned family gathering at Tvedestrand at Easter, in relation to making their engagement public. His eldest brother Fredrik and his wife would be coming from Kristiansand – Fredrik had just that year taken up the position of senior teacher at the cathedral school in that city. And Sophus' other brother, John Herman – Lieutenant Lie of Bergen – would also come, perhaps together with his wife. With regard to his sisters-in-law, Sophus commented that the lieutenant's wife, Petra Thaulow Kloumann was "one of the most charming of my relatives"; and in addition, the other sister-in-law, Amalie Nielsen, was "in all Respects a particularly estimable Person; although her Manner was not exactly pleasant, in any case, not upon first Acquaintance."

Sophus sent Anna "Tennyson's Poems" and an English grammar book, but the textbook "One Hundred Hours of English", that she seems to have requested, was temporarily sold out. However, at the Cammermeyer Bookshop he had seen "a Whole Section of English Novels by Well-known Writers", and in the next letter he would give her a number of titles such that she could tell him which she desired:

"believe me, my dear precious Little One, I know of no higher Desire than to bring some Joy to You."

For her part, Anna sent greetings from her girlfriend, Miss Zwilgmeyer, and Sophus asked Anna to greet her on his behalf, and say, as he hoped, could they "soon see her in our House?" The question mark that followed "our House" was a clear signal that he favoured a swift wedding. But this was a very sensitive theme that they did not find it easy to agree upon. Already now in mid-February 1873, Sophus expressed his fear that Anna would be "all too harsh" when he began to speak about the date for the wedding, but, he explained:

> You do not know how intensely I long to be able to conduct You to Your future Home. Believe me, my most Precious One, I shall do Everything, whatever is in my Power, such that You shall thrive.

And a little further on in the same letter: "Do You take me amiss when I urge You to think a *little* of training You as a future Housekeeper? I imagine that You have not exactly done any such great practical Preliminary Studies in this Direction?" Again he mentioned that in the summer she – at Moss and later at Ringerike – could get the chance to learn "Something of these Ways", and he added, "But have You not in Your own Home, quite naturally, taken care of a little Housekeeping? You are not becoming angry at me though? For my part, what dictates my Words is as much the Thought of Your, as well as my own, future Happiness." As a clarification as to why he wrote what he did, he mentioned: "One hears so often of the Cares and Worries that young Wives have, because they did not at the right Time involve themselves in Things." And he concluded, following a sudden thought: "In any case I have cleared my Conscience, after writing the Last. I pray You therefore, my dear Anna, tell me in Your next Letter that You have not become angry with me."

What and how she answered is not known. Sophus, on the other hand, opened his next letter by announcing right away that it would certainly be "a bad Letter" that she received from him this time – for the very moment he was sitting down to write, he got the news that a professor from his student days, Professor Christie, had suddenly and unexpectedly dropped dead at the age of only forty-two. Sophus wrote: "Indeed I have never really been in any close Relation to Christie, who on many Occasions has certainly given me powerful Support; but on such an Occasion as this, one sees so appallingly clearly how uncertain even one's own Calculations are."

This time Sophus spoke neither of "Matters of Housework" nor of "Housekeepers"; rather, he thanked and praised Anna in a profusely formal manner for having written to Mrs. Motzfeldt, as he had asked her: "To be sure, I have always felt that You were the most charming and sweetest Girl that I have ever met; I admit, nay more, repeat: once again You have given me new Proof of this."

Sophus was beginning to prepare for the mountain tour that they were to undertake during the summer. But, according to Sophus, Mrs. Motzfeldt wanted "directly or indirectly" to try to prevent the tour. She maintained that Anna was not physically strong enough for such rigours, and listed for him all the sorrowful cases where ladies, certainly from their own imprudence had contracted serious illnesses on hiking tours – "and including life-threatening Diseases." Sophus had reminded Mrs. Motzfeldt that her own son, Ernst, had taken his wife, Else, with him on several mountain tours without incident – and Else was "certainly no Titan". Moreover, Sophus had planned to take along *one* horse as well, upon which Anna could ride when the weather was bad. And after all, his sister, Laura, who was a "passionate Tourist" would be going too. Sophus also told Anna that he feared Mrs. Motzfeldt would frustrate the plans for the summer tour. "That would be *a great and almost irreplaceable Loss for Us Both.*" Thus, to help Anna appear "sensible and thoughtful" so that this would not happen, Sophus wrote in detail and encouragingly about the trip: "So long as we have only good Weather, You shall see that this will be the most enjoyable Journey You have ever taken." The plan was to travel "by Rail and Steamship through Ringerike and the lakes, Randsfjord or Spirilen. From there, they would take the great trunk road up through the districts of Valdres and eastern Slidre, and from the top of Valdres they would "wander for two or three Days in among the High Mountains, that here show themselves off in their full Glory." And for the further journey he made the following forecast:

You must not imagine Yourself having to become some rigorous Athlete. Our Mountain Tour will come to consist of: $\frac{1}{2}$ Mile of flat Mountain, then the first Cabin. 2 Miles along Lake Bygdin, where there are 4–5 remarkable Cabins with considerable Amenities; $\frac{1}{2}$ Mile flat Mountain to a new Cabin; $\frac{5}{4}$ Mile along Lake Tyen where there are several summer Farmsteads: $\frac{1}{2}$ Mile flat Mountain to the Highway. And as mentioned already, we shall have a Horse with a Lady's Saddle, and of course, a Guide.

He reassured Anna that he himself was a very experienced mountaineer, and that no one was as worried as he that "anything could happen" to her. Then he continued to describe the trip further: "Down to Sogn and the County of Bergen we will hopefully settle into a little Tranquillity at one or another of the many lovely Places along Sognefjord."

Anna seems to have appreciated Sophus's plans and care. In any case, he began his next letter as follows: "My dearest Anna, A thousand thanks for Your good and dear Letter. You must understand that it pleases me quite extraordinarily, that You have just as great a Desire as I that our great Summer Tour be realized." He was also happy that she would not be afraid of "those small Rigours that to be sure could not be avoided" – when she came to Moss she would be able to practice riding: "It is extraordinarily enjoyable, it seems to me, to sit a galloping Horse."

The time for making public the engagement began "fortunately to pull nearer", and he asked if she would not give him permission to tell some of his closest friends about it; for example, Axel and Barbra Motzfeldt, who were as well close family – and in any case he wanted to invite them to the theatre the following week and tell them about it. There were also a number of other friends – Axel Bruun, Axel Lund and Otto Aubert were named – whom he, "by Observation of the necessary Forsight", was eager to tell. In any case, Anna could not really have anything against him ordering a number of visiting cards with her name on them, such that when they should communicate the betrothal to relatives and friends, they had only to put her visiting card and his in an envelope together: "Certainly though, it is quite common to announce an engagement in such a Manner? Naturally at the same

Time we can of course write to them in detail, as we feel obliged, or so have the Desire." But what did she feel about it? Could he put the visiting cards in envelopes and send them off at the beginning of April, just as he was leaving Christiania to meet her in Risør? And anyway, could she send the addresses for those of her friends she wished "should get the official Communiqué of the Engagement"?

Anna must have proposed ordering one visiting card with both their names printed upon it, and Sophus replied: "I shall be so gallant as to set Your Name first." He would post them the moment he was about to board the steamship in Christiania on the way to Risør and Tvedestrand – on the afternoon when the ship arrived in Risør, Anna should come aboard, and after her departure from Risør, her friends in Risør, too, would get the letters with "the Information concerning the Engagement". There was certainly very little possibility that the ice would have gone out of Tvedestrandfjord by Easter, and therefore the ship would not be able to call at Tvedestrand. If that were the case, they would have to drive from Risør, and he reminded her that certainly from now on he would pay for all her travel disbursements. But should it happen that he had to go ashore in Risør, then she must "*definitively not* meet me at the Wharf, that is, if, as I consider extremely understandable, You feel in any Way embarassed. I can well appreciate that if You wait at the Wharf, Your Fellow Townsmen would, in one Way or another, plague You with their Indiscretion." Instead she could rest quite assured that not so many moments after the ship had docked, he would be up at her house. But he had a little request: "Should there be bad Weather, and, poor Wretch that I am, I get severe Pain in the Head, as unfortunately frequently happens to me at Sea, You must bear with me. I have been created a Landlubber: there I am no greater Wretch than most others." And if conditions were such that they had to drive from Risør to Tvedestrand, he asked her to think a little about how to obtain a horse and wagon – indeed, in such a large town as Risør it must be possible to get "a better Conveyance than the usual Post-Chaise"?

During the last four weeks leading to Easter, Sophus wrote four or five letters to Anna, and when he did not receive replies at the expected time, he became worried: "It is certainly stupid of me, I do not know whether I am plagued by Nervousness; but I do not quite come back into Equlibrium again until I have been soothed by a Letter from You." Anna reported from Risør that her father was in an unstable mood. Sophus expressed understanding of how distressing it must be to witness "a Father's Health and Spirits being so unreliable", and added, "But *in these Matters* certainly People cannot help at all." From the couple of chats Sophus had had with her father he would certainly not have had second thoughts about talking to him about "any Subject under the Sun", but he deferred to her to decide what now ought to be said to him and what ought to be kept quiet in order to spare him unnecessary worry. That her father did not like the fact that Anna herself was away from home for short periods, made Sophus worry about what his reaction would be when he led her to the altar. He freely admitted that he envied those of his friends who had already settled down – in domestic comfort, "the Comforts of Home that I, in that

Sense, have never known, but that I at the same time of course am certain I shall find at Your Side."

For Sophus this was "a clear Issue" and his "highest Desire" was that the wedding could take place as soon as possible. In any case, they had to speak about the date for the wedding when they met at Easter, and later when she came to Moss they had to "think of the Acquisition of Everything required for a new House." Since he did not understand anything about such matters, and wrote that he "had to rely upon You" – but he consoled himself with the fact that he had always managed to be involved in "Everything that was necessary." He was astonished that they were "in agreement in Principle about the Acquisition of Household Effects and such". As he saw it, newly married people all too often lived beyond their means; they let "Vanity run away with them", and thought only about what could be seen as "magnificent and elegant in the Eyes of Guests". What he himself would like to be able to afford that many would call "a Luxury", was namely the ability to travel: "Travel, especially in Summer is a Benefit for Women and for Men, for the Married as well as the Unmarried." If they "held the wedding in the course of the autumn (and it could not be any later!!!)", they could subsequently spend the first part of the winter in Christiania, and then travel – "under the Assumption that Parliament would grant Money for the Abel Works" – to Paris, where they would remain throughout the spring: "Spring is the loveliest Time to be in Paris . . . Nothing is prettier than to see the lovely fresh spring Green sprouting forth in the Middle of the great City."

From Risør Anna reported on tiresome whisperings and flutterings about whether it was true that she was engaged – the postmaster had also intimated something in that direction. Sophus urgently insisted that "these ugly People" should not prevent her from writing letters to him, and felt that "the enormous Interest in *Others'* Engagements" came from the fact that "People in common are so mentally impoverished and have Little or Nothing in which to be interested."

But now in the capital as well, there circulated unpleasant rumours about their engagement. The oracles from which these rumours stemmed were explained by Sophus: "An Acquaintance of mine, the Historian, Ernst Sars, has just spent some Weeks in Halden, where he had been invited to give a Series of Lectures for the educated General Public of that Place", and one evening when Sars was at a party, an unknown lady had come up to him and begun to speak "a great Deal" about Sophus Lie without Sars really knowing why. However in the end this lady, whom Sophus, from what Sars had said, thought must be one of Anna's girlfriends, told with conviction that she knew that Sophus Lie was engaged to Anna Birch. Sars had then thought that the engagement was public, and brought the news with him back to Christiania – to the Student Society and "The Green Room", and when precisely that day Sophus had not eaten dinner there, it was said that he had already left for Risør. In this letter to Anna, Sophus wrote: "Of course the next Day it was an easy Matter for me to lead them to other Thoughts; and I thought this Rumour had been quelled at Birth" – they all allowed themselves to be taken in by my honest Face." But if it really were such that it was one of Anna's girlfriends who had spoken to Sars – and if Anna thought she had to write to this erstwhile friend in Halden about this – then Sophus would urge her to be "as mollifying as possible": "To be sure it is of course shameful to abuse your Confidence, but People are after all *weak* on this Point."

Later Sophus got to know that it was Louise Zwilgmeyer who had spoken to Sars in Halden.

Otherwise, from the capital, Sophus was able to report that the middle of the month of March was wonderful "beautiful Winter Weather with warm Sunshine", and he recounted enthusiastically about a happy outing with the sled: First they had "Ladies and Gentlemen, in all, 15–16 Pairs" who gathered around twelve o'clock and ate breakfast at the home of Assessor Bruun, and from there they had gone in ten or twelve carriages about five and a half kilometres to a "great Hill up in eastern Aker County." And in "the most lovely Sunshine" they had been sledding for three or four hours: "It must be admitted that all did not go remarkably well, for there was far too much Snow. But when one goes sledding with Ladies, it is as a Rule best not to go too fast, especially when one like me is no Master of the fine Steering Mechanism." Sophus continued: "We enjoyed ourselves perfectly, and just think! not once did a gown or any finery come to be damaged." Both before and after sledding they had been into a farm belonging to some of Bruun's relatives, and been treated to "Wine and Cake": "So You can well understand that we fell into an animated Atmosphere." That evening, when they were back in the city, everyone had gone his way to make himself presentable for dinner and an evening dance. But Sophus, for his part, had "run up to the University and given a Lecture"; to Anna, he commented, "If the Truth be known, I am strongly inclined to suspect that my Audience found a bit of the Sledding Expedition's Mood in my Lecture, which at least was none the worse for it."

Then after the lecture, Sophus too took part in "a truly delightful Ball, a truly youthful Ball" (he was the only gentleman present over the age of twenty-three), and he related to Anna that one of her girlfriends – "Mrs. Nissen (née Prebensen), was one of the ten wives present. And precisely because she was Anna's friend, Sophus tried to make himself "as charming as possible in her Eyes," and he thought that he had "succeeded well."

When Anna did not comment on all that Sophus wrote about, Sophus interpreted her "Silence" to his "Advantage"; and in one way, what she felt about his letters, made him – as he wrote to her – "Certain that You have not misunderstood their Basic Tone." But all his talk and pressure for a rapid wedding had led Anna to express her "distress", and thus he replied promptly:

> I will truly allow You to have full Freedom. I shall completely give up Thoughts of Marriage until Autumn, if You do not change your Opinion on the Matter. [. . .] In such an important Matter, I would not exert the least improper Influence on Your free Self-Determination.

For all that, he went on to tell that Anna's aunts in Christiania – Mrs. Motzfeldt and Jacobine Birch-Reichenwald – who had on one occasion brought up the subject of the engagement, had said that they felt it was unnecessary to wait until winter for the wedding. He had answered both her aunts, saying that if Anna herself felt that she was too young, he would not "utilize too strong Persuasions", and he reiterated in his letter to her: "If You, when we talk about this at Easter, express it as Your Desire, that we wait until Winter, *then I shall yield*."

Jacobine Birch-Reichenwald now wrote Anna a letter in which she expressed her great joy over what had happened: Anna should give thanks to God for being allowed entry into such an amiable and intelligent family. Aunt Jacobine told about Sophus' sisters, who were completely rare, charming, bright, and good human beings who could help and support Anna in everything, and who had much they could teach her. Anna must be pleased to be able to follow Sophus and Laura to Mathilde in Tvedestrand at Easter. This was not the first time that Aunt Jacobine gave advice to Anna. Three or four years earlier Anna had told Aunt Jacobine that she wanted very much to leave school in Risør and be confirmed at once. On that occasion Anna got back a clear message that it was absolutely too soon, and a sincere wish that she must try again to recapture the desire to learn more. The reason that Anna had lost all desire for further schooling was, as far as Aunt Jacobine was concerned, that Anna read far too many novels, and she had thereupon urged Anna to speak to Mrs. Zwilgmeyer on the matter.

Easter was approaching. At the end of March, Sophus wrote that he felt it was "infinitely strange" that in only one week's time he would be meeting Anna again, and that then they could finally talk together without fear that outsiders would "see this and make their Observations thereupon". He wrote to her: "I am happy to the very Core of my Soul about this beautiful Time. I am so very pleased to bind You ever closer to me."

Somewhere along the way Sophus had his expectations fulfilled.

But in the first letters he wrote to her after Easter it is evident that various disagreements had come into being between the newly engaged couple. And once again this had to do – in any case, on the surface – with the timing of the wedding:

Dearest Anna!

First, my deepest, innermost Thanks for the last Week which, thanks to Your Amiability shall always remain a cherished Memory for me. Thanks, my own dearest One, for all the Signs of Your Love that You gave me, and forgive me, if You really feel that I occasionally was much too importunate and audacious in my Desires and Longings. But also, do really think over the Things upon which we were not quite in Agreement, really it was so absurd on my Part, as indeed You once felt. So, my own dearest and best Friend, if You arrive, after *mature* Consideration, at the same Result now as at Easter, then even though I find it so difficult, indeed, painful, I shall try to become, as we jokingly called it, more *humble*. In any case, write to me, my dearest, most cherished Friend, that at any rate You do not think less of me than before this. Write and tell me that You feel that I, in all major and substantial Respects, am such as You believed and desired me to be. If You do this, my dear Anna, then You are truly a sweet kind Girl.

Anna had stayed on for a while at Tvedestrand after Easter, and after Sophus had returned to Christiania. She enjoyed herself at the home of the Vogt family, and she had also come to an understanding with Sophus' sister Laura. Anna wrote to "her mother" in Risør that the Vogts had "a lovely House, it too is in the Swiss style, but larger and prettier than ours." She explained that the children "ceaselessly" called her "Aunt Anna" – "they loved me so much." During the whole time, Anna had shared a room with the eldest daughter, Lella, and the two became "terribly good Friends". According to Sophus, Lella (Eleonora), who was sixteen and about to be confirmed, had already expressed the desire to come and live with them in Christiania when Sophus and Anna were finally married.

For his part Sophus had stopped in Risør on his way back to Christiania, had spoken with Anna's mother, and come to understand that she was in agreement with him in respect to "the Time for our Wedding", but added again: "But naturally, my own dear One, if You maintain Your Decision, then, as painful as really it would be, I must remain at the Mercy of Your Will." He also expressed the hope that they might speak further about this when she, in the near future, came to Moss, and he hoped she, as before, would write "well and often" to him: "Do not be afraid to allow a little *Affection* to come out once in a while. Were I to have had the requisite Permission, I would have sent You Kisses by the Hundreds in the post, and demanded the Same by way of Reciprocity." And he concluded: "In any case, do not be afraid of me because I so utterly long to have You constantly close to me."

Before Anna had received this letter she had sent one to Sophus. And in his next letter he consequently had to admit it had been kindly done. But he had ceased to wonder at her kindness, as it seemed to be something that "came by itself". When it came to the question of the wedding, he seemed still to feel she was being absurd: "You are a strong little Lady; but I hope that Your charming Qualities will become more strongly developed than your Strength. He thought it was almost "shameful" that she had not admitted that she missed him, but of course he believed she wrote that way so that he would not have an impossibly inflated sense of himself. But she well knew he was certainly not humble in many matters, but she had to be aware of the fact that in relation to her, he was "the humblest Man in the World."

There was something else that caused him to worry. She had written that she had enjoyed a *journey by carriage* from Tvedestrand to Arendal, and Sophus reminded her that she had not been willing to drive together with him to Risør because she felt "it was so unpleasant" *to drive*. And he commented: "You see, I am catching You in a Self-Contradiction; I ought really to sentence You to a Fine, in a Manner which I believe You know, when You come to Moss. But I shall be as kind as You – *as a Rule* – tend to be, and believe that there were respectable Grounds, that caused Your Reluctance to drive with me."

It was as though Sophus had a new life in the capital after Easter. Earlier he had been "a Hater of Visitors" – "Visitors" were something that he had only considered as an external matter of form – but he was now sincerely pleased by the many who visited him and wished him every success with his betrothal. He had come to hear from several friends that they had long before given up all hope that he would ever become engaged – and, he explained to Anna:

Indeed, the Age of 30 is really not so very old for a newly-engaged Man. But the Thing is, I was already at the Age of 16 or 17 so terribly fully developed mentally and physically. Thus a long Time has passed since I became a full-grown Man, and therefore one is ready to believe I am considerably older. Even at the time I was 18–20 Years old, I was such; in any case, when I was at Moss or out on my Travels, I was as youthful and merry as anyone. At that Time I danced, and indeed, I courted, to my poor Ability, one or another of my old Lady Friends.

> But then came a Number of Years when I, so to say, neither dealt with Balls nor Amusements; when at least I never paid any Attention to young Ladies. And thus my Friends gave up the Belief that I would ever fall in love and become engaged. Thanks to You, my dearest One, Things have turned out otherwise. And it is my firm Hope that You shall see that in my strong Nature there still remains a not so little Fund of youthful Desire, of youthful Power. Dear Anna, You get to take me in Hand and correct me whenever You feel that I begin to be old and boring, then You shall see, it will all work out well.

Making public the engagement, with the subsequent visits and parties had disturbed him "a Bit", and he wrote to Anna that he therefore had to work "all the harder" in all the time remaining. First and foremost there were some papers to prepare – that he had promised to have already edited before Easter – for the Academy of Sciences in Christiania.

Other tasks that laid seige to his time were those involved in his collaboration with Sylow for the publication of Abel's collected works. Sylow lived at Halden, and they met at the end of April at the halfway point; that is, at Moss, to set up provisional plans for the work.

An incident had put him in a low Humour, and in the midst of things, caused him to write a rightly stupid Letter to his dearest Anna; this situation was a conflict that had flared up around one of his best Friends, Amund Helland. Helland was someone that Anna too, would later come to regard as a friend. Sophus now explained to Anna that Helland was a young Candidate in Mineralogy, who, with remarkable Skill, had always produced valuable Works in his field. But unfortunately Helland had entered a Sort of scientific Feud with Professor Kjerulf, and in this feud, according to Lie, Helland had "perhaps not observed all the due Respect, that narrow-minded People felt a young Man owed a Professor." Be that as it may, this led to a situation wherein conservative elements in Christiania had placed obstacles in the way of Helland's career in science. Among other things, the dispute had to do with the view on how glaciers had formed the topographical surface of the country. According to Helland, it was water and the erosion by glaciers that, through the ages, had formed fjords, valleys and lakes. To those of the more established scientific milieu around the geology professor, Theodor Kjerulf, the fjords, valleys and lakes were more likely cracks in the crust of the earth. Sophus Lie was one of those who strenuously supported Helland, and Lie seems to have identified his friend's struggle for new scientific insights, with his own. He told Anna about "a great Battle in the Academy of Sciences" before Easter, when he, in a discussion, had won "a brilliant Victory", adding:

> But our Opponents work in Silence with insinuating Innuendos, such that Things certainly appear doubtful. Nevertheless, I shall win in the End. *Forward march!* has always been my motto. I have won many Victories here in this Life, but almost always they have begun with a Defeat; therefore, if a Defeat antagonises me so much, it simultaneously urges me on.

He was also able to report to Anna that he had had great happiness on another front. "You should know that my scientific Papers are steadily meeting with more and more Success Abroad, most particularly in Germany, where personally I am most known."

Anna had her birthday on April 24th, and in relation to this, Sophus sent her three letters in all, with the "deepest of Congratulations" – and at the same time he asked her if she knew of anyone else who wrote so many letters of congratulations. He was convinced that she would be happy to get so many letters from him on her birthday, and wrote:

> So reward me by writing a really affectionate Letter. Tell me that You are so happy in Your Betrothal, and that You are thinking with such Joy of the Future at my Side. If You do this, then You are my own Sweet Girl as always. Do You feel that it is mistrustful on my Part to ask You to do this? You must certainly not feel this. Believe me, all Men who are so smitten by their Lady Loves, like nothing better than to get such Letters, always, again and again.
>
> I have an ugly Suspicion that all my Letters have a surprising Similarity to them. But in any case, it is certainly not an unpleasant Similarity. Is that not so?

He was very happy that she was now nineteen years of age: "18 Years Old sounded so frightfully young, when I at the same Moment must confess to my 30 Years. 30 and 19, it seems to me, go quite well together, although I would not have begrudged You a younger Man. But indeed, I am not going to let You go for that Kind of Reason." Quite the reverse; he felt it was quite remarkable how very much younger he felt himself to have become "both in Mind and Skin, since I got such a dear, lovely young Sweetheart." For a birthday present she would get "a little Gold Bauble", but since he was not sure exactly what she liked – "You have better Taste than I; I know nothing when it comes to such Things" – she would receive instead, ten speciedaler with which to buy herself whatever she wished when she came to Christiania.

Even now with the public announcement of the engagement, Anna's suitors in Risør were not completely out of the picture. Sophus wrote: "It makes me happy that Your Worshipper, Pedersen, has fallen into a state of Apathy. It would bring me the utmost of Pleasure were the same Thing were to happen to J. P. and M. P. But this is certainly not likely to occur."

Sophus loved every letter he received from her: "How very much every single amiable loving Word from You is a Ray of Light along the Road for me" – and he hoped that the time would soon come when he would always be able to "obtain Light and Hope from Your dear, good Eyes." And as a P.S., he wrote "Will You send me a Kiss in Your next Letter? A thousand happy Wishes, and thousand Kisses, Your Sophus."

The plan was that they should meet again in Moss at Pentacost, and perhaps in Christiania before then: "Do You also not long just a little tiny Bit to meet me again? Indeed, You well enough know that I have sworn in all Ways to be as amiable as it is possible for me to be."

Sophus tried to keep up the same tenor of correspondence throughout this waiting period, and he wrote letter after letter even though he did not always know for sure what he would say: "The Thing is, my Thoughts constantly wander to You and it is always my great Need to tell You this." He urged her to be indulgent toward "the Thinness of my Letters", and also admitted to "an egotistical Ulterior Motive": by writing often he could also hope for more frequent letters from her. And he expressed surprise that he had written "a whole little Letter" when, as he began, he only intended to write "a Couple of Words to her: "Is this not remarkable. What do You think is the Cause of it?" At any rate, "Now You are certainly no longer in any Doubt that I am so very fond of You, just the way others are of their dearest Loves. You are indeed, Child?"

Otherwise, he was busy and had much to do – editing and proofreading his own mathematical papers, and as well, the collaborative work with Sylow on Abel's works was very demanding of his time. In addition he helped Professor Broch with advice concerning the professorial position at the University of Lund in Sweden, the position that he himself had applied for the previous year. Broch wanted to know what Lie thought, since one of the applicants worked "in Things that have some similarity to Things that I myself am engaged in" – as Sophus explained to Anna. But at the same time he reassured her that he put writing letters to her above all else: "And yet You still speak as though I put Mathematics before You. Do You have any Doubt but that these can be well reconciled? Or do You think that at some Point in Time You will become jealous of it?"

In the course of reading one of her letters, he was as well "quite proud of what a learned little Wife" he would be getting. In particular he recalled the occasion when they spoke about a French baron who was coming to Risør and she had put a "quite correct Accent over his Name", and, Sophus continued: "You certainly have not been complimented very often about your Learning, so I hope my compliment pleases You."

With the Help of the Lord, we shall become two justly happy Married Folk. I feel so secure in this Respect. And think I am beginning to believe, that only terribly rarely shall there be Occasions when I become intense and nasty. That I shall always come to love You is of course certain.

But Anna certainly wanted to have no discussion on the wedding and marriage, even now, and she seemed also to have reacted to his keenness to get her to Christiania. Be that as it may, he began his next letter:

Dearest Anna!
You quite certainly know that I am not exactly enraptured by Your Letter, (as cherished as it is in and for itself, as always to me), in which You report to me Your Parents' Desire for You to remain a While longer in Risør. My first Thought was to telegraph You and implore You still, all the same, to come, but indeed, upon reflection, I must admit Your Parents have Demands upon You, that I shall not the least deny.

He had thought to say in the telegramme that she had to act according to her "own best Judgement", but that he "longed powerfully" for her. But in the end, for all that, he came to the conclusion that there was "really Nothing to do", and about this conclusion, he wrote:

You know – I feel You have to regard this as a Sign I am almost too *devoted*, that I find myself so patiently suffering this tough Fate. (I also manifested a Sign of this superordinately high Degree of Devotion on that Occasion at Christmas. You were *cruel* enough to forbid me to set my homeward Course via Risør.) In any case, my dearest Anna! It was a great Sorrow to me that so much Time must pass before I get to see You once more. But then, when we do meet, will You, indeed, be really quite charming and *kind*?

Only think, I feel it is almost a *little* ironic that You, after having given me such miserable Tidings, then express the Hope that in the Future I shall always be in Bubbling Spirits. In other respects, it is to the greatest Degree up to You to keep me in a constantly good Humour; one of the first Stipulations is that You come soon to Moss. [. . .]

Anna, my Dearest ! Are You not so terribly naughty with Respect to the Timing for our Wedding? Do You not know that a young Girl who is engaged, should be pliant and kind in this regard?

He again explained that he had put at their disposal all that they needed by way of "Money for Equipment and Furnishings and what not, all together", and she should be aware that "the Arrangement of Things should not be Cause for any Difficulty." He reminded her about how he had earlier in a letter clarified how devoted he had been on various occasions, and maintained:

Now You too, be so good as to reciprocate devotedly and indulgently on the Issue of the Determination of the Time of our Wedding. You know that You by your Nature are predisposed to being kind and charming. And that, indeed, You are also quite willing to be. Do not be angry with me, dearest Anna, when I always return to this Thing. After all, in this You must only see Signs of how very much I love You and how intensely I long for that lovely Time, when I shall always have You near me.

He urged her to write and tell him more about herself and what she was up to, and how she was: was she in good health again after her cold? Did she go out often and visit people, and was she "able to go out for a Stroll"? As for himself, he wrote: "I myself can never write much of such Things, for my Life goes along so monotonously. My Health never fails at all; I meet the same People everyday; I seldom go out to Social Functions (on this Point I can see myself becoming another Person, when I become a happy Husband); I never go to the Theatre, and so on." The only Social Life that with any sort of regularity he took part in, was once a week when he went to a little Amusement Club that a number of his friends had established – and in the course of the spring they took "small Excursions out around the islands, which is extraordinarily enjoyable."

Sophus asked Anna to convey his greetings to her friend, Miss Zwilgmeyer and apologized for not having had the opportunity to make her acquaintance: "I am conceited enough to feel People get a better Impression of me by talking to me than by merely seeing me." In a subsequent letter he commented upon Anna's relationship to Miss. Zwilgmeyer as follows: "I shall not delve into the Secrets You have with her."

Apart from this he was able to report that, together with his cousin, Anatomy Demonstrator Lie and his wife Amanda, he had thought of buying a house, that his brother-in-law Vogt, would come and give him advice on the matter, but a final decision could not be taken until Anna had come to the city. He concluded: "Write me a properly charming Letter; write that You are coming soon, and that You are thinking a little about coming around to my Point of View with Respect to the *Autumn*."

Anna at least must have answered that she would be arriving in Moss the week before Whitsun, and Sophus rapidly thanked her for her "dear Letter":

How pleasant it shall be to stroll together out in beautiful Nature and talk about the Future and our Aspirations. I think that since I became engaged to You, that the Future promises to stretch out in front of me, bright and clear. When we meet at Moss, we will certainly go walking together many times at Gernerlund, a lovely big Grove with Pathways and Benches; of course it is located a bit of a Stretch from the City: but of course we have the Advantage that not many People meet there. I shall do my Best not to beg for so very many Kisses. But, oh my Anna, I would so much like You to reciprocate and try to be even a little more generous in this Direction than You, curiously enough, have until now wanted to be. You shall see, on this Matter we shall become quite in Agreement. If only You would once in a while let it occur to You to give me a Kiss, then we would be in basic Agreement. But I must very much fear that when You read these Lines, You will clap your Hands together in Surprise that I even dare think the Time shall come, that You do not have so terribly much against letting me kiss You; and even that You should allow yourself to kiss me back. See to it, my own dear Anna! Such equally strange Things have happened many Times before now.

He reminded her that when he had proposed, she had felt "only that the Matter of being engaged, was quite horrible." Now she had gotten over that, but he feared that the next thing would be that she was "cruel enough to say" that it was best they "waited and always waited with the Wedding". Nevertheless he hoped fervently that after the stay in Moss, would "this Thing about forsaking Home" and living with him forever, fade away "as somewhat less appalling" than she for the time being was feeling.

And *à propos* the purchase of the house: Perhaps here as well it would be best to wait until they were married, and in addition, he knew just as little as she about being a house owner – but he had his plans and expectations:

I am extremely curious, about how it will go with our Housekeeping. Unfortunately I have no Sense of Order; but I console myself that I *can* learn Everything that is essential, and in a House, Order is the first Requirement. Fundamentally I have a very great Energy, once I have decided to do Something. But You, my dear Anna, are of course very neat and tidy? Are You not a little of the Opinion that at the Beginning You shall have to be a bit lenient with me, since I promise Improvement on this Point also. You see, I have the most noble good Intentions.

Otherwise, had she read anything in English recently? How was her instruction in French progressing, and should he send her some books? Had she perfected herself "as a prospective Housewife"? In any case she should not tell herself that there was no haste about this! She ought rather tell herself – as was indeed half true – that she was already half-educated! Thus in the end she must not answer that his letters were boring. Did she not recall the occasion in Tvedestrand when he had urged her not to think less of him because he was not "capable of being more amusing". Then she had answered that he did not have to be amusing. Sophus summed up as follows: "And I shall try to be as little boring as it is possible for me to be."

The days he and Anna were together at Moss before and after Pentacost seem to have been good, despite the same incongruities regarding the wedding, and so forth. In his first letter following these days together, he thanked her "for all the Courtesy" she had shown him. He urged her to take care of the cold she had contracted – even though a cold was not "such a dangerous Thing", they both would suffer if her "Cold were to develop its Dimensions", and he reminded her that she was not "some Titan of either Health or Physical Constitution; and Life demands Strength among Women as among Men." Therefore she ought to use the summer to build up her strength for the future, and he added, "You certainly must feel that I am boring with my Moralizing; but your Reason will tell You that I am right, and that You can quite feel that I am guided by good Motives when I write as I have done."

In a later letter he commented upon Pentacost, thus:

> You feel certain that I am graceless because I always ask You for more than You can give, because I always speak about the beautiful Future which is coming, that I am sure of, for there is such intense Sympathy between us, that the Wishes of the One immediately and unconsciously become the Other's.

He was thankful for all the signs of love that she had given him, and maintained: "And moreover it is now after all, Nature's Order that Man should ask, and Woman give."

With regard to the timing of the wedding, he seems to have made "an almost inconceivable Concession": it had been "an endless great Sorrow, to give up Hope definitively to have been permanently together with You this autumn. Admit that I have kept my Word honorably, that You would get your full Freedom."

They had agreed not to discuss the wedding any more for a while, but Sophus did not manage to refrain from writing: "You have filled my Soul, that almost all my Thoughts and Expectations for the Future are based upon You, and in any case, tied to You."

He wrote about the household furnishings they would be getting. It was his opinion that 1,000 speciedaler was more than sufficient, and cited examples from his circle of friends: Axel Bruun, who had a great number of rooms, a piano, and costly furnishings, had spent 1,400 speciedaler. Axel Lund, who also had many rooms and elegant furnishings, spent 1,100, and these sums also included his spouse's furnishings. Sophus commented that both these families were "Rich Men's Children". How much Ernst Motzfeldt used, he did not know, but he thought that inclusive of Else's furnishings, Motzfeldt had spent 1,000 spd. – and, Sophus wrote to Anna: *"In any case You must not be afraid that I shall be too niggardly*. I consider it only correct to be sober in this Regard. Believe me, there is a greater Danger that I shall fulfil too many rather than too few of your Wishes in this Field." He had also looked into providing the security of a widow's pension – perhaps 800 speciedaler would be sufficient, but they would have to see as time progressed, and he added, "Unfortunately most or almost all of my Friends are in better Postings than I. Will it bother You that we have to live more economically?"

If Lie did not feel himself at the top of things economically, with regard to mathematics he was certainly on the way up. He wrote to Anna at the beginning of July: "Taken as a Whole, Everything is going extraordinarily well in my scientific Work. I am meeting with more Appreciation than I could have expected." He told her: "about this pleasant Surprise" in a letter he had just received from Darboux in Paris with the information that he had been chosen for membership of a scientific association of that place, the *Philomatique*, and, in parentheses, mentioned that he had quite certainly told her before Christmas that he had been made a member of "the famous Academy of Sciences of Göttingen". Darboux had written in his letter that he had also been eager to have Felix Klein made a member of *Philomatique*, but that there remained ill-will at all levels in France toward the German nation. And with a thought to Lie's possible visit to Paris, Darboux informed him that since the war the cost of living in Paris had increased by nineteen to twenty percent, and that prices were constantly increasing. Darboux also reported that he had still not found time to write anything about Lie's "Berührungs-Transformationen". This was a twenty-seven-page treatise that had been submitted a month earlier to the Academy of Sciences, but that Lie, because he felt things moved too slowly, had gotten published at his own expense.

The summer of 1873 arrived, and it was time for the great tour into the mountains: from Christiania, up through the County of Valdres, and into the high mountain country of Jotunheimen, then down to Sognefjord – together with Anna and sister Laura. In the last letter Sophus wrote to his fiancée before she forsook her home in Risør to meet him, he wrote: "Keep well, my best and dearest Friend, and I hope that our next Meeting and the whole Foot Tour shall not be obscured by Clouds, that unfortunately we encountered last Time."

There are very few traces of how things progressed and the details of the travel route Sophus followed. But that they reached the mountain wilds and overnighted at the Norwegian Tourist Association's cabins at Tyin and Gjende, seems quite certain. (The tourist association, DNT, had been founded only five years earlier, in 1868). A dispute broke out between the keeper of these cabins, Ivar Beito, and Sophus Lie. This dispute consequently resulted in Sophus Lie having to meet later in the autumn with the Tourist Association's executive "to be confronted by Ivar Beito", and about the front line of this skirmish, he then commented to Anna: "Pleasant it will not be." But what the divergencies of opinion were, there is no mention. But perhaps it had something to do with a debate that broke out some days later during that autumn in the Tourist Association's annual general meeting which, by the way, was held at the Student Society. There were two points which provoked particular discussion at this meeting: should members have preference for lodgings "in Circumstances of Competition" with non-members, and should members even be required to pay for staying overnight at the Association's cabins? After a very basic exchange of words, it was decided that a charge ought to be "laid on every traveller", but that the Association's members ought to "be given preferential Treatment in obtaining places."

Sophus reminded Anna in a letter some months later that the red-bearded man who had rowed them across Tyin Lake and conducted them to Nystuen – the man had appeared in the capital that winter with a wood grouse that he, according to Sophus, would have given to Anna had she been there.

Another "travel memory" quite certainly from the summer of 1873, seems to have been Sophus and Anna's stay with Amund Helland – in a later exchange of letters, Sophus sent both photographs and frequent greetings to her from Helland – about her "Friend Helland" Sophus constantly had something to tell Anna. And Helland's enthusiastic descriptions of the mountain wilds of Norway seem to have been something that both Sophus and Anna identified with. In the most recent yearbook of the Tourist Association, Helland also wrote "About the Mountains" in a way which resonated with the public's view of the landscapes inherent qualities:

The Mountains open a view of Time, like the Stars in Space; one does not measure the Vastness of Sirius with a Tape Measure, and the Age of the Mountains demands a greater Unit of Measure than Years. There is much Wisdom to be obtained from the Mountains; it is a long history over which light is cast. [. . .] Wherever there is an open outlook over a region, there from time to time one can simply get intimations of the history of the formation of the mountains in that region; as with a human face, they can very often be read. Every single stone builds the land according to its nature: flat or undulating forms most frequently imply the stratified, slow accumulation of the ocean's sediment building up over long stretches; but suddenly there rears up a steep, mighty, mountain wall, that towers over this flat land such that you have a foreboding of volcanic stone and the catastrophe. At the highest and boldest of these mountain forms one often finds the suggestion of revolution. When one is on the high peak, it feels as though one is having great thoughts that prefer to come forth through revolution. When a mountain is being born, things certainly do not always proceed gently. A place must be made for everything that demands to advance; the old stones must be cast asunder, bent and cracked; poisonous acids, metallic steam, and other infernal devilries burst forth with the same, and many an innocent animal or lovely plant that probably deserves to live, goes down the drain; but when that peak is standing there, thus is it one more beauty that has come into the world and shall go down through the centuries.

"My Life's Good Fortune"

Following their long alpine tour, Anna, Sophus and his sister Laura were back in Christiania by the middle of August. The two ladies put up at a hotel, and a few days later continued on to Moss. After a short sojourn in Moss, Anna returned to her parents in Risør.

Sophus, in Christiania, was busier than he had been in a long time. For at least six hours a day he examined school matriculants who hoped to enter university, and for two hours a day he worked with Sylow on the new edition of Abel's collected works. In addition, he had to correct proofs of what he called his "ongoing Affairs". In the course of the month of August he delivered a paper to the Academy of Sciences in Christiania, entitled *Über eine Verbesserung der* JACOBI-MAYER*schen Integrations-Methode.*

His correspondence with Anna continued much as before, although with not quite the same frequency and regularity – but in any case, they wrote to one another once a week. When in the course of things he apologized that he was "a little remiss" in his correspondence, it was always because he had been occupied by mathematical work. But according to Sophus, this did not mean that their "Amiability" had diminished, but that he felt only more certain of "keeping" her, and added in his customary style: "But You my Dearest, You do not indeed think that I am a rather lukewarm, indifferent Lover, do You?" And he assured her that whenever so desired he could make a flying visit to Risør, and give her an encouraging kiss: "For I think that I am vain enough to believe that You would not be in the least unhappy were I to come unexpectedly, and give You one, or perhaps a Pair of Kisses. You would not, would You?"

Over the course of the summer Sophus and Anna had come to the agreement that they would marry in March of the following year, and he, in any case, considered this agreement to be binding. Apart from this, Sophus made only a couple of allusions to their tour through the mountains, and the time he spent together with Anna that summer. In the first instance, and apparently in response to her having reminded him, he wrote: "Indeed, I too feel sincerely that we had the truly most delightful Times together over the Summer, and not only during the last Part of the Tour."

Apart from the time they spent together with Amund Helland, Sophus had also had Anna with him on visits and at social gatherings with relatives in Christiania, to his old Uncle John and Uncle John's wife in Øvre Vollgate, and their son, the anatomy demonstrator at the university, Prosector Johannes Lie, his wife Amanda, and their two sons. An enduring friendship developed between Anna and Amanda – Amanda, as indeed Sophus had earlier described as an uncommonly pleasant Lady, was the daughter of the well-known Swedish writer, translator and folklorist, Arvid August Afzelius. In the course of that autumn Sophus regularly conveyed greetings from Uncle John and his wife, who always asked after her, as well as greetings from Amanda and Johannes – although Sophus did not always feel that they raised their children in the proper manner. He told Anna that Amanda allowed the children to go to school in wooden clogs, and he commented: "They do not look so bad when they are covered with Leather, but those poor Boys, their Friends play them for Fools. It is frightful to think of the Children in wooden Clogs and the Mother in her Silks and golden Finery!"

Sophus had actually considered changing his lodgings after his summer vacation. Social relations in the Motzfeldt house had not become any better, but his plans to wed in the spring and establish a home together with Anna seem to have made an interim move less pressing. Nevertheless he avoided Mrs. Motzfeldt "out of Fear of Scenes", but as well, in response to hints from Mrs. Motzfeldt, he also bought Ida a birthday present that autumn – "otherwise", he wrote to Anna, "it would have given rise to ugly Scenes." However, Ida, whose room was next to that of Sophus, had had a girl friend visiting for a period, and had mounted "a shameful Spectacle late into the Night."

While they were in Christiania in the course of the summer, Anna and Sophus had visited the photographer, and Sophus sent her a number of copies of the two of them together, pictures that she could distribute among her girl friends in Risør. He also sent pictures of her "friend Amund Helland", and of his friend Klein – about the latter he commented that the picture was only "a poor Compensation for the Person, whose Amiability You know I have always praised to the Skies." He also had photographs from Tvedestrand, in which Anna was together with the Vogt children, and as well he had received photos from his brother John Herman and family. He felt that all these, together with photos he had received from friends in Germany and France, would soon allow him to "fill a Couple of Albums." He would be giving a large portrait of Anna to her parents for Christmas, and it was in this connection that Sophus made his second allusion to the summer's tour through the mountains:

> During the Summer we had quite a Number of Struggles which should now Disappear. As I so often said, and which was so undoubtedly correct, the Thing was that we both held too fast to our respective Independence. An engagement, and even more, a Marriage, imposes on both Partners, certain Bindings that one must get used to. And I was quite often bad in the Demands I made upon my own sweet Lass; but You, my very best Friend, shall see that when all is said and done I am really not so bad.

Now, when he was thinking about the future, he announced his desire to be always and constantly near her. And it astonished him that:

> Is it not strange that fundamentally I cannot rightly understand my own being, in so far as I have so completely fastened myself to You. But I believe that You will also bear with me when, from Time to Time, Intensity takes control of me. Believe me, dear Anna, if only You understand how to handle me correctly, then little by little You can get this Intensity to leave

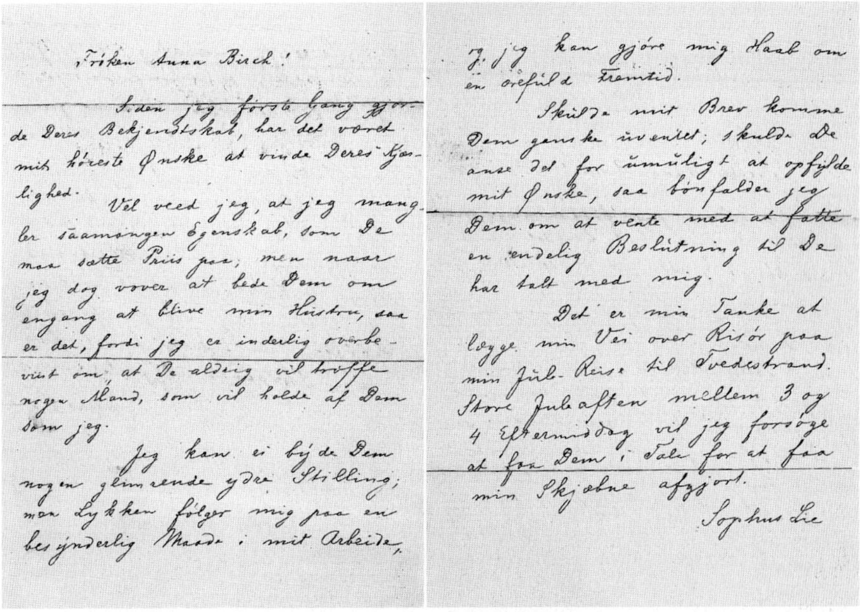

Fröken Anna Birch!

Siden jeg förste Gang gjorde Deres Bekjendtskab, har det været mit höieste Ønske at vinde Deres Kjærlighed. Vel veed jeg, at jeg mangler saamangen Egenskab, som De maa sætte Priis paa; men naar jeg dog vover at bede Dem om engang at blive min Hustru, saa er det, fordi jeg er inderlig overbeviist om, at De aldrig vil træffe nogen Mand, som vil holde af Dem som jeg.

Jeg kan ei byde Dem nogen glimrende ydre Stilling; men Lykken fölger mig paa en besynderlig Maade i mit Arbeide, og jeg kan gjöre mig Haab om en ærefuld Fremtid.

Skulde mit Brev komme Dem ganske uventet; skulde De anse det for umuligt at opfylde mit Ønske, saa bönfalder jeg Dem om at vente med at fatte en endelig Beslutning til De har talt med mig.

Det er min Tanke at lægge min Vei over Risör paa min Juul-Reise til Tvedestrand. Store Juleaften mellem 3 og 4 Eftermiddag vil jeg forsöge at faa Dem i Tale for at faa min Skjæbne afgjort.

Sophus Lie

Sophus Lie's letter, see page 169, to Anna Birch (photo) – of proposal, to which he received a reply when he went to Risør for Christmas Eve, 1872.

Some of Lie's mathematical friends.

Top:
Albert Viktor Bäcklund, professor at Lund.

Bottom right:
Hieronymous Georg Zeuthen, professor in Copenhagen.

Bottom left:
Luigi Cremona, professor in Milan.

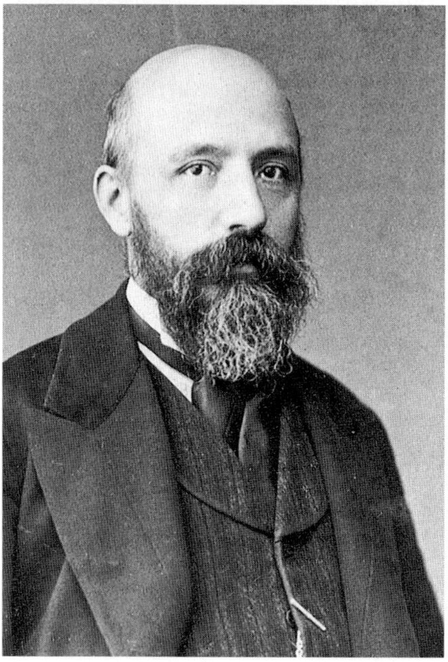

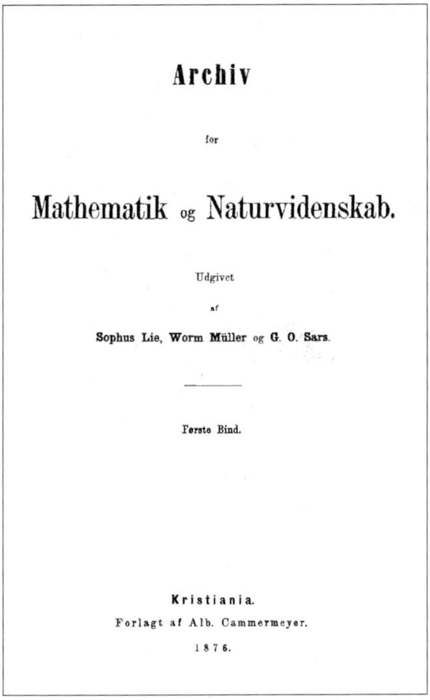

Archiv

for

Mathematik og Naturvidenskab.

Udgivet

af

Sophus Lie, Worm Müller og G. O. Sars.

Første Bind.

Kristiania.

Forlagt af Alb. Cammermeyer.

1876.

De realstuderendes læseværelse

ACTA MATHEMATICA

ZEITSCHRIFT	JOURNAL
HERAUSGEGEBEN	RÉDIGÉ
VON	PAR

G. MITTAG-LEFFLER

1

	STOCKHOLM	
	F. & G. BEIJER.	
BERLIN	1882.	PARIS
MAYER & MÜLLER.	IMPRIMERIE CENTRALE.	A. HERMANN.
38/39 FRANZÖSISCHE STRASSE		8, RUE DE LA SORBONNE

Top left:
The new journal of the natural sciences that Lie, together with G. O. Sars and Worm-Müller got started in 1876.

Top right:
The first issue of Acta Mathematica *with Professor Gösta Mittag-Leffler (photo, left) as editor came out in Stockholm in 1882. It quickly became one of the foremost periodicals and continues to publish. It is counted among the most prestigious publications in the mathematical world.*

No. 15. | Bladet udgaar hver Uge med et Nummer paa 8 Sider og koster Kr. 3,30 pr. Kvartal, Porto og Ombringelse heri iberegnet | Kristiania, 11. April 1886. | Avertissementsprisen er 20 Øre pr. Petitlinie for indenlandske og 30 Øre for udenlandske med 10 % Rabat for 3 Ganges Indrykkelse. | 13. Aarg

Professor Dr. Sophus Lie.

Ved Professor Lies Kaldelse til Lærer i Geometri ved Universitetet i Leipzig er der vist ham den største Udmærkelse, som overhovedet kan vises en Videnskabsmand. Naar et af Tysklands største Universiteter med Forbigaaelse af mange udmærkede tyske mathematiske Forskere henvender sig til en Udlænding, da indeholder denne Henvendelse fra aller kompetenteste Hold en Erklæring om, at den indkaldte Docent er den ypperste Repræsentant for sin Gren af den mathematiske Videnskab.

Marius Sophus Lie, hvis Billede idag pryder vort Blad, er født paa Nordfjordeid den 17de December 1842. Hans Forældre var Johan Herman Lie, død i 1873 som Sognepræst til Moss, og Maren Mette Stabel, død i 1852. Sine Barneaar tilbragte Lie paa Moss; han blev i 1859 Student, og i 1865 Kandidat i Realfagene. Som Student lagde han sig specielt efter Mathematik, ligesom han ogsaa med Iver beskjæftigede sig med andre Grene af Naturvidenskaberne. Da han var færdig med sine Examener, fortsatte han sine Studier, ligesom han virkede kraftigt for Udbredelsen af Kundskaber i Mathematik og Astronomi blandt Studenterne; han holdt saaledes en Række af Foredrag over det sidste Fag i Studentersamfundet, idet han illustrerte Foredragene ved Modeller, som han selv forfærdigede. I 1869 erholdt han Stipendium til en Udenlandsrejse for at studere Mathematik, og besøgte da Berlin og Paris, ligesom han foretog Fodvandringer i Frankrig, Schweiz og Italien. Den tysk-franske Krig brød ud under hans Ophold i Paris; paa en Vandring til Fontaineblau blev han arresteret som mistænkt for at være preussisk Spion; hans Breve, i hvilke der forekom mathematisk Tegn, mistænkeliggjorde ham, idet disse blev anseede for Zifferskrift. Hans Pas var ikke ordentlig viseret af Legationen i Paris og han maatte tilbringe hele 30 Dage i et elendigt Fængsel. Hans Stilling var i Virkeligheden kritiskt; thi Tyskerne nærmede sig, og Stemningen var ophidset. To Dage før Tyskerne kom, slap han ud.

Lies Arbejder i den moderne Geometri begyndte imidlertid at vække Opsigt inden den Kreds af Mathematikere, der beskjæftiger sig med denne Videnskab. Professor A. Clebsch i Göttingen henvendte i 1872 en Skrivelse til Professor O. J. Broch og omtalte den store Betydning, som Lies Arbejder havde for Mathematiken. Andre overordentlig hædrende Udtalelser fremkom fra fremragende Mathematikere i Berlin, Göttingen, Milano og Paris. En Del indflydelsesrige Thingmænd, nemlig Kildal, Riether, Daa, Steen, Sverdrup, Essendrop, Danielsen og Sæhlie fremsatte da Forslag om Oprettelsen af et extraordinært Professorat i Mathematik for Lie, hvilket Forslag Thinget vedtog, og disse Mænd har for den væsentligste Del Æren for, at Lie blev knyttet til Universitetet itide.

Som Lærer er Professor Lie meget yndet af sine Elever, og det har lykkedes ham at lægge Studiet af Geometrien ved Kristiania Universitet adskilligt højere end tidligere, saa at hans Bortrejse vil være et stort Tab.

Sine udenlandske Forbindelser har Professor Lie vedligeholdt ved Rejser. Foruden i 1869 foretog han Udenlandsrejser i 1872 paa egen Bekostning til Göttingen og Erlangen, i 1874 i Anledning af Udgivelsen af Abels Værker til Paris, i 1877 til München og i 1882 til Paris.

En ikke ringe Fortjeneste af vor Videnskab har Professor Lie derved, at han har grebet Initiativet til Grundlæggelsen af Archiv for Mathematik og Natuvidenskaberne; Adgangen til Offentliggjørelse af videnskabelige Arbejder er derved i høj Grad lettet, og den videnskabelige Literatur her i Landet fik en ikke ubetydelig Tilvæxt, efter at Lie sammen med Professorerne G. O. Sars og J. Worm Müller havde grundlagt dette Tidsskrift. Ogsaa Forslaget om en ny Udgave af Abels Værker kom i Virkeligheden fra Professor Lie, og Udgivelsen af disse blev iværksat af ham og Overlærer Sylow med Bevilgning fra Thinget.

Lie har altid hovedsagelig været optaget af sine videnskabelige Arbejder, og han har været fri for offentlige Tillidshverv, hvormed mathematiske Professorer ellers tildels blir stærkt belemret. Dog er han valgt til Medlem af Bestyrelsen af Turistforeningen, et Hverv, hvortil hans Interesser for Turistvæsenet og hans respektable Bedrifter som praktisk Turist gjør ham fortrinsvis skikket; han var for en Del Aar tilbage utvivlsomt en af Landets første Fodvandrere. Vort Lands politiske Udvikling har han fulgt med Interesse; paa Valgdagene findes han paa »venstre« Side.

De videnskabelige Sandheder, som Professor Lie har bragt frem i Dagen, vil være kjendt af forholdsvis faa; thi der gives ingen Kongevej til Mathematiken. Den mathematiske Videnskab vil paa sine øverste Højder altid savne den populære Anerkjen-

Professor Dr. Marius Sophus Lie.

Front page of Ny illustreret Tidende *on the occasion of Lie's departure for Leipzig. The unsigned article was written by Amund Helland. (Part of this text is cited on page 322.)*

me. But You will certainly still be fond of me, won't you, even if once in a while I am a little bit naughty?

The year 1873 became uniquely significant and decisive in the life of Sophus Lie. Whilst, during the course of the spring and summer, he had been completely fastened to his dearest Anna in a way that he scarcely understood, during the autumn of that year his future mathematical tasks seem to have become clear to him. The mathematical work that Lie began that autumn would come to be something that occupied him for the rest of his life.

In recent years he had worked on partial differential equations and he had also published a number of papers on this theme. His contact with Mayer in Leipzig was based on the fact that they both worked on such equations, and the correspondence between them dealt with the major issues in this field. But behind the integration methods – that is to say, the methods of solution that Lie now developed – he discerned much more extensive general theories. His studies and theories would come to have decisive significance in areas he himself could scarcely have foreseen. Rather late in life he himself summed up some of his own mathematical contributions (*Aftenposten* 25.11.1896), wherein he established "a Theory of *continuous* Groups" and pointed out "that this Theory is a natural Foundation for Differential Equation Theory and for Differential Geometry." And about this same time he announced (in the *Leipziger Berichte* 1895): "the Theory of Differential Equations is the most important Discipline in modern Mathematics." More than a hundred years later it is a fact that what Lie gave rise to has become its own mathematical discipline – modern Lie Theory with all its bifurcations and applications is fundamental to a whole series of fields of mathematics and natural science. Many, like the mathematician J. Dieudonné in *Gazette des Mathématiciens*, went so far as to say, in 1974, that it seemed there was scarcely any significant mathematical work possible without the use of Lie groups.

As early as approximately 1700, Leibniz and Newton – the founders of integral and differential calculus – solved simple differential equations which were used to describe and explain the planetary system. The differential equations quickly acquired a definite significance in physics, and in many ways physics first became a science only after these differential equations had been introduced, and after the discovery of mathematical methods of solution. It consequently became possible to verify changes and processes in nature that we perceive as continuous, due in part to the ability that now existed to build abstract mathematical models of them, and these were far better models than had previously existed. All changes of physical size – for example, the movement of bodies in free fall, or a planet's orbit around the sun – these movements and orbits could be described with differential equations.

While in an ordinary equation one is hunting for an unknown quantity x that fits the equation, in a differential equation it is a function $f(x)$ that one is looking for. While in a normal equation the given is, for example, a relationship between different powers of the unknown x, in a differential equation what is given is the relationship between the unknown function $f(x)$ and some of its derivatives: $f'(x)$, $f''(x)$, etc.

But even though processes, changes and phenomena in nature could now in principle be described and modelled with the help of systems of differential equations, there were still in most cases large problems of finding such solutions and demonstrating that solutions existed to the ensuing differential equations. Solutions to the relevant differential equations could be proven to exist in only a few instances; moreover, integrations were rarely found. As well, the few equations for which solutions had been found, seemed to be special cases that happened to occur here and there. With the lack of a general theory, various approaches and special sophisticated tricks were employed to integrate these equations.

During the course of the 1700s mathematical analysis gradually became synonymous with the study of differential equations. But in the study itself, in the theory of differential equations, the mathematicians groped around – their specific answers appeared almost to be solutions to puzzles, yet without any uniform relationship. The field was full of conjectures and experiments. Eventually it would be Sophus Lie who would find ways out of this indistinct and enigmatic landscape.

In the first models of the planetary movements, the mass of a planet was considered to be concentrated in its centre of gravity, and the planets were consequently seen as mathematical points. In this situation Newton's gravitational theory gave rise to linear differential equations of the first order. In subsequent studies, the basic concept – an accelerating force independent of time and an attracting centre – was held to be universal, but consequently, the models and the methods for understanding the processes were gradually improved. The processes that are dependent upon several variables: for instance, those containing quantities that are dependent upon *both* time and position, led to partial differential equations – equations that contain unknown functions in several variables, and the derived functions with respect to these variables. Partial differential equations are fundamental to all physical theorems: everywhere in the field of physics – indeed, in all of the natural sciences. Partial differential equations are used to formulate the laws of nature, laws that admit testing by means of experimentation. It is partial differential equations that describe how heat or sound spread through solids, liquids or gases; it is partial differential equations that plot out how electrical charges generate an energy field, how oil or other liquids flow through a reservoir, etc. Behind mechanical, optical, meteorological, astro-physical, electronic and elastic phenomena, behind waves and explosions, one finds the "glowing aura" of partial differential equations.

While Adolf Mayer in Leipzig had proceeded purely analytically in his studies of differential equations, it was Lie's work in 1872–73 which introduced more conceptual considerations. "Unfortunately we speak different languages in mathematics, or in any case, we reason in completely different ways," Lie had written to Mayer in December 1872, when he also for the first time, formulated what, much later, would come to be called Lie algebras.

In his doctoral work Lie had plotted out some extraordinary transformations, and with the help of these transformations he had demonstrated that straight lines in space could be mapped to curved surfaces in another space. The results that emerged had uncovered new and interesting properties concerning some of the geometric surfaces that were being studied at the time; for example, what was called the Kummer surface. When during the winter of 1872–73, Lie turned to the common theory of partial differential equations, it was these geometric transformations that he took as his starting point. And already in the work that was published in Christiania during May of 1872 he had found a new and simpler method for the integration of ordinary partial differential equations of the first order. Lie's method required far fewer and lower integration operations than those that had been developed in the fundamental works of mathematicians Cauchy, Jacobi and Pfaff. It was also a further working out of these theories with which Mayer had concerned himself.

Twenty years later Sophus Lie gave a short account of this historic development. In 1893 he wrote a three-page foreword to a textbook on partial differential equations, a book that rather soon became a standard work in mathematical teaching literature. This dealt with the work of the French mathematician Édouard Goursat, which then in 1893, two years after the publication of the French edition, was to come out in German. Lie had been asked to provide a foreword, and he undertook the task not because Goursat's work really required a foreword, he wrote, but because he could perhaps contribute to making the work better known and used. The first mathematician Lie focused upon was Lagrange, who in the 1770s had succeeded in reducing every partial differential equation of the first order to a system of ordinary differential equations. The next step had been taken by Monge who, by means of geometric reflections, and by introducing the concept *characteristic* as a structuring term, explained why Lagrange had succeeded so well. But Lagrange and Monge, and other mathematicians of the time obviously worked only with ordinary three-dimensional space, although they had for a long time considered some spatial concepts extended into n dimensions. It was therefore a great step forward when Pfaff generalised the theory in 1814, and showed how each partial differential equation of the first order, in two variables, could be converted into a system of ordinary differential equations (in three-dimensional space). A couple of years later Cauchy, following the route of analysis, reached the same result. Then, after Hamilton had shown that many problems linked to dynamic processes were connected to partial differential equations of the first order in a subtle manner, there came Jacobi, who managed to express these relationships still more clearly. Jacobi proved that the answer to each dynamic problem was a question of partial differential equations of the first order, and he developed what came to be called the Hamilton-Jacobi integration method. This method was criticised from various quarters, but it was gradually revealed that all the actual methods could be united with the appropriate conceptual developments of Cauchy's method. Jacobi came forward again with a new method, which, according to Lie, gave him a certain celebrity, but which in reality was very close to Pfaff's method. Jacobi had quite certainly made essential improvements that allowed one to avoid certain integration operations, but like all earlier methods, this one also contained a whole raft of exceptions. According to Lie, Jacobi's most important contribution to this field was the introduction and application of the concept of *functional determinant* and what was called the *Jacobian identity*. But the problems had been illuminated in a still sharper manner in, and with, the introduction of the concept of an infinitesimal contact transformation, a concept Lie mentioned being used by, among others, W. Thomsen, P. G. Tait and R. Lipschitz. And Lie himself appeared in the next stage of the development, recalling what happened at the beginning of the 1870s. The actual progress had come into being and found its place as follows: not only were the shortcomings in the old integration methods revealed, but also the means of remedying these shortcomings were found. And when Lie wrote his foreword to Goursat in 1893 he was also able to point back at some of his own work – it had been possible to provide a deeper basis for the ordinary theory of contact transformations, and thereby a generalization of the concepts, by combining the analytic theory of Pfaff and the geometric ideas of Poncelet and Plücker. Exceptions could now be handled easily, and problems that earlier were considered insoluble had now, quite simply, become trivial. Lie stressed that the simplification of integration that had been achieved had its explanation in the doctrine of differential invariants and infinitesimal transformations. These were new, complex mathematical concepts that had given the theory its true dimension and a thorough-going rational form. And to all this, Lie concluded, the work of Goursat gave a better introduction than any earlier work. One third of the space in Goursat's canonical textbook is concerned with a presentation of Sophus Lie's works.

But in 1873 Lie was still having difficulties trying to give his conceptual reflections a satisfactory form, and that autumn he therefore decided to translate his conceptual development in the language of mathematical analysis, and hoped thereby to become better understood both by Mayer and all the other mathematicians. In the papers that he published in Christiania he tried to develop such an analytic form, and still more in the articles on partial differential equations that he would gradually publish in *Mathematische Annalen*. Lie threw himself into this work and he allowed Mayer a free hand to undertake minor changes in the articles before they went to press. But this did not succeed completely in satisfying the analysts, and remarks were made to the effect that in his accounts, Lie based his work on synthetic reasoning that merely dressed itself in analytic clothing. This was a view that long hindered the spread of his ideas. For his part, Mayer as well seems to have not followed Lie's work beyond partial differential equations of the first order, and moreover, Lie did not inform Mayer in detail about his group theory discoveries.

The epoch-making theories that Lie now developed had their starting point in his discovery that the concept *symmetry* could be used to integrate; that is, to show that solutions to such differential equations do exist. The integration theory that he developed for differential equations (with known infinitesimal transformations) was the point of departure for the study of structures in other areas (finite dimensional continuous groups). Lie now embarked upon the task of determining all finite dimensional continuous groups of point transformations in the plane. There is not the least trace in either his letters or his notes as to how he got this idea, and how he worked out the first results. But, according to later descriptions from October 1873, he was certainly on the track of a theory of continuous transformation groups. This was a tremendously daring thought whose purpose demanded enormous endurance in terms of the colossal number of calculations this work involved. And, as he himself would come to express it, in terms of mathematics, for the next three years he lived completely in the realm of transformation groups and integration problems.

According to him he had little social contact with others, and he went out less than usual. However, Lie's student, Elling Holst, twenty years later, would recall a walking tour he took together with Lie that autumn. On that occasion Lie had spoken of some new ideas that were "of greater breadth and of more far-reaching applicability than anything else that he, up to that point, had discovered. He had succeeded in creating within general mathematical thinking, the counterpart to an ingenious principle worked out for equation theory by one of Abel's followers, the much too untimely deceased young French lad, Galois, and which was named for him, *Galois' Theory* or *Group Theory*."

Around this group theory or symmetry method that Lie developed for differential equations, would there come to be an impressive structural framework for continuous and infinitesimal groups: the superstructure of these groups which again can be seen to constitute modern Lie Theory, with concepts like Lie Groups and Lie Algebras. And again, these concepts are indispensable to the description of continuous symmetries. That the laws of physics are based on continuous symmetries is something that has become more and more clearly understood. The most modern theories of elementary particles could not be formulated without Lie Theory.

The fact – that during the autumn of 1873 Lie was concentrating on mathematics in a way that differed from his earlier work – is revealed as well in letters to his sweetheart Anna. On one occasion he guiltily reported that he had actually sat down in readiness to write her a letter: "But then exactly at that moment I received a Letter from Mayer in Leipzig which interested me to the most intense Degree in so far as he had considered pointing out a Mistake in one of my Works. I felt I had to explain the Matter to him on the Spot and whilst I was engaged therein, I lost all Consideration of the Time and it became too late to write to my dearest Sweetheart." He went on to implore her: "Do not become angry at Mayer because this Once he has gotten in Your Way. He has had much more Reason to be angry at You. You are aware that we had an extremely lively Correspondence over the Winter and during last Year. Over the Summer I became remiss in Terms of my Responses." There were several letters from Mayer that he had not answered, and in September and October neither of them wrote to the other, but then, at the beginning of November, he had, for three days in a row, sent off "long scientific Letters". On the fourth day, even before Mayer had received his letters, he received a letter from Mayer, and, as though hinting that in mathematics as well, a kind of telepathy was to be found, confided to Anna: "Just think, we had written simultaneously!"

Lie received yet another letter from Mayer, and when he finally got around to replying one day in November, he wrote:

> My most heartfelt thanks for Your two letters, which nonetheless I am not going to answer today. I want rather to clarify why I am not able to write more extensive letters for the time being. This is due to the fact that I am concentrating all my mental powers on a mathematical investigation that I find exceedingly interesting. I have achieved the most highly interesting results, and I expect many more. This has to do with an idea that has its source in earlier work done by Klein and I, namely on introducing concepts from substitution theory to the theory of differential equations.

And after having briefly clarified some introductory propositions and expressed the hope of finding a common method with which each partial differential equation could be brought into a "canonical form", he concluded: "At the moment I am in a great state of exaltation; I hope I do not become too disappointed in my expectations."

With Klein too, Sophus began to correspond over the course of the autumn, in a manner "somewhat more lively", as he put it. Klein, who had been on a long trip to England, talked in detail about the mathematical milieu he had encountered there. And in a letter to Anna, Sophus commented that indeed Klein, too, would have reason to complain that his (Lie's) "Amiability had slipped away": "I, who for so long wrote several Letters a Week to Him, have now become actually irresponsible. But partly this is Your Fault. But You, of course, are forgiven your Amiability." Meanwhile, Klein was planning to come to Norway during the summer, and Sophus, who now calculated on being well and truly married, wrote to his prospective wife: "Will it not be pleasant to have a Guest? You shall see that You shall grow fond of him. I suspect that already You have a Bed and Bedclothes for him. But if worst comes to worst, I will have to pay for it."

In order to show Anna that mathematics had not completely overtaken him, Sophus reported quite expansively on the few social events in which, despite everything, he took part. The first was the public "Harvest Festival" which indeed he felt she had read about in the newspapers. There had been much to see, and he had not nearly enough time in the couple of hours he had attended. What had amused him the most was a circus with two kinds of horses, and he explained: "That is to say, it is rather a Man who is both Horse and Rider. Or also it is two Men who form one Horse." In the first instance it was a man who wore around his "Middle a Gadget made of Basketry that made him look like a Horse", and in the other instance, the horse was composed of two men, one forming the forequarters and the other the hindquarters, and over them they wore a horse costume. One of these men, who probably for all his life had been practicing to whinny like a horse, was of such short stature that he could only be used as the hindquarters and his whinny consequently sounded like an act of "ventriloquism", commented Professor Lie, and he concluded: "It is inconceivable that they can look so life-like, rather like real Horses and Riders."

He had also been able to report on a good social gathering for the annual celebration of the Realistforening. All the professors of the Faculty of the Natural Sciences were invited, and because it had been Sophus Lie who, five years earlier, had gotten the association on its feet again, he was toasted "on three separate Occasions" to ceremonious acclaim, and to Anna he asked without a question mark: "Are You not proud on my Behalf." While the other professors had left at midnight, he had remained right up to two o'clock in the morning: "Among other Things, we improvised *stevje*[26], and several of the Students were remarkably witty, such that we had a delightful Time."

Apart from this, it happened that he attended the "Amusement Club" of which he was a member, but he was a very irregular member, and this was something he would like to improve, he confided to Anna. Meanwhile, one day he had been visited by Professor Ole Jacob Broch, whom he described to Anna as "a charming and good-tempered Man". Broch had been in Paris and Berlin and could therefore convey greetings from many of Sophus' acquaintances. From her "friend Helland" Sophus was constantly conveying greetings, and about the fact that Helland had been rejected as a member of the Academy of Sciences, and the subsequent outcry that led to almost the whole leadership of the Academy being defeated in the autumn election, he commented as follows: "Anything of this nature has scarcely occurred before. Consequently there is a terrible Turmoil in our Sciences Academy, which does not need exactly such a Reputation." As to his own status as a man engaged to be married, he reported the following:

> Otherwise, parenthetically speaking, it is remarkable how much more accommodating all the young Ladies have become since I became engaged. Basically I think highly of many of them. I am trying to make You jealous, but the Attempt catches in my Throat; I feel that You are the Sun, compared to all the Others!!! Think what it would be like if You wrote such a Thing to me! But many Years can certainly go by before You have come so far.

During this autumn in Risør, Anna was learning French, something that Sophus seemed to be very happy about and which promised well in relation to their planned tour to Paris. He expressed his pity for her, regarding her French teacher, who, from what she had said, was so strict, but then he added, "But think, I do not have the least Fear that You will allow Yourself to be intimidated. From my Experience I know that You possess the necessary Courage."

Sophus also encouraged Anna in the direction that his sisters Laura and Thea had been leading her, namely into knitting and sewing, and in addition, he sent her a sewing machine. The starting point seems to have been the great industrial exhibition which that year (1873) had been held to the west of Christiania, in the city of Drammen between June and September. It was a common feature of the times to arrange such exhibitions – both to promote trade and to educate the public by displaying new things and new technology; gradually the major industries' marketing demands became a driving force. The first World Exposition took place in London in 1851, and was followed by one in Paris. The first Norwegian industrial exhibition was arranged in Bergen in 1847. The exposition in Drammen had been

[26] Norwegian "stevje" are improvised songs, much like Caribbean calypso. They are composed spontaneously, usually between competing performers, and performed to traditional melodies. They are composed either of rhyming couplets, or as in the Middle Ages, rhyming second and fourth lines.

opened by King Oscar II, and attracted 80,000 people. Among the almost 1,500 displays, more than half were in the fields of crafts and industry. And here the Assistance Association of Moss, whose prime-mover was Laura Lie, won the silver medal for its knitted gloves. These "Moss gloves" quickly became a sought-after item, in whose distribution both Sophus and Anna became involved: "I shall be Thea and Laura's Commissioner to the best of my Ability; but I do not know if Laura will be content," Sophus wrote to Anna. He went on to advise his sweetheart against being too extremely diligent: "Only You should not be too dutiful. Just imagine that my little Sweetheart should be in Danger of becoming some Paragon of Virtue."

In any case he hoped that Anna had not read the serial that was published in *Aftenbladet*. It was a story that in a very unpleasant manner reminded him of their engagement – but, he recounted to her:

> The old Man and the young Wife were fighting with such Abandon, and just to think, without any Basis, because they had become jealous of one another. It is really remarkable how stupid Human Beings are to create for themselves unnecessary Strife. On this Point though, we have been more clever. But in other Respects I am afraid that we have comported ourselves just as stupidly, viz: for I am afraid that when once in a while we happen to fight, this had been due only to Misunderstanding: for when two People care for one another as we do, then one should, with a little Caution, be able to avoid all Disagreements. So next time we get to try to be more careful and deliberate: think how we shall have the most lovely Time together, when we no longer disagree. For whatever You now say, I fundamentally believe that we shall sometime soon achieve as complete a Harmony as two born, human Beings can.

Some months later however, he could tell her that "things went extremely well in the end" in the *Aftenblad* series: "Think of the serial, where Man and Wife actually cared enormously for one another, but both plagued themselves with the Thought that the Other did not properly care for him or her."

Sophus also commented in his letter to Anna on the autumn weather, and he implied that his health could be affected by the weather: "It is amazing how the Weather has an Influence on my Wellbeing. I really need Sunshine, at this Time of Year in any case; in Summer there can be too much of it." On another occasion when the weather was bad he expressed his intense desire that she could have been with him to cheer him up, but then perhaps she would think of earlier episodes "when in bad Weather I allow my Annoyance to spill over on You because You do not have any Desire to go outside in bad Weather. But my Child, You should not be afraid that I in the Future will plague You in this way."

According to Sophus, outdoor life and fresh air were the best remedies for most things. In the course of the autumn he was plagued by colds – "a violent Cold" he called it on one occasion. When he had a cold he was "not up to anything, not even to write Letters to my dear sweet little Love." But the reason for his cold was always presumably that he had been too little out of doors: "That is to say, I have been working too hard the whole autumn and too little out in the fresh Air. It is remarkable how sick and tired one becomes under such Circumstances." But in a gallant expression of regret to a planned dinner party at the home of his friend, Doctor Axel Lund, he managed to write:

My dear Axel!

I am sorry, but an awfully intense Cold, which in equal Degree has attacked my Head, my Humour, and my Wellbeing in Total, makes it impossible for me to appear tomorrow.

At the same Time I should inform You that a Number of Matters which have in part presented themselves in a new Light, and which partly have changed fundamentally, are forcing me to step back.

I hope You will do me the Service of making the Ladies my respectful Apologies for my Condition in this Matter in its Totality. I assure You I took this on with good Will and in good Faith.

Not only colds, but also dental problems had plagued him that autumn, and when he had finally got himself to the same dentist that Anna had visited during her stay in the capital, it went much better than he had expected. It had not been "particularly painful". The two largest cavities were in "the Wisdom Teeth, which seldom are fit for anything" but a third was in an "ordinary" tooth. And the dentist, who had praised "the Strength of my Teeth", and said as well that Anna had "extremely good Teeth", took only twelve shillings for his three cavities and Anna's one. (In comparison, postage for an ordinary letter was three shillings at the time.)

December 17th was Sophus' birthday. Anna sent him her congratulations and a pair of slippers. Sophus thanked her and commented, at first playfully about her exertion of domestic suzerainty over him – saying that this could imply "You are already making the Calculation that You will have me completely submissive to the Slipper." But otherwise, he was paralysed by her letter. She suddenly wanted to postpone the wedding indefinitely. He reckoned that she was influenced by her father's "Irresoluteness", and was not seriously able to believe that she in her "own Head" had been able to arrive at such a decision. Had she been able to do so, he would have believed that she was not fond of him. But "Faith in Your Love", that conviction which had been for so long the "Joy of my Life", was something from which he could not shake himself free. That it was a heartfelt blow to her parents to lose her, he could understand. For his part he had promised last Christmas to be carefully aware of her young age: *"And I have kept my Word."*

I promised You not to exercise any undue Pressure over You with Respect to the Determination of the Date of our Wedding. And may God be my Witness, I do not know anything about which I have more correctly kept my Word so manfully and honourably. Thus it was that I abandoned my Thinking about the Autumn of 1873.

But I have Your Promise that You will marry me in March 1874. You have promised me this many Times. Half a Year has gone by without You having retracted. Were You now, without any pressing Reason, to break Your Word, this would be the greatest Offense, the greatest Humiliation You could bring me. It would cast a Pall of Misfortune over our Future.

My dearest Anna, You may feel that I am hard, but it is my Duty to speak my Mind. And I cannot release You from your Promise. [. . .] I see the Reason in your Letter in that You are a good Daughter; but You know: the Woman shall abandon her Parents and follow her Man. [. . .] It will be a Misfortune for us both, if You break your Word. [. . .]

I have been so happy in the Certainty that I could soon call You mine. This infinite Grief that You have awakened with your last Letter, let it not be of long Duration.

Marriage at Last

Again, upon the next meeting between Sophus and Anna – the Christmas holidays at the home of her parents in Risør and at the home of his sister and brother-in-law in Tvedestrand – after the initial joy of reunion, this too, was an occasion marked by both harrowing disagreements and sincere love.

Their great "Bone of Contention" as Sophus called it, was the actual date for the wedding. And once again in this dispute he had to give in, the result being a new postponement. In letters written over the course of the winter and spring he licked his wounds and recalled that the relationship between them was based on his "Compliance and great Love", but admitted as well that it would not have gone quite so well if Anna had not been "so good and kind", and if he had not so clearly discerned that she cared for him "almost as much as he could ever desire". The wedding, which would launch their life together, would now take place in Risør at the beginning of August 1874.

He felt remorse and fell into "an ugly Humour" over the thought that he had plagued her "so shamefully during the Course of Christmas", he wrote. But then when he thought how often they had had such a jolly time together his spirits rose and he fell into a good humour. And the signs of their love were many. Never had she appeared more charming than now; and if it seemed to her that *this* sounded like a self-contradiction, she had to remember that "the human Heart is sometimes a curious Thing."

In addition to the "wedding theme", Sophus now also pointed out another reason for misunderstanding between them, and this was the feeling of what he called "the Sense of Dignity". When they attended the Christmas ball together at Risør he had wanted to go home before the end but she had wanted to stay and dance late into the night. The reason that he had then become "a little sour" was not as she probably thought, that he did not want to wait and watch her dance – quite the reverse, he assured her, "his greatest Joy at Balls in the Future" would be to watch her dance. As well, he himself delighted in dancing, but he maintained that no men of his age were enthusiastic dancers. No, what had spoiled the dance in Risør for them was something else, and he went on to explain: "I do not feel that You were justified in considering it obvious that I should be left waiting for the End. But, no, that was the case. It was Dignity that was the Culprit." As time passed, he rather, more and more, valued his "Dignity" less and less, and he hoped: "May it please God that You feel the same way." He added that he had been experiencing this "Sense of Dignity" as a veritable strait-jacket, but that he could now triumphantly announce: "Indeed, this Dignity, this Dignity, You must believe, it is about to receive its Passport in the very near Future."

Now in Christiania at the New Year what caused him worry was how to explain to relatives and friends that the wedding had been postponed from March until August. Sophus' major explanation was the state of affairs in Anna's home. Anna's father had been going through a very bad period. Whenever he was not sitting on the sofa and sleeping, he was expressing his great anguish that Anna was about to leave home. This was the same explanation that Anna's stepmother gave. Indeed, Sophus' relatives in Christiania, Moss and Tvedestrand were satisfied with this. The "Tvedestrand People" had even found it praiseworthy that Anna "did not find it easy to get married, as do young Girls in General". Only Sophus' sister Laura understood better, and more, because Anna had expressly wanted to know what Laura thought. And when Laura, by way of Sophus, came to know that Anna *regretted* that she, of her own hand had decided to postpone the wedding, Laura said that indeed no "irreparable Catastrophe" had occurred. This was also now Sophus' feeling, and he even said that he did not believe it was strange that Anna felt too young to get married, when barely a year earlier she had felt too young to become engaged.

However, Anna's aunts in the capital, Mrs. Motzfeldt and Mrs. Birch-Reichenwald felt that there must be other reasons underlying the postponement. In the words of Sophus, the two aunts "ripped open old Wounds" – namely they felt that the real reason indeed was that Anna "was not properly in Love" with him. Provoked by such suspicions he had written what he later remorsefully called "that ugly Letter" – one of the few letters Anna seems to have thrown away.

But the dissension between the two of them caused them to write to each other more frequently than ever, with Sophus arguing in the following manner: "Do You not believe that when *dangerous* Disagreements arise between People in Love they customarily express themselves by writing less frequently. Can we not try to put an End to this Impasse by considering that our Disagreement was not as dangerous as it was intense." He admitted that it had been he who had started it, but alleged that she was "really quick to take up the Issue". "Indeed, You are not quite perfect, but You are becoming so, right enough," he wrote, and in the following manner, summed up for her what had happened between the two of them.

He wrote that in the first period of their engagement he had been "awfully bad" toward her. He admitted that at that time they both made one another behave inappropriately, but felt that on this occasion he could not comprehend her behaviour as other than that she was "keeping her distance" from him:

> I feel that I have done Everything I could to make You happy; but yet You gave me no Sign of Your Love. I think You set more store in the Company of Others than in mine, that my Desires have so little Weight in Your Eyes.

Since in the meantime he had been coming, little by little, to another interpretation, he had found a "Basis of Clarification" that was not "so devastating" for his "Love", nor "so infinitely wounding" to his "Pride and Self-Esteem". Indeed, he now thought that she had cared for him during the first period of their engagement, and that with time she had come to love him more, but maintained that there was still a considerable distance between this and the kind of love that "penetrates Your whole Soul and which I sincerely hope one day shall be the Case". She must know and herself be "clearly conscious" that whenever he had been harsh toward her, this was not out of a lack of love – indeed, it could only be impossible for her to doubt that he always had cared for her with his whole soul – no, when he had been unpleasant, it had been because his "Love had been rejected". But now there was no longer any bitterness in him, and he hoped that when they next met – at Easter – "full Sympathy would hold sway" between them. He was convinced that they were constantly getting nearer and nearer to one another: "above all, we both have gradually learned to have *greater and greater Trust in one another*." And should "a little Friction" still come between them, in any case it was his hope and firm conviction that their future happiness would not be swayed in the least. She also had to know that in most other engagements as well, "the Sun was not always shining", but he believed that the "Obstinacies of Those in Love" manifest themselves differently in most others, especially when it came to jealousy, and he added: "I have a fundamental Admiration for myself in so far as I have such Trust in You that I do not become jealous of You." He was well aware that she was still invited out by two or three young male friends in Risør, but he reckoned that she behaved herself well and sent him good thoughts even when she went out for a stroll with them. Sophus was convinced that he and Anna already had "overcome the greatest Difficulties: It does not take even *one* Day's Labour to join two Wills into one," he concluded, and ended the letter by bringing to her attention that he was enclosing a speciedaler for her "Adventures", and pointed out that when the letter had become so long, this showed again how "exceptionally amiable and loving" he was, and asked, teasing: "is it not tiresome that I praise myself thus?"

Sophus returned time and again to what he, for his part, had had to renounce due to all the letter-writing and preoccupation with her – sitting in Christiania he felt himself "lonely, abandoned" and "almost unfortunate" because he had been drawn away from "old Friends and old Pursuits". But, he added, it had been particularly his "scientific Correspondence with the World Abroad" that suffered from their engagement, and he added: "But believe me, I have more than full Compensation in thinking about my own, dearest Girl."

It remained important to Sophus to go on reporting about exciting social contacts in the capital city and in any case to describe everyday life as something other than work-related. He reported that he utilised every opportunity to be "out on Skates", and it was not true what she had hinted, that he gave no thought to his "Health": "Quite the opposite. But this Winter, what with Abel and other ongoing Commitments that are beyond my Control, the Issue is that, for the Present, I have had too much to do." And on another occasion, in response to her question about why he had not taken part in something or other, he wrote that it was due to "the Fatigue of Overwork". In future he would arrange matters differently with respect to his work. He wrote: "Ever since I began with Abel I have constantly had an enormous Amount to do." Probably this indicated that from now on he would have less to do with the task of editing Abel's works, something also indicated by later statements to the effect that it was Sylow who did most of the work and ought to receive the greatest honour for the work. Nevertheless, in the coming years much time would be devoted to this work – it was the tracking down of outstanding materials, being constantly in contact with Sylow and the printer, and later there were many proofreadings and corrections. During that spring (May 1874) he reported to Anna: "We have had enormous Difficulties at this Time, Sylow and I, in concluding an Agreement for the Printing of Abel's Works. At last it now looks as though we shall really have an Agreement in place."

Sophus recounted for Anna the evening gatherings to which he had been invited. These were at the homes of Ernst Motzfeldt, Otto Aubert, Uncle John, and a large party at his cousins', Johannes and Amanda. One day, in addition, there was a retirement party for Professor Rasch. This had been arranged by professors and a number of the students: "It was a little overly ceremonious; otherwise, it was rather enjoyable – your friend Amund Helland was also there." Once again Sophus recounted that he had been "strongly engrossed in the Helland Battle, which still continued with great Energy within the University." This time it had to do with Helland's position as a university research fellow. According to Sophus, the geology professor, Kjerulf, had felt himself "personally offended by Helland's Works and opposed his Employment with Tooth and Claw." Consequently, Sophus thought it doubtful that Helland would get any post at all, but he added that there was scarcely any doubt that Ernst Sars, whom Anna had also met, would become professor, "although he too had Non-friends who worked energetically against him." About his own professorial appointment two years earlier, Sophus wrote: "It went more smoothly for me. At that Time I had no Enemies; now, I dare say, after I have begun to get myself mixed up in various Struggles – while only scientific ones – I too would have encountered Opponents." Sophus also reported that Ernst Sars' brother, the zoologist, Georg Ossian Sars, would probably also be made professor in the near future.

On another occasion it was the carnival. Sophus reported that he had defied Mrs. Motzfeldt's desire that he take out Ida, but when he chanced to meet her "good female Friend, Julle Wedel", he thought "it would have been pleasant to be her Knight". And he went on to comment: "It is fundamentally remarkable that I have never been in love with Julle, for I have always considered her to be a splendid and lovely Girl. But it arises from the Fact that I have always felt she had too little Understanding." When later that spring the wedding of Miss Wedel was to be held, Sophus wrote to his fiancée: "You can imagine that now on the Day, she [Julle] looks lovely, and above all, she looks the Picture of Health, more powerful and rosy-cheeked than is seldom enough seen among Peasant Lasses. I do not think You will look quite so powerful as a Bride, but I have made the secure Calculation that You will truly look blossoming when I Lead You to the Altar."

The date for the wedding was settled for August 11th. As early as March, Sophus had written to Professor Bjerknes to exchange examination duties with him, such that Bjerknes took over the *examen artium* (in August), and Lie the Secondary Exam (which would be wound up before the summer holidays), and, Lie explained:

"The Issue is, I have thought to have my Wedding in August. A number of Circumstances would make it particularly inconvenient for me to be in Christiania during August."

At Easter time he was astonished that Anna had not expressed any longing to see him, but he forgave her, and believed the reason was her headache, which arose from the spring air, or perhaps it was the "Chlorosis" or "green sickness" that many young girls in Christiania suffered from – in any case, he urged her to be a "clever Girl", so he could be proud of her "red Cheeks" and "blossoming Appearance".

The only thing he worried about before they were to meet at Easter, was the sea journey to Risør. In return he hoped that she appreciated that he "in this Manner, Time after Time, undertook such unpleasant Journeys" in order to spend some days together with her – "I become quite agitated in myself, when I think of all my Trips to Risør and Moss. But I also make precise Calculations that I shall receive rich Recompense both now at Easter, and what is more, throughout Life. Do You not agree with this, my Child?"

She must have written back quickly and told him that she was very happy about his arrival, and he answered immediately: "Not that I consequently have had a Shadow of Doubt in this Respect, for I have had a more than good Sign that You feel my Company is really pleasant; but I feel it is so good that my little Sweetheart is breaking herself free of being shy, to rather often tell me that she cares for me and that she longs for me."

He felt that like earlier holidays, they should spend it with time divided between Risør and Tvedestrand, something he considered was most pleasant for both of them, and for her parents: "But in the end You must be a sweet Girl in Tvedestrand and not once again make me jealous of Johan. Meanwhile I assure You quite seriously that I was quite jealous of the Children in Tvedestrand, whose Company last year You seemed to find more pleasurable than mine. But on that Occasion You were so shy, poor Child."

What role "shyness" and "dignity" played between them that Easter, Sophus made no comment, apart from thanking her for having been "so infinitely kind and sweet at Easter". In the course of the Easter holidays she must have expressed her fear that he – with time, when she would no longer look so presentable – would come to regard her more coolly. He attempted "to speak some Words of Comfort":

> To be sure I must begin with the Admission that perhaps in the short Time we knew one another before the Engagement, I would not have been so captivated by You if You had not appeared so lovely, and above All, so innocent and unspoiled. Because of this, it may seem reasonable to state in Advance, that, with the Passage of Time, and even though You may no longer look the way You do now, I would think no less of You. But you see, my Child, if You actually ever lost any of Your Appearance, then in the meanwhile Love would have made me blind to Your Shortcomings. Think, Child, that were I so evil as to consider this Possibility as the Basis of my Thinking at Easter that You were so perfect, that You should put this partly at least to my blind Infatuation. Do You not think that this is shameful?

When they parted after the Easter holidays, and he had travelled by steamship from Risør, a rather odd situation arose. He had stood on the top deck and in every manner conceivable tried to make himself visible to Anna below on the quay. However she had not been able pick him out of the crowd. He recalled: "You looked constantly in the other Directions, at other parts of the Steamship. I consoled myself that this was the way my poor little Sweetheart looked for me; only think: You looked almost a little sorrowful because You had been unable to find me anywhere."

Upon his return to Christiania he found letters waiting from Mayer and Klein. Klein, who thought the wedding had taken place in March, still planned to visit Norway in the summer – staying for a month from the middle of August. He was excited about it and asked to which degree knowledge of the Norwegian language was desirable. Sophus wrote back and explained the amended wedding plans and urged him to postpone his tour to Norway. Following the wedding on August 11th, Sophus and Anna planned to travel to Paris, both for a honeymoon and because Sophus had commitments in Paris in relation to the task of assembling the works of Abel. As well as looking up Darboux and Jordan, Sophus also wanted to track down Abel's missing Paris Treatise in the French archives. Sophus thus urged Klein instead to journey to Paris and meet them there. He described Klein to Anna again as "an exceptionally worthy Person" and gave the information that he hoped they would also be able to meet Mayer during the honeymoon tour: "He [Mayer] has perhaps not acquired such a winning Personality as Klein, but he is also exceptionally engaging. I am very proud of the fact that my Friends are more dependable than Your Girlfriends." This latter was an allusion to the ties linking her to Miss Louise Zwilgmeyer, which were certainly hanging by a thread at that moment.

The next thing he had to remember was to honour her birthday on April 24th. This year he sent her three letters and in addition, a telegram: "You certainly cannot complain that Security has cooled me down. Indeed, can You?" He hoped that the engagement had contributed to the past year being a happy one for her, and he stressed:

> I believe now in Reality – as in other respects I remain so modest – that You have actually been particularly lucky with Your Engagement. For I am now comparatively a rather respectable Person, and in any Case, I am extraordinarily kind. I am rather too kind rather than not kind enough.

He sent her best wishes on her twentieth birthday. "May the Lord bring You a happy year, beginning today. Thank the Lord for the important Step which You in Your Year shall undertake to make it truly your Best." His present was twenty speciedaler that he considered might be a contribution to obtaining a pianoforte "for our House". He had begun to look for suitable apartments and asked if he could go ahead on his own accord if he found something he felt was good and had to be decided quickly: "In any case, my Child, You would indeed have to be content with the fact that I did not ask You for advice." He also sent her a book for her birthday, something she had asked him for, and while he had waited for the post office to open – the one which dealt with packages – he began to browse in the book: "I must otherwise confess that it does not do anything honourable to the enhancement of Your Taste, that You think so highly of this from what You have read about it. By my Assessment it is a terribly silly and insignificant Novel, but I have certainly not really read it, so my Judgement is perhaps too strong." Earlier, in response to her enquiry, he had sent her the book *My Wife and I*, a popular story by the Danish writer, Carl Henrik Scharling. Unbidden, he now sent her a book by Jonas Lie. In other respects there were many who considered that Sophus and Jonas Lie were relatives, and perhaps Sophus thought so too. In any case, Jonas Lie seems to have been in many respects a kind of favourite author, and Sophus recommended his novels to Anna on several occasions.

In the last birthday letter he wrote that he felt it was strange to think ahead to her next birthday: "Then we shall be two dignified old Married Folk," and he thought consequently of when "the Shyness and Bashfulness that still leaves its Tracks somewhat on You, will have completely disappeared. And You shall then, with full Confidence, without any Reticence, back me up. Do not You also think so?"

The age difference between them, as she seems to have been preoccupied by, was completely without significance, he argued. When he was twenty-three he felt just as old as he did now, and he did not think that his "Character or Mode of Thinking and Speaking" had suffered great changes from those times:

> If You are now in proper Perfection, then You will see that You too little by little find that we get along very well no matter what our Ages are. God knows that I have indeed wished the difference was a little less [. . .] Believe me, Child, other young Girls *who marry are no more experienced than You*; and hence it happens that most are much less perceptive than You [. . .] As soon as You have been married a Couple of Months, then You shall astonish Yourself over how earlier You felt You were too young to marry.

Sophus was sincerely convinced that she would become "as happy a Wife as any", and gave the following attestation: "Nature has blessed You with many good Gifts, and I have met few (I know none other than Ernst Motzfeldt) who Destiny has decreed to be happy to this Degree. Here You must not believe that I am thinking that You yourself have got such a kind Sweetheart. No, I am thinking rather of your happy Temperament."

Through the spring they enjoyed themselves mutually with letters from one another – she had only to excuse the fact that his handwriting had become so thick; for he had begun to use a quill pen and had still not learned to sharpen it. Sophus remarked that a long time had passed without them having any disagreements:

> Child, You can well believe that these shall become much reduced than what, for example, a Year ago we both imagined. How infinitely good it is how we won out over what at that Time could have separated us. It is so strange, it was basically nothing other than a Misunderstanding, but it was, indeed, perhaps the Standard Rule in these Matters. Therefore let us in the Future be as open as we have been until now, and You will see, my dear Girl, that things shall go as famously as in any Novel.

Letter-writing, according to Sophus, was "a tolerably good Thermometer for the Warmth of the Love of two Sweethearts", and he was always trying to sum up their relationship, the twists and turns of what had happened, and to establish premises that would secure a happy marriage and a bright future for them both.

What had made him "irritable" earlier had been transformed and gone away: "For if You in the Beginning were somewhat independent, everything now seems to be too submissive, rather than the reverse." He was certain that she would come to be "a little Paradigmatic Example of a Wife". And if she did not feel certain enough that he would become "a Paradigmatic Example of a Husband," then that too was no reason for worrying: "Be calm, Child, I also believe with Certainty that at some Time in the Future I shall obtain Your Praise".

She was not at all pleased that he was always stressing how "kind" she was, but he maintained that her kindness was "a Quality of an extremely pronounced Degree." He simply managed not to describe her as "kind", for his firm conviction was that she was "the kindest best Girl under the Sun."

When in the most lovely of spring weather he took hikes over the fields and out into the woods along Frogner River[27]: "I know nothing more Pleasant than to walk along a cheerful River and see the Sun refracted through the Trees" – he wished only that she were with him. "Tell me, do You sometimes long for me? A proper Bride does so, but You think of course that a proper Bride does so indeed, but does not say so?"

When one day he received a letter that he characterised as "a thorough-going little Sorrow, like Resignation", he felt that it was because she had had a headache: "If any Other had read it, then he would scarcely have obtained the Impression of it having been written by a happy Bride, because we are indeed in agreement that that is what You are."

In her next letter Anna reported that she had had the mumps. Sophus immediately conferred with his uncle, and reported back that Doctor John Lie rather considered the illness to be benign, but without due care it could become "extraordinarily dangerous". Anna quickly returned to good health. Moreover her father became better, and in any case, Anna thought, he wanted to be well for the wedding. But both he and her mother in Risør felt it very sad that Anna should be leaving them so soon.

It was now clear that Klein would not come and meet them in Paris, and Sophus was "plainly sad" that Klein would neither be coming to Norway nor meeting them in Paris – "the Germans are so hated there," he explained to Anna, and added, "You cannot have the slightest Idea how wretched it is that Klein and I cannot be together this Summer. Such Gatherings are crucial to our Work for Years." In his letter, Klein had also registered his great disappointment that they would not be together on a daily basis: "I would now have loved to have come to visit you, and again to penetrate to the soul of your works," he wrote – but a quick meeting in Paris, due to "the unsavoury political situation", would be of little satisfaction. Sophus insisted to Anna that there was a possibility that they would meet Mayer, either on route to or from Paris, and he hoped she would not have much against such a meeting: "In 3 or 4 Days Time we can certainly come to speak a frightening

[27] Today these woods are parkland in the middle of Oslo, and the vestiges of "Frogner River", which has since been piped away underground, are the ponds visible from the main bridge in Vigeland Park.

Amount about Mathematics. But he is a lovely Man and You will certainly have good Feelings about him." Apart from this, he had come to hear that one of her aunts, Miss Birch, was going to LeHavre to bathe, and that she would certainly then go on to Paris while they were there. Sophus hoped it would not be for too many days, but felt that for a short time a visit would perhaps be good for Anna, while he was searching the archives for Abel's manuscript.

He asked Anna to send him her baptismal certificate which he needed to "set up a Widow's Fund Deposit Account": "It is strange how Everything is pulling in on us now. The Wedding is now approaching with such rapid Strides, that even my Impatience is somewhat mollified."

They would spend the wedding night at a Risør hotel. The following morning they would go on by boat to Larvik, and the plan was that the day after, they would take the steamship to Fredrikshavn in Denmark. He wanted to reserve a stateroom, and he explained: "A little Room with two bunks for us. The one Bunk is above the other. I shall certainly take the upper. In any case I shall have to take care that my Head is over your Feet." But were there to be inclement weather that day in Larvik, Sophus took precautions: in that event, he wanted, despite the fact that it would cost more, to travel overland via Christiania and through Sweden. But he added: "Otherwise I trust our Fate to my lucky Star, which is usually willing to follow me."

On this much they were in agreement except that she would have preferred to spend a couple of days in Risør after the wedding. The only thing they disagreed about, was the time for the wedding announcement in the church – the publishing of the banns. Anna wanted to postpone this announcement for as long as possible, and Sophus felt that he could adapt to her wishes, and added: "But I do not want in the Future to promise You to conform to unreasonable Desires. Aren't I awful, and aren't You having second Thoughts?"

Early in July Sophus reported being extremely busy with Abel's works. He told about his friends who had begun to leave the city: Amund Helland on his usual summer tour; Axel Bruun and his wife to Switzerland; his sister Laura and a friend who would go on a foot tour through Telemark to Tvedestrand; the Motzfeldts were making themselves ready to travel to the Sandefjord baths; Ernst and Else planned a tour to Valdres but would otherwise spend the summer vacation in Asker, outside Christiania. "As for myself, I intend to take my usual Foot Tour," Sophus reported, and rattled off a series of parties and visits that first had to be wound up. There was an evening at the home of Mrs. Bruun, and a party at Doctor Lund's summer place, and he had to look in on the Birch-Reichenwalds and Mrs. Blytt, and of course he had to visit Ernst and Else in Asker "to where the whole World is travelling since the Railway came into use." (The railway line from Christiania, via Asker, to Drammen, was opened in 1872).

Sophus' "usual Foot Tour" including part of the route along the same trails that he had hiked twelve years earlier with Ernst Motzfeldt and Kristofer Janson: via Kongsberg to Seljord in Telemark, over through Vinje and Haukeli to Odda, then by boat across Hardangerfjord to Granvin, then up through Voss to Gudvangen and Lærdal in the fjord country of western Norway, and back over Hallingdal and onwards to Gol on the high plateau of Hardangervidda, and from there to Numedal and back to Christiania, where he arrived on August 3rd. He had then been away a good three weeks, and there was only a week until he would be married in Risør.

He wrote to Anna: "Just think, I am sincerely overjoyed that our Wedding is going to occur completely in Peace and Quiet." He was particularly happy that none of Anna's aunts would be attending, but he appreciated that Birch-Reichenwald was expected.

He wrote his last letter to her before the wedding, on August 6th. He explained that he would be taking the steamship two days later, that he would be bringing with him several wedding presents from friends, that in Risør he would immediately take a hotel room, and that she could do as she pleased, and either meet him at the steamship quay or not. And he concluded: "It has suddenly struck me that this is in all human Probability the last Letter that You will receive from me during our Engagement."

Herr Professor mit seiner Gemahlin

Sophus and Anna married in Risør on August 11, 1874. The celebration seems to have gone as Sophus had wished "quietly and peacefully."

They embarked upon their planned honeymoon journey, arrived in Paris, and remained there until the middle of October. They stayed at 8 rue Bagneux (Vaugiraud) and seem, all in all, to have been "very satisfied". Sophus was particularly happy that the weather had been so propitious, as he wrote to his friend Ernst Motzfeldt after they had been in the great city for a month. Otherwise, about the honeymoon days in Paris he wrote:

> Actually Time here goes by in a highly uniform Manner. Part of the Time I sit and read or write, either at home or in a Library. Then we utilise 2–3 Hours to go out visiting one Sight or another. Then I visit one or another Acquaintance, and finally, once in a While I attend a Party. Anna has also been invited, but when however the Gathering is only for Gentlemen she does not have the Desire to attend. If we are invited to a Couple of married Friends of mine, who are not here for the time being, then I take her along with me.

This time too, Motzfeldt back in Christiania was asked to send money: even though Sophus thought his normal "Monetary Calculations" were accurate, he urged Motzfeldt "for Security's Sake" to send 120 speciedaler from Sophus' August-September salary. The plan was now to remain in Paris until the middle of October, then spend a short week in Germany, and return to Norway and Christiania at the beginning of November.

Anna, too, reported on life in the big city. After a good month in Paris she wrote home to her "Mama" in Risør in glowing terms about the great woods with their tall acacia trees through which they had driven on their last day into Paris, about the house they lived in, and about life out in the big wide world. She had become acquainted with two English ladies in the pension where they stayed. She reported that she managed quite well speaking English; moreover, every other day she took French lessons from another lady who lived in the house. Apart from this she sewed pillowcases and looked forward to returning to Christiania to get to see all of her "lovely Things". Anna sincerely hoped that "Mama" could visit them as soon as they were back in Christiania and "look at Everything in our House". She was infinitely thankful that her stepmother had kept all her clothing and linen in such good condition, and had looked after the wedding presents – there were several down duvets, there were tablecloths and serviettes, and handsome kitchen furnishings – "and to think, getting an Oleander!" Anna wrote, reminding her stepmother that the latter now had no need to give away her own oleander plant.

It is infinitely pleasant here, as you can imagine. I see so many strange sights. This Evening (Saturday) we are going to a great Circus. [...] We have also been to the Theatre. This Afternoon Sophus and I strolled around a Palace called Luxembourg, and walked in the Garden, which was a marvellous Delight. It seems very strange not to be eating Dinner here before 7 o'clock, but we are always buying Wine Grapes and Cakes to tide us over [...] On Saturday we went to the Bois de Boulogne and watched the Horse-Racing. It was tremendously entertaining. All the Fashionable of Paris were there, and consequently it was a Place of great Elegance and Finery.

The next day, Sunday, she continued:

Last Evening we had a tremendously enjoyable Time at the Circus: there were so many lovely Horses and such skilled circus Riders. [...] In the Afternoon we had been to Versailles and seen the huge enormous Palace. It is the largest one here, and they have many lovely Paintings. It was vastly entertaining. We journeyed there on a little Railway. Sophus is fervently sweet and charming, as you can imagine. He does not know what to do to keep me entertained, so we are having a splendid time. [...] We also have very cosy attractive living quarters.

In addition, Anna was able to tell her stepmother in Risør that her Aunt Elise had arrived from LeHavre to visit them, and that the aunt had accompanied her on an outing and they bought silk for a gown: "we purchased the light green; it really is lovely. We bought 30 Alen[28] so that I can make a Number of additional Embellishments. The Light Red does not last at all well; I believe you feel the same about it. Do you think I need any new Woollen Frocks for Winter?" And a little later in the letter: "In any case, I think I shall buy a Wool Frock, for I must have it to use when I am invited out in Christiania to small Parties. Sophus is giving me Money for Fur Trimmings, that I am quite certain about, and that will be quite grand."

Lie had received a number of promises of permission to go into the archives of the Academy of Sciences in Paris in order to study the original manuscript of Abel's famous Paris Treatise. It was of particular interest to Lie and Sylow in relation to their work of compiling a new edition of Abel's works, to examine this manuscript and compare it with the published version. While in Paris in the autumn of 1826, Abel had submitted this work to the French Institute, but it had been left lying on the great Cauchy's desk – and forgotten. Right after Abel's death it had been found, and it had been decided that it would be published, but the events of the July Revolution of 1830 had once again sent the manuscript down the wrong road, and thus the first edition of Abel's works came out in 1839 under the editorship of B.M. Holmboe, without this paper, the Paris Treatise. The manuscript did not come to light again until 1841, when it was printed in Paris. At this point in time, eveyone took for granted that Abel's original manuscript was to be found in Paris.

In this same period while Lie was in Paris, Sylow was in Germany hunting down information about Abel's works and manuscripts in that country – and to a certain degree, Lie and Sylow tried to coordinate their efforts. That September from Paris, Lie sent two letters to Sylow who was in Göttingen. Lie continued to assume that he would get to see Abel's original manuscript, but everything became slower and more difficult than he had expected. To Sylow he reported: "Here one must

[28] Alen: one alen measured two English feet, or 0.6277 m.

observe a Diversity of Formalities. What particularly gives rise to Delays in that the Committee that gives Authorization to see the manuscripts can only be brought together once a Month. It seems hardly possible that we might get the Manuscript home." Were he only to see it, he would "get to check painstakingly all the doubtful Passages, as well as get a dependable Transcript made." Apart from this, Lie reported that he was diligently studying Abel, that he thought he understood the difficulty with the Paris Treatise such as it stood currently, and he remained firmly optimistic: "When once again we get started, I think that my Participation in the Work shall become fully effective." Lie had received a message from Weierstrass that there were some discrepancies between Abel's original letters, and the printed versions. Lie had written back and said that Sylow would look him up in Berlin. While in Göttingen, Sylow received a letter from Professor Bjerknes – who for his part had begun to write a lengthy biography of Abel – with the report that one of Abel's journals or workbooks had been found. It had been Mrs. Holmboe, who at Bjerknes' request, had searched the attic and any hiding places for Abel papers, and had found this journal, which was in extremely miserable condition after having survived both fire and water. The thing was, that in his time, Professor Holmboe had assembled at his home all of Abel's remaining papers, but in 1850, the same year that Holmboe died, his house had been damaged by a great fire. And all three who were now working on Abel's life and his oeuvre – Bjerknes, Sylow and Lie – hoped that more Abel material would show up from the fire-damaged Holmboe house. But it was obvious to Lie that in the forthcoming edition of Abel's works, despite the various complications, it was necessary to distinguish between "Things Abel himself had published, and Works that he had held back and possibly would himself have described as unpolished."

Lie discovered in Paris "after much Nonsense and many Contradictions", as he wrote to Sylow on October 17th, that the manuscript had gone missing. Abel's Paris Treatise was not in the Institute's archives, and had probably not been there since the printing of the treatise in 1841. No one knew what had become of Abel's work, but the suspicion was widespread that the publisher on that occasion, the Italian Libri, had stolen the manuscript. Not until 1952 was Abel's original manuscript located, by the Norwegian mathematician Viggo Brun, in Florence, Libri's home town. Thus, Lie did not get any further in his work with Abel's Paris Treatise, but he was able to report from the French metropolis that their friend Jordan, had read through both volumes of the first edition of Abel's works, the one edited by Holmboe, and corrected errors of language: "In any case", Lie commented, "in this we have a secure Point of Departure", who had also had his friend Darboux negotiate with the bookdealer, Gauthier Villars on the price and sale of the new Abel edition. On the basis of a forty percent deduction on the sale price of 30 francs, Gauthier Villars wanted to have 150–200 copies of the Abel edition to sell in France, Italy and England. Darboux had also advised them to have the Abel edition printed in Paris. Lie felt that this was certainly not something to be thinking about now, but that they could confront the book publisher Grøndahl in Christiania – if he did not do respectable work – with the knowledge that everything could be done better and considerably cheaper in Paris. This would go on for another seven years, with much work for both the editors and the printer, in terms of both the mathematical content and the French text, before Sylow and Lie's edition of the works of Abel was finished. Lie in Paris sent Sylow warmest greetings from Jordan who explained that he had found much that would be of particular interest to Sylow. "You can well believe that I have made a Diversity of useful Visits," Lie concluded, saying that he had twice in Paris happened to encounter Professor Broch.

Lie wrote three letters from Paris to Mayer in Germany – one of them concerned mainly the work with the Abel edition and plans for when and how they should meet in Germany. In another letter dated October 5th, Lie reported that the day before he had discussed the three-body problem with the mathematician Puiseaux. In other words, what they actually discussed was the prize offered by the French Academy for a paper illuminating the three-body-problem, or what was known as *perturbation*: when the position and velocity of three bodies or astral entities which mutually attract one another, are known at one point in time, how can the position and velocity of these three bodies be calculated for another point in time? Earlier in their correspondence Mayer and Lie had commented on this competition, and they were both convinced that their own respective works provided the prize-winning answer to the question posed by the French Academy. The problem was however, they had not been aware of the deadline set for the submission of answers. Lie had found out from Puiseaux that no prize had been awarded, that it was somewhat uncertain whether the deadline for submissions had been extended, and that their works did not directly answer the problem. But Lie continued to feel that the works that he and Mayer had published on partial differential equations simplified the problem considerably, and that their works could therefore have passed as answers. In any case, when Puiseaux came out a good half year later with the final report on the perturbation competition (the three-body problem), Lie felt that the whole thing was "quite absurd". He said in a letter to Mayer: "Have You read Puiseaux's report on the three-body problem? It is quite absurd. Now, when that which these Gentlemen have desired for so long, comes true, then they almost ignore it."

At the beginning of October in Paris, Lie also got a couple of letters from Klein. Since there had not been a Norwegian tour, Klein had journeyed to Italy, but would be back in Germany in the middle of October, and he also very much wanted to meet Lie. He proposed that all three – Lie, Mayer and Klein – should meet on the evening of October 18th, in Cologne, and the following day proceed to Düsseldorf, where Klein hoped they could stay with his parents, who of course wanted very much to see Lie again with his new wife.

This went as planned. They met in Cologne and journeyed on to Düsseldorf, whereupon they found that Klein's sister was awaiting the birth of her fourth child, and therefore there was no room for Sophus and Anna at the home of Klein's parents. Thus the Lies, together with Mayer, stayed at a hotel called "Zum Breidenbacher Hof", which according to Klein was really the only hotel in Düsseldorf for one who was travelling with his wife. Klein did not have all the time in the world. He had to be back at Erlangen by October 26th at the latest, and on the way he also had to postpone some meetings in Leipzig, yet Lie, Klein and Mayer had five or six days together in Düsseldorf. The joy of being together again must have been considerable for them all, but especially for Lie and Klein – how Anna felt in this company is not known. In one of the first letters he wrote to Lie following their reunion, Klein wrote: "When I think of you I so often have the feeling of a long, indeed almost hopeless disconnection from my better self." The romantic conception of twin souls seems to have lived on in Klein.

The days they had together in Düsseldorf would also give rise to a significant report: Klein wrote down, from Lie's verbal account of his most recent ideas for a paper, a study that Klein also edited for Lie, and a couple of months later was published in *Göttinger Nachrichten* under the title "Über Gruppen von Transformationen". This was Lie's first publication on group theory – eight pages – wherein parallels were also drawn with Klein's geometric methodological approach as outlined in the *Erlangen Programme*.

Sophus and Anna left Düsseldorf on October 24th, and seem to have stopped for a few days in Berlin and Copenhagen on their way home. They had returned to Christiania at the beginning of November, and in the next two weeks they were – uninterruptedly, according to Sophus – getting ready their new home, with the address: "Hægdehougen [Hill Garden], Mrs. Blytt's Property". Mrs. Blytt was the widow of Professor Mathias Blytt, and as well, the mother of Sophus' friend Axel Blytt, who was a university research fellow in botany. The result of all Sophus's efforts to find lodgings for them had simply been to move from one friend's mother to another: from Mrs. Motzfeldt to Mrs. Blytt. This was also a time, of course, when it was difficult to find apartments in the capital city. Christiania – and the official form of the name was now to be spelled Kristiania – was a city of 80–90,000 people, and with a geographical expansion of the city a few years later, the population reached 100,000. With the growth of commercial and industrial enterprise, more and more space in the city centre was taken up with offices, stores and workshops. In a year, there were only about 400 new apartments available to let in Christiania – and this continued to be the situation, as it had for some years, when Sophus and Anna established their home. (Sophus, who was conservative in his manner of writing and never used the new orthography, always wrote "Christiania", and indeed the two forms of the name for the capital substituted for one another for a long time.)

The wedding presents and Anna's things from home now arrived by boat from Risør. Sophus also seemed to have gotten some furniture (dining room chairs) from his family home in Moss, but the most important contribution from that direction was that they got to take over the serving girl, Stina Pettersen, who had been in service in the home of Sophus' sister Thea in Moss. For many years to come this Stina would prove to be an invaluable support to Anna in all her household duties. Otherwise, Anna seems to have had a flair for fixing up the apartment and making their new home comfortable, yet she never found herself completely at ease in these rooms in Mrs. Blytt's building. They would only live there however for a little more than half a year, until the summer of 1875. When Sophus wrote to Anna (who was home visiting her parents in Risør while he was on his annual tour in the mountains), he said: "You can rest assured that we will never again come to live at Mrs. Blytt's."

A short time after their homecoming from France and Germany, and after the two weeks of uninterrupted work to set up his new home, Lie wrote to Mayer that the stay in Düsseldorf would become decisive for his coming work. After having printed three or four lesser papers in Christiania, he wanted to recapitulate everything in a long article for *Mathematische Annalen*, and he wanted at the same time to develop both synthetic and analytic aspects of his points of view. In addition he planned to give out a book over the next couple of years, a book which brought together his account of partial differential equations. He thought the comprehension of his work was made more difficult by virtue of the articles being published and read individually, while in reality they had reciprocal relationship with one another, building on one another. For some years, Lie held firmly to these plans to bring out a book with his collected works on partial differential equations, and Klein wrote that the prestigious publisher, Teubner Verlag in Leipzig, would, with recommendations from him and from Mayer, certainly publish such a work.

Lie resumed his university teaching commitments in Christiania, something he never considered very demanding. That autumn the university calendar announced that he would "lecture 2 hours weekly on Differential and Integral Calculations applied to analytic Geometry." In earlier university calendars he had only been put down as "Will announce his Lectures at a later Date". He also continued the task, from Christiania, of hunting down missing Abel material. What in particular he was looking for was a paper that at New Year's 1824 had accompanied Abel's request for a travel stipend as it travelled back and forth between the various government departments. Abel's title for this paper had been *Integration af Differentialformler* [The Integration of Differential Formulas] and promised that it might contain extremely exciting ideas. But Lie received a negative response to his request from the Church and Education Ministry. Abel's paper had disappeared without a trace – and to this day it has still not reappeared.

Now, a short time after their return in November 1874, one of Lie's most important students also travelled abroad. This was Elling Holst who, with private support from a well-off wholesale merchant in his hometown of Drammen, left for Germany. In June of that year Holst had passed his public service examination in science – it was still called the science-teacher examination – and on his own trip abroad Lie had prepared for his student's stay. In Düsseldorf he had spoken with Klein, as well as with Mayer about Holst, and in the course of things, a study programme was tailored specifically for him. During this seven-month stay abroad, Holst received several letters from Lie with advice and encouraging comments, and in Christiania Lie tried to ensure that Holst would get one of the available research stipends. Holst's work and mathematical future were also discussed in the exchange of letters between Lie and Klein.

"If there is any Way at all to do so, You should go to Erlangen right away. I will pay 5 Speciedaler to meet part of the Extra Expense." This was the first advice and promise Lie gave to Holst. And when someone else had advised Holst to go to

Berlin, Lie maintained: "Believe me, there is nothing for You in Berlin unless You were to be as lucky as I and meet another Klein there."

Holst followed Lie's advice, and Lie wrote: "In all human Probability, You will not regret it". Lie maintained that Klein in Erlangen was "equally eminent as Researcher and Teacher" and a "brilliant Mind", which was attested to as well by the fact that he had just been appointed professor at the technical college in Munich. Lie felt that Holst could, "with full Confidence" go in whatever direction Klein suggested. He prepared Holst by saying that Holst, with his background in Christiania, "where the University's Lectures were still conducted exclusively in the old Classical Domains", would quite often come to feel "despondent" out there in "one of the Flashpoints of modern Research". But he must only refrain from giving up: "Do not lose Courage. [. . .] Suddenly one day a Light will go on for You." And in the following words Lie expressed what he ardently hoped would probably happen to Holst: "That Klein will do what he once considered doing, that is, lead You into my Things, such that when You come home we can work together in Common. You have to believe that we could really accomplish Something."

In relation to his life in Christiania, Lie recounted only that he suffered a degree of domestic distraction: "Otherwise, as still newly married, I am starkly distracted. And as a result my Letter-writing abroad has become completely shelved. Greet Klein for me. But I shall be writing to him right away."

For his part, Klein wrote in a letter to Lie that he thought Holst's talent lay with geometry, and that he had carried out a series of small independent works, but that he still lacked knowledge of modern mathematics, and therefore it was desirable that he remain as long as possible in Germany, and that he would gladly take him along to Munich. In a later letter to Lie, written in May 1875 from Munich, Klein wrote that it was difficult to form the future course of Holst's studies. Holst was too independent and held too many different interests, Klein felt – he had long ago given up trying to teach him modern analysis, and now after various attempts he had now given him a pure geometry assignment, that had a certain connection to something Klein himself had worked with, namely determining structures on surfaces of the third order. Apart from this, Klein informed Lie of the various posts and positions in the German mathematical milieu, and he asked Lie to greet his wife, and announced that he too now – encouraged by Lie's example – was going to get married, and that then, together with his spouse, perhaps could come to Norway for a visit next year. (Nothing came of this plan for a tour to Norway in 1876).

Lie in Christiania did not succeed in getting Holst a research stipend, or not yet in any case. In this connection he oriented Holst in the intricacies of the tug-of-war between faculties in the Christiania milieu: there were many who were now strongly opposed to allowing an adjunct stipend to go at this time to a mathematician. Lie mentioned Professors Waage and Schübeler, and said that Professor Bjerknes too did not desire to see any new scholarships in mathematics. But Professor Broch would warmly support Holst – "He knows how to grab hold of a Matter," Lie commented, and in the following words, characterised his own abilities in such matters:

I consequently am not fit for getting Anything Going, as You know I have little Talent for socialising with Folk. [...] It is going to be a painful Transition for You from Germany to Christiania, whatever the Row in Science Circles is about. Down there You are used to talking to People who have Faith in themselves, who believe in You; here You scarcely meet People who in general have any Faith in the Future of Science, and who in any case do not believe that any Other than themselves ever do Anything. There is a cold Fog invading the scientific World up here, as far as I know it. And what is fatal is that I am so diametrically different from Klein with Respect to the Ability to be able to get into the Thinking of Others.

A lively correspondence also began to blossom at this time between Lie and the Swedish mathematician Bäcklund. Albert Victor Bäcklund was three years younger than Lie, and also in the autumn of 1871 he had been among the applicants for the vacant professorship at the University of Lund, but was actually too young to have been taken seriously. As far as Lie was concerned it had been this application for the post in Lund which had resulted in the rash of sympathy that, a year later, had broken out in Parliament, when Lie himself had been appointed professor. For Bäcklund too, the application for the professorship was to have fateful consequences. Namely, one of the friendly academic secretaries had allowed Bäcklund to read the works that Lie had submitted with his application, and this led the young Bäcklund into mathematical fields that he would work with for a long time, and where he also created for himself a posthumous reputation, for *Bäcklund Transformations* became a well-known concept in the mathematical literature. It was Lie's first work on partial differential equations that inspired Bäcklund, and in the correspondence that developed between them, their new papers were exchanged and commented upon. Lie would come to regard Bäcklund's work as the most important Swedish contribution to mathematics. The contact between them was particular intense now, from late autumn of 1874, and for the next couple of years.

Bäcklund's background was in astronomy and mechanics – in his doctoral thesis of 1868 he determined latitude at the newly established observatory at Lund; a year later he became associate professor of astronomy at the University of Lund, where he later became professor of both mechanics and physics, and finally, rector. Much of Bäcklund's work gradually came to deal with mechanical and physical theories, but it was his work in pure mathematics that outlived him, and it was in this field that contact with Sophus Lie was important. Stemming from Lie's work on the well-known point transformations, another class of transformations were discovered – namely, contact transformations – with the property that each differential equation could be converted to a new differential equation, and with this there arose the question: is it possible to find other transformations with the same properties. Bäcklund would come to demonstrate that this could not be so, but he also showed the existence of infinitely many new transformations that could transform a given differential equation, or a given system of differential equations, into other such systems. And it was this type of transformations, and their properties that gradually received the name *Bäcklund Transformations*. The most important example of such a transformation is the one he found in 1883, and which gives the recipe for how a surface with constant negative surface curvature can be transformed into a new surface with the same properties. Lie was always preoccupied with creating the best working conditions for Bäcklund , and tried to help him as well as he could with statements of support – apart from pure mathematics their correspondence addressed the fillling and shifting of various academic positions and personal relations within the milieux of science in the Nordic countries. Afterwards there were many who lamented that Bäcklund through all his posts and teaching duties did not find the opportunity to dedicate his whole life to pure mathematics.

After an intense exchange of letters during the winter and the following spring, Bäcklund came to Christiania to visit Lie at the end of June 1875. Anna was away for a summer visit to Risør, and in a letter to her, he told about his Swedish friend's stay: "We have had an extremely pleasant Time and have talked an awful lot about Mathematics." Bäcklund also thanked Lie for a lovely week in Christiania – and their correspondence continued, with a discussion of transformation problems, designing of papers, and Bäcklund's struggle to get a decent professorship.

Lie left for the mountains at the beginning of July, and for a month that summer he was hiking together with Ernst Motzfeldt. They went up in the high mountains, in Jotunheimen, and passed along the great alpine lake of Gjendin[29]. On their homeward journey they visited Bjørnsterne Bjørnson, Norway's leading poet, who earlier that summer had moved to his new home north of Lillehammer, at Aulestad in Gausdal. Motzfeldt, who had indeed sat on the executive of the Student Society together with Bjørnson, stayed several days longer at Aulestad than did Lie, who hastened back to Christiania. Anna was still in Risør, and according to plan would return a little over a week later, in the middle of August. "You can well believe that it is not very pleasant for me here alone," Sophus wrote to her. He sent her five speciedaler, saying it was horribly hot in the capital, but that her flowers were surviving well – "Stina, it seems, has done Things well." He himself had "a tremendous Amount to do" – he was an examiner of the *examen artium*, and concluded: "I long terribly for my little Wife and hope I get to see her again very soon."

[29] Gjendin is a lake in Jotunheimen made famous by Henrik Ibsen's lyrical play "Peer Gynt".

In the Lee of "the Modern Breakthrough"

From the autumn of 1875 Sophus and Anna's residence was the Tandbergs' løkke[30] or summerhouse estate on Drammensveien, and they would continue to live here for over four years. They took an apartment on the second floor of the house of the Tandbergs. The husband died a couple of years later, and in the census, "Housewife Charlotte Tandberg" was redesignated as "Landlord". The well-off citizenry owned their rows of houses and summer places out along Drammensveien, which was the main arterial road for traffic in and out of the city. Much of the new construction and expansion faced other streets and areas. This autumn was to see the first tramlines in the streets of the capital. The cars went from Homansbyen (in the region of today's Bislet Stadium) to the downtown Stortorvet (Main Square) and on to (old) Oslo in the east, on elevated tracks. These cars were drawn by horses. The tram line concession had been granted on the understanding that they would function with horsepower, and not steam, for which also an application had been made. The tramline's inauguration on October 6th was an event that caused a sensation, and drew a large crowd. The cars were fully loaded. During its first year the tramline conveyed one and a half million passengers. Horse trams were a step in the direction of modernity, and this improvement in communications naturally led to urban development a couple of years later, which in turn brought the population of Christiania to over one hundred thousand inhabitants.

Lie used scarcely 10–15 minutes to walk from home on Drammensveien – passing by the Royal Palace, where Brynjulf Bergslien's monumental mounted statue of King Karl Johan was unveiled that autumn – to his lectures at the University, and as earlier, it appears that lecturing cost him little in terms of preparation. As a rule he stood up without a manuscript and impressed his listeners with his superior knowledge of the material. In the university calendar for that autumn there was the notice that Professor Lie would later announce his programme of lectures, something that probably meant in practice he, as in the two previous semesters, would continue to lecture three times a week "on Differential and Integral Calculations applied to Analytic Geometry". Not until the following autumn would the university calendar listing be changed, and then it merely announced that Extraordinary Professor Sophus Lie "will continue his Lectures on analytic Geometry 4 Hours weekly". The content of these lectures was probably not so very far from "the old

---

[30] The better-off families of Christiania repaired to their summer properties in the countryside on the edge of the city. Such summer residences, or *løkker*, were both productive gardens and places of relaxation.

Classical Domains" that he himself, in his letter to Elling Holst, accusatorily argued was the only form of lecturing in Christiania. But the classical mathematics certainly had to be taught before one could speak of anything else, and the syllabus for the public service examination in science, "the science teacher's examination", took as its aim, first and foremost, to educate science teachers for the higher schools. It was these students Sophus Lie taught. He had no colleagues and no students who could assist him in his own mathematical research. His only contact with his own field was by letter, and above all with Klein and Mayer in Germany.

During the summer of 1875, Elling Holst had returned to Christiania from abroad. His wedding, in which he married Inger Skavlan, sister to Lie's friend, the professor of literature, Olaf Skavlan, was held in August. Holst became a teacher at Aars' and Voss' School in Christiania, and a year later was a research fellow in mathematics at the university – a position for which he had to thank, not least of all, Sophus Lie for his persistent pressure. Lie seems still for a time to have had expectations that Holst could become a mathematician he could collaborate with, and someone who could contribute to Lie's own mathematical research. As a fresh research fellow Holst received a public grant to study in Copenhagen, and he won the university prize competition, receiving a gold medal for his work entitled "On Poncelet's Significance for Geometry. A Contribution to the Developmental History of the Ideas of Modern Geometry." This was a work to which Lie gave his official stamp of quality by having it published under the university programme in Christiania. Holst later managed to study in Paris, published several works on geometry, and took his doctorate in 1882 with a dissertation entitled "A Pair of Synthetic Methods especially for Use in the Study of metric Properties". Holst came to be known as an unusually inspiring teacher, both to those who studied mathematics only for the secondary examination, the improvement tests, and for those who went further toward "the science teacher examination". When, in 1895, Holst was finally made associate professor in mathematics it was "with the particular duty to teach the common school mathematics, treated scientifically".

But despite the support and encouragement Lie gave Holst, he realized rather early on that Holst would never be what he had hoped: a collaborator and support in the new pioneering mathematical work. As early as New Year 1876, Lie must have had clear indications about how Holst would and could contribute to mathematical research, and he commented about this in a letter to Klein, who indeed also knew Holst's mathematical capacity. Klein answered Lie: "I am quite in agreement with your assessment of Holst. You must nonetheless give him my greetings when you see him." Nevertheless Lie kept in close touch with Holst through the years, through both letters and meetings – and after Lie's departure from Christiania it was Holst who took over part of his teaching responsibilities. It was also Holst who worked to get Lie back from Germany, and who wrote the first biographical articles about Lie.

Externally speaking, it was accordingly the correspondence with Klein and Mayer that continued to inspire Lie in his work. After Klein got married in August 1875, and had returned from his honeymoon, he confided to Lie that he now felt both

Artist and mountaineer Catharine Hermine Kølle painted this picture about 1840, on the way between Grungedal and Haukeli, one of the major routes between east and west in southern Norway.

The Norwegian landscape, which painters, writers and researchers in the natural sciences had long studied and enjoyed, gradually began to come to the attention of other groups. To hike in the wilds of the Norwegian mountains during the summer months became a fond pastime, not least for students. Sophus Lie hiked this route through Telemark several times, the first being in 1862 together with Kristofer Janson.

The Land Route at Risør

Commissioned by royalty, this was painted by Jacob Munch in 1821 – and such was the national highway and surrounding landscape in the 1870s when Sophus Lie journeyed to and fro on visits to his fiancée, Anna Birch, at Risør. On one of his journeys along this road, Sophus wrote to Anna: "The Weather was so good and in the Evenings I had the most lovely Moonlight. People think that I have very strange Tastes; but this strengthens the Muscles and Nerves, and besides, I find myself with such good Feelings as I roam the Countryside as a Hobo. By Habit, I am always deep in my own Thoughts, and building Castles in the Air for the Future."

Romsdalshorn, *painted by Johan Fredrik Eckersberg in 1865*

Sophus Lie was a most avid hiker and mountaineer, and was admired for his long and gruelling expeditions in all directions across the country. He was active in DNT, the Norwegian Tourist Association from its inception in 1868, and for a time was on the executive. Lie was among the first to extol the magnificent mountain landscape of Romsdal and the interior of Nordmøre. He wrote home to his wife, probably in 1879: "From Molde we went to Eikesdal, which as I have said before, runs parallel with Romsdal. It was here that Bjørnson wrote his famous Song 'Over the high Mountains'. Particularly at its upper End, the Valley is especially surprisingly narrow. There are many magnificent waterfalls."

Swiss landscape, *painted by Knud Andreassen Baade in 1850*

After Lie arrived in Leipzig at the age of forty-four, in 1886, one of his dearest wishes was to get to know the mountains of Switzerland, Austria and northern Italy as well as possible, and he used every opportunity to do so. As early as his first autumn in Germany, he wrote home to his friend Motzfeldt: "I have now been out for fourteen Days on a Foot Tour in the Tyrol and Salzburg. I am determined, so long as I can hold out in this Manner, to pursue my Plan of going on a Foot Tour every Summer, until I turn 60. It is still going fine but I find that with the Rucksack on my Back some of the greater Climbs are more difficult."

in a better mood and in a better condition to work than he had for a long time. During the recent years his ideas had been much too dispersed, Klein wrote, and reported that once again he was working with "Analysis situs" (topology) – and added that it was really not something Lie himself was interested in ["die Dich ja Nichts angeht"] – and that he would also try to write an article he had long been intending to write, on partial differential equations. The article would not be original, but rather, instructive, Klein wrote, and he asked Lie to send back the letter he had received from Klein at the end of July, where he had written down a number of formulas which Klein now needed so as not to have to do the work again. Klein would also be grateful if Lie wrote what, according to Lie's assessment, it was in Klein's work that ought to be included in the article. It seems that Lie neither answered this question nor returned Klein's letter; in any case, the original is to be found among Lie's surviving papers.

Before the summer vacation Lie had sent his article on partial differential equations to Klein for publication in *Mathematische Annalen*, and now in October, Klein could report that three sheets were already printed, and that the journal would also be printing a work by Bäcklund on his special transformations, together with a note by Zeuthen on singular points. Klein apologized that he for the time being had not managed to write, and nor had he found the correct form for his critical comments on Darboux's work. Otherwise, the existence of this critical organ, a journal where one could freely express one's meaning, was something he had often longed for.

Some months later Klein was able to report that Darboux had written to him for the first time in two and a half years, and asked for references from articles published in *Mathematische Annalen*. Klein was pleased to receive this enquiry, for indeed he had several interests in common with Darboux, as he wrote to Lie, and he would now set aside his projected critique of Darboux. Klein was otherwise able to report that he was working tremendously well with his study of plane curves, that his work with *Mathematische Annalen* took up a great deal of time, and that he had had an unexpected visit from Ernst Motzfeldt, and that the mathematician Ferdinand Lindemann wanted to include in his textbook everything that Klein had written through the years on contact transformations and ordinary differential equations.

Sophus and Anna in Christiania kept in close contact with their families, or at least with parts of their respective families. Anna wrote regularly to her stepmother in Risør, and in the capital they were in the habit of exchanging visits with the Birch-Reichenwalds and the Motzfeldts. On Sophus' side there seem to have been regular contacts with Uncle John and his wife in Øvre Vollgate, with Cousin Johannes, his Amanda and their children. Doctor John Lie, the congenial, seventy-five-year-old, had in the meanwhile aroused public notice when he applied for the vacant public position of medical brigadier. The Department had unequivocably considered him too old, but then the charming Lie showed himself off in person before King Oscar, in order to prove he retained the energies of youth. Naturally enough, the king did not want to override the Department's decision, but ordered that the honourable and gentlemanly John Lie be appointed Knight of the Order of St. Olav "for his worthy Contributions to Public Life". He had already, at an earlier date, been awarded the Wasa Order.

It seems they also maintained frequent contact with Sophus' sisters, Laura and Mathilde, and Mathilde's family in Tvedestrand. Encouraged by the good results of the industrial exhibition in Drammen in 1873, Sophus' sister Laura was working on plans to establish a school in Moss for perspective serving girls. These plans did not come to be; instead, she applied for a position as girls' teacher in the capital, but the first time around she was not successful. Sister Laura moved to Christiania in 1875, and a short time later began to work as housekeeper for a well-to-do businessman at Dronningens gate 13, who had four children to educate, and after a couple of years as housekeeper for this businessman and his children, she was appointed a teacher within the Christiania public school system – with an annual salary of 720 kroner, first at Kampen School[31], and then seven years at the Ruseløkken School in another working class quarter of the city – before, in October 1886, getting down to what would be her great life's work, as the housekeeper of the Eugenia Institution in Christiania, a position she held for more than twenty years. Her considerable ability in the literary fields, as well as her practical activities were highly prized everywhere, and by all reports, left the mark of her extraordinary pedagogical abilities within and without the school. What was particularly noted was her ability to get children *to work*, and during the cold winters she helped both the children and parents in her district with food and clothing. Sister Laura remained close to Sophus and Anna down through the years, and to their three children she became a much-loved aunt.

Sophus seemed to have a more sporadic relationship with his brothers. There was little or no contact with his eldest brother, Fredrik, who was senior teacher at the Kristiansand Cathedral School. Later however, Fredrik's son, Sigurd Lie, who was born in 1871, and who would consequently become a well-known composer, for a certain period was a frequent visitor at his uncle's house. But Sophus and Anna had more contact during these years with his other brother, Lieutenant-Colonel John Herman in Bergen. John Herman even went so far as to write a letter, in December 1875, specially to Anna, in which he expressed his hope of initiating a correspondence with his sister-in-law, since she was now "drawn into the most intense of all Associations by settling down at the Side of Brother Sophus in the Role of his Better-Half." John Herman taught mathematics at the Bergen grammar school which taught the sciences, as well as working in the city's military office. His wife, Petra Thaulow Klouman, the daughter of General Klouman of Bergenhus, eventually had ten children, two dying at a young age. Lieutenant-Colonel Lie had only his officer's salary with which to provide the means of living and schooling for his children, and he complained about the cost of living and high prices. But

[31] Kampen was then a residential area for industrial workers and their families, and today, in the twenty-first century, is a renovated oasis of compact nineteenth century wood-frame houses surrounded by Oslo's twentieth century modernity.

in the field of shipping, the economic downturn was somewhat lighter, and John Herman had good contacts among Bergen's shipping merchants and shipowners. And after Vicar Lie had left a sum of money to be divided among his children, John Herman had got Sophus and Laura to join him in purchasing "shipping shares". Laura gradually contributed 1,000 speciedaler, a sum that was now the equivalent of 4,000 kroner, and as for the two brothers, they probably discussed still greater amounts. In order to minimise risk they decided to purchase shares in three different ships, Although it was John Herman who informed them what was for sale, it seems that it was Sophus who had the best overview in the early years, of the size of the sum each of them had invested in each of the three shares, and how much each of them had profited. Later it was to be John Herman who informed the other two on how much they had to the good. In the first half of the 1870s there had been a strong upturn in trade and commerce, particularly in shipping. Norway was then the greatest shipping nation after England and the USA. Even when the economic cycle turned, precisely during these years of 1874–75, the profitability of freight transport remained high, for the time being. In 1876 John Herman was able to report to Sophus that one of the ships in which they had shares, paid dividends of nine percent – a time when average dividends in other sectors were only half as high. The ship that had returned this good dividend plied the Baltic Sea with cargoes of timber; another ship, which sailed the Atlantic, paid lesser returns. The siblings Sophus, Laura and John Herman seem to have stood together on these "Shipping Shares" for over ten years. It was not until 1886 that John Herman thought that the times had been so bad that they ought to consider selling their shares and buying bank bonds instead – at that point Sophus received a payment of 3,000 kroner on this investment in the shipping business.

For the time being, it seemed as though the on-going work on the Abel edition cost Sophus little in terms of effort and worry. From time to time he replied to some enquiries from the publisher, Grøndahl, but the real protracted work came several years later when the corrections and proofs had to be checked. Meanwhile, it was Sylow who carried out the consistent day-to-day work on the Abel papers. In December 1875 Lie received a letter from the German mathematician and editor, Carl Wilhelm Borchhardt, with questions about dates of publication of the different volumes of *Crelle's Journal* between 1826 and 1829. These were precisely the years during which Abel had published his epoch-making works, and after his death there broke out a certain discussion on whether the German Jacobi had taken something from Abel and published it as his own. Therefore it was very important to get to know the precise publishing dates for the various works that both Abel and Jacobi had brought out in this period. This question as to which of them had been first with the discoveries, and therefore had priority was also something with which Professor Bjerknes was strongly preoccupied as he worked on his biography of Abel. In Bjerknes' Abel biography too, this question – the struggle to award priority to Abel or Jacobi – was given pride of place. After having studied the dates of letters and journals, Bjerknes' conclusion was clear and unambiguous: Abel had come out first with his discoveries, and Jacobi had, in illicit ways, taken the honour for some things that belonged to Abel. In any case, according to Bjerknes, it was shameful that Jacobi had not once in his works referred to Abel. Lie, at this point in time, seems to have been little preoccupied with this question, and it appears unclear as to whether he was able to give Borchhardt any satisfactory answers. As far as it goes, this also seems remarkable that Borchhardt himself did not have information concerning these dates, for indeed, it was Borchhardt who had taken over the editorship of *Crelle's Journal* after Crelle, and had edited the journal for the past twenty years. When later, in 1880, Bjerknes' biography of Abel was published, Lie was extremely critical of Bjerknes' preoccupation with this question of priority – remarkably enough, since this was, or would become a type of question in which Lie himself, in relation to his own works, was, and would continue to be deeply absorbed.

Contact with the Swedish mathematician, Victor Bäcklund lasted throughout 1875 and 1876. By means of letters and reports, they kept one another up to date in relation to their foreign contacts, and they discussed problems in their fields, first and foremost Darboux's works, and in particular, his integration methods.

At New Year's 1876, Lie received word from Dresden in a letter from the mathematics professor, Leo Köningsberger, who had requested references for Lie's writings, that it would be an honour to have Lie's name, together with a list of his published writings. Lie answered promptly and received a letter of thanks in return.

In 1876 Lie also received the considerable esteem arising from being one of twelve foreign honorary members of the *Mathematical Society* of London.

An important incident in the academic milieu on the homefront that year was the foundation of a new scientific periodical, *Archiv for Mathematik og Naturvidenskab*, which in a short time led to an almost doubling in the increase of the annual publication of scientific works in Norway. Sophus Lie was one of the three who took the initiative for this, a journal in which he would come to publish a series of articles. It was important to Lie to get his "group theoretical" research printed: the writings of the Academy of Sciences came out too seldom, and in the foreign journals he felt he had to bring out works of greater and more synthesising nature. But there were also deeper ideological reasons for this periodical. Many were very critical of how difficult it was for new insights and new research to come onto the scene and find space in the established milieu. The new periodical was founded in stark and conscious opposition to the country's leading journal of the natural sciences, *Nyt Magazin for Naturvidenskaberne*, which was the successor to Norway's first science journal, *Magazin for Naturvidenskaberne*, which was founded in 1823, and where, among others, Abel made his debut as a mathematical writer.

The geology professor Th. Kjerulf was central to the editorship of the well-established *Nyt Magazin for Naturvidenskaberne*. Professor Kjerulf was among those who campaigned energetically against Amund Helland being allowed to have the floor to profess new scientific theories and outlooks. Lie, who had struggled discretely for Helland in the Academy of Sciences, had realized as a result of this strife, that the country's scientific environment was too narrow. There was however a *Polyteknisk Tidsskrift*, founded in 1854, and which was closely linked to the scientific world, but this was oriented more toward experimental developments in science and technology. Helland had been locked out, censured and slandered, and forced to turn to Sweden to get his works published, with the support of Professor Nordenskiöld. Helland came to symbolise the struggle of the country's young scientists against the established ethos of public officialdom. Many felt it was high time that new channels for publishing were set up. Furthermore, many were also clear that the factions, whose members were visible and went into scientific production, also fissioned into political left and right, a process that was breaking out in all institutions, private as well as public.

On different fronts, the various academic and scientific disciplines and their milieux each had their own struggles against the establishment. Apart from Sophus Lie, the editorial board for the new journal, *Archiv for Mathematik og Naturvidenskab*, included the zoologist, Georg Ossian Sars and the medical intern, Jakob Worm-Müller.

In the milieu of the natural sciences some of these contradictions were explicitly around Amund Helland and his point of view of geological change and development. Helland's work was attacked both verbally and in print. In opposition to the time-honoured view that valleys and fjords were formed as a result of basic geological faults, Helland argued in favour of the strong erosive workings of glaciers, and consequently their ability to hollow out valleys and fjords. As support for his point of view, among other things, Helland had measured a number of mud flows carried by rivers coming out from under Norwegian glaciers, and in 1875 he had been in Greenland and measured the speed of glacial movement. The measurements he made in Greenland aroused excitement as he found that a glacier could move up to twenty-two metres in a twenty-four hour period, and this was a new overwhelming proof that the action of glaciers was sufficient to account for the furrowed landscape.

But great and grave dissension now broke out in the realm of scientific theory and outlook. This took the form of serious debate about how human beings and all living things had developed from earliest time onward. It was the Englishman Charles Darwin with his Theory of Evolution, who created seismic splits and tensions not only in the scientific, but also in all the academic milieux, as well as in all cultural camps, and indeed, among broad sections of the people. Darwin's epoch-making book, *The Origin of Species*, came out in 1859, and many Norwegian scientists came to know Darwin's works and theories in the course of the 1860s.

In Norway at that time, the person best prepared to evaluate the new theories was Professor Michael Sars, who had mapped the dispersal of marine species not only along the Norwegian coast, but also the coasts of the Adriatic and Mediterranean Seas. He had studied the development of marine species and proven the link between existing marine species and fossils from bygone times. Despite an initial opposition to Darwin's theory, Michael Sars, through his own extensive studies, came little by little to support the new theories of Darwin. But not until 1869, only six months before his death, did Michael Sars express this support unequivocably. In a lecture he held at the Academy of Sciences, "On Results Gained from Researches of the Ocean Deeps", one finds (in his manuscript): "I, for my Part, must openly state, that I, prejudiced by the old Preconceptions, at the Beginning was not inclined to take up Darwin's Theory; but, indeed, the more I have thought it over, the more I, especially led by my later Studies of Nature, became drawn to the Recognition of this Theory's Truth."

Upon the death of Michael Sars, his son, Georg Ossian Sars took over his father's place on the editorial board of the establisment journal, *Nyt Magazin for Naturvidenskaberne*, and it was here that young Sars published his own works, until he now accordingly helped found *Archiv for Mathematik og Naturvidenskab*, and he continued on as editor of this new journal until his death over fifty years later.

Georg Ossian Sars himself was one of the most diligent contributors, with almost sixty articles. He continued as well to follow professionally in his father's footsteps. He was a marine zoologist and undertook several research expeditions in Norwegian and foreign waters. He gradually withdrew more and more from the reality of human life and devoted himself to crustaceans, for the knowledge of which he became world-renowned – the greater part of the 8,000 pages of material he left behind at his death, decorated with his own neat and tidy drawings, dealt with the crustaceans. Much of this material was published in *Archivet*, as well as in many foreign journals and books. But in the 1870s he was a clear-sighted and fearless representative of Darwin's Theory of Evolution, and all that that entailed.

In 1872 and 1875 he had published two university programmes of lectures under the common title, "On Some Remarkable Forms of Animal Life from the great Depths of the Norwegian Coast." (The first was partly based on his father's surviving papers, while the other presented new discoveries about *Brisinga coronata*.) This was the first time G. O. Sars wrote in English, and was indeed closely intertwined with his view of Darwin's revolutionary insights into biology, and was a homage to the science of zoology, expressed through the medium of the English language. Perhaps it also served as a hint to Kjerulf, the editor of *Magazinet*, to reconsider his predominantly German-speaking praxis.

To G. O. Sars, Darwinism was a theory about how new and complex phenomena could occur and develop. But the debate around Darwinism and the Theory of Evolution was for the most part a discussion on religious and non-religious science, and many complaints were lodged with the Collegium concerning G. O. Sars' lectures. This resulted in him not being given permission to teach the young and impressionable souls who came to the university to prepare for the secondary examination. G. O. Sars lectured only to those enrolled for the science teacher public examination, and to medical students.

Through the 1870s, Georg Ossian Sars, together with his brother Ernest Sars – who maintained that much of history could be understood in the same way as the world of nature (and in the Norwegian context the historian became more famous than the zoologist) – became the most notable representatives of the new theories and the new attitudes that were asserting themselves everywhere and causing debate. The year after *Archivet* was founded, Ernst Sars became the driving force behind the journal *Nyt norsk Tidsskrift*, which Olaf Skavlan would join five years later, when it was then under the title, *Nyt Tidsskrift*.

Archiv for Mathematik og Naturvidenskab was Norwegian science's answer to the new currents within European science. The climatic shift, which in the Nordic countries was referred to as "the modern breakthrough" here found its mouthpiece for the natural sciences. During these years its methods of thinking and its attitudes were a melting pot which supported a positive and optimistic belief in the future and the possibilities for a better life. In literature, this was the time that realism blossomed forth – the kind of literature that would set actual problems "for debate", and was thereby functionally useful to both human and social life.

A polarization had clearly come to the surface in the academic milieu when Ernst Sars was appointed professor of history in May 1874. This appointment had occurred in the same manner as the professorship of Sophus Lie – it had been undertaken by Parliament, and several of the same initiators supported it. But, as Sophus commented to Anna, when Ernst Sars was appointed by Parliament, it was against strong and outspoken opposition from parts of the established university community, which was still the bastion of the hegemonic outlook of the public servants. Furthermore, the reaction from university quarters came later that year – while Sophus and Anna were on their honeymoon in Paris – and one of the main opponents of Ernst Sars and all that he espoused, was Professor (of Medicine) Ernst Lochmann, who specialized in medicines, poisonous substances, and hygiene. When Lochmann gave a speech to the incoming students at the university's assembly hall in September 1874, he fought back vehemently against what he considered to be dangerous and contagious tendencies both in the worlds of science and of Norwegian spiritual life. His theme was natural science and its position in contemporary thought. Having briefly outlined the highly progressive benefits that the natural sciences had undoubtedly brought human kind, Lochmann settled into a relentless thundering speech about how the triumphal forward march of the natural sciences had to be stopped. That is to say, the natural sciences had taken licence in areas beyond their competence, had expounded upon, and set about rectifying fields in which they absolutely did not belong. All in all, concluded Lochmann, this mighty man of Norway's medical faculty, it was the contagious spirit of materialism that had broken out. Lochmann maintained that when only physical and chemical energies and cellular structures were recognised as real, this turn of events led many scientists to cease believing in the immortality of the soul. Rather they took the position that the only thing capable of perception and recognition was the steadily improving and developing human brain, and in this spirit of the natural sciences, Professor Lochmann consequently saw a close relationship to the positivism of the day.

When mathematics, physics and statistics were applied to the conditions of human life, as well as to the realm of the spirit, then there was no longer any place for God, but only great natural laws which in increasing degrees men would uncover and understand. In the eyes of these ungodly ones, humanity was a current in which the individual's will was constantly attenuated, and freed from personal responsibility and blame. With reference to the French writer, Émile Zola, Lochmann stated: just as a liquid like vitriol was innocent and could not be blamed for its corrosive properties, and that sugar was sweet and ought not to be praised for its sweetness, now too should the human being not be held responsible for his deeds. Lochmann scoffed – should crimes, suicide, mental illnesses, and indeed even religion and morals, likewise now be treated as the results of natural laws, climate and soil! And that such conceptions were necessarily associated with the extreme left in politics, should certainly not astonish anyone. Positivists and materialists could certainly, on the surface of things, have a dazzling character in modern culture, but bore within them the embryo of society's downfall. Lochmann's more dire prediction was that without faith in the eternal, without a thought for judgement and responsibility, both the individual and society would disintegrate. There was talk about a contagion, which was already exploding into epidemic proportions. Thus, with professional gravity, Lochmann brought his arguments to ground: in the same way that modern medicine showed contagions could be confined and stopped, so too could the infection of literature and the soul be prevented by means of proper hygiene. Moreover, one of the most important tasks in this struggle was to keep modern schools of thought out of the university – as in previous generations the university must be built upon a Christian foundation. With Professor Gisle Johnson as its greatest exponent, the Faculty of Religion was pointed out as paradigmatic. Even though this faculty had not actively opposed the professorship for Ernst Sars, it was to be lauded for not having the monstrously dangerous, and contagious ideas now found in the Faculties of History and Philosophy, and Mathematics and the Natural Sciences.

But even within Lochmann's own scientific circles there were some who felt that the milieu was so suffocating that they too wanted to strike a blow for a new and freer type of research. The doctors and medical faculty had had their own professional journal since the 1820s – at first *Eyr*, and then from 1840, *Norsk Magazin for Lægevidenskaben* – but not all of them felt that this was to their satisfaction. The third man on the editorial board of the new journal, *Archiv for Mathematik og Naturvidenskab* – who joined Lie and G.O. Sars – was the man of medicine, Jacob Worm-Müller.

Worm-Müller was eight years older than Lie and three years older than G.O. Sars. Following his public service examination in medicine and some years of a busy medical practice in Christiania, he had decided to devote his life to science. For six years, beginning in 1864, Worm-Müller lived abroad, mostly at his own expense. He studied physiology at the leading universities in Germany and Austria; he gave lectures at gatherings of naturalists, and he published several articles. His first great work was an enquiry into the influence of heat and chemical materials on the electro-kinetic energy of muscles and nerves. He had then become a research fellow at the University of Christiania in 1870 – this being the same year that G.O. Sars also held a research fellowship and the year before Lie received the same much sought-after adjunct stipend, that is, the graduate scholarship.

Worm-Müller became extraordinary professor of physiology in 1873 and directed the Physiological-Chemical Institute, which he himself had founded. He published a series of papers in German, not the least of which in *Pflügers Archiv für die gesammte Physiologie*, on "The Mechanics and Valve Contractions of the Heart", on "The Function of the Papillary Muscles", and he submitted fundamental analysis of diabetes based upon a series of chemical-physiological inquiries, showing that the substance bromide of potassium worked on diabetes and quantitatively determined the presence of grape sugar in the urine. Worm-Müller was said to be a restless and extremely energetic man of science who was tireless in carrying out and checking upon his experiments. His constant good humour made him a popular lecturer about whom the students were always telling anecdotes. But a debilitating asthma gradually prevented him from working, and he died in 1889, only fifty-five years of age. Some years later, Amund Helland took over the position of the third member of the editorial board of *Archiv for Mathematik og Naturvidenskab.*

Studies and researches into the present-day and the actual were now on the agenda. The theories that depicted and mapped out the development of human and animal life on earth and in the sea, as well as the formation of the landscape, were now a sign of the times. There was discussion everywhere about the objective representation of reality, a realism with clear didactic intentions: the belief in progress was an important factor in intellectual life. From Denmark came Georg Brandes' idea of free-thinking, free from the constraints of all socially and traditionally determined contexts – including Christianity – thoughts that appeared so radical that Brandes was denied the right to hold lectures at the University of Christiania that year (1876).

The new faith in development and progress was raised against the religious and romantic dream of a new millennium. In literature, psychological absorption was a vital quality, and in opposition to romanticism's cultivation of self, there was now an insistence that through an individual's specific traits it should be possible to glimpse the social type. The play between the individual and the social type was one of the demands that realism now placed upon character delineation. It was important to characterise "the species", observe the relations, the milieu – the actual place to which the characters belonged. There were patterns to be found; there were also invariables in the human psyche – this applied to the uncovering of laws of intellectual life and behaviour as well. Within Norway there was also talk about the distinctive "folk character" of various parts of the country, and everywhere "the two cultures" were set up against one another – a nationally determined peasant and public servant culture against an international, imported culture.

Philosophy Professor Marcus Jacob Monrad was at the pinnacle of his academic power and prestige during the 1870s and he also played an important role in the struggle against Darwinism and positivism both in the university and in society. Even though Monrad – Sophus Lie's old teacher of Norwegian and philosophy – had not at any time had the qualifications to understand the contents of Lie's mathematical work, he had nonetheless been among the first to support Lie's cause, and he contributed to getting his "Imaginary Theory" published in 1869. And now, in his argumentation against positivism, he used mathematical concepts. Monrad declined to side with the dogmatic theological standpoints of Lochmann and his like, yet he reacted strongly against the proximity between human beings and animals toward which Darwinism tended. Monrad maintained that the natural laws revealed themselves as conceptual laws, and that therefore in principle there was little or no difference between speculative philosophy and empirical research. But the factual ought not be lined up together with the ideal, and the natural laws must never be substituted for the laws of ethics, to which, according to him, there was a tendency in the thinking of positivists. The laws of society could not be plotted by taking as one's point of departure human conduct, passions and feelings – the reexamination that the positivists insisted upon was at best imperfect.

Following Michael Sars' lecture to the Academy of Sciences six months before his death, and his public presentation of his own meticulous enquiries which had been constructed according to the theory Darwin had formulated, the Academy of Sciences set out to throw light upon, and to discuss, both Darwinism and positivism, by organising a series of lectures. It was Professor Monrad who supported these considerations, which also resulted his book in 1874, entitled *New Currents of Thought in Recent Times, A Critical Survey,* and which was translated into German five years later. In this work he used mathematical concepts to demonstrate that not all true comprehension and knowledge can be based on experience. The leading positivists – and he mentioned the names of August Comte and John Stuart Mill – had namely, "the most unmathematical Thoughts, assuming that the Points of Geometry, Lines, etc., are physical Realities". They claimed that the Point is *minimum visibile,* the Line is without width, etc., and maintained as well that basic premises of mathematics too, were founded upon "experimental Truths, Generalisations drawn from Observations." As far as the positivists were concerned, then, even Euclid's fundamental plane geometry was conditioned by, and founded upon, human experience.

According to Monrad, the consequences of such a mode of thinking was meaninglessness, and explained as follows: "The Statement that two straight Lines could never enclose any Space was an Induction made completely on the Basis of Experience – this Experience was made partly by real external existing Lines, and partly by such Lines as we have construed internally." Moreover, the positivists did not preclude "anything as *unthinkable*"; thus, this was none other than "The Possibility that sometimes we could make use of what Experience contradicted" – namely, the complete nonsense that at some point in time it could be demonstrated that two straight lines could enclose a space. Monrad maintained that it was quite likely that experience and observation could have provided historical and psychological bases for human beings having arrived at this and at many other "Common Truths", but that "when once these have been found, they do not rest upon these Experiences, but are evident in themselves, since the Opposite shows itself to be unthinkable." Monrad gave the following example: "Let us assume that one perhaps first by means of adding up Apples or Nuts, arrived at the Insight that $3+1=2\times 2''$, but thereafter and for all time this would endure of necessity – experience will never be able to show anything different. To take "the Unthinkable only as a psychological Fact, and not a logical Condition" was an attitude completely in opposition to basic mathematical principles – and thus, should the certainty of these basic principles thus only "agree with our most ingrained Preconceptions"? No, asserted Monrad, there were truths that existed beyond experience – "the straight Line is a Concept which has certain *necessary* Properties", and that "two straight Lines cannot enclose any Space" is an eternally valid truth.

Sophus Lie never commented on this line of argument made by his old teacher and well-wisher. Were he to have entered the argument and used his mathematical insights, he would have most certainly complicated the picture to a degree beyond recognition both by Monrad and his opponents. In modern geometry, Euclid's old parallel postulate was no longer the only one possible, nor was it external experience that was the basis for the new research and insight.

It seems that in other countries as well, mathematicians did not try to advance their professional views in this debate about empiricism and positivism. To be sure, in England non-Euclidian geometry had for a period been regarded as equally outrageous and blasphemous as Darwinism, and subject therefore to attempts at exclusion from tradition-laden teaching institutions. But most mathematicians in the 1870s and later, felt that studies of the history of mathematics were an unnecessary waste of time, something of little worth and almost deserving of contempt. If older points of view were correct and had validity, and were embraced, then so too were the newer mathematical theories; if however they were erroneous, it was thus obviously misleading to study and be preoccupied with them – such was the attitude. Instead, one fastened on to a certain "superstitious faith" in the power of differential equations to solve the enigmas of the world, and the most practically-inclined believed that in function theory – and above all, in Weierstrass function theory – they had found the perfect tool. And rather than the everyday common philosophical debate, the mathematical world allowed itself to be astonished by the publication, in 1876, of works left behind by Bernhard Riemann at the time of his death. Riemann had been dead for ten years, but his works revealed themselves to be full of epoch-making ideas, although inaccessible to most.

The scholarly historical aspects of mathematics were definitely standing in the shade. Perhaps Sophus Lie's later newspaper articles about the teaching of mathematics might imply that he had preserved a certain interest in the subject? But he probably thanked his lucky stars that he was involved in a science where he could not be monitored and supervised, where he could involve himself peacefully in his research, where he could not be held responsible by society for his points of view, and did not have to pull back from his research, and defend himself against attacks – whether in speeches or written texts – made by interested observers of the issues of the day. For Lie seems never to have allowed an inherited belief in God to come in conflict with his scholarly and scientific research. Undistracted by actual debates, he could devote all his mental powers to building a new mathematical discipline. It was more difficult to muster a similar concentration in fields in which the struggle over positivism was raging and throwing down intellectual impediments to the work. In such fields, this battle was making research much more of an act of keeping one's balance, than it was a free act of enquiry.

Lie launched the first volume of *Archiv for Mathematik og Naturvidenskab* with the first parts of his theory of transformation groups, and in the course of the next ten years he published in all, forty articles in this journal. The essential features of a new mathematical discipline were thus formed.

The issue of *Archiv for Mathematik og Naturvidenskab* for March 1876, came out with its first issue, and Lie's "Theorie der Transformations-Gruppen (Erste Abhandlung)" extended to thirty-nine pages. In the next number, which came out in May, he had two papers, and in September he followed up with a fourth paper. Altogether in the first volume he contributed 120 pages. One of the reasons why Lie now concentrated on problems of group theory was that he had realized that in order for integration problems of differential equations to be taken further, the properties of the relevant groups had to be clarified. In the course of the autumn of 1876, he in fact succeeded in finding and formulating a new general integration theory.

His life together with Anna and the atmosphere in their home at the Tandberg Estate seems also at this time to have provided Sophus with good working conditions. Anna seems to have slid rapidly into the role of professor's wife in the capital city, was satisfied with the social life in which she took part. The serving girl Stina was accomplished in all the practical domestic duties, and among other things, looked after all the heating chores, which for the most part centred around coal.

Anna's father had been admitted once again to the Gausdal Asylum, and he remained there for several years. According to the journals he was "extremely sensitive and irritable" and was often "offended by every Insignificance", although he could also be "calm" and "quite sensible" but would "only reluctantly dress himself properly or allow himself to be washed and combed". It was also noted at Gausdal that Father Birch had been invited to visit his son-in-law "several Times, but on the Way he abandoned the Outing because he had to put himself a little in order, because the Invitation came too late, because the Return Home was too early, etc." In his good periods during these years, Father Birch did visit Sophus and Anna down in the city – it was once reported that he had strolled into the city without permission: "When a Guard was sent after him he was on his Way back to the Asylum."

For Anna, her father's illness had become part of everyday life that she could handle calmly. Worse, however, was the upsetting news from Risør that Anna's stepmother had fallen ill; she had a tumour in her breast and Anna was very worried. At some point during the spring, Mrs. Birch came to Christiania to have the tumour removed. The operation seems to have gone well, as expected. Consequently, as summer was approaching, Anna journeyed to Risør to be together with her "Mama". It gradually became evident however that Anna's stepmother had cancer, from which she was to die four years later. They tried to keep this diagnosis secret from the patient.

Sophus finished his lectures, which this spring had consisted of four hours a week "on analytic Geometry", and he looked forward to his annual trip into the mountains. He wrote to Mayer in the first half of July that it was about 30 degrees Celsius in Christiania, and it was high time to get himself off to the mountains "to seek, for my Constitution, a more suitable temperature". But first he wanted to write about his new theories, which had been printed in Christiania, and he reported that the article for *Mathematische Annalen* could not be ready before September. As to his other plans, he told about how he would work with contact transformations and partial differential equations over the coming winter, and he estimated that this would be a work of about twenty-five pages. His goal was to produce his own works in context and according to a uniform plan. At the same time he also wanted to communicate other important material: above all he wanted to give Mayer's theorem – if Mayer now gave him permission to do so – a prominent place side by side with the Poisson-Jacobian theorems. The following autumn he would like very much to come to Germany, and as well he hoped that Mayer and Klein again, as previously, would help him. The only thing that emerged from these plans was his trip to Germany.

In a similar vein he maintained his correspondence with Klein. Lie wrote about how he had generalized two of Jacobi's theorems and showed that one was a special case, and that they were not two different theorems as Jacobi had described.

Sophus left for the mountains, and Stina the serving girl, looked after the house. On July 23rd Stina in Christiania wrote to Anna in Risør that everything was in good order with the flowers, and that it was quiet in the apartment as well as in the capital city.

Anna and Sophus were back together again in Christiania at the end of August. Throughout the autumn he continued his lectures "on analytic Geometry" four times a week. As to his own research, he reported to Mayer that he was working "awfully keenly", that he was applying his ideas on transformation groups and was full of expectations. In October he worked over the proofs of his work on the new integration methods of what was referred to as the Monge-Ampère equation, a work that came out in *Archivet* at New Year's 1877.

There exist no letters or other documents to indicate how Sophus "got on" with his Anna at this time when her "Mama" was ill. In a letter to Mayer he characterized it as an unfortunate incident in his wife's immediately family, a misfortune he for his own part had turned into something positive.

Early in the New Year, Lie had injured his shoulder and for a time was prevented from working. In March he sent a letter to Mayer in which he wrote that he had far too many plans "Ich habe zu viele Pläne!" Lie's group theory insights seem to have led him to better and simpler solutions to a series of problems that he had posed earlier. In most cases where accurate solutions existed for a differential equation, this was based upon one or another symmetry property of the equation. Lie now found methods for how these symmetry properties could be found. He wrote to Mayer that in his studies of partial differential equations he would now also treat the Pfaff problem.

In April 1877, he received a letter from Munich with an invitation to go there and give lectures in September. The letter was signed by D. L. Seidel, but Klein had probably had a finger in the game as well. About this same time Lie wrote to Mayer:

> Ridiculous as it is, I still have difficulty writing due to my shoulder, which apart from this was not in the least seriously injured. For all that, this detail has not hindered my mathematical speculations. Quite the opposite. And it is actually quite comical: in recent years I have always made my discoveries when, as it were, I have been exposed to one mishap or another. In the spring of 1872 I had injured my eyes, and at that same point in time I found the integration methods. In January 1873, my father died, and I worked on group theory. During spring 1876, several misfortunes occurred in my wife's immediate family and precisely those same days, I found my new integration theory. In January 1877, I injured my shoulder such that I was unable to continue working at my usual desk, and the same evening I got a brilliant idea about minimal surfaces, an idea that has gradually given me many joys. As something notable in this same direction I can mention that the main theme in my paper "Über Komplexe" came to me one evening in Paris when I was, so to say, asleep. A short time afterwards, while I was sitting for a month in prison at Fontainebleau, I had there the best serenity of thought for developing my discoveries of the time, which in other respects have given me the greatest joy.

Sophus and Anna were expecting their first child in May. Would this also be something that waylaid and tied him down in a manner that would give inspiration to new works?

A Steady Stream of Works

Sophus and Anna had a daughter on May 22, 1877. The newborn was christened Marie, and was most often called Lillemor and Maia, later Mai.

In the middle of July Anna took the child and the serving girl, Stina, to Risør. About the same time Sophus departed on his annual mountain tour. On July 17th Anna wrote to Sophus that they had had a fine boat tour from Christiania, the baby had slept like a stone, they had arrived in good condition and well received in Risør, and Mama felt that the little one was just like Anna when she was small. Enclosed in the letter was also "an itsy-bitsy Kiss from Maia".

The letter had been sent to Grandalen Sanatorium where Sophus should stop according to plan when he was between the town of Hønefoss and the alpine country. The next letter from Anna in Risør was addressed to Lom in the Jotunheimen mountains. Anna was able to report that Maia had a voracious appetite, regular digestion "2–3 times a day", and weighed all of 24 marks (about six kilograms). "Mama" could sit all day holding the baby, and every day Stina took longer tours with Maia. Otherwise, Stina seemed to feel it was very good to be in Risør. Anna also explained that they had both hens and pullets, so she was eating eggs morning and evening. "Mama" earned money otherwise by having between five and eleven dinner guests every day. In a later letter – addressed to Røsheim Alpine Hotel in Lom – Anna also explained that she had salmon for dinner every day, that she was going to visit the Zwilgmeyers, and that she was "so intensely happy to be home again".

For his part, Sophus too seems to have been very much at peace with the condition of things. He was happy that it had gone so well for Anna and the child, happy too that the annual mountain tour was very much up to expectations, and he thought that certainly he would now be able to settle himself into new mathematic efforts. In a letter to Mayer – dated "In den Gebirgen, 27 July 77" – he wrote that naturally enough due to the birth of his daughter he had not done so very much work in a long time, but that he had thought a great deal about his book, which would be a collection of all the theories linked to differential equations and the existence of diverse integrals. In this work he would aim as far as possible to simplify things, through editing, and even if he now for the time being despaired at the thought of this work, he hoped that in one way or another it could come about if Mayer and Klein would help him.

Following his hiking tour in the mountains, Sophus had planned to travel almost immediately to Germany in time for the natural sciences meeting in Munich,

Norwegian mountain peaks in autumn, *painted by the German artist, Walter Leistikow ca. 1897*

In the industrialized Germany of the Kaiser at the end of the nineteenth century, the Nordic countries and their nature were of particular interest and admiration. Some of Lie's German friends – Felix Klein and Adolf Mayer and their wives – often spoke about travelling to Norway, and over the years, Friedrich Engel became a warm friend of Norway. Engel taught himself Norwegian, and even wrote an article in Germany about Norway's battle over language.

Leipzig University, 1890

Dresden aside, Leipzig was the most important city in the Kingdom of Saxony. The university was one of the oldest in Germany, founded in 1409, and was in all ways richly equipped with library, reading rooms, institutes, seminars, laboratories, etc. The university and the city experienced a great flowering in the latter half of the nineteenth century – both culturally and scientifically. Leipzig was a centre of music, and for a long time the city had found itself at the core of the German book trade. Now too, the city was in the process of becoming a leading industrial centre; the population increased, the city established new suburbs. The university was closely linked to prominent researchers and men of science in a whole series of different fields. In the course of his twelve years in the city, Lie formed "a school" and he became a central figure in European mathematical milieux.

From Paris, *painted by Frits Thaulow in 1879*

Many Norwegian and other Nordic artists – painters, writers and musicians – stayed in Paris and both sought and received inspiration from the great city during the last decades of the nineteenth century. For Sophus Lie too, and his science, Paris represented what was most advanced and innovative. Lie went to Paris for the first time during the summer of 1870, four years later he was back on his honeymoon, in 1882 he gave an epoch-making lecture, and he was in Paris on several other occasions in the 1890s to be honoured by the mathematical elite of France.

Melancholy, *with the coastline at Åsgårdstrand, was painted by Edvard Munch in 1892*

Lie would be very depressed at various periods, and beside himself with anxiety. He considered himself a "Bellwether Person", and pointed out: "As a rule, I have had the greatest Self-Confidence, but when Self-Confidence shows itself to be wrong, the Game was over for me." He wrote this to his friend Motzfeldt in the autumn of 1889 after he had been "seized by a limitless despair", and stayed for nine months in a psychiatric clinic. Upon Lie's discharge, the doctor's diagnosis was still "melancholy not cured". But Lie literally walked the disease out of his system – his own prescription being long tours on foot. During the summers in Norway during the 1890s – he also stayed at Åsgårdstrand, but it was particularly Anna and the children who stayed there, while Lie set off for extra hiking tours in the wilds of the mountains.

where he would also meet Klein and Mayer and spend a couple of months with them. His agreement with Anna was that she, together with the child and serving girl, would remain for the most part in Risør. Right before Lie left for the mountains he had received a very accommodating letter from Klein, who had arranged all the official papers for Lie's temporary residence in Germany and invited Lie to live with him for the first while. They could then arrange things, Klein wrote, exactly as they were when they stayed together in Paris during the summer of 1870: they could work together as often as they liked, and when either of them wanted to be alone, they had only to close the door. During the actual meetings of the natural sciences, Lie could find a room in the city, for Klein had already promised accommodations to his brother-in-law, the physicist Lommel, and his wife. And interposed among the practical information and commentaries, Klein had written: "What you write about strictness particularly interests me." In the context in which this is embedded, it almost indicates that Lie, in using the term "strictness" was referring to something in the direction of "well formulated".

Sophus was back in Christiania from his mountain tour by the middle of August. Anna wrote and expressed her delight that he had had an excellent trip, but she worried about his lengthy tour abroad: "May God deliver us safely into November when we are all happily together again, although indeed Time goes by at a furious pace." Anna told him about the baby – affectionately referred to as Lillemor [Little Mother] – who now had red roses in her cheeks, and laughed aloud at the capers of a kitten. Anna thanked him for the bottle of port and the half speciedaler he had sent to purchase clothing for the little one, and reported that she would buy a woollen shirt for the three-month anniversary of Lillemor's birth on August 21st. Otherwise, Anna was able to report from Risør that "Mama" was feeling well and working fully, but that Mrs. Zwilgmeyer and Mrs. Carstensen, who were both between 70 and 80 years of age, were both going to Christiania to have abdominal operations.

Anna wrote a letter on August 23rd that Sophus would receive in Copenhagen. She thanked him for two letters, a postcard, money and "an elegant book" he had sent her. The book dealt with child-growth and child-rearing, and Anna wrote that it made her "anxious about how infinitely difficult it was, the bringing up of Little Ones, and how infinitely many Considerations there were, that it certainly became impossible to carry the Task through. Do You also not feel the same?" What Sophus answered is not known – nor is there any trace of what he replied to her question about whether he envied Klein "who has a Son, and You only a Daughter?" Anna was able to report that Lillemor had spherical cheeks; she sent him a small lock of the little one's hair, and wished him "a really pleasant and successful Stay in Munich".

Anna and Sophus regularly exchanged letters during his stay abroad. Anna began her letters with "Dear Sophus", later "Dear Husband" and "My dear Husband", and on October 10th, she called him "My own dear Husband". He responded with "Dearest most precious Anna!" and "My dearest Anken. You are my brave kind precious Wife." About the little one, she reported on her "sweet small Arms with their Dimples – they are so round and plump." But one day something happened that had made Anna sad: she had had a visit by the Zwilgmeyer sisters – Dikken, who was home from Christiania, and Louise – and Louise had said that Maia had such small eyes because her cheeks were far too chubby. Anna thought this was a frightfully ugly thing to say, but had managed to hide her feelings and even to say yes to a promenade with the Zwilgmeyer sisters later that evening. What had been worse was that Louise had invited herself to come and live at Anna's, once Anna, together with the baby and Stina, had moved back to Christiania. According to the plan, Sophus was expected back from Germany during the first week of November, and Anna wondered what he would say if Louise were still to be living with them in the apartment upon his return. Louise's errand in Christiania was a visit to the dentist, and even though Anna felt it was impertinent to invite oneself in such a manner, she could not bring herself to say anything other than yes to Louise. To Sophus she said, "Dear sweet Husband, if only You were here, then I should ask You what I should do." Had she done the right thing? Louise had said that they who could not invite themselves to stay for some days were not really friends!

Anna also received a visit in Risør from Sophus' sister Mathilde, and his brother-in-law, Dr. Vogt, from Tvedestrand. The reason for this visit seems first and foremost to have been that at the beginning of September, "Mama", Mrs. Birch, had taken a bad turn for the worse. Vogt advised against a new operation, and felt it was best if Mrs. Birch was not informed about her disease, which had now been definitely diagnosed as cancer. Anna felt it was awful to have to answer "Mama's" questions about the disease: had she spoken to Dr. Vogt, and what had he said, and so forth. Anna was in "mortal Anxiety" for fear that others in the town would tell her stepmother about her disease, and what, in that event, she should reply. What she wrote to Sophus was: "Your Letter has been my only Comfort and Joy in Days", and she explained that after having received this diagnosis, in light of "this great Sorrow" she had been unable to bring herself to go out strolling with her friends, and that she might well return to Christiania somewhat earlier, probably around September 22nd. When she sat holding the child in her arms however, she could not really feel unfortunate: Maia smiled and laughed, so intensely delighted, the poor Little One." Otherwise Doctor Vogt felt that Maia was "absolutely like" Sophus, and Anna too, felt that the baby resembled her father when she slept.

Sophus wrote from Munich that the "sorrowful News" had moved him deeply "both for Your poor Mother's Sake and for Your and Maia's Sake". He felt it was correct that her "Mother" [stepmother] not be persuaded against her will to have a new operation. Such surgery, he felt, would give her more suffering than help – "The previous Operation has indeed certainly not improved her Condition," and did Anna not also think that her "Mother now particularly had seen that the first Surgery had not done its Job." Sophus had written about this both to his sister Mathilde and his brother-in-law, Dr. Vogt, and he had received positive letters in return, and in addition, he had written to Ernst Motzfeldt since he believed the Christiania relatives would press for a new operation. Should another operation nonetheless be conjectured, Anna must immediately send a telegram and he would come right home. In any case, were the operation to go ahead in Christiania: "It would thus be my Duty to stand at Your Side at such a trying Time," and he would not permit, both for her sake and for the child's, and for Mrs. Tandberg's, to: "let a Chloroform-Operation be performed in her House", but would provide for a dignified room with good care and attendance. He wanted to send her money, and urged her to telegraph if anything should happen – "Under such Circumstances it is not appropriate to save for a rainy Day."

> God save, protect and give You strength in this distressing Time. And let us hope that All will go better than it looks for the Moment. So many Times up to this Point in Life has Help come when it is most needed. And thus do I have good Expectations this Time too.

Sophus was happy that Anna had decided to return to their apartment in Christiania and he reckoned she could find the keys where he had left them: "The Key to our Bedroom is stuffed into the black Sofa. The Keys to the Secretary [a writing desk, usually possessing many drawers] are stuck down by the Head-rest of my Bed. The Key to the Entryway is lying in the Drawer of my Washing Table. The rest of the Keys will be found in the Secretary."

Anna seems to have been dismally unhappy back in the apartment in Christiania. She wrote to Sophus that it was awful to know that her stepmother was "all alone in her great Pain", and the first night back in Christiania she had cried herself to sleep. She recounted the terrible parting scene that had happened at the quay in Risør. They had both decided to be "brave and not cry on the Quay", but when they were standing there and the steamship had come in, then "People came to put a Ship's Owner, Jørgensen, on board." He had been quite mad and was to be sent to Gausdal Asylum, but "He fought against boarding and his Wife was pulling back on his Arm." It had been appalling to watch – "Mama began to sob compulsively and almost fainted. Everyone on the Quay became completely alarmed." Anna too had felt "completely miserable", but worse still was that Mrs. Birch could only cry and cry, worse and worse, until in the end she had to be taken in hand and led home by one of the women who had appeared on the scene. Anna had had to board the ship "in infinite misery" over knowing that her stepmother was so very sick and "so intensely distressed that I should be leaving." Anna wrote: "Dear Sophus, I am so endlessly distressed about Mama – when I was down there she instilled great Courage in me, by herself being so cheerful, that I never believed there was any Danger. Oh, dear Lord, that my own beloved Mama should suffer so much. She who had never had it easy, at least not since she met Papa. And how sweet and understanding she has been all these Years toward us. I cannot bear to think about it." Anna confided to Sophus that when she was alone, she could often not hold back her tears.

Other that this, she reported that there had been five or six cases of smallpox in the capital, and Anna wondered if Maia should be vaccinated, and who in that case she should go to. The question of whether Mrs. Birch would be operated upon or not, had still not been clarified. Doctors Hjorth and Scheen in Risør disagreed, and Sophus expressed satisfaction that Hjorth agreed with Brother-in-law Vogt that the operation was not advisable. But Anna reported that his cousin Johannes was in favour of an operation: "One can never say that it is too late, he says."

For his part, Sophus believed in a cure without an operation, but Anna wrote back:

No, Sophus, You write about Hope of Cure, which I do not have now. How is Cancer to be cured when it is so widespread, as it certainly is in Mama, and when the wounds are not removed, it can certainly not go away by itself; that has never happened before. I too have heard about Cancer being cured, but that is when a little Tumour has come, which is removed right away. It has happened that in such Cases the Cancer Tumour never comes back again. But I do not believe this when the Wound is so large.

Sophus tried to comfort and encourage his "brave kind precious Wife": "It is my daily Prayer, that God provide that your poor dear Mother quickly gets better than she has been these last two years. If anyone should be so blessed, it is she. And I am always of the Hope that she receive her Reward also here in this World." He understood that it had been "unpleasant" for her to come back to Christiania, but he hoped that she would rapidly come to have "everything pleasant", and he added: "Hire all the Help You require. I shall be so happy to come back again to my own dear Wife and my little Maia."

A letter he had written to his friend Ernst Motzfeldt from the mountains before he left for abroad, seems to indicate however, that all was not so simple and happy:

Dear Ernst!

I see myself in need of appealing to your usual Willingness to be of Service. As perhaps I have indicated before, the Costs of my Mother-in-law's Illness have caused my Bank Balance to be overdrawn to the sum of ca. 20 Spd. beyond the Maximum that I can apply thereto. You would do me a great Service if, as I believe You are willing to, to get these 20 Spd. paid back by her Relatives. For all that, the Expense that I have come to bear is so significant that I shall feel the Sting of it for a long time. [. . .] If You speak with Mrs. Birch in relation to this Matter, then I would tell You at the Same Time, that for the time being I am terribly irritated by the Manner in which she treats Anna. Mrs. Birch allows herself to nag at her for what she or I do not do which Mrs. Birch feels we ought to do. When I did not accede to Mrs. Birch's Request that I come down there before my Travels, the Reason was that I am awfully sure that had I done so I would have given her a sharp little Speech. And I do not really want to do that, when indeed at Heart she means well.

Sophus now wrote to Anna from Munich and reported that he was having "a very good Time". Klein's wife was "remarkably amiable" and Klein's son was a "fine strong Boy", but that Mayer's wife was ill and was keeping to her bed. Both Klein and Mayer spoke about "some Time in the Future making a Journey to Christiania", and it was not inconceivable that Mayer would come next summer "for as You know he is very wealthy". Sophus was looking forward to encountering "a whole Number of mathematical Friends" at the Meetings of the Natural Sciences, that was to begin some days later. He thanked Anna for being so diligent in her letter-writing and asked her to give his greetings to Amund Helland and to urge him to write and explain how things were going.

Lie gave his lecture at the Meetings of the German Natural Sciences in Munich – "Der 50. Versammlung deutscher Naturforscher und Ärzte" – on September 19th, and the title of the lecture was "Über Minimalflächen, insonderheit über algebraische", and he based it on the forty-page article he had sent for printing in May, in *Archiv for Mathematik og Videnskab*. Some of those at the gathering detected a mistake in Lie's presentation – which had to do with the expression of a formula in the event where the minimal surface became a double surface. And only three days later, Lie sent off a "Rectification" to the Academy of Sciences in Christiania, who consequently printed this later in the year. (Minimal surfaces can be said to be simply the least conceivable surfaces stretched between given points – the points here conceived as a framework that the surface must pass through. The surface that results when a ring is dipped in strong soapy water and held up in the air is therefore a minimal surface – and thus most surfaces in nature can be seen to be such – to the degree that nature's "indolence" prefers the lightest, most simple, and appears "to choose" the shortest, quickest minimal "solution" in its phenomena.) Once Lie had completed his official obligations in Munich, he thought he would stay a further month in Germany.

Anna reported from Christiania that eight barrels of coal had arrived at the house and that she had got them all inside the double doors for winter. She did not know whether or not she should also buy butter for 15–16 speciedaler to have for the winter. "You say so often that I should not bother too much with saving Money, but my Dear, I am indeed careful and do not want to pay more than things cost," she wrote, and recounted that she had not gone out quite as often as Sophus was always urging her to do: but Josephine, the little daughter of Cousin Johannes and Amanda, came often to look in on Maia, and these occasions were both pleasant and sociable. Maia now loved to lie awake and be entertained – this was probably due to teething, Anna thought, and otherwise felt that things were going quite well with the child. But with great alarm she recounted that she had found three or four gray hairs on her own head, not at the temples but on the back of her head: "I have kept them in my Pocket Book", she wrote. Louise Zwilgmeyer had arrived from Risør as they had agreed, and Maia had received a perfectly lovely dress from Dikken Zwilgmeyer. Anna wrote that they had had a lovely time together, and she repeated what Louise had told her about an aunt who had suffered from breast cancer for ten or eleven years, and had been operated upon five times. This aunt managed to hold the cancer in check with a plant juice that could be obtained in Germany, and Anna wrote: "If only You could obtain some of it." Whether Sophus actually purchased this plant juice is not known, but "Mama" Birch lived another two and a half years before succumbing to the cancer.

Regarding other events in their home, Anna wrote that she had one day had a visit from Ernst Motzfeldt, and on another day she had been visited by her father, who was terribly well for the time being. Sister Laura, too, had visited her. And also the nineteen-year-old Johan Herman Vogt from Tvedestrand had come and stayed with her for some days – he was having his eyes tested, and Anna felt that he was in good form and his eyes were good. Johan Herman – the nephew Sophus had taught to swim in Tvedestrandfjord eleven years earlier – had become a student the previous year, and after spending one year at the Dresden Polytechnikum, he was now back in Christiania.

One of the actual topics of conversation that fall in the capital was the great flood of Akerselv, the river bisecting the city into east and west. The river overflowed its banks everywhere and did such damage to the unsuspecting and innocent citizens that the obstinate owners of the river, who for long had struggled to defend their private ownership, now had to admit that a communal control was necessary.

Sophus was on his way home on October 15th. He reported from Berlin that all was "remarkably well". And Anna wrote back that she was looking forward to going down to the railway station to meet him in a very short time.

Sophus seems to have arrived home at the end of October or in the first days of November. Some books and a number of letters lay in "the Secretary", waiting for him. There were a couple of letters from Bäcklund who requested support in relation to an application for a professorship in Lund, and there were letters from the German mathematicians August Weiler in Mannheim, Lebrecht Henneberg in Freiburg and H. A. Schwarz in Göttingen with thanks for papers which Lie had sent them. Schwarz also hoped to meet Lie in Uppsala in relation to a meeting to which the Royal Academy of Sciences had invited him.

Lie began his lectures at the university; according to the official calendar of lectures he was lecturing five times a week "on Analytic Geometry and Stereometry". Perhaps he began these lectures earlier than expected. In any case, his sister Mathilde in Tvedestrand commented in a letter to Anna that it was "lovely that Sophus had got home this early". Otherwise Mathilde was thanking Anna and Sophus for their "Kindness toward Johan", and she continued: "(Johan) writes that he had an enjoyable time at Your Place, and I feel naturally enough that it is so lovely to know that from time to time he is with You; it is evident that nothing good comes, particularly to Johan, from such a far too free Student Life, where they live, each in his own Digs and show no Consideration for anyone."

Shortly after returning home Lie received a letter from Mrs. Klein, and in his reply, written in December, he wrote that upon receiving her letter he had immediately decided to send her husband a picture of Abel. Since now he had had the good fortune to lay his hands on such a picture, he was sending along with it his Christmas greetings. Sophus sent greetings from his wife Anna, and explained that both she and his daughter, who for quite some time had had two teeth and had another two on the way, were doing well. He asked Mrs. Klein to greet her husband and tell him that Lie was working energetically and well on transformation groups, and that he would soon be writing to Felix as well.

But several months would go by before Lie would send anything to Klein, and then, only a couple of weeks after receiving a letter from Klein in Munich. When Lie finally wrote – probably at the end of May 1878 – he also let it be known that he, "for the whole winter through, had lived almost the life of a hermit". About his tour abroad in the autumn of 1877, Lie said that apart from the encouragement and stimulation he had received from Klein and Mayer, he really had been disappointed with the trip. A considerable reason for this had no doubt been the nerve-wracking reports from home about his mother-in-law's illness, but a still deeper reason lay with the irritation and exasperation aroused within him by the mistake he made in his minimal surface calculations. ["Vielleicht noch mehr lag die Ursache in meinem Ärger über den Fehler in meinen Minimalflächen."] Moreover, Professor Schwarz, according to Lie, was "a downright openly moralistic kind of person", who told everyone wherever he had been, in Germany, Sweden and Norway, about this mistake – and some expressions from Brill had also irritated him. In summary, Lie wrote: "On the whole I have been rather irritated and out of sorts. And I could only regain my equilibrium by working out some mathematics, which has given me back my full self-respect." And Lie explained that when he had been unable to obtain quick results with the minimal surfaces, he had taken himself back to his transformation groups, "which in all likelihood will prove to be my life's most important achievement." Lie went on to say – in this letter from the first half of 1878 – that he had begun the enormous task of determining all point groups in space and that through frightful [schändliche] computations, he had succeeded in this. But the main reason was that he had achieved new points of view that stood out as fundamental to his theory. He wrote: "My major result, which applies to all dimensions, is that the linear group is the only one among the infinitely small, that possesses the greatest possible transitivity. And with this, consequently, my old conjecture is successfully realized."

Again he had begun to work with minimal surfaces, and had achieved good results: "I am quite certain," he wrote to Klein, "that it will be of great utility for this theory [that is, the theory of transformation groups] to develop the synthetic treatment, side by side with the analytic." He now published his discovery in *Archivet* – in the first two issues of 1878 he had four articles, and in the next two issues he had two new articles – on transformation groups and minimal surfaces. Other than this, Lie was able to report that he had begun to edit a couple of articles on minimal surfaces for *Mathematische Annalen*. The first of these, dated Christiania, 20 June, 1878, was published a year later.

At the beginning of May Lie had again received a letter from Lebrecht Henneberg, who was now in Zurich. He thanked Lie for having sent his articles and for finding that some of his own theories had been applied in such a fine manner in Lie's work. Henneberg sent his latest articles to Lie, and at the same time asked to be sent Lie's latest piece on minimal surfaces. Lie reported to Klein that he was glad that he had developed a friendly relationship with Henneberg – in reality the correspondence between them had begun following the meeting of natural sciences in Munich in the autumn of 1877. According to Lie Henneberg had "found some very beautiful theorems about minimal surfaces", and Lie maintained that his new investigations into minimal surfaces merged with Henneberg's works just as well as did his own earlier works. (Lie had written to Henneberg that in his works he was going to award Henneberg priority, but added that Henneberg's discovery only had influence on the way that Lie had discovered a mistake in his own investigations.)

As for Klein's mathematical work for the time being, Lie wrote that perhaps he too could work on some occasion from a geometric point of view concerning the theory of elliptic functions. And he added: "It is really remarkable how few people there are who have acquired a really audacious geometric way of thinking. We have learned that well from Plücker; in any case, we have grounds for being eternally grateful to Plücker."

At the university Lie seems to have felt himself freer than earlier when it came to fasten names to his lectures. According to the university calendar of lectures during the spring semester of 1878 he would lecture "3 Hours weekly on Space Geometry", but in the remaining two weekly hours he would lecture "on a Theme" which he later would "determine following Conference with those studying". It seems somewhat unclear what these two hours were used for, but probably this spring Lie began to teach more from his own research. As early as the following semester, it was announced in the university calendar, that Lie in addition to "Space Geometry" would lecture twice a week on "Contact Transformation". Elling Holst wrote then to Ludvig Sylow on November 2nd, that he was pleased Lie was lecturing on his own things, and Holst indicated that this was great progress, that Christiania now held lectures on current, ongoing mathematics. Holst was also able to inform that Lie's lectures were followed by an audience of four, and that in addition to his Contact Transformations, Lie had also begun on "his partial differential equations".

At home on the Tandberg Estate, little Maia was growing in the most "gratifying manner". Anna, on the other hand, had not been in the best of health over the winter and spring, but was happily regaining her health, Lie now reported to Klein in the pre-summer of 1878.

Lie had finished his great treatise on minimal surfaces by June 1878. It was twice as long as he had expected it to be, and would as well be published in *Mathematische Annalen* the following year. But shortly before he was going to send his work to Klein he had discovered some minor inaccuracies that meant that he would hold back the treatise and calmly go through it still one more time. Lie admitted that he was not quite in the habit of giving himself time to do this:

> I have always been of the conviction that my work would make a methodologically valuable contribution. As *for me personally* small minor inaccuracies are immaterial. But I cannot expect that minimal men [die Minimalmenschen], who in part will greet my new geometrical reflections with suspicion and mistrust, to have the same view.

In this last there was an open allusion to the unpleasant memory of the mistake that some had pointed out in his work – as late as two years after the meeting in Munich Lie asked Klein if anyone had found other mistakes in his paper.

Now in the pre-summer of 1878 Lie also gave an account of his next project, in which he intended to establish a common relationship between curvature theory for algebraic curves in space and the theory of minimal surfaces – and hopefully he would manage to produce this in an "elegant analytic form", but everything took longer than expected.

In Christiania it had also been terribly hot and this made it almost impossible to work – the month of July had begun, and Sophus was longing to get up in the high mountains. On July 7th he reported to Sylow that he was ready to set out on his annual alpine tour, and that he would probably be back in Christiania at the beginning of August, or at least he would be back by August 15th, "to hold the Artium", that is to say, his usual fourteen days with the *examen artium*. The main reason for the letter was to ask Sylow if he could free himself from the school, at least until some time in September, for the Abel work. The previous year Parliament had granted 5,540 kroner for "the gradual Publication of Mathematician Abel's surviving Works", and it was thus "the Fund for the Publication of Abel's Works" that he hoped would support Sylow. (Sylow had already had a leave of absence for four years, from 1873, to work on Abel's papers.)

From the Department of Church and Education in Christiania, there came the report that professorial salaries would be increased, and to Lie this was welcome news indeed. In recent years the salary had been completely inadequate, he reported to Klein. Lie was still "extraordinary" professor. Two years earlier Bjerknes had proposed Lie for an ordinary professorship, but according to the surviving protocols of its deliberations, the Collegium had, at the time, other plans. It would not be for another year, until 1878, that extraordinary professors would receive the same pay as the ordinary professors.

The summer holiday was much as it had been the previous year – Sophus left for the mountains, and Anna took their daughter to Risør. And they seem each to have enjoyed their respective venues.

Sophus was back at the university for the artium examination, a job he categorized as "boring, but not as taxing as in earlier Years". Anna had reported from Risør that little Maia had been sick, and therefore their return journey to Christiania had to be postponed. On August 22nd, Sophus wrote to Anna:

My dearest Anken!
I understand from Your Letter that our Maia is in Recovery. God provide that it goes well and surely ever- better with her. You must of course not come home until Scheen [the Risør doctor] determines that Travel is of no Danger. Otherwise You should not go out in bad Weather, particularly not when the Wind is southerly.

They had hired a new serving girl in Christiania – Karen – and Sophus was of the opinion that she was longing for the return of Anna and Maia, but other than that, she was willing and kind and did her best. Sophus asked Anna to order two bushels of apples – "That is, after all, the Amount we normally buy" – and he asked her to pay Dr. Scheen "4 Spd. or more, according to Circumstance". In the following sentence there was mention of paying a night nurse "5 Kroner [. . .] and more if she is more frequently up and awake over our Treasure." And Sophus closed with his familiar admonition:

Finally, You must look after Your own Health during These Days. As I have said before, eat and drink well, and go outdoors a great deal.

Greet your dear Mother warmly for me.

Now be careful, my Dearest. And may I soon have You and Maia back with me. That is when I feel the safest, when I myself can look after You. Give Maia a Kiss from me, and say that her own Papa, who longs for You Both, sends her his greetings.

Three days later he again wrote to his "dearest Anken", and he hoped it would be the last letter she received in Risør before she left: "I feel that I am an exceptional Husband for writing Letters. But You too, are my own brilliant Wife." The weather had been good in the capital and the wind from the north: "May the Lord grant You as good Weather on Sunday, so that You can travel. But as I said before, if it blows, then in the name of God, wait." And when they came to Christiania he would take shawls and lap rugs with him "down to the Quay; it gets cold in the Evening, so we have to be careful, especially with our dear Maia." And he urged her to give "the Girl on the Steamship a good tip so that she will help You and the Child." Apart from this he was able to report from the capital that Anna's father at Gausdal Asylum was going on an organised tour to visit some of his relatives in the countryside: "This is indeed very commendable of them," Sophus commented – and reported that on his hunting trip, Ernst Motzfeldt had shot both ptarmigan and reindeer, and "Now he will be proud and have much to boast about."

Throughout the autumn Sophus held his hour-long lectures five times a week, and in at least two of them he took as his point of departure his own research. These were the lectures followed by the audience of four, and which had been described by Holst as a happy and significant forward step in what the university offered by way of teaching.

In terms of his own research, Lie worked further with minimal surfaces, geodesic curves, surfaces with constant curvature, "translation surfaces", and transformation groups. Articles and studies flowed from him in an almost steady stream – published, above all, in his "own" *Archivet for Mathematik og Naturvidenskab* where he numbered the works: "Theorie der Transformations-Gruppen I, II, III, IV, V", "Sätze über Minimalflächen I, II, III", "Beiträge zur Theorie der Minimalflächen I, II", "Zur Theorie der Flächen constanter Krümmung I, II, III, IV, V", "Weitere Untersuchungen über Minimalflächen", among others. But he also prepared much other work for editing as well, bigger articles for *Mathematische Annalen* in Göttingen, in order to reach a wider audience. In his exchange of letters with Klein there was much talk about a work on transformation groups for *Mathematische Annalen* (a work that was finished a year later and began to be published serially in June 1880, under the title "Theorie der Transformationsgruppen I"). In the meanwhile he had made public two works on minimal surfaces, in *Mathematische Annalen*, in addition to nine articles that for the most part were printed in German in *Archivet*, and a couple in the Proceedings of the Academy of Sciences – one of these works, about how surfaces and the radiuses of their curvatures were connected, was also translated into French and published in *Bulletin des sciences mathématiques et astronomiques*. Lie thought that curve theory and the theory of geodesic curves were together a good field for applying transformation group theory. About his major work on transformation groups, he wrote to Klein early in 1879:

I am thinking a great deal about a publication in Annalen, of my transformation groups, which I can now generate better (more strongly and clearly) than before. It is my aim to give an account of the theory in such a manner that the reader does not need to be familiar with the newer works on partial differential equations and contact transformations. I must make haste with this work, for the ideas, which I contemplate as mine, are beginning to spread and be used (particularly by certain Frenchmen to whom I have sent my work), and considered new and used in particular forms. I do not know if it is better to begin with the general theory or with interesting applications. If my working energy, which for the time being is satisfactory, continues to hold up, then the general theory as well as applications by means of differential equations, etc., will get a great and comprehensive treatment.

Lie had undertaken a work on the sphere geometry of Reye – "a geometry that utilizes the sphere as a point" – and to Klein, Lie expressed his exasperation that Reye, while knowing better, did not mention that others, earlier, and at tremendous length, had developed a sphere geometry. And it was also terribly "comical," Lie pointed out, that Reye should have complained that he [Lie] and Klein had advanced Chasles, at Reye's expense, as discoverer of the tetrahedron complex.

It seems that family life at the Tandberg Estate was going very well. "I am living extremely well" ["Ich lebe recht gut"] is a standard formula in his letters. Sophus and Anna kept contact with their relatives, but there does not seem to have been any socialising beyond this. A harrowing event occurred in April 1879, when Uncle John Lie had been injured by a tram. The eighty-year-old doctor had had six teeth knocked out, a leg wound, a fractured lower jaw, and concussion. But the old one – who still considered himself a youth – and who was still held to be one of the "characters" on the city scene – kept a low profile for only a short while before he was once again carrying out his deeds among the city's poor, and in military circles.

When John Lie died two years later (in January 1881), all the officers stationed in Christiania made sure he had a vast funeral – it was said that in their eyes John Lie was the most congenial old doctor that one could ever imagine.

Once again Sophus set out on his annual hiking tour to the high mountains, but this time he was accompanied by his nephew, Johan Herman Vogt. Fifty years later, Vogt was to recount his experiences from this tour: one morning after having overnighted at a "sæter" (a farmer's alpine summer camp where the livestock was allowed to range freely), when the sæter boy, who had asked their names, learned they were Vogt and Lie, he had quickly retorted, if Lie had Sophus as a first name, then there was a song about him. And Vogt recalled only one verse of the long song – "Fifty miles in one shake/With rucksack and its ache/Is to him a piece of cake." Vogt commented that there was only one thing wrong with this: Sophus Lie's rucksack caused no ache because in general it contained very little.

Lie was back in the capital in mid-August for his duties with the *examen artium* – and he seems to have become a fair and just examiner. It appears that a certain dispute broke out that autumn[32] among the fifteen members of "the Deputation for the Examen Artium". Professor of Egyptology Lieblein put forward a proposal to discontinue the written exams in mathematics for those seeking the *examen artium* in classics. Now that they had a sciences line, this naturally enough was

[32] In high latitude Norway autumn is considered to begin in mid-August.

chosen by all those who had the "Desire and Ability for Mathematics and the Fields of Science", Lieblein reasoned, and he pointed out that it was unreasonable that those who otherwise, were gifted people, but absolutely lacked the ability to learn mathematics, should have to have their academic records ruined by a low mark in mathematics. In any case, the demands (verbal and written) for such subsidiary subjects – as mathematics was to the classical, or Latin line" should be reduced – and Lieblein got a majority of the "examination deputation" to support this demand. Whereupon Lie wrote a detailed account to the Collegium. Lie stated that in so far as he was in agreement that the written exam "in general ought only be lengthened for the Main Subject", he was nonetheless afraid that such changes would weaken the position of mathematics, and he doubted that "anyone possessing a one-sided Gift had to date been kept from the University by Reason of this written Mathematics." The Collegium sent on Lieblein's proposal to the Department, which, a year later, reported back that changes in the requirements concerning the classics *examen artium* were not relevant.

The work on the Abel edition had for a long time been left in Sylow's hands, but now the printing had begun, and Lie had to take on his share of the proof-reading. Also during this autumn of 1879, Lie eagerly promoted the view that Sylow had to take leave from his job as senior teacher in the town of Halden in order to devote all his time to the works of Abel. The printing was undertaken by Grøndahl's Publishing House in Christiania. In the middle of August 1879, the typesetter at Grøndahl had resigned his position; he complained about the miserable working conditions, and the much too inadequate pay for such a time-consuming work. He threatened to go to the newspaper *Aftenposten* with his story. For the moment Lie wanted to give the typesetter five speciedaler extra, and also managed to obtain these extra monies. Thus the printing continued, sheet by printer's sheet, and the reading of the first, second and third proofs. Meanwhile he continued his lecturing at the university – three hours a week "on analytic Plane Geometry" and two hours "on a Segment of higher Geometry".

In addition to the mathematics included in the exchange of letters with Klein, came also discussion about a new stay in Paris. Klein had plans for a stay in the French capital, and Lie wrote: "I cherish the idea that we should visit Paris together once again. Soon ten years will have melted away since we last were there together. In other respects, I had thought not to travel abroad until 1881, but perhaps I could travel as soon as 1880." And thereafter it seems that plans developed between the two friends to meet in Paris in the autumn of 1880. Lie wrote: "In Paris I would for the most part audit your talks and discussions with different mathematicians. In all respects this is a role that suits me fine." Lie mentioned that Holst, during this winter of 1879–80 would be staying in Paris, and added, "Holst's book on Poncelet is, I feel, meritorious, for it contains much that is historical."

Elling Holst had received a public grant to travel to Paris and England to engage in post-doctoral studies, and, from October 1879, together with his wife, stayed in Paris. Holst reported to Lie from Paris on his meetings with French mathematicians. Lie wrote back and expressed his pleasure that Holst's ideas had met with "Appreciation", and were considered new "by them who are competent to judge them." He continued:

As far as I am concerned, I do not have great Understanding of such Things for the good Reason that I am terribly ignorant of this Part of Geometry. [. . .] I can only commend You for pursuing Your Ideas to their utmost Consequences.

Apart from this, You ought naturally to utilize the Time in Paris to make Yourself familiar with the most important Works that are now being produced by French Mathematicians. Among the elderly You know I (apart from Chasles) particularly appreciate *Bonnet* in "L'Analyse appliqué à la géometrie". It is a good Thing about the Frenchmen, that their Works are so well edited. I can understand most of the French, but only very few German Works.

Lie also sent along from Christiania a pile of his recent papers, which he asked Holst to give to friends and acquaintances in Paris. Lie counselled Holst to visit the mathematician Mannheim first – "he speaks, as I recall, German," Lie wrote, and mentioned that he had as well written to Mannheim, and urged him "to advise You, particularly about which of the younger Geometers You best can seek out." Mannheim would certainly also introduce Holst to mathematicians in the Academy or wherever he now had to meet them. Lie continued in his letter to Holst: "In that Event, You could greet them on my Behalf, and say You have some Papers from me, and together with this, finally ask them if You can pay them a Visit and deliver these Papers." However, Lie added: "Nevertheless, I repeat what I said earlier, You must not impose upon the French Mathematicians' Time." In addition to Mannheim, Darboux, Lévy and Halphèn were mentioned by name.

Lie also asked Holst to undertake some more specific things: "On some Occasion, ask Darboux or the Others if anything has been written in France in recent Years on Surfaces of a constant Curvature." Lie also got a letter from Darboux about this time, recounting that he had met Holst, who, in Darboux's words, "seems to be a good Geometer and a very pleasant fellow." Otherwise Darboux had learned about some of Lie's work. Darboux had a great desire to refer to this work and requested information on where it was published. Could they not otherwise exchange the journals they respectively worked on – *Archivet* for *le Bulletin*? Darboux asked. And Lie promptly arranged the both the one and the other, and received a letter of gratitude in return.

Lie in Christiania had received some money with which to buy some mathematical models, and from Paris Holst had sent him price estimates for such models. In a letter written to Klein at the New Year's 1880, Lie mentioned that he had 200 marks with which to buy such geometric models, and thus he had already ordered "*Catalan's* polyhedre semi-régulieres", and urged Klein to come up with suggestions for how to use the rest of the money.

At some point during the winter of 1879–80, Lie with wife and child, moved from the Tandberg Estate to the Schønheyden Estate, which later had the address Drammensveien 84. The move seems to have come abruptly – and one of the stories that later spread about the eccentric genius of Sophus Lie may have had its genesis in this move. In an undated letter to Holst we learn only: "Having just moved, I have forgotten You for the Span of a Week, but You will have had enough to do." At the same time, Holst must have known more – in any case, it was one of Holst's friends who recounted a long time afterwards that Lie had been given notice. The reason for this was said to be that from time to time when he was preoccupied with one thing or another, Lie would go up and down through the rooms with a sword in hand, and in the course of things would stick the sword into the ceiling with such force that he could lift himself off the floor with it. The hostess was said to have observed this one day and immediately put a stop to it.

Now, at New Year's 1880, Lie reported to Klein that it looked quite certain that he would be able to travel to Paris in the autumn, and he was already looking forward to it. But whether the university would grant him the necessary money would not be known finally until late in the spring, and he added: "As for other Difficulties, hopefully there will be none to fear, except that a man with a family can never be certain." The fact that Anna was pregnant and awaiting the birth of their second baby during the summer seems to have upset his plans. Shortly thereafter there came another letter from Klein in which the latter conveniently proposed that the visit be postponed. Lie answered quickly that indeed a postponement suited him well, and announced that he now had definitely put off his travel until 1881. Lie continued to hope that they could travel together, and added: "I can hardly travel to Paris alone." After five years in Munich, Klein had just been appointed professor in Leipzig; upon hearing of this, Lie sent his hearty congratulations.

Lie had a great deal of work during the spring: "In a word, I am struggling with surfaces of constant curvature," he wrote to Klein. "It is moving forward. But every step involves abominable calculations which, for the time being, seem unavoidable." He said that he considered it a certainty that he would soon be able to demonstrate that a combination of his own, and mathematician Bianchi's investigations would find all possible surfaces with constant curvature. The proof also involved integration methods other than those found by Monge, Ampère, Darboux, Lévy – and Lie believed these new methods depended upon the existence of some singular transformations that an equation could give rise to. But perhaps he would never come to develop such speculations any further. Lie thanked Klein, and the whole editorial board of *Mathematische Annalen*, for the coming number in which Lie's ninety-page work on transformation groups would be published: "For me, it is a great encouragement. And I really need encouragement from abroad, for here in Christiania, where not a single soul reads my works, there is nothing to encourage me."

At the university in the spring semester, Lie carried on his lecturing "five Hours weekly on the Geometries of Plane and Space."

He admitted to Klein: "Since 1872 I have not had my full work energy, but despite this, I have become successively more at peace."

On June 9th, "Mama" Birch died in Risør.

On July 5th Sophus and Anna had their second child. She was christened Dagny, but in everyday life was called simply Dadi.

The Mathematical Milieu at Home and Abroad

There is absolutely no indication in the existing documents and letters as to whether Sophus took his usual mountain tour in the summer of 1880. Perhaps he stayed home as support to Anna and Maia, who was now three, and to be near little Dadi. Anna, who seemed to have a special eye for hair, this summer clipped a lock Sophus' head and folded it into a piece of paper. Then, on the outside, she wrote: "Sophus Lie, 38 years old". Unlike what she had found in her own hair, there was no grey in this lock.

Lie wrote to Klein at the end of July. He recounted that they had had a daughter and reported that he had taken a big step forward in his theory of surfaces with constant curvature. He hoped that during the autumn semester he would have a large contribution on this ready for *Mathematische Annalen* – in other respects he felt he was far in arrears with publications for *Annalen*, and concluded: "For the time being it is not possible for me to do otherwise. I am excited as to whether my transformation groups will be read, and similarly, whether they contain any mistakes."

For a couple of weeks from the end of August he examined the *artium* students as usual. At the university he continued his lecturing: five hours a week "on Space Geometry". Apart from this it seems that through the autumn he was mostly to be found at home, together with Anna and children at Skarpsno, Drammensveien 84. Whether there was correspondence with, or visits from, relatives, there is no record, and the telephone had just come into use. At this time in Christiania there were only 169 telephone subscribers.

Lie's studies continued to come out in *Archivet*, and right at the end of the year, he received a letter from Darboux, who wrote that Lie's works were highly cherished in Paris, and that the application to which Lie had put his surfaces had aroused special interest and goodwill. Darboux himself was ecstatic about Lie's elegant solutions, he wrote, and sent along a French translation of one of Lie's works that he had thought to publish in *Comptes Rendus*.

The work on the Abel edition was nearing completion. Much of the time-consuming proofreading and correcting was done, but there still remained some principal problems of language to sort out, and the foreword had yet to be written, and the title page to be set up. Abel himself had written most of the works in French, and the remainder, written in Norwegian and German, had been translated into French, and even those of Abel's letters, written in Norwegian, and which were available, would enter the edition clad in the language of France. But what about Abel's letters to his German friend and editor, A. L. Crelle? Should they be published in German now when everything else was in French? "It seems to me that is absurd," Lie wrote to Sylow. In Lie's view a number of pages in German, in a book written in French, would be astounding, and would give "the general Reader, who did not study the Introductions and Notes, the Impression that Everything was published in the Language in which it had been written". And as a postscript regarding Abel's works, Lie added, "Among other things, I have found a highly remarkable Integration Theory. It allows one to find an Integral containing as many Parameters as one wishes."

Sylow wanted very much to ask Professor Broch for advice in relation to the question of translation. Broch was both a master of languages and well-known in foreign circles. He had a German wife and now lived in Paris, as two years earlier he had been appointed director of the international office for "weights and measures" in Sèvre. From this honorary position, Broch worked energetically to get standards established for weights and measures in the wider international society – it was scarcely possible to achieve a higher reputation than Broch now held within the sciences in general. But for his part, Lie had little desire to ask Broch – Broch would "certainly not go into the Matter," Lie felt, and pointed out that if he himself were to explain things to Broch, he would have difficulty being impartial in his presentation. Lie had however asked advice of Broch – in the course of one of Broch's visits home to Norway – and he now reported to Sylow that Broch had not shown any "particular Sympathy" for any of the alternatives. In the end, the result was that everything was translated into French.

For about the same length of time that Lie and Sylow had been working on Abel's collected works, Professor Bjerknes had been working on his biography of Abel. And down through the years, particularly between Sylow and Bjerknes, there developed ideas and points of view about Abel's life and his work. The first parts of Bjerknes' Abel biography was published in Stockholm in the *Nordisk Tidskrift* in 1878 and 1879. Parts of the material were also made known through articles in Norwegian newspapers and periodicals. But Bjerknes gradually and increasingly concentrated on what had earlier been interpreted as "that noble Competition" between Abel and the German mathematician Jacobi, a "competition" which had resulted in them both being honoured for having discovered the theory of elliptic functions. This material, according to the editorial board of *Nordisk Tidskrift*, was too heavy-going for their readers – but the issue was solved by the whole of Bjerknes' Abel biography being published in 1880 as a supplementary issue of the periodical.

Now, at New Year's 1881, Lie commented on Bjerknes' book, in a letter to Sylow: "It [the Abel biography] contains Things, that are not advisable. A couple of audacious Hypotheses, were, so to speak, elevated to Truth, even though, from my Perspective, lacking in decisive Foundation." In his investigations, Bjerknes had concluded something that, far beyond mathematical circles, was considered sensational; namely, Abel's major work in this "competition" had already been in print for two months when Jacobi submitted his study, and Bjerknes had found it overwhelmingly to be the case that during these two months Jacobi had studied Abel's work. Bjerknes' conclusion was clear: Abel had suffered great injustice – first of all, because in his work Jacobi did not cite Abel on this decisive point, which represented a theoretical breakthrough, and second, because later Jacobi failed to give a clear message about this historic connection. Abel died only a year after this famous "competition", in 1829, and could not defend himself. Jacobi lived another twenty-two years – he founded a school of thought in Göttingen, which Klein too had been influenced by, and had left works that Lie as well had taken as a point of departure, but Jacobi had never committed himself beyond expressing his great admiration of Abel.

In a later letter to Sylow, Lie made concrete his objections to Bjerknes's book. What hurt him most was that "Bjerknes seems to want to divest Jacobi of Part of the Honour for his greatest Discovery: the Conversion of algebraic Integrals", and there were also other unfortunate expressions, like: "Jacobi trod on his [Abel's] Heels," and Jacobi's "Agressiveness".

Lie also reacted to the fact that Bjerknes had not mentioned that they, Sylow and Lie, had given him information to use in the book. According to Lie, it ought to have been of some consequence to Bjerknes that he had not mentioned receiving "Assistance" from those who were working on Abel's collected works. If Sylow had received some help from Bjerknes, had he not "received 10 Times more from us", he asked, and continued, "I do not consider that there are any Grounds for us always to accept the Impertinences of Bjerknes, and in Consequence, pay him unmerited Courtesy." For Sylow, who in the course of the work had often discussed matters with Bjerknes, and frequented his home, the matter does not seem to have been quite so clearcut – perhaps he was also thinking of his own future position in life. Sylow as still only a senior teacher at the Fredrikshald Latin and Grammar School, but both his own mathematical studies, and the work with the Abel edition certainly made him qualified for much higher positions.

At this time, Bjerknes was certainly one of the most talked-about scientists in Europe. In the summer of 1879, he had been in Paris, and before large groups of European scientists, presented experimental proof for his new hydro-dynamic findings. The following summer he had presented similar experiments at the meetings of the Scandinavian natural science researchers in Stockholm, and now in 1881, he was being invited to present and demonstrate his apparatuses at the international electrical exhibition in Paris.

Lie went on a trip to Uppsala in the spring of 1881, and in Stockholm he met for the first time the Swedish mathematician Gösta Mittag-Leffler, who, following successful years in Helsinki, had just returned to lead the Stockholm College, which would become the precedessor to Stockholm University. At this meeting with Mittag-Leffler, Lie came up with the idea for a Nordic mathematical periodical with Mittag-Leffler as editor – there was no pure mathematics journal in either Denmark, Iceland, Finland, Sweden or Norway. And Mittag-Leffler, who in addition to his mathematical work also had a great desire and capacity for establishing scientific milieux, seized on the idea. This meeting resulted in the first issue of

Acta Mathematica coming out a year later in December. This quickly came to be a leading journal; it still continues, and is considered to be among the most prestige-laden journals in the mathematical world.

It took scarcely twenty months to move from the idea for a Nordic journal to its launch – whereupon Mittag-Leffler was able to submit the first issue of "our Idea," as Lie called it, ceremoniously to the king. Lie and Mittag-Leffler corresponded frequently, and during this period Lie wrote twelve long letters in which he made proposals and commented on plans, as well as strategy and tactics, and from then on, the two of them remained in friendly contact with one another. It seems that from the beginning Lie had envisioned a journal that for the most part would publish works by Nordic researchers, and about Nordic contributors. He expressed his reflections to Mittag-Leffler on September 6, 1881, as follows:

> Here in Norway we shall participate to the best of our Ability; but we are indeed small, and to be sure, for the time being, can do less. I have still some Hope that Sylow will now become more productive, and so too, Bjerknes. Both could surely submit particularly valuable Contributions, if they only would. But unfortunately both are a little unpredictable. Broch hardly writes Mathematics any longer. Prof. of Mechanics Guldberg is indeed always producing Something; occasionally together with Mohn on meteorological Themes, occasionally together with Professor Waage on mathematical Matters. Prof. Guldberg is a gifted Man, but I know little about the Significance of his scientific Enterprise. And thus we have Dr. Guldberg, whom Sylow and I place rather low. Finally, Elling Holst, who does not lack Talent, and who indeed, on Occasion can present very good Works. For the time being, among the Students, there are few for whom I have any Hope. Unfortunately the latest Changes in our University's Provisions have spelled Damage to Mathematics while allowing Physics and the Natural Sciences to advance more than before.

According to Lie's thinking, it was realistic to expect "a Contribution of 8 ark [1 ark or printer's sheet = 16 octavo pages] annually from Norway – Half thereof from me." Sweden and Denmark "could easily present Double that", Lie felt, concluding: "After all this Calculation, we will, in any case, have got enough Material, and what is more, we can certainly count on Contributions from German, French, English and Italian Mathematicians." Lie asked Mittag-Leffler to take over the correspondence with Zeuthen in Copenhagen, who, when first approached by Lie, had expressed a certain scepticism about the idea.

Mittag-Leffler for his part seems gradually to have set his sights upon other and greater possibilities, and it also became quite forcefully clear to Lie that Mittag-Leffler was "1000 Times a greater Diplomat" than he, and several times when the topic of obtaining moral or economic support, Lie's comment was: "It has more Clout when You write."

That summer of 1881 Lie left for the mountains with Ernst Motzfeldt. On July 7th they were in Lillehammer, and on that occasion they did not go west toward Jotunheimen as they had done so many times before, but to the Rondane region. Had they gone west they would of course have to pass through Aulestad, and Motzfeldt, it was said, had little desire to meet Bjørnson again. This was due in part to the fact that following his homecoming from America in May, Bjørnson had been involved in a heated agitation against the monarchy, and in favour of a

Norwegian republic, and in part, due to Motzfeldt's having been, the previous year, retained by Bjørnson to sell his estate of Aulestad, and to date, having found no interested purchasers.

According to Motzfeldt's diary, the tour to Rondane was relaxing for them both. They were back in Christiania at the beginning of August. A month later Lie wrote to Mittag-Leffler:

Upon my Homecoming from my Mountain Tour ca. 12 August, I found a Number of trying Businesses had mounted up; in particular, the Corrections to the [Abel] Edition and Archivet, and moreover, our Student Examinations, which are especially rather arduous.

Lie and Sylow had still not "placed the final period" on the question of whom they should acknowledge and thank in the foreword to the Abel edition. Lie stood unflinchingly firm on his decision to strike Bjerknes' name, but left it up to Sylow, with the following strong request: "Now, in God's name, You get to decide."

Lie wanted Bjerknes to be cited only for his Abel biography, and added (September 11, 1881): "Bjerknes has meanwhile full Wind in his Sails these Days. At any rate, this seems to have arisen from the science Dilettantes in Paris following his Experiments with Interest." He went on to point out: "It has certainly always been my Position toward Bjerknes to avoid Conflict. But he is expecting too much." About the same time Lie wrote to Mittag-Leffler in Sweden:

Here the Norwegian newspapers are frequently reporting that Prof. Bjerknes' Apparatuses arouse splendid Attention at the electrical Exposition in Paris, that indeed they might be considered the Pearl of the Exposition. Can one believe that such a Thing is true? Have You heard anything of this Nature *directly* from Paris? Bjerknes who for years was as unassuming as Sylow, is still not quite free to engage in a little Humbug here in his homeland. He is now struggling to have founded a hydrodynamic Institute at our University, with a Per-Annum of 4000 Kroner.

Bjerknes's experiments were being discussed by many in Paris, and among others by the British journal *Nature*, as the most interesting display in the whole exhibition. Bjerknes received the exposition's highest distinction, with, among others, the Americans, Alexander Graham Bell for his telephone, and Thomas Alva Edison for his electric light bulb, the German, Werner Siemens for his dynamo machine and his electric railway, and seven other exhibitors.

The question of what the title page of the collected works of Abel should look like was also the object of discussion. Lie wanted the title to appear without information on the editors and which scientific associations of which they were members: "This is typographically ugly. And why indeed does a Reader need to know whether either of us is in this or that Position? We stand as Norwegian Mathematicians. That is the Issue." Lie stated that on his calling card he had only his name, without a title, and this could perhaps "be construed as Arrogance", but then he only gave it out to those who were somewhat acquainted with him, he maintained. But which name, Sylow or Lie, should come first on the Abel edition? Lie was in a state of doubt – it was normal to put the names alphabetically, and "L", "unfortunately" came before "S": "But on the other Hand, You have done the much greater Work on the Edition than I have," wrote Lie, who would thus necessarily stand in second place: "This was the Reason that a Year ago I suggested that You take over the whole Edition." Lie's suggested solution seemed now to be that in the foreword they explain more about the working relationship between them, and he brought to attention that each of the notes to the studies was signed – and here Lie had far fewer and shorter comments than had Sylow, whose notes in many places had almost become independent papers. As a final point, Lie added, that everybody certainly knew who had done the most – at least all of those, both at home and abroad, with whom he and Sylow had been in contact over the past seven years, and most certainly as well, all those with whom Bjerknes had been in contact, he added with sarcasm.

Lie was concerned about Sylow's position and his future. He now tried, without luck, to get Sylow to apply for the vacant professorship in Helsinki – it was "stupid of him" not to apply, Lie commented to Mittag-Leffler, and continued: "He [Sylow] has the Basis for considering himself hard done by here in Norway, however his own Modesty was partly to blame" – and: "It is remarkable that his Erudition and Discernment are tied so little to the Urge for Creativity. He is devilishly bright, that is for sure." To Sylow himself, Lie wrote: "As far as it goes, I myself feel convinced that You, in reality, if not completely, will get what You deserve." What Lie did to assist Sylow in this direction is uncertain, but fifteen years later, when Sylow was sixty-four years old and still a senior teacher in Halden, Lie in any case put down his views strongly in a feature article in *Aftenposten*, entitled "On Abel, Évariste Galois, and Ludvig Sylow". This had to do with securing Sylow the situation in which he "could devote all his Powers to working out his mathematical Ideas." Two years thereafter, Sylow too was made an extraordinary professor in Kristiania/Christiania, where he was an inspiring teacher right up to his death twenty years later.

Lie and Sylow also had to deal with the problem of the sub-title to the Abel edition during that autumn. Broch wrote from Paris that it ought not to stand as "Second édition" but rather "Nouvelle édition", as Holmboe's edition of 1839, had not been complete, and it ought to be stated that the edition had the support "du Gouvernement Norvégien," and not "de l'État Norvégien", but about this, Lie could certainly speak with Professor Storm, Broch felt. When Abel's collected works were finally submitted for printing – Volume One being 621 quarto pages, and Volume Two, 338 – they were ceremoniously launched in Christiania before the Academy of Sciences, on December 9th, 1881 – the title page read: *"Oeuvres complètes de Niels Henrik Abel. Nouvelle Édition publiée aux frais de L'État Norvégien par MM. L. Sylow et S. Lie."*

Lie now devoted great efforts to getting the Abel edition out to mathematicians all over Europe, and he received many letters of gratitude in return. He took care that the work, through Gauthier Villars in Paris, was for sale in France, Great Britain and Italy. Over a two-year period, Gauthier Villars purchased 250 copies; in Germany, at the beginning, it was a hard slog to find an appropriate channel for sales. Lie took care that above all, the work be presented formally and with ceremony. In Germany it was launched by Weierstrass, in Paris by Broch, and Lie asked Mittag-Leffler if he could get the book launched in the Swedish Academy of Sciences. Klein in Leipzig helped by having 100 copies sold through the well-known Teubner Verlag. A little while later, Lie summed things up for Mittag-Leffler:

> It is basically remarkable how little Abel is read in the Original by the great Majority of Mathematicians, instead of knowing him second Hand. How many have, for Ex., read Abel's Paris Memoir, which indeed still, apart from the Abelian Theorem, contains what is called the "Riemann" *p* in nuce. To be sure the Treatise is not a little heavy-going.

Some months later, Weierstrass sent to the University Library in Christiania all Abel's letters and manuscripts that had been in the possession of the Berlin Academy. In consequence, and directed through the university, Lie suggested that, in addition to thanking him, they ought to reciprocate by awarding one of the greater Royal Norwegian Orders to Weierstrass. But nothing came of this, and to Mittag-Leffler Lie explained, the reason "most certainly" was that "the Governing Party looked askance at me, who had become Professor at the Initiative of Parliament, and as *eo ipso*, to listen to me was to listen to the reddest of the red (without any Grounds therefor)."

In the foreword to the Abel edition many were thanked for their suggestions, help and support. Among the Norwegians were Broch, and of course the deceased Holmboe for his work with the first Abel edition – but there was no acknowledgement of Bjerknes. Toward the end of the foreword one finds only that Professor Bjerknes, basing himself on comprehensive studies and materials gathered for the current edition of Abel's works, had recently published a detailed biography of Abel. The hope, that the biography would rapidly be translated to a better-known language, was expressed, but also consideration was drawn to the fact that they, the editors of the current edition, did not share all the points of view of the biographer, even though they too awarded Abel the main honour for the discovery of elliptic functions.

Bjerknes was prepared for the fact that his Abel book – because he had exploded the myth of "the noble competition" – would, for national reason, attract a degree of unpleasantness. But the fact that his points of view were not greeted with full support, was a great disappointment to Bjerknes. And without mincing words, he let Lie and Sylow know that the foreword to Abel's collected works was the least fortunate place to make reservations about one who struggled to see that justice was done for Abel. Lie reported to Sylow: "Bjerknes is utterly distraught about us declaring ourselves in disagreement with certain of his Ideas and 'even to do so in a Main Sentence rather than a secondary one'." (Using a book's foreword clearly to explicate the development of one's meaning was something that Lie also would come to utilize again. Meanwhile, Bjerknes' book was published in French four years later, translated by the mathematician Houël in Bordeaux, with whom Lie also had a degree of correspondence.)

Lie was relieved that the Abel edition had been completed, and expressed this in letters both to Klein and Mittag-Leffler. While the work was still ongoing, Lie did not want any payment for it, but now that it was finished and it became evident there was a sum of about 500 speciedaler (1 speciedaler = 4 kroner) left after all the expenses of the Abel edition had been paid, he reported to Sylow that he wanted to have 400 of these, not as an honorarium, but for the losses he suffered during the years between 1874 and 1878. Due namely to the work on the Abel edition he had had to relinquish a subsidiary income from teaching, and in these years he had had only "the Wage plus Cost-of-Living-Supplement" of 950 speciedaler, which "did not stretch to cover my Family's Maintenance, including the Widow Pension Fund" – this had consequently led to the fact that he must, "given this most embarassing Economy, go into Debt."

He wrote to Klein now as well that the work on the Abel edition in the recent years had hindered his own work, and that the editorial board of *Mathematische Annalen* hopefully would notice that he now had more time for his own studies. The first piece he had ready to go was a work on geodesic curves, and then again there would be the planes with constant curvature. Lie asked Klein to give his greetings to Mayer, and say that he would soon be in contact: "For years I have held myself back, not really consciously. But rather it has just happened that way," he wrote, and some months later to Klein again: "I believe that my level of working energy has reached a minimum in the autumn of 1881. I hope that many years will go by before it [the trajectory of his working energy] reaches its next maximum."

The year 1882 was also busy and eventful for Lie. During the spring semester he lectured two hours a week "on Cartesian Plane Geometry", two hours "on a simple Segment of the new Geometry", together with one hour "on elliptic Functions". In April he sent off to Klein and *Mathematische Annalen* his study of geodesic curves, and with this, he wrote that he, with the greatest of interest, although only in its general contours, was following the "remarkable and stunning progress" made in function theory. "The view of the future that you presented in your [Erlangen] Programme is becoming more and more fulfilled," he wrote, in a state of wonder, to Klein. What Lie seems to have put off mentioning to Klein for the longest time, was the lively correspondence in which he was now engaged as well with Mittag-Leffler.

Mittag-Leffler had at this time had far-reaching contact with French mathematicians, first and foremost with Charles Hermite, and from early on he had been attentive to three unusually talented students of Hermite: Paul Appell, Emile Picard, and above all, Henri Poincaré. Already by the autumn of 1881, in a letter to Lie, Mittag-Leffler had placed Poincaré among those who had advanced mathematics the most during the current century, and Lie then answered:

> I am, in a certain Manner interested to know if he [Poincaré] is Right in his Claim that he can integrate any homogeneous linear Differential Equation with rational + resp. algebraic Coefficients. For, about the Integration of such Differential Equations, I have for the longest time fed the Solution of all infinitesimal Transformations, by any ordinary or partial Differential Equation, back into themselves. If Poincaré's Claim is correct, then I can, among other Things, integrate any Differential Equation $y(n) = f(x y\, y'y'' \dots y^{(n-1)})$.

Similarly, other Classes of algebraic Differential Equations could be shown, which permit *unknown* inf. Transformations to be integrated. Halphèn occupied himself with the Integration of Differential Equations that permitted (infinitely many or even all) linear Transformations. Valuable as these Inquiries most certainly are, it still remains to be pointed out that it is perhaps immaterial that the Transformations in question are linear. The main Issue is that the Differential Equations in question still admit a Transformation Group.

And he concluded: "I have ordered Poincaré's Memoirs but unfortunately they are sold out." In a letter to Mittag-Leffler, Lie also mentioned that he and Sylow had received a long letter from Hermite, who had communicated his satisfaction with the reservation in the foreword of the Abel edition in relation to Bjerknes – and Lie could not now refrain from commenting: "It is remarkable that a discerning and prudent Man like Bjerknes could expose himself as he has done on a Couple of Points." Lie knew that Mittag-Leffler too was working on a biography of Abel, and he expressed great interest in the forthcoming book, but added, "Be a little careful if You confer on this Matter with Bjerknes. He will certainly try to get You on board." Bjerknes and Mittag-Leffler too engaged in an extensive correspondence with one another, but Mittag-Leffler's book on Abel was not published until the year Bjerknes died, more than twenty years later.

Lie's reaction to Bjerknes' account of Abel's having priority of discovery seems also to have had its bases in the situation of the day. He wrote to Mittag-Leffler: "If we Scandinavians, who indeed disappear in Numbers in Comparison to Germans, are to claim any Glory from the Rights of Abel, then we must first and foremost be vigilant against Injustice to Jacobi."

The work on the Nordic periodical – *Acta Mathematica*, "our Idea" – more and more came to be directed by Mittag- Leffler, and Lie seems to have been satisfied with this. Mittag-Leffler was unsurpassed in his ability to solve organisational and economic problems, and moreover, had the best contacts in the political and economic milieux of Sweden, not to mention, in international scientific circles – and within Sweden, he utilized his friendship with the mathematician and cabinet minister, Carl Johan Malmsten. But some of the relations in the international journal situation created difficulties. In 1880, two of the great names in mathematics, Kronecker and Weierstrass – following the death of editor C.W. Borchardt – had now begun to edit the prestigious *Crelle's Journal* in Berlin. Weierstrass was Mittag-Leffler's former teacher, and a good and trusted friend, but Mittag-Leffler had doubts about the administrative abilities of the two German mathematicians. Moreover, it was said that Kronecker and Weierstrass did not get along with one another. Mittag-Leffler therefore feared that the renowned journal would now fall into decay, and that a competing new Nordic journal would hasten such a decay, something that in turn would create bad will in Germany toward the Nordic initiative. For Lie, the greater considerations were Klein and *Mathematische Annalen* in Göttingen, where more and more, Klein was becoming the major editor. He mentioned several times in letters to Mittag-Leffler that he would continue to write for *Mathematische Annalen*, and he wrote that he could not imagine that Klein or any of his friends would oppose the idea of a Nordic periodical, but nonetheless, he added: "Another Issue is that the Editors of Math. Annalen may not be happy about an Undertaking that deprives them of a considerable Amount of Material." The precondition for the Nordic journal was that it, in the words of Lie, "*only* take in Papers from the great Languages of Culture. This Claim must be satisfied if one is to compete on the World Market."

Lie and Mittag-Leffler also sent their own mathematical works to one another. Lie commented (in February 1882):

> Your great Discoveries in Function Theory and the Calculation of Integrals have my greatest Admiration and Sympathy, and according to my Abilities, and as far as Time allows, I have tried to form an Idea about the Essentials of Progress in these Fields. Unfortunately I must regret that in my Youth I read so little about Function Theory, and now it is a heavy task to complete these Lacunæ.

About his own work, Lie wrote: "I believe that my Transformation Groups (about which I first wrote in the Göttinger Math. at the conclusion of 1874) will come to have special Significance in the Theory of Differential Equations of two Variables *if Poincaré's Assertions continue to stand*." And a little later: "There is an essential Difference between Your and Poincaré's Inquiries into Functions and Differential Equations, and my more modest studies of Transformation Groups and Differential Equations - there Precision plays a rather different Role for You, where on the whole you study analytic Functions, than it does for me." Lie then formulated for Mittag-Leffler his "first Idea", as: "I can see myself transferring to Differential Equation Theory the Ideas which, for those concerned with algebraic Equations, have given rise to the Doctrine of *Abelian Equations*."

Throughout that spring Lie tried to get "Norwegian rich Men" to make economic contributions to the Nordic journal, but found out rather quickly that attitudes were "hardly particularly promising". He conferred with Bjerknes about the matter, and Bjerknes felt that Professor Aschehoug would be the best person from whom to ask advice – and Lie reported to Mittag-Leffler that Aschehoug was "precisely the Man". He was interested in science, "and not only in Mathematics", and above all, he had great influence in "The Conservative Party, that is to say, the Party that has the Wealth." In Sweden, Mittag-Leffler had always managed to get great contributions for the journal – and even the King supported the initiative – such that Lie commented: "Had our political Situation not been so desperate, perhaps we too could have gotten something from Parliament. But just at the present Time it so happens that there is a sudden tragic Change in the Liberality which hitherto Parliament has shown toward Science." Nonetheless, the Norwegian government gave its promise in 1883, to give 1,000 kroner to the journal, corresponding to the contributions of the Swedish and Danish governments.

The first editorial board contained only Nordic mathematicians. Lie was no doubt involved in deciding who would sit on the first editorial board, but the periodical was mainly run by Mittag-Leffler, who now first and foremost got in touch with his mathematical friends in Paris. Mittag-Leffler had arrived at the view that they found themselves in a time that resembled the 1820s, when elliptic functions were discovered, and precisely as young Abel had done at that time to make the newly-founded *Crelle's Journal* well-known, so too now would the young Poincaré secure the success of the Nordic journal, *Acta Mathematica.*

Mittag-Leffler wrote to Poincaré at the end of March 1882, and explained openly the role he envisioned. Poincaré's response was positive and he began right away to send his works to Stockholm. Lie wrote to Mittag-Leffler: "If it really is so, as You make the Calculation that by greatly expanding the Contributions made by the Frenchmen You mention, so indeed is it quite certain that this is sufficient to establish the Journal's Reputation." They got it right: it was first and foremost by getting Poincaré linked to the journal that *Acta Mathematica* so rapidly won itself international renown. Goodwill from the German mathematicians was secured by, among other things, having Weierstrass, Kronecker, Kummer and Schering honoured with one of the Royal Orders of Sweden.

As early as the spring of 1882, it seems to have been clear that Lie would, in the coming autumn, quite certainly get a leave of absence and a public stipend for a tour abroad, and Lie now turned back to the thought that he and Klein could again meet in Paris. He wrote:

> Do you not also think about a trip to Paris? It would be strange to meet in Paris again. Then once again we could go to Sceaux [Parc de Sceaux, south of Paris] and drink coffee among the trees, and even once more admire the hippopotamuses in the zoological gardens, and perhaps even meet once more for an evening at Closerie des Lilas. Think of it!

The reservations that nonetheless seem to have lain behind these hopes and expectations, were due first of all to the fact that Lie had come to know Klein's health was not good – he was suffering from asthma – and Lie commented: "With me, it is my nerves that are not in order," and he added, "For me there is nothing so essential, both for body and for soul, as a journey abroad. What drains me is sitting for hours at a time and writing." And right before the end of the semester, when he had received a firm official response about the travel grant, he wrote that he could not give up hope that Klein too would came to Paris:

> With regard to your health, I hope so much that it is going better. For one thing, I believe that you would have difficulty to find a better travel partner; around me you will always find *unconditional* sympathy and friendship. Therefore, perhaps a trip in my company would prove to be less disquieting [aufregend] than not to travel. For me it is a major issue to get together with you and be in your company.

In Paris Lie would make it clear to Halphèn that several of his theorems were only corollaries of those he (Lie) had published long before on infinitesimal transformations. He also mentioned to Klein that he had several points of contact with the works of the mathematician Laguerre, particularly regarding his investigations into invariants of linear differential equations. And Lie asked Klein, when he had the time, to write to Poincaré, and tell Poincaré that Lie wanted very much to meet him in Paris. Lie added: "I believe that my transformation theories can profitably be combined with his magnificent enquiries."

This was a sore point. Klein's relations with Poincaré was undergoing considerable strain – in addition to his unsatisfactory health, his relationship to Poincaré was quite certainly another reason for Klein's refusal to meet Lie in Paris.

Klein had been working for the last couple of years – inspired by Riemann's studies – on a geometric approach to function theory, and during the autumn of 1881, he had also made public a more substantial work on a new type of algebraic functions, a work that he certainly thought would be the best and most important of his mathematical career. But as early as half a year before, in May 1881, he had seen that young Poincaré in Paris was beginning to publish the main points of a new theory for a new type of functions – what came to be called automorphic functions, which Poincaré also called *fuchsian*, after the German mathematician Fuchs. Poincaré's studies were highly tangential to what Klein was working on. Klein had thus rapidly made contact with Poincaré, and in the beginning it had been a friendly correspondence, and a "competition" in which they both tried to formulate a major theorem that would simplify and be fundamental to the theory. Klein also managed to formulate such a theorem and a strategy for proving it, but this "competition" with Poincaré would nonetheless drive him into a state of mental and emotional exhaustion, which had consequences.

Poincaré would also come to play an important role for Lie as well. With a thought to his forthcoming trip to Paris, Lie also wrote to Mittag-Leffler before the summer holidays (1882):

> I had thought to ask You to introduce me to Poincaré. Perhaps at some occasion when You are writing to him, You can tell him that I intend to spend October and November in Paris and that it is extremely important for me to make his Acquaintance. [. . .] Perhaps Paris will become the Centre of Mathematics again soon. It seems to me to promise to do so in many Branches of Mathematics. I believe it is an Advantage for the World if the Frenchmen come once again to be at the Forefront, for it is certainly far more pleasant to read French than it is German Works.

For most of July, Lie hiked in the mountains as usual. How Anna, with their children aged five and two, spent the summer seems unclear, but she probably stayed home for part of the time, at the Schønheyden Estate at Skarpsno, and perhaps had already begun to spend part of the summer at Åsgårdsstrand[33], something that in later years seems to have become customary. For his part, Sophus hiked over the mountains to Nordfjord. In the middle of July he was at his birthplace, Nordfjordeid, and went on further, to Stryn, eastward, over the mountains again, and down to Lillehammer. He was back in Christiania in early August, in time for the *artium* examinations. He remained in Christiania for about a month before he left for abroad. In the university calendar of lectures for that autumn one learns that Prof. Lie: "is staying Abroad on a public Stipend." In Copenhagen he met Zeuthen and, with considerable expectations, they discussed the Nordic periodical. Lie thought he would contribute one of his larger pieces on surfaces with constant curvature, and it was important to him to get Zeuthen to submit a contribution. Lie had pointed out to Mittag-Leffler:

---

[33] Åsgårdsstrand, on the western shore of Oslofjord, would also become the summer residence of the expressionist painter Edvard Munch. See also "Melancholy", the Edvard Munch painting reproduced at p. 263.

With regard to Function Theory, the Journal will strike a blow and receive a dominant Position. But when it comes to Geometry (which represents a great Number of the reading Public), it is of capital Importance to get the Danes, and in particular, Zeuthen, with us. It is certainly the Case, that for the Time being, there is more Movement in Analysis than in Geometry; but nonetheless, to my Way of Thinking, it is necessary that the Journal also make its Mark in Geometry.

From Copenhagen, Lie went on to Klein and his wife in Leipzig, and he reported home to Anna from there that he had "had a long and interesting Discussion with Klein. Tomorrow we are going out to visit Mayer who lives in the Countryside." Otherwise things were "remarkable', and he hoped it was the same with her and the children. He asked her to greet everyone, "particularly Laura", and he reminded her of his further itinerary and addresses where she could reach him. For his part, Klein reported, on September 19 (1882), to Poincaré in Paris that he had been paid a visit by Lie for a couple of days, and he prepared Poincaré that in a month's time he would be meeting Klein's "friend S. Lie" – and he wrote that even though Lie was not working on function theory, he was still "vitally interested in the progress that had recently been made in function theory."

Following his visit to Leipzig, Lie was off to the Swiss towns and cities of Pontresina, Andermatt, Zermatt and Geneva before he would arrive, according to his calculations, in Paris at the beginning of October. It seems that he kept to this itinerary and schedule. "For me, as a Norwegian and an old alpine wanderer, Switzerland is of particular interest," he wrote to Klein. Lie hiked for a couple of weeks in the Swiss Alps, and came to Paris in good health and good mood. He moved into the same boarding establishment where he and Anna had spent their honeymoon, at 8 rue Bagneux. He wrote home to Anna and told her that the landlady, Madame Maurvais had died, but that the boarding house continued, that as usual there were people of many nationalities there, mostly English, and mostly ladies, but so far, he had "almost not spoken a Word to Anyone". The price was reasonable, six francs, but he felt that he got so little food that he would soon have to move: "For 8 fr. per Day I shall undoubtedly get a good Pension." He went on to tell Anna that he had already been to visit one mathematician, and had an invitation to another, and that for the time being, Professor Broch was in Paris.

Sophus asked Anna to send him her glove size, and to try to find out if the sizes in Paris were the same as in Norway, or else she could clip off "the uppermost Edge of the Glove (the Wristband), which shows the Circumference of the Wrist." Thus Sophus had already begun to think about what he would buy to bring home, and he speculated on what he would buy for the children, and asked Anna for advice: "It does not have to be so very cheap. Perhaps I shall lump the Paris present and Christmas present into one." He praised her for having remembered to send a telegram for Ernst Motzfeldt's wedding – Motzfeldt had lost his first wife, Else, the year before, and now, in September, had married her sister, Margrethe. Otherwise, Anna had only used the money "after the best Consideration, approximately in the Manner that we usually do." He reminded her that he had money to the good at Aars [Latin and Grammar School], money that she could have disbursed to her if someone would certify who she was. If letters came from Gauthier Villars or Teubner, she must open them, and were there anything about Abel's works, she must go to Department Secretary Horn; if they were of a private character, she must send them, "and so too, above all, *all* Letters from Abroad, and all Printed Matter," with the exception of the large American packages that contained *"American Journal"*, edited by Sylvester.

Sophus wrote home diligently from Paris; he also thanked Anna for writing so often to him – her 'dear Letters" were a great joy to him, particularly everything she wrote about the children. In addition to Anna, Holst, Klein, and Mittag-Leffler all received letters from Lie with information as to what had happened to him in Paris. In his first letters to mathematical friends, Lie accentuated Klein's position and contributions. He was very clear about the rivalrous relations which existed between German and French scientists and mathematicians – he knew well what he was now involved in.

He wrote to Holst: "Klein is and is becoming a brilliant Mathematician par excellence. His Health however is not quite what it ought to be. But anything Else is not to be expected, the Way he works." It was Klein's time-consuming teaching, his extensive editorial contributions to *Mathematische Annalen*, and his professional "competition" with Poincaré, that Lie cited here. In one of his letters to Holst, dated 7 October 1882, Lie wrote as well: "Shall we get together this Winter and form a Mathematical Sciences Association? Principally in the geometric and purely analytic Direction. No Spheres in Water! Or in any case, only in Moderation!" This last was a pointed reference to the experiments of Bjerknes.

Lie reported from Paris to Mittag-Leffler: "Naturally enough, Klein wants to stand on the most friendly Footing in regard to the Periodical," and added: "Klein is a far more remarkable Mathematician that the Majority of the German Function-theorists would admit, or perhaps, understand, if only his Health does not Shackle his Working Powers too severely." And following the first meeting with French mathematicians, he himself wrote reassuringly to Klein: "Your stocks are high in Paris!" But then he added: "All the same I say, along with Abel: C'est beaucoup plus difficile de gagner l'intimité des Français etc." ("It is a great deal more difficult to come into close acquaintance with the Frenchmen [than the Germans].")

Lie wanted to get to know the French mathematicians, and above all, make his own work known – and he succeeded, even though he perhaps did not get as far as he hoped. After the very first meetings in Paris, he wrote to Mittag-Leffler:

> It is curious that without Comparison, my most original and far-reaching Works meet with much less Acceptance than my less important Works. But such is the Way of the World.

His "far-reaching Works" were the theory of transformation groups, and at first his thinking was to arouse interest in these by having discussions with Hermite: "I am counting so much on You [Mittag-Leffler] on Occasion, to recommend me as well to Hermite," and Lie continued: "Picard, who indeed is evidently quite a Devil, is Someone with whom I have fewer Points of Contact, since even in his doctoral Dissertation there are Things included mindlessly, that I had long ago published."

The first time Lie had appeared at the French Academy he had met Halphèn, Darboux, Poincaré, Lévy and Stephanos – they were, one and all very accommodating, Lie reported to Klein. He told Klein that while at the Academy he had also spoken with Chebyshev, a marvellous old gentleman who, vital as a youth, had spoken about a number of things he had published, from subjects that had now been drawn from Abel's surviving papers, to differential equations in which the multiplier could be established when one knew an infinitesimal transformation.

In a letter to Holst, Lie expressed particular happiness to have met Stephanos: "He [Stephanos] seems to have found common Appreciation as a talented Mathematician." Hermite, however, had promptly disappeared from the Academy the moment Lie arrived. Poincaré was able to report that he had already received the proof pages from Mittag-Leffler, of his work for the new Nordic journal, which would certainly be delayed due to Weierstrass' contribution, which had still not been submitted. But otherwise, Poincaré had a little speech impediment that made it difficult for Lie to understand him.

To Mittag-Leffler, Lie wrote that he was working on a paper for the journal, but:

> Before I send You my first Piece, I want to secure myself by having a Conference with Darboux and others, to ensure that the Content is new. I certainly always have enough Material. But the Issue is that my first Work should be short, easy to read, and of value. These three Claims limit my Choices to a great Degree.

Lie did not come to decide the subjects he would write about. In fact, he neither now, nor later, ever came to publish anything in *Acta Mathematica*. But as Mittag-Leffler wrote later (in *Acta* on the occasion of Lie's death): "It was he [Sophus Lie] who first understood that the time was ripe (l'époque était venue) to publish a substantial Scandinavian journal."

In the middle of October he wrote home to Anna that even though not much had happened, he was nonetheless able to report that things were still going well with him: "Since I arrived in Paris it has been almost continual good Weather. Yes, just think, here it is still rather oppressively warm, and even in the Evenings after the Sun has gone down. I still have had no use for my Dust Cloak." He wrote about the streets and the places they had been together – "about a lovely Tour out to Parque de Vicennes where I had been once before together with You". "In the Café de la Régence I now get both Morgenblad and Dagblad; so now You no longer need to send me the daily news." He was also quite certainly aware that the Norwegian writer he earlier had described with great sympathy, Jonas Lie, had now, in October, arrived in Paris. A couple of months later, in December 1882, Bjørnson also arrived in Paris with his family, and they would live there for five years. As for Jonas Lie and his wife, Paris would be their home for the next twenty-four years.

The most exciting novels of the period were French; moreover, the best actors hailed from France, and French journalism and criticism were considered to be both the most entertaining and the most thoughtful. But the Norwegian poets – and later, the Norwegian painters – were fascinated by what they considered the lack of morality adhering to the ontology of the French. During this whole period, Norwegian literary eyes looked askance at "French Morality". In a letter written to Ernst Sars, Bjørnson, for example, compared Norwegian and French

"Consciousness of Sin" in the following manner: "They [Frenchmen] do not hold themselves back, the majority of them, from their conquest of women, and certainly it wounds neither them nor their morality as it does with us, who consider it a sin, even though we do it." In one of the first letters that Sophus wrote to Anna about social relations at the pension, it appears that he was alluding to these activities when he wrote: "Here there are few Gentlemen, so I could certainly find Appreciation."

Lie had several meetings with the French mathematicians, not only in larger gatherings and in the Academy, but also individually, and he was invited to give a lecture to the *Société Mathématique de France* on November 3rd. Lie had noticed that his geometric works were better known in Paris than in Germany. He was constantly hoping to find "an open ear" for his transformation group theory, but he could see clearly that it would be by means of works with a more practical application that he could impress and demonstrate his mastery. He wrote to Klein about his forthcoming lecture to *Société Mathématique*: "On that occasion I must try to make advertisement for my integration theories."

During recent years, the French mathematicians Halphèn and Laguerre had worked with investigations of differential invariants and related integration methods of the projective group of a surface. In light of this, Lie seems now to have brought forward his earlier studies of integration theory to a complete system with known infinitesimal transformations in order to find simpler and better methods.

With Hermite, who had disappeared so suddenly at the time of Lie's first meeting, he later talked "on all sorts of things". Lie wrote to Klein that he felt Hermite had a very friendly and amiable way about him; yet he was not completely sure how deeply Hermite had gone into what Lie told him. The most remarkable thing he got to hear from Hermite – who was said to have been incapable of reading a word of German – was that he, Hermite, had been told by Mittag-Leffler that the German mathematicians *hated* the French mathematicians – and Lie's subsequent protests went unheeded. Hermite was also known to be a very devout man, but had declared himself amazingly forcefully and energetically on the nationalist question, but Lie considered that Mittag-Leffler's statement had probably been somewhat vividly coloured. Hermite was consequently very eager to hear about sparks and disagreements between French and German mathematicians, and he presented conditions in Paris as idyllic.

"It is certainly no better in Paris than in Germany," Lie commented to Klein. He wrote that Hermite naturally enough praised Weierstrass, but complained that his presentations were difficult to read, and pointed to Lazarus Fuchs as the German mathematician who had in the clearest manner managed to edit his theories. Hermite had also spoken, with great familiarity, about Klein's results, but in such an indeterminate and approximate manner that Lie felt he must report directly to Klein that he was not sure how well, in reality, Hermite knew Klein's work. "Is it not such that Fuchs is Klein's *bête noir*," Hermite had asked, and Lie had then answered that Klein certainly was familiar with Fuchs' achievements, but that Klein certainly considered that in France, Fuchs was attributed honour for things that others – Riemann for example – had done. "But that we know well," was Hermite's reply.

Behind these insinuations lay some of the contentious argumentation to which Klein was now succumbing. Early in the correspondence and the "competition" with Poincaré, Klein had come out with bombastic expressions regarding the name of the new functions – a discourse over naming that Poincaré for his part seems to have considered completely immaterial. Poincaré's first innovative work was published under the title *Sur les fonctions fuchsiennes* – and his point of departure was something that he had grabbed from the work of Fuchs, who was a leading exponent of the Berlin School of German mathematics. Klein felt it was completely wrong to use Fuchs' name, both for purely factual reasons, and certainly also because he did not want to establish any Berlin-Paris connection. When Poincaré, without any warning, began to use the name *fonctions kleinéennes*, Klein became even more upset, and held fast to an old-fashioned terminology: "invariant functions with respect to a group of linear transformations"; later he used the description "automorphic functions". Fuchs and others got themselves involved in the discussion. Klein's preoccupation with the name became more and more a struggle over priority. Klein seems to have been backed up into a corner, rather than crowned with the laurels that he at first had hoped would be the result of this work. Well, wrote Poincaré, call the functions instead "Schall und Rauch", citing with tolerance a line from Goethe's *Faust*.

Lie seems to have wanted to encourage Klein from Paris by referring to what the French mathematicians felt: Halphèn, Darboux and Stephanos all spoke with the greatest familiarity about him, Klein. Both Stephanos and Darboux thought of visiting Klein in Leipzig. Poincaré had said on occasion that "all mathematics was a group history" ("dass alle Mathematik eine Gruppengeschichte war") – and Lie had then explained about Klein's Erlangen Programme, which Poincaré did not in fact know about. Poincaré had said that in the beginning he felt that it was difficult to read Klein, but that now it was going quite easily. Darboux and Jordan had also said that Klein placed great demands upon his readers when he frequently did not elaborate his proofs sufficiently.

Lie had spoken at greater length with Jordan. Jordan's mother had just at that moment died, and Lie had been invited to her funeral. Otherwise, about what he saw and learned of the French mathematicians, he wrote: "Picard is Hermite's son-in-law. He quite naturally makes a very intelligent impression." But according to Lie, Hermite was almost always "far too polite" and he was not so "agile in the head", as for example, Picard. It was certainly also son-in-law Picard who gave Hermite references to the most important German works, which Hermite subsequently was unable to read for himself.

About his own works and their reception in Paris, Lie was rather satisfied. He was impressed by how scrupulously Darboux had studied his works, and more and more was lecturing on this material at the Sorbonne. But it was unfortunate that Darboux still pursued and "plundered" Lie's work. To Klein, Lie wrote: "he [Darboux] makes immaterial changes and publishes this without mentioning me.

Now he has begun on the surfaces of constant curvature. I must therefore, as quickly as possible, prepare my Christiania pieces for submission to *Math. Annalen*."

For Halphèn – whom Lie characterised as "extremely friendly", and otherwise as someone who knew something about his work on minimal surfaces – Lie tried to explain the relationship between his integration theory and Halphèn's invariant differentials. It was this on which he would later give a lecture, and already as early as October, he reported to Klein that this was a theme that he had also discussed with Hermite, Poincaré, Picard, Jordan and Chebychev – and the latter two in particular had encouraged him energetically. Poincaré, on the other hand, had paid strict attention to him during the meeting, and likewise Picard. Therefore he feared that Hermite's familiarity with his work was fundamentally only phraseology. "Hermite is a man of the world in conversation, but seems now to have become a little old," Lie added.

Lie also commented on his meetings with the mathematicians Bonnet, Lévy, Bouquet and Mannheim. Bonnet was very friendly and vigorously interested in Lie's geometric matters, but "colossally thick". As to Lévy, he said only that he was a good mathematician who knew little about Lie's integration theories. Lie expressed his gratitude to Bouquet, who together with Briot, had so beautifully acknowledged Abel. What Hermite had said about Abel and Jacobi in his lectures at the Sorbonne, was however "remarkably incorrect". According to Lie, Hermite had declared that Euler had already inverted the integral, and where Abel had the priority, Hermite mentioned that Abel and Jacobi were simultaneous, and where Jacobi had priority, he did not mention Abel at all. (It was still Hermite who was to have said about Abel that he had given mathematicians enough material to work on for several centuries!)

Together with Mannheim one day, Lie had gone out on the town and drank beer, and they had had with them the editor of *Liouville's Journal*, Rescal. This well-known journal was however for the time being, lightly regarded in Paris – or so Lie was able to report, and added that he could well understand this after having heard Editor Rescal speaking about the direction of the newer function theory, which he quite obviously did not understand. Mannheim, as always, was friendly – "he is really a good human being, and he warned me against Darboux, as he had all good reason to do." But Lie also had to add in his letter to Klein: "I must nonetheless speak with Darboux, for he understands me best. And I must pay something for the enjoyment of it. In any case, science demands it."

According to Lie, Darboux in addition had the bad habit of talking scornfully about most other mathematicians, but Darboux made an exception for Klein – as did Halphèn who, in other respects, had the same bad habit. In another letter to Klein a little later, Lie wrote that Darboux had married a poor girl, and Lie explained that when he and Anna had been on their honeymoon tour in Paris eight years before, Darboux had been terribly impoverished – but since then he had subsequently become Chasles successor, and had as well, other sources of income.

The month of October passed. Lie was satisfied with the days, but then it began to become colder in Paris and Lie bought himself an overcoat, hat and umbrella for approximately 100 francs. On November 3rd, he impressed his audience at the *Société Mathematique de France* with the presentation of integration methods, and his fame increased by several notches.

A few days later he wrote home to Anna with great satisfaction, and reminded her that he wanted to give her a gift to the value of 100 francs – partly as compensation for having been so long alone. Now I have to do things well, so that You are satisfied with what I purchase." He indicated that this might well be something in gold: "It is really what I too appreciate the most" and moreover it would provide them both with the most enduring "Memories of the Paris Tour". Sophus thought he would buy a brooch and a locket: "It is much better to buy a few handsome Things than several middling ones," he commented, and explained that he had been "tremendously economical". He nonetheless had to ask Anna to send him money, which the whole time he had meant she should, namely thirty francs to Paris and one hundred francs to Copenhagen, where he would receive it on his way home:

I am so looking forward to coming home. I count the Days until I leave. And yet things have gone marvellously for me here and I shall no doubt come to speak many times of a new Tour abroad. Next time it will presumably be to Germany, to Leipzig.

While I was in Switzerland I speculated many Times on whether it would be possible at some Point in the Future that we and the Children to travel to Switzerland together. I think that Mai and I could walk together, and You and Dagny, wherever one is unable to go further by Railway, one must spend a Couple of Days here and there in one of the many lovely Pensions, which are located in the most beautiful Places. Who knows! If Fortune smiles upon us, so perhaps this Dream can be realized.

It began to be cold and bitter in Paris, and difficult to get the rooms warmed: "I generally sit double-clad, or with my Overcoat on and a Blanket over my Knees. In this way it is tolerable," he wrote to Anna, and added, "I have bought a Number of Pictures for the Children and shall try to buy some Silk Ribbons." It was easier to find pictures for Maia than for Dadi, and he therefore asked Anna about finding something to give her "from Papa" – otherwise there was an agreement that Anna would buy things for the sum of two kroner for each of the girls: "Show Dadi at the same time my Photograph, and tell her not to forget her Papa." Then he admonished his spouse to be on guard against draughts assailing the children at any hiatuses between upper and lower body clothing: "Dress the Children properly. Make sure that they have adequate Underclothing such that they cannot become cold in the Discontinuities."

On November 14th, he received a dinner invitation from Poincaré – the following Saturday at seven o'clock, 66 rue Guy Lussac. Lie replied the following day, in German, that it would be a great pleasure for him, and at the same time explained that he had in fact been around to Poincaré's house and found that he was not at home. That same day, Lie sent a mathematical "Communication" on the integration of a system of differential equations, to the Academy of Sciences in Christiania. But it seems that most of his remaining time in Paris he spent hunting for a brooch or a locket for Anna. (In later letters, both to Klein and Mittag-Leffler, he regretted that he had been too late in contacting Picard and Appell.) In the windows of the shops he saw displayed "a Number of Lockets; but scarcely any Brooches," he wrote to Anna, and asked, "Could it be that they have gone out of fashion? Perhaps they have Brooches in the Back of the Shop! As far as Brooches are concerned, it is not precisely necessary that they have a Pin on the Reverse, are oval and *without corners*, is it?" Sophus seems to have had another solution in mind. He wrote: "When You can get beautiful Things for 90–100 Kroner in Chr.a, it is not really worth me buying here unless I find really beautiful Things. You can count on me. In any Case, I shall do all in my Power."

He told Anna that he was still on the move, that he had invitations both to breakfast and dinner, and that he was out in "société mathématique". "But nevertheless, it feels a little lonely here," and again: "I am heartily looking forward to being home again soon, which still is the best." According to plan he would leave Paris on November 22nd in the evening, reach Copenhagen on the 25th and visit Zeuthen, then go on via Lund to visit Bäcklund, and finally be home with Anna on December 1st or 2nd.

Whether he came home with a locket or a brooch, is unknown. But he did arrive in Christiania at the beginning of December, and for the rest of the month seems to have been much at home in the apartment – perhaps he was trying to adjust to everyday life with Anna and the girls; in any case, he tried to sum up some of what had happened and was happening in the mathematical world around him.

Mittag-Leffler launched the first issue of *Acta Mathematica* with due pomp and ceremony in Stockholm on December 12th. In a letter to Klein, Lie expressed his fear that Mittag-Leffler placed too much emphasis on a grandiose style: "When he [Mittag-Leffler] begins to tire, I fear a decadence. But until then, I expect all will go well for some years. The timing is really especially well-chosen, for the discoveries of recent years are quite securely epoch-making in the history of mathematics." About Mittag-Leffler himself, Lie wrote that he was not only a prominent mathematician and a remarkable social entrepreneur (Gesellschafter), but also "a good human being", and at the same time, that he was too much the "intriguing diplomat" – so that from the point of view of an undiplomatic person like himself, Lie, things would absolutely not go well if they were to work together.

About Klein's own work, Lie wrote, "I do not doubt that you, in your considerable preparatory work for Poincaré's discoveries, will get the recognition you deserve. With all your pupils you have an army that represents great power." But Lie did not presumably know all that had happened to his friend in Leipzig. Acute asthmatic attacks, in addition to overwork and the "competition" with Poincaré had driven the thirty-three-year-old Klein to a breakdown that would be followed by years of physical as well as emotional weakness. Klein himself later wrote about this autumn of 1882: "My really productive labour power in theoretical mathematics was decimated in 1882. Everything that would follow later, in that it did not deal with the elaboration of earlier works, only dealt with minor issues. Therefore Poincaré had a free field and published up to 1884 in *Acta Mathematica* his five great treatises on the new functions." Following his breakdown, Klein suffered from depression for several years, and thereafter he gradually become more highly esteemed and well-known, not really as a researcher, but as an exquisite lecturer, a good editor and a scientific organiser.

Lie wrote a letter to Mittag-Leffler on December 24, 1882:

Overall, I am extremely satisfied with my Stay in Paris, even if I (like Abel) associate on a much more intimate Footing with the Germans. In Paris, as courteous as they are, one always has the impression that these Gentlemen are extremely preoccupied. There is great mathematical Life in Paris now for the Time Being. I have the Feeling that the Germans' Sovereignty in Mathematics is beginning to weaken. And despite the good Friendship I have with many Germans, this indeed is dear to my Heart. For I certainly by far prefer to read French, as opposed to German Mathematics.

In Paris I had the most Points of Contact with Darboux, and to some extent with Halphèn, Laguerre and Lévy, who are all prominent, talented and original Mathematicians. I am most grateful to You for introducing me to Hermite and Poincaré. It certainly interested me to talk to Both, even though my Knowledge is far too restricted in those Fields where they shine. Poincaré is difficult to understand, as he often speaks indistinctly. Although I understood him to the extent that on several Occasions I got a lively Impression of his unusual Mental Powers.

Picard and Appell both made an extraordinarily good Impression on me and I regret that I waited so long before seeking them out.

Almost all the French Mathematicians were talking about Your Journal. As a Rule, they are all at Work on a Paper for the Same.

Picard in Paris wrote for *Comptes Rendus* about Lie's work on transformation groups, and he predicted that these theories would in the future be of great significance. These words motivated Lie. He would strike the forge while the iron was hot: he would make ready two comprehensive articles for *Mathematische Annalen*: on differential invariants, and differential equations and continuous groups.

"It is Lonely, Terribly Lonely"

According to the prospectus of lectures in the university calendar, Lie lectured in the spring term of 1883 for "two hours a week on Cartesian Space Geometry together with three hours a week on higher Geometry."

The correspondence with Mittag-Leffler continued, and the first item it dealt with was securing support for the new journal, *Acta Mathematica*. Lie outlined the political situation in Norway, and was active in trying to get the granting authorities involved in talks. First however, he had to kill the impression that had also come to Mittag-Leffler's ears – namely that Lie, due to the parliamentary vote which ten years earlier had appointed him professor – "would have a special Influence with the Left's [Liberal Party's] leadership." Lie wrote: "On the Occasion that Parliament found to make me Professor, I had literally *never spoken a Word to a single individual* Member of Parliament." Later, of course, he had on a couple of occasions "approached Men of Parliament to advance Matters" in which he was interested, and he repeated that he now also, in regard to *Acta*, would try to get support privately." Apart from this, Lie did not think that the political situation would have any bearing on the question of granting support to the new Nordic periodical. "The Liberal Party has no Basis for looking askance at either Broch, Bjerknes, Sylow or me, and our Rightists [Conservatives] always vote for Grants that the Government proposes." The fact that Bjerknes had thrown in his lot with the Liberal Party in the last election, also augured well, Lie felt, and even that Broch had given "his Name" would be a strong support. Lie communicated what was the prevailing view; namely, that many in the Liberal Party felt that "an acceptable Way out of our impending Discord" was to have Broch become the country's new Prime Minister:

> As far as Broch and I are concerned (You will excuse the Juxtaposition which for the Moment is absurd, but the World sometimes is absurd) we somehow agree in Politics, as we both think our two Parties are both engaged in a competition of Stupidities.

The parliamentary election of 1882 – also called the "Impeachment" election – that had accordingly taken place while Lie was in Paris – was the beginning of a new step in the country's political history. From now on, the struggle between the two parties – a Left (Liberal) and a Right (Conservative) – which for so long had been struggling with one another over state power now permeated every sphere of social life. At this moment, in January 1883, the new Parliament met – 83 from the Left and 31 from the Right – and barely three months later they had agreed to an impeachment indictment against a sitting minister, Selmer, a process which the following year concluded with Norwegian parliamentarism being established as the system of government.

The foreman of this parliament's Budget Committee was one of Lie's old friends from "The Green Room", H. E. Berner, who had also been editor of *Dagbladet* when the vehement struggle was raging around Amund Helland. Another person from the same circles, Lars Holst, had subsequently during the spring of 1883, become the new editor of *Dagbladet*.

In order to be completely sure of receiving the Norwegian grant to *Acta Mathematica*, Mittag-Leffler had also asked the historian, Ernst Sars, to write a recommendation. Lie felt this was a good idea and reported that Sars, "on the other hand", had sought information from him about the project. The only thing that was uncertain, according to Lie, was if *Acta* would be seen as hindering, rather than facilitating contributions to other Norwegian institutions and organisations, but Lie felt that certainly Sars would recommend *Acta* without any Reservations.

At the beginning of March, when Lie wrote to Mittag-Leffler, he said that the reason why he had not made good his promise to send "a more substantial Paper" to *Acta Mathematica* was partly due to his tour abroad, and partly that he had been preoccupied with his own "Investigations", but in addition to this:

> I knew that for You the Major Question was to get Works from the most celebrated Writers, and by that means to bring Fame to Acta. I can certainly imagine that it is desirable for You to have some Norwegian Works as soon as possible; but I have Reason to think that You will find Yourself better served with Works from another of my Countrymen.

Mittag-Leffler seems not to have been completely satisfied with Lie's efforts on behalf of *Acta*. For his part, Lie thought that it was unnecessary to advertise the periodical in *Archivet*, where indeed all the readers knew about it. In *Nyt Tidskrift*, which Ernst Sars and Olaf Skavlan edited, *Acta* had been reviewed by A. H., which surely stood for Amund Helland. After he had declared that this could scarcely be considered "a competent review of mathematical works" – "for no royal road to mathematics is given" – he gave unstinting praise to the initiative and all the prominent mathematicians who took part. The only reservation that A. H. had, was about foreigners who might be confused about "Norway's constitutional relationship to Sweden" when Norway should render contributions to "a Swedish journal". Lie had to soothe Mittag-Leffler afresh with the reminder that most of the influential members of the Budget Committee would quite certainly support the grant and pointed out again that the Left wanted to expand its "Goodwill in favour of Science", and the Right "for less pure Motives". Mittag-Leffler's disquiet must be attributed to "a Misunderstanding of the Statement, which easily has been coloured by momentary Bad Feelings." And Lie continued:

> I urge You to be aware that I, not alone, am following with Astonishment the brilliant superlative Manner wherewith You lead Acta, but also, parallel with this, that I am working in the Manner I consider wisest, for the Attainment of the Norwegian State Contribution.

About his own mathematical achievements, Lie wrote:

Concerning my own Contribution to Acta, I want first and foremost to say that I know very well that my Works can be read by very few Mathematicians. My Treatises are considered to be badly edited. Under these Conditions I myself truly have no great Desire to write for the first Issues of Acta. At the same time my Aim has always been to submit a Contribution. Only it is difficult for me to agree, myself, about what I should send You.

And no matter what Mittag-Leffler and the French mathematicians had flattered him with, it seems that Lie had a level-headed view of the matter. He wrote:

As for the rest, I want to say that I quite certainly have my own Share of Vanity and Conceit, but yet I understand well that one must apply Subtraction and Division to any unwarranted Praise.

On May 23rd, Lie was able to report to Mittag-Leffler that 1,000 kroner per annum had been granted to *Acta Mathematica*, the service in return to be thirty copies of each issue of the periodical, but Lie added: "Yesterday our Academy of Sciences was deprived of one-half Part of its Contribution." The Academy of Sciences in Christiania had received 2,000 kroner, and this support had been set up against contributions to professional publications, which in total amounted to 3,000 kroner, of which *Acta* received 1,000. That Lie used the words "was deprived" in reference to the Academy of Sciences contribution being cut in half in comparison with previous years, might indicate that he was not completely on board with the current thinking about the circumstances of science held by his good friend Amund Helland and others from their old circle of friends.

The debate about the contribution to the Academy of Sciences had been long and bitter, and most of the argumentation for cuts in the contributions had been formulated by Helland, who, in several newspaper articles, had accused the Academy of Sciences of "Coquettishness" and "the Closed Door of the Guild Hall". Others had grasped at the name Christiania Academy of Sciences, saying that it indicated it was "a Society of more local Nature". By virtue of the cutting of state support and therefore of forcing the Academy of Sciences to find, by its own hand, alternate means of support, many had hoped that the Academy would open up in such a way that whoever was interested in science could join by paying an annual subscription. Such a system would generate an income that far outstripped the sum by which the state had now reduced it, and above all, it would smash the out-dated guild system and make "Entry to the Academy's Deliberations and Discussion" completely and equally accessible to all. It was a democratising move, in tandem with many of the era's contemporary attitudes. However the Academy of Sciences itself did not hold with such an attitude, and the halving of its contribution from the state led to, among other things, restrictions being placed upon its publishing activities.

Throughout the winter and spring of 1883, Lie corresponded with Mayer and Klein in Leipzig, and with Poincaré in Paris.

In February he reported to Mayer that in the course of recent months he had succeeded in making a great step forward in the theory of transformation groups: "Earlier I restricted myself in the main to groups with a *finite* number of parameters. Nevertheless there are also groups with *infinitely many parameters*. Now I have also mastered these groups."

In March, Lie wrote to Poincaré – first thanking him for the remarkable goodwill he [Lie] had been shown in Paris, and urged Poincaré also to convey greetings to his good wife. Then Lie thanked him for the brilliant works that Poincaré had been successively sending him and reported that he had read Poincaré's studies in *Acta* with "pure delight" ["wahrem Genuss"]. Beside the generality of the results, Lie expressed special amazement at the simplicity with which Poincaré had treated such difficult problems. Lie wrote that the mathematical world waited in a state of excitement for his next work, and he expressed the hope that he himself, with his limited knowledge of function theory would manage to penetrate the theories: "Through your functions, the doctrine of functions has, in my eyes, taken on a whole new concrete content." About his own work, Lie wrote that he very much wanted to direct Poincaré's attention to "les groupes continus", and he thought also that Poincaré would come to play a role in the theory of differential equations. Lie communicated that he was treating continuous groups with an infinite number of parameters just as easily as those with a limited number of parameters – in particular, that he had determined all the plane's infinite and continuous groups. Lie sent Poincaré a picture of Abel, and stressed that this was something that he had wanted to do for a long time, long before Mittag-Leffler had decided to publish an Abel portrait in *Acta Mathematica*.

Poincaré answered immediately, and sent along a photograph of himself, for which Lie gave his hearty thanks. Lie then answered a question about a certain equation system with two random analytic functions by, in his terminology, determining an infinite continuous group that allowed two extrinsic differential equations to be invariant. In his letter of reply, Lie also showed, in eight pages, how he prepared the way for the principles of his transformation theory, and he sincerely hoped he had succeeded in giving Poincaré some idea of his investigations. And since for the most part the letter had dealt with his ideas and works, Lie wanted to conclude by saying that Poincaré's last piece in *Acta* – on Fuchsian functions – had aroused his admiration. It would, however, be several years before Lie and Poincaré again exchanged letters.

During the spring, Lie reported to Klein that on the whole, the winter had been good, and he continued to be satisfied with the force of his own labour, and in the course of the summer he expected to be able to send him "a large integration theory". Otherwise, it did not seem that Lie wanted to go into, or that he even understood, the extent of Klein's breakdown and illness. Lie commented on his own work and that of others, as usual. He hoped that in the course of a few months he would have this piece ready, his "large integration theory", a continuation and development of the work in *Göttinger Nachrichten* from 1874. He discussed Halphèn's research and expressed the fear that Halphèn would not do him justice. He commented on Picard's latest work, which interested him particularly, and he mentioned that Bäcklund's determination of geodesic curves on surfaces with constant curvature was really a simple corollary of Lie's own works. He explained that he was working "with full energy" once more on transformation groups and differential equations – after working for some years mostly with geometry – and he underlined the fact that, as each year passed, he became more and more certain that the theory of transformation groups would place the theory of differential equations onto a completely new track, if he now could only get all his results together and edited. He concluded: "Hopefully things are going well for you now."

It is quite probable that this summer too, 1883, Sophus Lie left for a month's holiday in the Norwegian mountains, and judging from everything, it seems that Anna, as during the previous summer, was home at Skarpsno with the two "little Girls", Mai and Dagny, six and three years of age – and was probably also out of the city for part of the time. Where Sophus went is uncertain, but in a letter to Mittag-Leffler he recommended, and in any case gave a persuasive account of "the most interesting Overland Route from Trondheim to Veblungsnes (near [the town of] Åndalsnes) in [the county of] Romsdal":

By Railway to Støren. Overland to Opdal, through Sundal to Sundalsøre (which is the only Place in Sundal where one can get good overnight Accommodations; although on the other hand, the Stations between Støren and Aune in Opdal are better). From Sundalsøre a magnificent *Hike* of ca. 4 Hours to Litledal and Litledal Lake, first by Boat 1–2 Kilometers, thereafter overland almost the whole way to the Lake. From Sundalsøre by steamship to Eidsøre. From there, overland to Eidsvåg. (From here an utterly interesting Trip by boat and overland to Nøste in Erdsfjord and up the Valley *halfway* to *Eikesdal Lake*, which *is counted among* Norway's first-rank Sights.) [. . .] Eidsøre by steamship to Veblungsnæs, from where one can make the most utterly interesting Trips to Isterdal or to Hen and onward to the far side of the mountains Vængetinderne and Romsdalshorn. Veblungsnæs is a good Station, either one makes the Effort to reach Ladested or Åndal (also called Næs) [Åndalsnes], or *Åk*, which is the most well-known.

A shorter and equally beautiful though not spectacular Route is the following: from Trondheim to Ørkedalsøren [Orkanger] by local Steamship. Overland through Ørkedal and Surendal to Stangvik. From there by boat and overland to Thingvold and Eidsøre, and from there, as above.

The normal Route from Veblungsnæs to Sondmør is by Steamship to Vestnæs and thereafter by Highway to Ørskog.

The greatest Sights in Sondmør are Norangsdal, *Hjørendfjord* and Geiranger.

Lie received the new issues of *Acta Mathematica* from Mittag-Leffler, gradually, as they came out, and in addition, Cantor's book, for which Lie thanked him heartily, and studied with "the greatest Interest", and wrote back: "I truly think that it is a Fundamental Work."

Lie never came to complete any papers for publication in *Acta*, and gradually his relationship with Mittag-Leffler became more reserved. During the autumn (September 1883) Lie wrote to Klein: "I am soon sending a larger work to Math. Annalen. I am terribly displeased with Mr. Mittag-Leffler." He continued:

I hope you are in good health. It is now precisely one year ago that I visited you in Leipzig. The next time I come to Germany, I shall stay longer. It is lonely, terribly lonely here in Christiania where not a single human being understands my works and my interests. Moreover, for the time being we have an absurd situation in Norway. An indescribable stupidity and fanaticism in *all* directions.

This latter was an allusion to the Right-Left polarisation that had spread and made independent initiative difficult everywhere. As well, the work he hoped soon to have ready for *Mathematische Annalen*, seems to have been the thesis on differential invariants, which would not be published for another year. But Lie could now also report from Christiania that he was busy with the printing of a very important work on continuous groups with an infinite number of parameters, and that thereby the theory of transformation groups had made a great stride forward – the study "Über unendliche continuirliche Gruppen", fifty-six pages in length, was published by the Christiania Academy of Sciences.

On this occasion again, and certainly now as a kind of consolation and encouragement, Lie repeated for Klein what the French mathematicians had said about him, and Lie suggested that perhaps a trip to Paris would bring joy and inspiration to Klein. As to his own condition, Lie wrote that he actually had points of contact with many mathematicians in Paris, but added: "If only once I could acquire the art of asserting myself, that indeed would be fine. But up to now it seems it will never happen." The correspondence with Poincaré had come to a stop. Lie commented: "Poincaré impressed me on several accounts; however, I did not succeed in entering into real scientific contact with him."

Lie praised a paper by Stephanos published in *Mathematische Annalen*, and it was now in September 1883, that Lie also commented on a paper that Klein had sent him – by a young mathematician by the name of Friedrich Engel – in the following words: "Such a man as Engel ought to study my work on transformation groups in *Math. Annalen*." Whether it was this reaction that gave Klein the idea of Engel as a possible assistant to Lie, is uncertain, but in any case, Klein came to mention, a little later that autumn, the possibility that the twenty-two year-old Engel could come to Christiania to assist in the writing up and editing of Lie's theories, an offer Lie then seems to have thought was so insubstantial and airy that he did not at once comment upon it.

In December 1883, Klein received the honour of an offer of a position at Johns Hopkins University in Baltimore, USA, as the successor to J. J. Sylvester. This offer served both to uplift and inspire Klein, but he was in doubt about what he should do, and he also asked Lie for advice. Lie's reaction to this offer to Klein was spontaneous – he certainly felt that Klein ought to say yes, even though, as he expressed it, for him personally of course, this would be "a great Loss", were Klein to go so far away: "Indeed, in Leipzig one can always meet you." Also his students in Leipzig would certainly suffer, but Lie's foremost worry was for the journal. How would things go with *Mathematische Annalen*, for which Klein was the driving force, and Lie could not imagine who would be able to replace him: "For namely, in my eyes, such a job is exceptionally difficult, and demands many abilities whose combination is seldom found." By and large, however, Lie felt that a stay in Baltimore of five to ten years would be useful and important for advancement; moreover, it would do him good to get away from the rivalries that existed between Klein and the Berlin School, which according to Lie, had for a long time hardly accorded Klein's "brilliant geometric works" the justice they deserved. And now, when Klein's health was not in a satisfactory state, Lie felt he could leave with "honour". "The victory is yours. [...] As for the rest, I think your renown will mature rapidly with a move to Baltimore." And still once again, Lie reminded him that Klein's stock was high among the young mathematicians of Paris. Both they and the British would follow his activity from Baltimore all the more eagerly. As well, the eyes of the Berlin School would bug out when he arrived in America. Lie expected that Klein in Baltimore would have more time for his own work, although, whether the climatic conditions would suit him, he had no inkling. But his advice to Klein was clear:

> I believe that by and large it would be useful both for your physical, and of course, your mental health, to leave Germany for some years. Your renown would mature significantly by doing so – you would certainly get more time for your scientific activity, and at least you would be followed attentively by a wider circle. In addition, this comes at a very important moment, that is, that you, in America, would find an enormous territory for your activities. [...] But apart from this, your spouse can give you the best advice.

Lie was pleased that once again Klein would reprint in *Mathematische Annalen* the little piece on tangent curves in Kummer surfaces that they had written together and published in the Berlin Academy's *Monatsberichte* at Christmas time in 1870. He commented, "That era of the sphere, 1870–72, was for me a happy time, which I shall always look back upon with joy."

Klein had also been accommodating in providing a transcript of Weierstrass' lectures, and now, at New Year's 1883–84, it was sent off from Germany. For Lie this fitted his plans very well for his university lectures:

> I am trying successively, that is to say, to reform the lecturing in Christiania. Before I was appointed to the university, there existed *not a single trace of new* geometry and extremely little of the function theory presentations of Cauchy, Abel, Jacobi and their followers. (I am speaking of the university's *lectures.*) I have now successively (and with great caution) established some tolerably civilised, if modest, lectures in geometry. I have also begun to deliver (although I am followed by envious and outraged eyes) modern function theory, and with time it will turn into something. However, I must be cautious, on one hand, because function theory is not my field, and on the other, because the students always protest when it becomes more demanding. The prevailing mathematical climate is very poor, even though Broch (who now lives permanently in Paris), Bjerknes and Guldberg are talented men. The greatest hindrance is that for the examination, *the only question is one of reproduction.* Before I took my doctorate, there were almost no doctoral theses – only three in the time period between 1811–1869, and here I speak not only of mathematical theses, but also of doctoral works in total. In recent years it has become somewhat better. You must remember that Norway really did not exist before the present century.

Lie was very excited about whether Klein would go to America. He well knew that in Leipzig, Klein was encompassed in sympathy, but he continued to believe that a move to Baltimore would rejuvenate him: "You have talent (which I unfortunately lack) to excite sympathy. And I doubt not but, that in America you will find the same sympathy, whilst at the same time you are away from your enemies."

"Also for today, a little mathematics," Lie wrote, and maintained that the theory of invariants by deformation, which were developed by Gauss, Minding and others, was a special case of his ordinary invariant theory – and that his theory also embraced other special theories, considered by Lie to be "more an advantage than a drawback":

My main achievement in the area of *infinite continuous* groups lies actually within my definition of these concepts, which for me always seem vague and indeterminate. After I had come up with the definition, I found a little method for determining all infinite groups. At least this brought to light the general theory of invariants for infinite groups in a simple manner. At the same time, I noticed that while the existence of, and theory about, differential invariants for *finite* groups, both synthetic and analytic, is immediately evident without a trace of calculation, when one thinks of a continuous manifold transformed by a continuous group, then I can still not synthetically, but only analytically, understand that any infinite continuous group determines differential invariants.

And after some observations about curves of curvature on surfaces, differential equations of invariants, and that which in the terminology of today are called finite dimensional groups (Lie groups), he wrote: "As you see, I am still working in a confined circle. But still it is so spacious as to give me enough to do for my whole life."

Klein, for his part, had begun to collect his old writings for republication in *Mathematische Annalen*, and Lie asked him if he did not also want to publish here his "Programme Statement" from Erlangen: "It is certainly your most important work from the time period of 1872. *It would now be much better understood than at that time.*"

In evaluating the offer from Baltimore, Klein had certainly also thought about who should take over from him in Leipzig, and in these deliberations it seems that right from the first moment, Lie had stood foremost in Klein's mind. Early in January 1884, these plans were presented seriously to Lie, and he reacted with great joy – perhaps "much too sanguinely" he agreed in a letter to Klein written on January 26th. But were a move from Christiania to Leipzig really to be in the offing, then he would immediately come to Leipzig to talk over the whole thing with Klein, and above all, would, by personal inspection, find out "whether the summer heat and summer light in Leipzig would be injurious to my left eye". And because Anna was expecting a child in May, such a journey could not be discussed before June, July or August, Lie wrote, and underlined that in every way it was best that a rapid decision not be demanded, but that he quite certainly would greet such an offer gladly. "I cannot express myself any more definitively," he added.

In Christiania Lie tried to improve "his health and his nervous system", which he had neglected for a long time: "My corporeal essence has always been extremely strong," he wrote to Klein, but felt that the spectacles that had been prescribed to him in 1872, contained a not-insubstantial error. And through the improvement that was now going to be made, it was "not impossible, or even probable" that his left eye could be better equipped to tolerate strong sunlight: "In reality, I have always believed that my eyes were basically strong, even though my sight has not been strong."

At this point in time (January 1884), he had not talked to Anna about an eventual move to Leipzig, but he assured Klein that she would not "set down difficulties". Quite the opposite: for a long time she had proposed, half in jest, that he took a year off and took her and the children for a trip to Germany or France: "And besides, in Leipzig moreover, one finds (as far as I know) there are always many young ladies from Norway who are taking their musical education there."

All the same, the number of lectures and the amount of examination work in Leipzig – as Klein had portrayed it – had dismayed Lie. But the fact that Klein in his teaching had stressed curvature theory and themes associated with it, suited Lie well, he wrote, adding: "For Monge (after Poncelet and Plücker) has always been a shining example for me." And almost as an application for the position in Leipzig, but also in answer to a question by Klein, Lie undertook to explain which mathematicians he knew and had read, what income he received, and duties he fulfilled as a professor in Christiania, and he also gave a short autobiography up to the time when he had first met Klein in 1869.

In his reeling off of the mathematicians he knew and had met – after Monge, Poncelet and Plücker – he wrote:

> I know Steiner unfortunately only second-hand; likewise Hesse and Staudt; on the other hand, I know Möbius and Chasles well. Of the newer geometers I know, apart from you and Darboux, there are precious few. For example, I know only individual works by Clebsch, Zeuthen and Cremona; likewise of Beltrami, Reye and almost nothing of Dini, von Mannheim, Sturm, Schubert. *Weingarten* is particularly *prominent*. Enneper is not bad, nor are Hoppe, Bäcklund and Bianchi, and *Stephanos* has especially done great things as a geometer. Bianchi's latest work is very good, and Bäcklund's, still better. Most other so-called geometers, f. ex. Brill, Noether, Lindemann, actually do not deal with what I would call real geometry.

On the subject of salaries in Christiania, Lie wrote that an ordinary professor began with an annual salary of "4500 Kroner = 5000 Marks", after five years he got "5000 Kroner = 5500 Marks", and after still a further five years, "5500 Kroner = 6111 Marks," and finally, after a further five years, "6000 Kr. = 6666 M." In addition to this, about twenty of the oldest professors got a cost of living remuneration of about 600 kroner. The extraordinary professors had not claimed such a living bonus – a juridically doubtful matter, Lie commented, but for the past five years, they had otherwise had the same salaries as ordinary professors. Consequently, for the time being, Lie had 5,500 kroner as his annual salaried income, which in 1887 would rise to 6,000 kroner. He mentioned that whenever he liked, he could, if he so desired, become an ordinary professor – this had also been proposed by the faculty – but when it had so little economic significance, he would scarcely make any gain in that direction. He was also a little freer as an extraordinary professor, and particularly now, if there were to be a question of seeking leave of absence for a certain number of years. While the ordinary professors were duty-bound to lecture five hours weekly, the requirement for the extraordinary professors was formulated by a very unclear method, and they usually lectured from two to four hours per week. In addition, Lie also pointed out that most of the professors in Christiania only applied a portion of their time to their teaching and their scholarship and sciences, that otherwise they were busy with different ventures that brought in additional income. In the course of his twelve years as professor, he himself had earned about 3,000 marks in this way, the majority of which dated from the work on the Abel edition.

In his summary overview of his life and scientific development, he felt there was little new to tell Klein. But nevertheless, it was here that he informed Klein that during these last years of study he had become "melancholy and eccentric" and he thought he had lost his "mental abilities" ('geistige Fähigkeiten"). And he recounted that as early as the autumn of 1868 he had seriously begun to study geometry, and at the beginning it seemed difficult to him. "Not a soul here [Christiania] had any idea about geometry". But after he had made his "imaginary theory" in December of 1868, he no longer had any doubts about his calling ("meine Ruf"). "Broch, who later would become Cabinet Minister, took an interest in me," Lie continued, and concluded: "The rest, you know."

In Leipzig, Klein gradually came to the consideration that he would say no to the esteemed offer from America; the reasons that were decisive seemed to be those of health and economics. Lie commented: "Well, perhaps this is best for you, even though I have not made any strong judgement about it." As for his own position: "There is much that draws me away from Christiania. Nevertheless, man has a certain inertia, wherein what is, is gladly accepted. In many ways I have a remarkable life here."

Over the winter and spring Lie also had a degree of correspondence with Reye. Reye sent Lie his latest work on linear and quadratic ray complexes, but had also expressed a degree of dissatisfaction with the manner in which Lie and Klein, in their joint paper on curves and surfaces, had referred to Chasles in relation to tetrahedral complexes. Lie admitted that this work – published in Paris in the summer of 1870 – perhaps due to its concision, lacked certain things, but nonetheless, these could easily be remedied, and he acknowledged Reye's worthy achievements in this field. But when Lie now got to see that in Reye's work he had barely mentioned Klein, and had not even included Lie's name, he became "unaccustomedly annoyed", and indicated to Klein: "The issue is that Kummer made him [Reye] aware of our things, which he then studied careful, and thereafter did not even mention me, and you only fleetingly. [...] To me it looks like he [Reye] dares to ignore me, something that is impossible when it comes to you." And Lie told Klein that Darboux in Paris had also expressed himself critical of Reye, and in fact, mentioned him as someone who only plucked things from the writings of others and put them together again without any independent point of view, and in effect was no more than a compiler.

Lie also now had a critical word against Mittag-Leffler, and the reason was first and foremost that Mittag-Leffler should have refused to publish an article that Klein had submitted to *Acta*:

> Mittag-Leffler is also a peculiar man. He has in any case taken me by the nose in a very polite manner. What you tell me is quite remarkable and annoys me to the highest degree. But he regards himself as sole editor. Despite that he really can thank me a great deal for the fact that Norway and Denmark have gone in together with Sweden to publish a journal, to which annually Norway contributes 1,200 Marks, he still, on several occasions and in the most sensational manner wounds me. He is colossally arrogant. For him the four great mathematicians of today are Weierstrass, Hermite, Poincaré and Leffler, and the two latter are the greatest of the future; to him, most of the others are zeros.

Lie also commented on this event in a letter to Sylow, dated February 1, 1884, after having told about the rejection of Klein's work, and that Mittag-Leffler seemed thereby to have been following in "the Footprints of the Berliners", Lie wrote: "Leffler is intelligent enough, but I do not set an unqualified Price on his remaining Abilities."

The only message Lie sent to Mittag-Leffler this year was a card, communicating in haste to a query. Lie wrote: "It would please me if Madame Kovalevsky were to join the Editorial Board of Acta."

In a letter written later that spring, Lie also took up his scientific and personal relationship with Klein. First, he described his study of differential invariants, which he soon would have ready for *Mathematische Annalen*, and which he hoped Klein and Mayer would comment upon before it went to press, and then he described another paper *"on the determination of continuous groups by calculating with differential invariants"*, and the application of this theory to a series of problems. Whereupon he hoped in the course of the spring to send Klein two large works whose points of departure were treatises on differential equations and continuous groups, works that were further workings of things he had done in 1874 and published in *Archivet* in 1883. And then he went on to take up personal relationships:

> In my forthcoming works I am going into my old relationships to you, and my more recent ones with Halphèn, Poincaré (and Picard).
>
> What my relationship to you concerns, as I have had to insist, is that my work on differential equations and yours on transformation theory are not, as individual mathematicians have intimated or believed, some refinement or working out of your ideas. My position is somewhat troublesome. To the highest degree I feel that my relationship to you has been of the greatest significance to me. And similarly, it goes without saying that already, before I came to Germany, through Plücker, albeit in an indistinct form, I was involved in the study of contact transformations, and which opened up great, if somewhat indistinct vistas before me.

On April 24, 1884, Anna gave birth to a son, and just as their eldest daughter was named after Anna's mother, now the newborn son was christened Herman, after Sophus' father.

Lie wrote to Mayer a few days later: "My wife has newly had a *son* . . . With that it so happens that certain parts of my work are not to a sufficient degree worked through." It was the work on "differential equations that admit a continuous finite group" that was referred to here. Lie reiterated that he had difficulties with mastering the demands that an analytic form of presentation placed upon the editing, but pointed out that the present work was not his last word on this "extensive theory, which shows itself to have so many analogies with the *Abel-Galois* equation theory". The aim of this work, Lie wrote to Mayer, was first and foremost to secure for himself priority: "For a work in Math. Ann. will indeed not, as in a Norwegian [journal] be ignored." A good month earlier, Lie had sent off his work on differential invariants, and this was read and assessed in Leipzig by Mayer and Engel, something that Lie was extremely grateful for.

The reality of plans to have Engel come to Christiania seems now to have become taken more seriously by Lie. Engel would be heartily welcome in Christiania, Lie wrote to Klein at the end of May, and promised that he would allocate much time for Engel, if he really now were to come. Whether Engel would be satisfied with the stay was something else: "For Christiania is not a good place for mathematics, and from the lectures he will not be able to profit, when they are, needs be, so very elementary." And as to younger mathematicians, there were few: Holst was certainly a fine fellow, and his last work on metric invariants contained the kernel of something good, but he was far too busy with music, art and politics. Nor would Engel be able to profit greatly from Bjerknes, Lie felt, and wrote that Bjerknes' thoughts went in two narrow directions: "The one is Jacobi's injustice toward Abel, and the other is 'spheres in fluids'. And both parts have, at least for me, long since become boring, and it is seldom that new ideas are found around him." He himself, he could devote an hour or two, either in the mornings, or after he had completed his lectures, to Engel, and in any case, the afternoons: "We shall not lack for material. The theory of transformation groups must be our theme." And then he rattled off a series of more concrete work tasks that he could, with Engel's help, continue with.

At the end of June, Lie wrote to Engel:

Dear Herr Engel!
When at the end of 1883 Klein first mentioned plans about You coming to Christiania, I dismissed the thought as so incredible that I did not even reply to it. When however he later returned to the plan, I grabbed at it with both fists. For me and my investigations this would be a great advantage if the plan indeed is allowed to be realized. I certainly know Your competence, not only from Mayer's and Klein's laudatory references, but also from Your interesting, independent works, which I hereby thank You heartily for, as well as for Your valuable remarks about my last pieces, which momentarily I shall be sending to Leipzig. Whether You will be satisfied with a stay here, is certainly another question. I can only promise to do my very best. I shall particularly, if I get to maintain my usually not so very strong powers of labour, place much time at Your disposition. [...] For the last while politics has captured all attention here. Exactly during these days an amazing thing is happening, which also lays siege to me completely, even though I certainly also apply some hours to editorial work.

In a letter to Klein, Lie wrote that he had become "deeply moved" by the foreword that Klein had written to his work on the icosahedron ("Vorlesungen über das Ikosaeder"), where he gave Lie more honour than he felt he deserved. Lie was overjoyed to be able, once again, to come into contact with Klein's ideas, and wrote: "It is a pity that we did not manage to follow one another better." He went on to say that he could also present his work in approximately the same simple manner as Klein. Otherwise he discussed Klein's use of the term "isomorphic" where he himself had written "gleichzusammengesetzt". He expressed apprehension about Halphèn who constantly used his ideas without acknowledgement, and he felt it was shocking that Darboux had once, almost blatantly expressed the view that treatises published in Christiania were beyond respect.

Lie commented as well on the dramatic political situation in the country at that time (June 1884): "Big things have been happening in the last few days in

Christiania." There was a government crisis, and Broch had been called home from Paris to conciliate and form a government that transcended the dogmatic lines of parties. However, the centre elements were too weak; in any case, Broch did not succeed. But through these "June negotiations" nevertheless a fundamental shift in the times came to reign over the country's political life – the country turned from a system with a public servant government to one with a parliamentary popular government. Consequently, Johan Sverdrup now built his government on the centre position, and concentrated all power in Parliament, much as Broch had attempted to do. Lie summed up the situation: "It is now the first time that Norway has formed a government with members from the opposition. Passions are high and everything is politics. Hopefully we shall now have more peaceable conditions."

It seems that this summer, Sophus's mountain tour was shorter and other than planned. At least he expressed it this way later, without explaining where he had been, or what had happened. Anna wrote a letter to him at the end of July – addressed to Sogndal in the western part of the country, but had it returned – in which she complained that she had not heard from him for a week, and seven year-old Mai had written her own letter to her father. From their home at Skarpsno, Anna related that the children were well, that they had strawberries often, but that she had not been able to bathe for three days due to the high water, which was up almost into the garden and was inundating the bathhouse. She reported that the "little Girls" were "awfully good during the daytime, and Little Lad exceptionally sweet." She reported that there was a mountain of post for him – from the Ministry, from Mayer, Engel, Teubner, *Mathematische Annalen*, from France and Italy, and that his brother Fredrik was in the city in connection with his son's artium examination. Anna's only worry was that the servant girl, Josephine, whom they had had for five years, might take a post somewhere else. Anna was sorry that she had already once again "engaged a Girl" – it would have been better to keep Josephine, who also was now "so attentive to Little Lad".

Summoned to Leipzig

Friedrich Engel arrived in Christiania in September of 1884, and he stayed for about nine months. He would later become Lie's closest collaborator, and came to devote much of his life to working on Lie's mathematics. It was Engel who was the driving force behind the publication of Lie's *Gesammelte Abhandlungen* in seven volumes – the last volume being published in 1960, nineteen years after Engel's own death.

On several occasions Engel described his relationship to Lie, and about their first meeting, in Christiania in the autumn of 1884, he wrote:

When I came to know him Lie was at the height of his powers, a pure Germanic gigantic presence who quickly gave one the impression of an unusual physical strength and endurance. In truth, he was still famous among his countrymen as a hiker, but it ought to be pointed out that at that time in Norway it was far more comprehensible to evaluate the hiker than the mathematician. [...]

Already by 1884 his body had a tendency toward plumpness, but without a trace of heaviness. His broad countenance was adorned with a light blonde full beard. His blue-grey eyes looked out calmly from behind thick spectacles that he wore due to his myopia. [...] His voice was strong and powerful, but when he spoke loudly his voice timbre seemed not to correspond to his powerful frame.

His manner was direct and open, and he was opposed to any pretension, such that he instinctively instilled confidence. From the first moment, one felt that toward him one ought always to be open, that he presented himself exactly as he was. He was forthright and completely frank in his expressions about people with whom he had no sympathy; above all, he hated every trace of intrigue, and therefore frequently used the designation of intriguer, on the whole, he had a great propensity for sniffing out intrigues. [...]

The goal of my journey was twofold: on one hand, under *Lie's* own guidance, I should become immersed in his theories, and on the other, I should exercise a sort of pressure on him, to get him to carry on his work for a coherent presentation of one of his greater theories, with which I should help him apply his hand.

When Lie began to realise that perhaps Engel would indeed come to him in Christiania, he had stressed, in his letter to Klein in May 1884, that his treatise "Theorie der Transformationsgruppen I", which had been published in *Mathematische Annalen* in the summer of 1880, was a good piece of work:

even though individual segments were also too broad in so far as I did not dare to work through my line of thinking synthetically (that is, my conceptual operations), but had to dispose of it by analysis.

I can now, with the help of Engel, continue with what was long ago commissioned by, and printed in Annalen: 1) The determination of all finite groups of contact transformations in the plane, which do not allow themselves to be treated in groups of point transformations.

2) The determination of all the plane's *infinite* groups. 3) In addition, there is a third general work on diverse extensions of continuous groups to an n-dimension manifold.

Another series of works which for the time being I have partially edited, will deal with invariants of a finite or an infinite group which generate a surface, and the utility this theory has for differential equations.

As well, various geometric investigations would be able to furnish us with material.

In a letter Lie had written to Engel, in June, when it had finally become clear that Engel would in fact come to Christiania, he stated:

In my investigations of differential equations that admit a finite continuous group, the analogy between substitution theory and transformation theory has always hovered in my mind. I have always operated with such concepts as subgroups, invariant subgroups, *transitivity of the infinitesimal*, primitivity, etc. etc. When I call my method of detection synthetic, I thereby understand that I, on one hand, am operating with the notion of manifolds; on the other hand, that I am completely operating with concepts. One can demonstrate that certain integration problems can be reduced to certain auxiliary equations of *definite* order and with definite properties, *wherein a further reduction generally is impossible.* How far the analogy with algebraic equations can be taken, I cannot say, for the good reason that I am little familiar with equation theory. In this direction I am waiting for You with great anticipation. I do not really know anything about what the analogy is to modern function theory.

Friedrich Engel was twenty-three years of age when he met Lie in Christiania, and he received, in his own words "the very best reception". Like Lie, Engel was also the son of a vicar of the Protestant faith. But when Engel was four years old, his father had become the teacher of religion at a higher school in Greiz, eighty kilometres south of Leipzig, and here Engel went to school until he was eighteen and seriously began to study mathematics – for the most part in Leipzig, but also in Berlin, where he came to know Weierstrass and the Berlin School. But it was in Leipzig at New Year's 1883, that he had taken his state examination in mathematics, and it was in Leipzig six months later, that he took his doctorate under Mayer, on work in the theory of contact transformations. After a year of military service in Dresden, he returned to Leipzig in the spring of 1884, to attend Klein's mathematics seminar with the aim of writing a more extensive work, which would qualify him for a scientific posting, a so-called competence dissertation. At this stage Mayer and Klein had intervened to get him a stipend from the University of Leipzig (Kregel von Sternbach's Stipend), and support from the Royal Society of the Sciences of Saxony – and this enabled him to travel to Sophus Lie in Christiania.

About his nine-month long stay in Christiania, Engel also wrote:

Lie had certainly been thinking for a long time about writing a more extensive work on transformation groups, and while he had still not come to the point of doing so, he had actually made a beginning therein. Without any apparent support, which my coming provided, this would certainly not have come to anything. It would have met the same fate as a work on partial equations of the first order, which he had planned out in the seventies, but which got no further than him setting up a detailed outline and some possible chapters.

Lie now really made a resolution, to write a work on transformation groups. It would not become an easily understood introduction to the elements of the theory. It would not be a popular book, but still, it would be something that was extremely close to my heart: he wanted to give a comprehensive, systematic and powerful presentation. Moreover, it should be a seminal work, which could account for the greatest possible variation. Therefore we began, as Lie used to say, right off with "mit voller Musik" ["with full music"].

We got together twice a day, in the morning at my rooms, and in the afternoon at Lie's. Little by little, to direct my thinking, he developed the individual chapters verbally, the way they should be carried out according to his plan; in order to give me support in working out the preliminary draft, he gave me a short sketch, so to say, a skeleton, which I could provide with flesh and blood. In this way, I early on had the very best guidance in group theory, of which I had a partial and superficial knowledge when I arrived in Christiania. For everyday I was astonished anew by the splendid construction Lie had built up alone, by himself, and had in his head. I noticed that his treatises gave only a very weak idea about the extent and depth of his conceptual world.

At Christmas time of 1884, the first preliminary draft was finished and Lie worked through the whole thing himself, for now he would work out the plan for the final draft. From the end of January 1885, we again got down to working through it, the finished chapters were revised and new ones added. Up to the end of June 1885, when I left Christiania, a mound of manuscript pages had accumulated, which according to Lie's calculation would have come to about thirty arks [1 ark = 16 pages]. Upon the eventual conclusion of the work, which would require a further eight years and some months, these thirty arks would have expanded to one hundred and twenty-five, which at that time neither of us foresaw. [. . .]

But it must be added that with only some few exceptions, all the theories that one finds in the three volumes of *Lie's* "Theorie der Transformationsgruppen", were already in *Lie's* possession at the time I arrived in Christiania. In reality, the task was only to see that some details were more elaborated, and in general, presented with a systematic cohesion.

When Engel returned to Leipzig in July of 1885, he continued to work in the spirit of Lie. He fulfilled his academic qualifications with a treatise entitled "Über die Definitionsgleichungen der continuirlichen Transformationsgruppen", a work which he had, for the most part, written in Christiania. In addition, Engel gave his examination lecture on "Anwendungen der Gruppentheorie auf Differentialgleichungen". And when he immediately began to teach as a private docent at the University of Leipzig, he began with a series of lectures on the theory of differential equations of the first order.

During Engel's stay in Christiania, Lie had commented, in several letters to Klein, about the collaboration with Engel, but there was never any full-fledged description of it. Shortly after Engel had arrived, Lie wrote: "I am quite in agreement with you and Mayer that Engel is very talented. He thinks scientifically, and at the same time possesses originality."

For the whole time, Lie expressed his satisfaction with the editing process – until about Christmas 1884, he had expressed the hope that the work would go even faster in time. After this work, which he consequently thought would be about thirty arks (480 pages), was finished, he saw for himself a future work on differential equations, and as a more distant possibility, a third work stood before him, on geometric applications of these theories. During the spring of 1885, he reported that the programme that he had set for himself in the autumn, was, in essence, successfully completed "with the valuable assistance of Engel." The "theoretical difficulties" were not many, he reported, but they were located "naturally enough in the fixing of the basic concepts" – and otherwise, Lie explained that in the course of the year he had come to see a greater generality in his theories, and there would certainly be enough to do in the future.

In the course of the work, Lie had also kept Klein abreast of Engel's own work – his qualification dissertation. Lie extolled the many beautiful applications that Engel had reached by means of the new common method of determining all continuous groups:

> On the whole, the work gives the impression that Engel's real skill is more in the analytic direction, than in the purely conceptual. About this it is still difficult to judge, as his labour power has been essentially sequestered by me. Overall, I have a *very favourable impression* of him. And given all circumstances, one can expect many useful works from him. And there is a good possibility that in time he will become a truly prominent mathematician.

Engel became a professor of mathematics, and above all was known as the great connoisseur of Lie's mathematics. Engel's research contributions and interests would, to a large degree, deal with different aspects of Lie's theories. When, almost forty years later, Engel looked back on what began in Christiania that autumn of 1884, he wrote:

> The three-quarters of a year that on that occasion I managed to spend working together with *Lie* is still unforgettable to me. I reckon them as the happiest times of my whole life, as was the long series of years I was granted to be *Lie's* co-worker and possess his unqualified faith.

Beside the intense and extensive collaboration with Engel, Lie held his lectures at the university. In any case, during the autumn of 1884, he lectured three hours weekly "on projective Plane Geometry", plus two hours in which, according to the university calendar's list of lectures, he would announce at a later date. In the following spring semester he then lectured four hours a week "on projective Plane Geometry together with one Hour weekly on partial Differential Equations." In the autumn semester of 1885, according to the catalogue of lectures, it was "five Hours weekly on projective Space Geometry," and in the spring semester of 1886 he should "lecture 6 Hours weekly on projective Space Geometry."

One of those who would begin to attend Lie's lectures during this autumn of 1884 was Axel Thue, who later became professor and a prominent mathematician. In his first years as a mathematics student, Thue made very accurate notes from Lie's lectures, and from the clearly-written notes made by Thue, it seems that Lie's lectures appeared almost to be a textbook. But according to other verbal reports, Lie's lectures were not always equally structured. However, all were in agreement that Lie's lectures were quite different from all the others – his lectures were always full of life and an immediacy that was completely unlike the usual lectern performance – his clear ideas and speech, his lively interest, and his powerful personality, all combined to make him a master of the art of lecturing, it was said. In a brilliantly lucid manner he could throw light upon the internal relations between a series of different mathematical facts, and by posing counter-questions to the students, he got them actively to take part in the solutions to the problems. Sometimes his lectures could develop into a convivial chat, and at other times they would take greater swings and be invested with solemnity. He never wrote down his lectures in advance, and he never once had notes with him. When a problem had been solved and discussed, he would formulate a resumé, recapping what had been gone through. He could be radiant when it was something that really touched his soul; but it also happened that when once in a while a solution did not work out, that he would break off the lecture by tossing a piece of chalk at the board and storm out.

And Axel Thue, may have had some difficulties in writing up the lecture notes; anyway, in the summer of 1885, he reported to Elling Holst that he had "written off some of the least interesting lectures on the higher geometry."

Apart from mathematics – the lecturing and the working together with Engel – Lie was still involved in the science milieu of the country. During the spring of 1885 the tumult broke out once again around his friend, Amund Helland. On March 2nd the Mathematical Association was founded.

Lie seems to have had little discernment of personal animosities, and could invite around him people who were not on speaking terms with one another. Anna later told her son about a terrible dinner party at their home at Drammensveien 84. Together with Amund Helland among others from the Left, he had, on the other side, invited Ernst Motzfeldt, Axel Lund and others. The result was that the representatives of the two diametrically opposed political tendencies had not exchanged a single word all evening.

Where Lie's mountain wandering took him in the summer of 1885 seems again to be unknown. But in any case, that autumn new outlooks on the future opened up to him.

That is to say, in August Klein was offered a position at the University of Göttingen, and even though Klein quite certainly now again was in doubt about what he should do, the thought of being a successor to the likes of Gauss, Riemann and Clebsch, gradually served to make up his mind. Klein said yes to Göttingen, and as his successor in Leipzig, he wanted to have Lie. After a certain period of ifs and buts on both sides, this was indeed what came to pass.

But Klein had to struggle to create a place for Lie at Leipzig. First of all, the position had to be secured – that is, it would be easy to leave Klein's chair empty, but he strove to awaken in the majority the fear that thereby the field of mathematics would weaken in the city; and on the other hand, others had to be convinced by arguments that the foreigner, Sophus Lie, was the right man for the position. Through his faculty, Klein had the argumentation sent on to the Ministry of Culture of Saxony. In this document it was pointed out that among the many geometers who, through quite well-rounded studies, had arrived at important results, there was one man who, to an unqualified degree, loomed above the others. This was Sophus Lie, who with his "independent ideas, the strong consistency of his work, and his high standards" was the only one who "in the power of his personality and his original thinking" would be in a position to "lay the foundation for an independent school of geometry". Further argument in favour of Lie was the fact that Engel had come back from Christiania "with a number of new points of view", and that Lie in earlier statements seemed to be positive toward a career in Germany, and that he, by such a move, would thus "have a possibility of overcoming the isolation" in which he was living in Norway.

One of Lie's toughest competitors for the position was the mathematician H. A. Schwarz, who had as his spokesman the powerful Weierstrass in Berlin. But Berlin was located in Prussia and Leipzig in Saxony, and the two German states could not involve themselves in one another's internal affairs. Nevertheless, Weierstrass wrote a letter to Schwarz (December 20, 1885), stating that had Leipzig been a Prussian university, he would have said that the foreigner, Lie – despite a deal of good work – "neither in regard to science nor as a teacher" was of such consequence that he warranted a passing over of actual "countrymen". And in conclusion, Weierstrass said: "Now they are beginning to say that he [Lie] is a new Abel, whom one must secure at any price. A beautiful beginning for a new era under Klein's presidency!"

The Ministry of Culture and Education in Dresden, which was the capital of Saxony, decided on December 18, 1885, officially to offer Sophus Lie the professorship.

For his part, through a frequent exchange of letters with Klein, Lie had already, within the period of a month, reconciled himself to the fact that in reality this would come to pass. At the end of November he had written to his friend Ernst Motzfeldt:

Dear Ernst!

You have no doubt heard that I am thinking of becoming Professor in Leipzig. Salary 7500 Marks should not in and of itself pull me away from Chr.a; but the large Surroundings, the different, larger working Circles tempt me. Our miserable Conditions for Science (particularly with such poor Wretches as the Majority of my Faculty Colleagues) have always been awkward for me.

That many personal Considerations held me back is self-evident; but the Result is that I am still going to venture out into the World and try my Wings.

From the first instant Lie had desired to accept the position in Leipzig. Anna seems to have supported him in this as well, even if, most likely at this time, at least in the beginning, she was not wholehearted about it, when it actually came down to talking about moving. Lie's considerations were that he wanted very much to be able to keep "the Right to march back" to his position in Christiania without having to apply or be appointed all over again. If all links to Christiania and Norway were broken off, he would have to secure for himself a pension in Leipzig, and this would cost money. The thought of what might happen were he or members of the family to fall ill preoccupied him – he also feared that his ability to work during the summers would be impaired due to the heat, but hoped he would rapidly get used to the summer temperatures, and in emergencies, he could use blue-glass spectacles against the sunlight.

He asked Motzfeldt for advice as to how he should request a "Leave from Service", and if he should seek this for a certain number of years or for an indefinite time, of course without pay. But whom should he apply to? Parliament, which had appointed him? And then there was the big question of whether the Ministry in

Saxony would accept such an arrangement – leave of absence without pay for a definite or indefinite period of time. How did one address men of stature in Germany, like a minister of culture? This he asked Klein, and added: "Indeed, since 1814 we in Norway have had half-republican, very middle-class with simple social relations."

Lie pointed out to Klein that the position in Christiania could in general be left to stand unfilled since at its inception it was given to him personally, not primarily as a lecturing position, but rather to engage in scientific work: "Consequently my position is in a manner similar to that of our poets: Bjørnson, Ibsen, etc., who received a *poet's allowance*, and are away for the most part in Paris, Dresden, Munich, Rome."

The poet's allowances that Parliament had granted, to three persons in 1885, were each for the sum of 1,600 kroner per annum – Ibsen lived in Munich, Bjørnson was still in Paris and for the most part lived on credit, which in these years had reached the sum of 15,000 kroner. Other than this, in the autumn of 1885, Alexander Kielland was for the first time refused a poet's allowance, and the arguments that some of the Members of Parliament then brought out into the public square, were considered by many to be evidence of parliament suffering a political hang-over – this, they said, was what occurred when the common people were elected to make important decisions. And Lie, following in the wake of the Kielland issue, seemed gradually to take a different view of "public opinion", and the ability of the public to draw the correct conclusions.

Following advice from both Motzfeldt and Klein, and after having written personally to Prime Minister Johan Sverdrup "in an as far as possible dextrous Manner", Lie managed to organise it such that he got his "Leave from Service" for an indeterminate period of time and without pay, and without this coming up in a way that would necessitate the Ministry of Culture in Saxony having to take account of the Norwegian scheme. It could otherwise appear as though Norwegian professors were still reasonably free to determine their own conditions. Sophus Lie's nephew, Johan Herman Lie Vogt, who had meanwhile that spring sought advice from his uncle on how best he could apply to be accepted by the Academy of Sciences, gave an example: "Here every professor did exactly what he wanted. The one professor of chemistry, namely Waage, had gotten not a few months leave to travel to the Holy Land and be a pilgrim to Jerusalem."

Lie was fortunate in that everything was arranged in the best manner, and looked forward to arriving in Leipzig. The official date for taking up the new post was April 1, 1886, but as early as the middle of February he himself wanted to take a tour to Leipzig to orient himself with respect to lecturing and seminars, investigate the prices and find accommodations – in order to be able to move with the family by the middle of April, in good time for the summer semester, which did not begin that year until April 28th.

One of the reasons why he was so keen to secure a pension and preserve the connection to his position in Christiania perhaps lay in the fact that he had seen the great difficulties his brother had had in this respect. Senior teacher Fredrik

Lie in Kristiansand had been sick and unfit for work during the autumn of 1885, and had himself to pay for his own substitute. To get away from his domestic situation, Fredrik had come to Christiania, and every day through the autumn he spent two or three hours at the home of Sophus and Anna. According to Sophus this was also a reason why Anna was a little out of sorts with the talk of an eventual move to Leipzig. She too had been ill over Christmas, and the main reason, as Lie reported to Klein, was probably "the heavy atmosphere that my brother's distressing illness" had left in the house. Sophus' other brother, Johan Herman in Bergen, wrote expressing his hope that their elder brother Fredrik still had "the Strength to pull himself together" – "because this [the recovery] depends in particular on a strong Will." John Herman reported that he too wanted very much to help Fredrik economically but that for the time being, the conditions of shipping were so bad that he had to evaluate this very carefully. In this letter from his Bergen brother, Sophus was given best wishes, that they were overjoyed on his behalf about the offer from Leipzig. And he gave thanks for allowing their eldest daughter, Augusta, who was eighteen and looking for a serving position in the capital, to stay with Sophus and Anna. Augusta had on that occasion written home and told what a lovely time she had had while she stayed with them. Her father reported that Augusta was now happy to have gotten an office job with an income of fifty kroner a month.

It seems that, after all the doubts had been tidied away and the formalities regarding the position in Leipzig had been carried out, Anna had become enthusiastic about their new prospects. From both acquaintances and those they did not know, who had lived in Leipzig, they heard only good words about the city. And in response to Lie's request, both Klein and Mrs. Klein sent information on conditions in the city – how much it would cost to rent the space they required; that is, a kitchen, a little room for the girl who helped with the children, who they were fortunate enough to be able to take with them, as well as a bedroom for him and Anna, two rooms for the children, a study, dining room, and finally a sitting room. Preferably the house ought not to face the sun, and it would be pleasant if it were near the countryside. Lie underlined that he and Anna were accustomed to living modestly – the prices of meat, sugar, clothing and common services were similarly raised and compared, and all in all, he concluded that in economic terms too, the move would be better for them – especially once he got to know as well that he would receive 2,000 marks toward the cost of moving. "A move from Christiania to Leipzig is a frightfully complicated story, but since other people have managed it, so must I as well," he wrote to Klein. And finally, there was discussion about taking over from Klein. Lie asked Klein what he had been teaching, and gave him information about his own possible lecture series, and informed him that he would prefer not to lecture before nine o'clock in the morning. And was he expected to give an inaugural lecture? If so, should it be a more general talk, or a formal lecture? He confided to Klein: "As to mathematics I tend in general to speak freely and easily without a manuscript. As for non-mathematical issues, I can hardly speak without a Script. I have no Head for non-mathematical Matters."

Mayer had sent him the edition of Möbius that Klein had published, and Lie thanked him: "Möbius was one of the classics, clear, simple and path-breaking." And a little later he wrote to Klein in a vein in which he expressed the view that the world would not, as well it might, appreciate and honour his work with transformation groups and their application to different disciplines: "As little as I dare compare myself with Möbius, yet I nevertheless rather believe the recognition, within certain limits, of my contributions too, with time, will increase."

Thus it was that Lie travelled to Leipzig in the middle of February. He stopped off at Lund to visit Bäcklund, and sat, weatherbound for three days in Schleswig due to snow, and in Göttingen he had a very friendly talk with Schwarz, his most indignant competitor for the position in Leipzig, and in consequence, he wrote to Klein that apart from Klein there was nobody to be found in Germany whom he understood better – yet Schwarz had also communicated some of the unpleasant "truths", among other things that Lie must not believe he was invited to Leipzig only on the basis of his scientific merits, but also because he was not involved in function theory.

While Sophus was on this trip to Leipzig, Anna and the children lived for a time with Ernst Motzfeldt and his wife Margrethe. Lie wrote home to Motzfeldt from Leipzig:

I am running up and down like a Maniac. I expect that this Month I shall have made many more Visits (ca. 70–80) than I have done in my whole Time in Christiania. Everyone here is very courteous. On the other hand, there are certainly one or two Mathematicians external to Leipzig, whom I have crossed, and who are irritated by me. As far as it goes, to the best of my Knowledge, there have been no other Foreigners, other than Abel[34] and I, appointed Professor at a German University. (The Swiss are out of the Calculation here.) It is rather amusing. In Christiania I have often felt myself to be treated unfairly, so I have truly achieved an unmerited honour.

To Anna he wrote as well that there was much to do and that he was "much on the Move", and he urged her: "Have Mai read aloud a little German every Day" and he reminded her that her final letter to Leipzig must be posted at the latest by March 10th, and that in the next three or four days it would be possible to reach him by *poste restante* in Copenhagen and in Malmö.

In Christiania on April 10th, a farewell gathering was held in Lie's honour, complete with speeches and songs especially composed for the occasion.

Bjerknes gave the speech on behalf of the faculty and the scientific milieu of Christiania. He began with a reminder that "we" belong to "a very small people on the fringes of the world", and that the scientific work which was carried out here "in our little country" under "difficult conditions", could not be expected to be other than a "modest, little-noticed contribution to the collective whole." "Our University" was young, whilst its access to "means and powers" was limited, nor

[34] Notice of the appointment of Abel to the university in Berlin reached Norway two days after Abel's untimely death in early spring 1829.

could it be expected to be at the top in all scientific fields that are now developing continuously. But from time to time, and often by unexpected means, conditions could nevertheless be right – and indeed this had also happened once earlier, in "the university's earliest days", and even in the same field of science to which Lie now brought such honour, namely mathematics, in which a Norwegian had found world renown, such that foremost countries wanted to have them "for the advance of their own science". And following a long comparison with Abel and his contributions – and after an interpretation of how mathematical teaching in Norway, not least at the university level, had become so much better since Abel's time – he also commented on Lie's student period, and above all, his further thinking and self-study. Bjerknes stated: "Anyone who manages to achieve something more, must to a certain degree also be a self-taught man." On behalf of his colleagues, Bjerknes expressed his gratitude for everything that Lie had "done for the future of science, and thereby for the honour of our country." As for the contribution he had made to the university: "The loss that it is suffering is great, because you are leaving it." He urged Lie to remember once he was abroad and had "a series of conveniences" and assistants at his disposition, and in addition could "debate scientific materials in numerous circles of professors, docents and doctors", that he had "climbed up to these glories from our dear old basement auditorium no. 33". In conclusion, Bjerknes said: "Therefore, do not forget *Universitas regia fredericiana*. Do not forget this country with its quiet study, the country with its great nature. It is a poor home, but it has fostered gifted sons."

And the song, which was almost certainly authored by the literature professor, Olaf Skavlan, went to the melody "Once there was so brave a man" ["Det var engang en tapper mand"]:

The Sciences, 'tis often said,
Stride down Republic's Path,
And at their trusted life-long Head
Who could they place but Math?
[...]
And he doth speak a unique Tongue
That few can read, and fewer write.
'Tis Greek to Folk both old and young
And shall scarcely set the World alight.
A Dialect did Sofus Lie,
Of this rare Lingo learn,
And joined the Tribe of Two or Three
Who, its Meaning could discern.

Yet, Sofus Lie another Tongue imparts,
A Language each one comprehends:
The radiant Idiom of our Hearts
That's won him such good Friends.
This is of course the People's Lingo
In each and every far-off Land
Where the X-Y-Alpha-Sigma Jingle
Is a Thing that Few dare understand.

Now Sofus Lie to Leipzig goes,
And shall of course new Friendships cast,
But old dear Friends at Home he knows
Remain sublime and unsurpassed.
One Day the Journey back he'll wend,
When Siren Longings beckon 'Home again,'
And seize our dear and precious Friend.
Leb' wohl, auf Wiedersehn!

Then, the day after the grand farewell gathering, the front page of *Ny illustreret Tidende* (see also p. 210) featured Professor Dr. Sophus Lie in both picture and story. The picture was a drawing signed H. G. Olsen, and the unsigned story text had been written by Amund Helland. In addition to biographical information and laudatory descriptions of his affairs – as a man of science both within and without the university, as a mountaineer, and as an engaged citizen "on the Left's Side" – it was also ascertained:

By being appointed Teacher of Geometry at the University of Leipzig, Professor Lie is being shown the greatest Distinction that, on the whole, can be shown a Man of Science. [...]

The scientific Truths that Professor Lie has brought into the Light of Day, would be familiar to comparatively few, for there is no Royal Road leading into Mathematics. The Mathematical Society will always, at its greatest Moments, lack the popular Recognition, which comes, for Example to be shared by the Discoverer of the Telephone or a lucky geographical Explorer. Nevertheless it is acknowledged on all Sides that the new Breadths of Vision that have seized the Field of Mathematics are the Most wondrous Possession of the Human Mind, and are a firm Foundation on which the other Natural Sciences are built. [...]

Now, as Professor Lie is forsaking his Fatherland, it is under the Assumption that down there, in Leipzig, he will have better and more propitious Conditions for his future mathematical Studies. We, however, have the firm Hope that he will not abandon the Norwegian University for good, and we will remain under the Assumption that, after not so very many Years, he shall wend his way back again, and thus we dare to regard his Summons to Leipzig University as an indivisibly joyous Occasion.

Before he set out on his journey, Lie was also named Knight of St. Olav's Order. Whereupon he asked Klein if he ought to inform the Saxon authorities about this, and added: "As democratic as I am, and as insignificant as this is in and for itself, under the new circumstances I do not regard it unkindly."

PART VI

Professor in Leipzig

Seeburgstrasse, 1890

The Coming Period: A More Difficult Balancing Act

Lie arrived in Leipzig with his family in the middle of April 1886. He was forty-four years old, Anna was thirty-two, and the children, nine, four and two. They moved into the ground floor of a house at Seeburgstrasse 5 – a few minutes' walk from the university – and they were to live there for the full twelve years they stayed in Leipzig. Mrs. Anna and the children would gradually come to thrive so well there among their friends, at social events and school, that it was difficult for them to return to Norway.

In the course of his years in Leipzig, Lie would come to be a leading mathematician and a central figure in the mathematical milieu of Europe. Lie formed "a school" in Leipzig, and talented students from the European and North American continents came to sit in front of his lectern and hear him teach geometry and differential equations, as well as discourse upon his own theories.

However, the main weight of Lie's work in Leipzig became the writing down and editing of his works. The collaboration with Engel continued. The major work, *Theorie der Transformationsgruppen* was produced by the major publisher of scientific works, B.G. Teubner of Leipzig, in 1888, 1890, and 1893 – three volumes totalling 2,000 pages. This was the treatise on transformation groups on which Lie had begun to work, with the assistance of Engel, while he still lived in Christiania.

Right away during his first year at Leipzig, Lie had the good fortune to find a most brilliant pupil in Georg Scheffers. Scheffers made precise notes of Lie's lectures, and later under Lie's guidance, worked these up into two books which gave the best introduction to Lie's integration- and group theories; these would later be followed by a book on Lie's geometric research. These three books as well – again totalling over 2,000 pages – would be published by Teubner Verlag.

During his years at Leipzig, Lie also wrote almost forty studies and treatises – to the total of more than 1,200 pages.

Lie became Professor of Mathematics and Head of the Mathematics Seminar – this seminar functioned almost as an independent institute, and at that, an institute for geometric studies. The Seminar was founded by Klein six years earlier and its prototype had been the Mathematics Seminar of Berlin, set up by Weierstrass and Kummer in 1861. (It had been at a session of this seminar in Berlin in 1869 that Lie had made such an impression when he came up to the blackboard and solved a problem that Kummer had been working on.)

Enrollment in mathematics at Leipzig had been great under Klein's leadership; indeed, far too great. Because so many had studied mathematics in recent years, the need for mathematical teachers in the higher schools had been met in spades – qualified graduates now had to wait six or seven years for placement. This led to the number of mathematical students beginning to fall off – from eighty in the summer of 1886, to around twenty students eight years later. At the beginning, this situation gave Lie cause for concern that his lectures on general topics did not draw in as many students as he expected. But he appears to have found consolation in the fact that his special lectures on his own theories attracted a satisfactory audience. Moreover it was certainly flattering when, a couple of years later, in 1888, and at the initiative of Poincaré, Picard and Darboux, students were sent to Lie from the highly esteemed École Normale Supérieure in Paris.

From 1890, Lie was one of five professors of mathematics at the University of Leipzig – the four others being Adolf Mayer, Carl Neumann, Heinrich Bruns and Wilhelm Scheibner – but of the fifty-six students who in the course of these years came to take their doctorates in mathematics at Leipzig, twenty-six were supervised by Lie, and several of these went on to do further work on Lie's ideas.

The Mathematics Seminar was divided into two departments, each with its own haunts, and each with its own assistant. The first department was located in the old city centre at Ritterstrasse 24, and here Engel worked as the librarian. Department II where Lie held most of his lectures, was located at Brüderstrasse 32, in what was called the Czermakian Spectatorium, which in addition to a large amphitheatre with a good skylight, contained a collection of models. Friedrich Schur was responsible for the care of the collection of models, but a couple of years after Lie's arrival, Schur found a position at Dorpat, and Engel took over the assistant position in Department II as well. Lie's residence in Seeburgstrasse was about midway between these two venues – a five- to ten minute walk in either direction.

Another seminal person in the work around Lie's life and mathematical endeavours at this time was Eduard Study. During Lie's first couple of years in Leipzig, Study functioned as a private docent at Leipzig, as did Felix Hausdorff later, from 1895, and both came to play roles in Lie's working circle. To be a private docent in Leipzig meant giving lectures without either pay or employment. In Germany, in order to have the right to be a private docent, one was required to have published a distinctive qualifying dissertation – neither the licentiate nor the doctorate was considered sufficient. Scheffers also functioned as a private docent in Leipzig, from 1891 until he was appointed professor and left for Darmstadt in 1897.

Right from the beginning it was a very arduous and demanding programme of work that Lie had to prepare for – ten to twelve hours of lecturing every week, in addition to supervising students, and his own intense and demanding work.

Many years later one of his students, Gerhard Kowalewski, wrote down his memories: "Lie liked to teach, particularly when it came to his own ideas, and he had lively contact with his students. [...] He had the habit of posing questions to us while he was lecturing [...] and did not take it badly if whoever had been asked a question did not know the answer, but rather, he would turn toward one [of us] from the elite. [...] Lie never wore a tie. A full beard adorned the place where a tie ought to have been. [...] And at the beginning of the lecture he had the habit of freeing himself of his stiff collar, while saying, "I cherish freedom."

But the congenial, conversational tone that he was known for when he lectured in his mother tongue in Christiania, was difficult to find in a foreign language, at least during the first years. At the end of his first year in Leipzig, Lie admitted to Klein that while he usually had improvised his lectures in Christiania, now, with greater demands upon him, he had to work in a completely different manner: "Now I work out the lecture, then, so to speak, I try to memorize it like an actor. But up to this point, I have not succeeded very well." The foreignness of his situation, he wrote, meant that he always had to think a little about language, and only then, about the mathematical content. Lie made these conditions concrete in a letter to his friend Motzfeldt in Christiania. He wrote that whilst he had used five minutes a day to prepare to lecture in Christiania, now, due to the competition and the German language, he used up to three hours in detailed preparation, and in addition to that, he had the supervision of students.

Teaching and supervising gradually consumed more time than he would have liked; nor did his personal plan for the stay on "the Leipzig plains" work out to perfection. Apart from the language, which he did not feel he had mastered as well as he would have liked, the social customs and the etiquette were new and different, and as far as the university's leading organs were concerned, he felt he never got his point of view across properly. He began to miss his friends in Christiania, and above all, he missed Norway's natural landscape. After barely a year in Leipzig, he wrote to his friend Ernst Motzfeldt: "I miss Chr.a's beautiful Nature and the refreshing Walks." And again, a year later: "It is remarkable, but every Year I become more and more attached to Nature. I miss it so much that I feel in Time it is what shall drag me back from Leipzig to Norway."

During these first years in Leipzig, Lie regularly read two or three Norwegian newspapers in the "newspaper and artists' Café Mercur", and he commented passionately to his friends in Norway about academic and political life in his homeland. He reported with pleasure on the rest of his family: "the Children are thriving remarkably well here" and "Anna is getting on fine, receiving frequent Gewandthaus concert tickets from our rich Friends. (I myself have come to the Realisation that I am not musical enough to sit for a Couple of Hours in the finest of Full Dress just to listen.)"

Lie found only partial outlet to his need for the outdoor life and long exhausting tours by travelling to the Tyrol and Thüringen, and excursions together with the family to places closer to Leipzig. His worry about the hot summer sunshine of Leipzig proved to be unfounded – Lie was surprised that he could tolerate summer heat so well, and that he was not plagued by the summer sunlight. "I believe that the Leipzig Smoke weakens the Rays of the Sun", he wrote to Motzfeldt after having been in the city for two summers. He returned to Norway for the first time in the autumn of 1888 – alone, but he had already begun to look forward to a new trip to Norway together with the family, two years hence.

But before things got so far, many things in Leipzig would have changed for the Lie family. There would be no journeys to the homeland before 1892. The pressure of work, problems of collaboration, and domestic anxieties made him sleepless and depressed, and in 1889 he had a complete breakdown. He became silent and apathetic in November 1889, and was sent to a psychiatric clinic near Hanover. One of the things he would later say about his breakdown was that he was "in the grip of a boundless Desperation" ("von einer grenzenlosen Verzweiflung ergriffen").

He was a patient at the psychiatric clinic for a good seven months. Here he was given opium and sleeping draughts, but maintained that in the end he literally walked the affliction out of himself with long hikes. When Mrs. Anna came to fetch him, he was discharged, but in the reception book of the clinic his condition at the time of release remained "a Melancholy not cured" ("ungeheilte Melancholie").

In a letter he had written to Picard a short time after his return to Leipzig in the summer of 1890, and before he felt strong enough to return to his duties at the university, he gave thanks for all that Picard had done for the theory of transformation groups. Then, for Picard's benefit, he explained his illness and its effects:

> I suffer from almost complete insomnia. This affliction, together with the remedies that are applied against it, have broken me down in such a manner that my scientific work has quite come to a stop, although my strong physique still certainly wants to go on living.
>
> Under these incomprehensibly sad conditions it has been a great pleasure to me that you have gone so deeply into group theory. Science has been the greatest joy of my life. It has been my good fortune to have been allowed to contribute to the development of science.
>
> To me, my life is actually incomprehensible. As a young man I had no presentiment that I was in possession of originality. It was only as a 26-year-old that I suddenly realized that I could create something. I read a little and began to produce. In those years, 1869–74, I had a mass of ideas that with the passage of time I have only managed to develop imperfectly.
>
> It was group theory and its great significance for differential equations that particularly interested me. But the publicising has been pitifully slow. I have not been able to edit, and I have always been afraid of making mistakes. Not these small immaterial errors, that my brilliant and nevertheless untalented friend A. Mayer is so often correcting for me. No, it is the deep-lying mistakes that I fear. I am pleased that my group theory, which is now available, does not seem to contain any fundamental errors. Certainly the foundations of function theory could be accounted for in a better manner.
>
> My conscientious but painstakingly slow assistant, Engel, has presumably postponed the publication of my investigations. The foreign conditions of Leipzig have also completely consumed my labour power.

Lie thought that he would never regain his full labour power. He wrote to Motzfeldt in Christiania that the fame he "hungered after" would probably never come to him, and he had already begun to console himself with the thought that his life's work would withstand the test of time and with "the Passage of the Years, become more and more appreciated, no Doubt about it."

Things did not go as Lie feared, in any case, not in relation to his ability to work again. He reported further to Motzfeldt right after his breakdown: "Here in Leipzig my Colleagues and Everyone else are showing me the most splendid Goodwill. People are doing everything possible to bring me back into equilibrium again."

Lie would come to continue his innovative mathematical studies with a series of articles. The second volume of his major work, *Theorie der Transformationsgruppen* came out while he was hospitalised; the first book of his lectures was published a year later, in 1891, and two years later, the third volume of *Theorie der Transformationsgruppen*, more or less simultaneously with the second collection of his lectures. Lie's works were spreading and increasingly more highly valued and appreciated.

His contact with the French mathematicians gradually increased significantly. The most evident expression of this came in 1892 when Lie was elected a member of the French Academy of Sciences, taking the place vacated with the death of Kronecker. In addition to his purely professional work, Lie was also regarded as the foremost bridge-builder and conciliator between French and German men of science, who still entertained a degree of mutual mistrust following the war a good twenty years earlier. The interest and admiration that the French mathematicians accorded Lie when he came to Paris during the spring of 1893, brought him joy, and inspired his further work. In a letter to Poincaré, Lie wrote: "Once again the centre of gravity for mathematics is Paris and France" – after having commented upon the foundation of geometrical thinking, such as had been formulated by Euclid via Lobachevsky and Riemann, and after having discussed points, space, surfaces and curves as basic concepts, Lie continued: "My group theory provides a good foundation for a geometry built upon real or complex manifolds of several numerical variables, but not for a geometry built upon the concept of space."

When in 1895 l'École Normale Supérieure in Paris celebrated its centenary with full pomp and ceremony, Lie was invited to speak about Galois and his significance for the development of mathematics – an honorific task that had been given to him in recognition that it was precisely Lie who most clearly had developed Galois' brilliant ideas to a further degree (see below p. 374).

But despite fame and good fortune, Lie seems nonetheless to have found that in his everyday life in Leipzig much had changed following his mental illness in 1889–90. Friends and colleagues gradually came to see changes in Lie's attitudes and his behaviour, and not least, encounter problems of working together with him – his mistrust ruined earlier friendships. It was said that Lie had almost become paranoid in his accusations and allegations that others were stealing, and equipping themselves with his ideas, in order to publish them as their own. In the autumn of 1893 this came to a culmination with the publication of the third and last volume of the great work on transformation groups. In a foreword to this volume, Lie commented upon Klein's geometrical works and the associated attitudes and points of view, which at that point in time seemed to be spreading through the mathematical circles, about the relationship between him and Klein. He concluded: "I am no pupil of Klein's, nor is the reverse the case, although it might be closer to the truth." (See below, p. 371.) This was a description that sent shock waves through the whole German science milieu. It was said that this was stark evidence on Lie's part of a "megalomania that had broken into full flame."

Naturally enough in Norway, Lie's international fame was clearly evident, and it gradually also came to be known that he was unhappy living abroad. Many came to feel, little by little, that it was high time Norway's great son be brought back home to the fatherland, and also to show what resources the nation could muster in the looming struggle over the union with Sweden. Fridtjof Nansen, Bjørnstjerne Bjørnson and Elling Holst, among others, worked so effectively to get Norway's famous man of science back, that Parliament as early as 1894, undertook to change Lie's title to "Professor of Transformation Group Theory", and offered him a salary of 10,000 kroner – almost double as much as was paid to an ordinary professor.

But despite all this goodwill and single-minded desire to get Lie back to Norway, almost four years would pass before he returned on a permanent basis. It took time to wind up work projects, book publications and duties in Leipzig.

In the Big City of Leipzig

Lie began to lecture even before the agreed date of his inaugural lecture and official reception at the beginning of the summer semester in 1886. The university year in Leipzig was divided into a winter semester and a summer semester. The summer semester began in the middle of April and continued to the middle of August. This was followed by a two-month holiday before the winter semester started in mid-October and continued until mid-March – when again there was a one-month holiday before the new summer semester began. This was an arrangement of the teaching year that Lie quickly came to like, even though in Christiania he was, of course, used to having time off during the middle of summer – between the Norwegian spring and autumn semesters – when he would leave for the mountains in order to avoid the summer heat. But he seemed to think it was both natural and good that the teaching in Leipzig continued through the warmest part of the year: "Here in July one certainly cannot take any Summer Tour, on foot at least," he wrote home to Motzfeldt – besides, the heat was less bothersome when one was working, and moreover, the houses in Leipzig were certainly designed such that the heat did not obtain any "real Strength". In addition, his own house in Seeburgstrasse kept the heat out, as he reported to his friend in Christiania, adding, "Had I such a House on Drammensveien [in Christiania/Oslo], I would certainly feel fortunate."

Engel later described Lie's choice of residence as "most highly unfortunate": "Whereas his house in Kristiania was light and friendly with a delightful view over Frogner Cove, in Leipzig, out of fear of the southern sun, he sought out a residence that lay far too much in the shade of the higher surrounding houses." Another reason why Lie chose this house in Seeburgstrasse was certainly that it was only a short distances from the parklands and his two workplaces, the two mathematical departments. In any case, the Family Lie moved into Seeburgstrasse 5 "with Furniture and all the Household Goods", together with their serving girl Andrine, whom they had taken along from Christiania. Due to construction in the area, their address changed in 1891–92 to Seeburgstrasse 33.

Lie had already been entered in the protocols – Personal-Verzeichnisse der Universität Leipzig – from the 20th of January 1886, as Professor of Geometry and Director of the Mathematical Seminar and Institute. Since the salary was paid out at the end of work periods of three months, perhaps this meant that Lie received his first pay upon his arrival. In any case it would easily have been useful: "It is such that in the Nature of Things, I have a frightful Number of Expenses" – most things were a little more costly than in Christiania, Lie reported, who anyway hoped that the household expenses would not be more than 200–300 kroner more expensive than in Christiania, even though meat cost twice as much. Clothing however was a little cheaper, "but here we must dress more elegantly, especially since an ordinary Professor is a great Man in Leipzig," so all in all he did not think that there would be any saving to speak of, when it came to items of expenditure – and in addition, the cost of housing was higher than in Christiania. But he certainly earned more than he had in Christiania, and in summary he told his friend Motzfeldt, "The Results are indeed that I am managing rather well" – and, just as he had done for the last four or five years, he had been laying "a little aside for the Children".

The title of the lectures with which Lie began the semester was "Anwendung der Differential- und Integralrechnung auf Geometrie" – from Tuesdays to Fridays between twelve and one o'clock, and on Mondays between 10 and 12 he held seminar exercises: "Anwendung der Theorie der infinitesimalen Transformationen auf Differentialgleichungen". According to the records there were generally fourteen students attending Lie's sessions during this first semester. During the years this number would vary somewhat, but during his last years it stabilized at over twenty. In addition to the fact that Lie was rather disappointed that so few came to hear his lectures at the beginning, he had difficulties keeping the same students over several semesters. German students generally used four years to prepare for the state examination, and this exam was not much below the level of a German doctorate, but which did not have the same value as a Norwegian doctorate, and far less than the French. In the Seminar, Lie tried to build upon Klein's teaching, and in the correspondence between them during that early period, Lie often asked Klein for practical advice. Lie wanted very much to find time, beyond the allotted hours, to work with five or six students who came to the seminar – and in the first years he supported many of the students to advance to their doctorates. To some degree Lie compared the teaching in the German seminar system with what was offered at l'École Normale Supérieure in Paris, and felt that it was a great acknowledgement both of his teaching work and his position as a man of science when, after a couple of years in Leipzig he began to get students from l'École Normale Supérieure.

At the time of Lie's arrival, the University of Leipzig had 3,251 students, or more than three times the number in Christiania. Of these, eighty studied mathematics, and fourteen of these, accordingly gathered for Lie's lectures. (A few years earlier the number of mathematics students had been up to 250.) Beyond this, the students were registered in three categories: those who were domiciled in Saxony; those who came from other German states; and foreigners. Scarcely half of the students at Leipzig were from Saxony, and such were the conditions now in mathematical studies that among the eighty students, six were foreigners (from Bulgaria, Austria, Turkey and North America).

Leipzig, next to Dresden, was the most important city in the Kingdom of Saxony, which was part of the League of German States, and a deep-rooted sense of competition existed between the adminstrative and cultural Dresden, and the more trade-oriented and industrialized Leipzig.

The University of Leipzig had been founded as early as 1409 and was one of the oldest in Germany. It was richly equipped with libraries, reading rooms, numerous seminars, institutes and clinics, chemical laboratories, and an observatory, a museum of mineralogy and a botanical garden – particularly in the last decades, since the 1850s, the university had experienced a blossoming. This occurred, not least, because the former King of Saxony, King Johan I, who in the course of his twenty-year reign from 1854, had brought together a number of prominent men into the cultural and scientific administration of the country. King Johan himself had been a brilliant philologist and historian – and he had distinguished himself particularly in Dante research with his verse translation of *The Divine Comedy*, complete with critical and historical notations. And mathematics too, had been given special treatment. King Johann had pulled in Carl von Gerber as Minister of Culture – and it was he whom Lie had met in Dresden when he came in February 1886 to look into the conditions and requirements of the professorship in Leipzig. One of those whom von Gerber had brought into the Ministry of Culture and Public Education was the mathematician, Oskar Schlömilch, who for many years had been putting out the journal *Zeitschrift für Mathematik und Physik*. It had been Schlömilch who had backed the employment of Felix Klein as professor at Leipzig in 1880, and it was at his initiative that Klein had been able to establish the research institute (the Mathematics Seminar). Now, in addition to having administrative responsibility for mathematical instruction at schools and colleges in Saxony, Schlömilch was also a mathematical consultant and advisor at the important publishing house of Teubner.

In every way, Lie had landed in a milieu that certainly valued mathematical research – Lie was treated in the best manner possible by Teubner's, by colleagues at Leipzig, and by the Saxon administration. He also wrote home about how "The Sciences stand in Honour" in Leipzig, and how much easier everything was where "the Regard for Science" always carried great weight. Lie felt welcome. He had gotten the best of conditions for publishing his works, and he wanted to form "a school" in the city.

Not only the university, but the whole city was experiencing a florescence. For example, in addition to the royal university clinics, the city had two state hospitals, a famous homeopathic hospital, and children's hospital, three institutes for the blind, and a facility for the deaf and dumb, an orphanage, and an institution of semi-religious nursing sisters. The city in 1886 had approximately 170,000 inhabitants – strong industrialisation was continuing to develop, the population was increasing, and a series of suburb communities were being incorporated into the city. When the Lies left the city twelve years later, the population was estimated to be well over 400,000.

The city had been a centre of money and trade right from the 1400s and 1500s – such that after only Berlin and Hamburg, Leipzig was the largest commercial city, with three large annual markets: at New Year's, Easter and Michaelmas. Leipzig had long been central to the German book trade, and in competition with Frankfurt, the city still had a leading position in the production of books. Now, as the nineteenth century was coming to a close, the city was in the process of transforming from a commercial centre to an industrial city. In addition to the paper and book-publishing industries, there were now metallurgical industries, with foundries and factories producing machines and equipment, electrical components and motors, foodstuffs and stimulants; and there were great breweries and factories in the fields of textiles and chemicals, among others. The city expanded outwards; smoke from the factory chimneys blotted out the sunlight – something that Lie with his sensitivity to light was thankful for, but he was also to see the industrial proletariat grow in strength and significance.

In the old city core of Altstadt there were whole series of monumental buildings and venerable churches – the houses were tall and the streets narrow, but the city also had many large open squares. Augustusplatz, where the old university was located, was considered one of the most beautiful in Germany. There was a magnificent city hall, several beautifully decorated theatre buildings – ornamented both externally and internally – as well as concert halls, the *Gewandthaus*, an impressive state courthouse, and a university library with abundant collections of books. There were also great book collections to be found at the *Reichsgericht* (the state courthouse), and at the bookdealers' exchange. Moreover there was an art academy, a crafts and industries school, several museums, parks, a castle, a zoological garden, and a palm garden in a greenhouse of glass and ironwork. The city also had a public race course and a great amusement establishment with a circus and a concert hall housing one of Germany's largest organs. There were nine daily political newspapers in Leipzig, and the city had a great number of associations and societies – one of them being the Royal Academy of the Sciences of Saxony, which had been planned out by one of the city's greatest sons, Gottfrid Wilhelm Leibniz, but it was not founded until 1846, on the two hundredth anniversary of Leibniz's birth. This science association, which was divided into two classes, that of mathematics-physics, and that of history and philology, soon had Sophus Lie as a member, and he would also come to publish a series of his investigations in the Academy's "Proceedings".

Early in the 1700s, Johann Sebastian Bach, in his roles as church organist and director of music at the university, had made Leipzig a centre of music. Richard Wagner was born and raised in Leipzig; Robert Schumann was central to the musical life of the city. Leipzig's position as one of Germany's most important centres of music had gradually been strengthened – and the fact that Felix Mendelsohn Bartoldy had founded Germany's first music school in the city in 1843 was not the least of the reasons for this turn of events. Moreover, only three years after the *Gewandthaus* was finished in 1884, the music school – which had now become the Royal Conservatory of Music of Leipzig – got its own concert hall, and together with the magnificent *Gewandthaus* it subsequently formed a truly musical quarter of the city. People came from all over Europe and America to hear, learn and study music. When the conservatory celebrated its fiftieth anniversary in 1893 it had 6,000 students; more than half being foreigners, including several from Norway. Twenty-year-old Sigurd Lie, Sophus' nephew, arrived in Leipzig during the autumn of 1891, to study at the Royal Conservatory of Music, and during the year and a half that he was there, he frequently dropped in on the Lies at Seeburgstrasse. Sigurd Lie had already contracted tuberculosis, and during his stay in Leipzig he had a strong bout of illness. While he only lived to be thirty-three, he managed to compose a series of orchestral works, and above all, he set the works of contemporary lyricists to music. While in Leipzig, Sigurd Lie studied composition under Carl Reinecke, who in his time had been a teacher of both Edvard Grieg and Johan Svendsen. Edvard Grieg, by the way, visited the city several times while the Lie family lived there, and in the course of those visits, Grieg and Lie met one another.

Many well-known poets as well had lived and studied in Leipzig down through the times, and linked their artistic work in various ways to the life of the city. Schiller had lived there when he was writing "Lied an die Freude"(Hymn to Pleasure); Schegel and Novalis, and several luminaries from the heyday of romanticism had been students in the city. The best known was perhaps Goethe, who had set parts of the action of *Faust* in one of the city's great wine cellars; later, among others, Theodor Fontane had spent the academic year in Leipzig.

Leipzig had had a highly honourable history in all ways. The great political incident which was perhaps most talked about was the great and decisive blow against Napoleon in 1813 – what was called the *Völkerschlacht* – in which 200,000 men went up against one another in the Leipzig plains. One, out of the innumerable who were wounded and died on the battlefield on that occasion, was the twenty-two-year-old poet and hero of freedom, Theodor Körner. The monument that was put up in honour of Körner's short but blazing patriotism, was also used, in Norway, as a model for a dignified commemorative monument to be built for Niels Henrik Abel.

Lie felt well from the first moments in Leipzig and could report to his friend Motzfeldt soon after his arrival:

> My Nerves, which were under considerable attack, for the last Period in Christiania, are at last beginning to recover again.
>
> I have now held my first three Lectures and think myself that they were creditably done. It is to be hoped that my Audience is not in too great a Disagreement with my Assessment.

With King Albert I of Saxony in attendance, Lie held his inaugural lecture on May 29th – the holding of such *Antrittsvorlesung* for a broader public was an old tradition in Germany. Lie's topic was "Über den Einfluss der Geometrie auf die Entwicklung der Mathematik" and thereafter, he took the oath of office. As an ordinary professor, Lie was now a member of the University Plenum, which was composed of sixty-four professors. On many occasions Lie would place great emphasis upon this disciplined and learned gathering to the detriment of the casual and often professionally weak relations with which he was familiar from the university in Christiania – even though in the end he would complain that he had never had any real influence on the leading organs of the university in Leipzig.

By virtue of his office, Lie was taken on a tour to Dresden; he had Anna with him when "following the Leipzig Custom, we had an Audience with His Majesty". And right away that first summer he had arranged to have a summer place outside

Felix Klein

Adolf Mayer

Georg Scheffers

Friedrich Engel

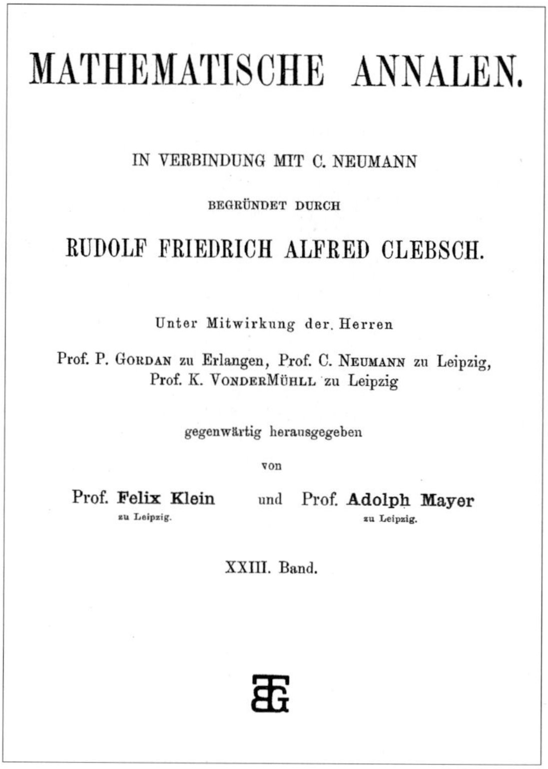

MATHEMATISCHE ANNALEN.

IN VERBINDUNG MIT C. NEUMANN

BEGRÜNDET DURCH

RUDOLF FRIEDRICH ALFRED CLEBSCH.

Unter Mitwirkung der Herren

Prof. P. GORDAN zu Erlangen, Prof. C. NEUMANN zu Leipzig,
Prof. K. VONDERMÜHLL zu Leipzig

gegenwärtig herausgegeben

von

Prof. **Felix Klein** zu Leipzig. und Prof. **Adolph Mayer** zu Leipzig.

XXIII. Band.

Mathematische Annalen, *the German periodical in which Lie published several of his important treatises.*

Bottom left:
The title page of the first volume of Lie's major work – the second and third volumes came out in 1890 and 1893.

Bottom right:
Lie's lectures. The title page from Lie's own copy. Lie's collection of books and papers were purchased by the state in 1900, and were placed in "De realstuderendes læsevarelse" [the science students' reading room] at the University of Christiania/Oslo (notice the name stamp). This reading room was inaugurated in 1908.

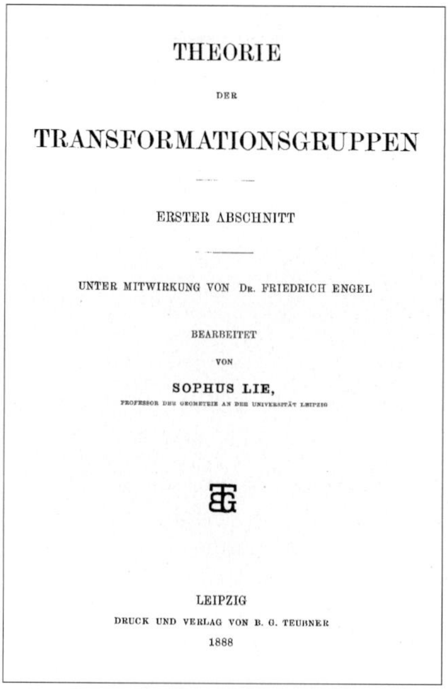

THEORIE

DER

TRANSFORMATIONSGRUPPEN

ERSTER ABSCHNITT

UNTER MITWIRKUNG VON DR. FRIEDRICH ENGEL

BEARBEITET

VON

SOPHUS LIE,

PROFESSOR DER GEOMETRIE AN DER UNIVERSITÄT LEIPZIG

LEIPZIG
DRUCK UND VERLAG VON B. G. TEUBNER
1888

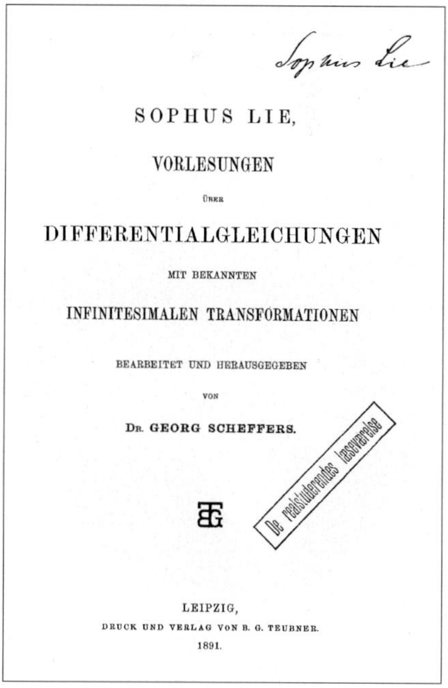

Sophus Lie

SOPHUS LIE,

VORLESUNGEN

ÜBER

DIFFERENTIALGLEICHUNGEN

MIT BEKANNTEN

INFINITESIMALEN TRANSFORMATIONEN

BEARBEITET UND HERAUSGEGEBEN

VON

DR. GEORG SCHEFFERS.

De realstuderendes læsevarelse

LEIPZIG,
DRUCK UND VERLAG VON B. G. TEUBNER.
1891.

Alfred Clebsch

Carl Neumann

Hermann Amandus Schwarz

Eduard Study

Wilhelm Ostwald, who came from Riga in Latvia, became professor of physical chemistry and was a friend and associate of Lie's during the latter's first years in Leipzig. He received the Nobel Prize for chemistry in 1909.

the city where he, and particularly Anna and the children, could - according to the practice and standards of the bourgeoisie – enjoy the countryside. This place, Berga an der Elster, was located 50 or 60 kilometers south of the city, and was a much-frequented summer place for the upper bourgeoisie of Leipzig. Lie had entered into a rental agreement in Berga for a house called "Angermühle" (Mill in the Meadow), something which perhaps might suggest a house at the bottom of a valley beside a river – later it appears that they also used the hotel in that same community. Lie himself referred to the place simply as lying "in the Countryside of Thyringen (Saxony Weimar)." And here, in the countryside, the family went in August after the end of the summer semester. Sophus generally stayed for a week in Berga before going on further for a two-week walking tour in Switzerland and the Tyrol. He wrote to Motzfeldt:

> I am anxious to see if, in this manner, I can keep to my Plan to go on a Foot Tour every Summer until I reach my 60th Year. Until now it is going well, although I sometimes find the greater ascents hard-going with a Rucksack on my Back.

Otherwise during their first summer in Leipzig the weather conditions were rather cold: "But the question still arises, how will it go during a normal, or more to the point, an abnormally hot Summer," he wrote to Motzfeldt, adding:

> We are indeed very well satisfied here. The Saxons are an extraordinarily pleasant Race of People. To Anna, as to me, they show every possible Goodwill. [...] Neither Anna nor I have noticed any real Homesickness.

Rightly enough it had been a little difficult for nine-year-old Marie, who at the beginning had had no children to talk to, and at school she understood very little, but after about four months in Leipzig, her father could report: "Now however she has become amazingly good in German and is doing well here." As for himself, Lie felt that it would take several years before he could "judge whether it was sensible" that he had moved to Leipzig: "There is certainly one Disadvantage, that the Lectures and the other teaching work require much more Time than in Christiania, where as far as it went we Mathematicians had easy Days." On the other hand, he hoped that the teaching load in Leipzig would "bear Fruit in quite another Sense than in Christiania," but he admitted, "It costs me much more than I had expected, to use this foreign Language." He was "too old to become any real full-blooded German." But he was still "extremely well pleased" with his German lectures, and hoped the students were of the same opinion. Lie reported to Klein that he felt his own theories also "showed themselves to be pedagogically suitable".

Lie was invited to Berlin at the end of September 1886, to the Meetings of the German Natural Sciences, which was still the great assembly point for all researchers in the natural sciences from medicine to mathematics. Four years later mathematics hived itself off, and formed Deutschen Mathematiker-Vereinigung, DMV. Lie looked forward to meeting Klein in Berlin, and one of the reasons that Lie attended, seems to have been to enjoy Klein's presence. As he wrote to Klein before the meetings: "There are many things that I so much want to talk over with you. It is all too long-winded to write about." Lie stayed at the Grand Hotel, Alexanderplatz. He felt himself active and on the move, and wrote to friend Motzfeldt: "Should I accomplish what I desire, then I shall advance both my Person and my Ideas." At the meeting in Berlin (21 September, 1886) Lie gave a lecture entitled, "Tatsachen, welche der Geometrie zu Grunde Liegen" ["Problems Lying at the Foundation of Geometry"] – a sharp commentary upon, and almost a critique of, a work by the famous German natural scientist, Hermann von Helmholtz. Lie demonstrated that the new theories of transformation groups could give a better and more comprehensive explanation of geometry's basic problems than von Helmholtz had come out with in his famous 1868 paper.

Still, the main people in Lie's professional work were Engel and Mayer, and gradually Scheffers as well. The correspondence between Lie in Leipzig and Klein in Göttingen was still frequent, and of both a professional and personal character. In Berlin and other places, Klein, Lie and Mayer were regarded as a group – referred to derogatorily as the "société thuriféraire". The French "thuriféraire" was actually a word used in the ecclesiastical sense that had to do with swinging the censer in the mass, and as applied to Klein, Lie and Mayer, "société thuriféraire" also had a figurative meaning in the general direction of "a gang of fawning lackeys".

Mayer had renounced his ordinary professorship in Leipzig, in favour of an honourary professorship – in order to make room for Lie. The twenty-five-year-old Engel still would put in a couple more years as a private docent – and his collaboration with Lie was underway again, although things did not go quite as Lie would have liked. In a letter to Klein he expressed his desperation about Engel's dilatoriness – he, Engel, after a month, had not advanced as far as he had in one day in Christiania. To Engel himself, Lie pointed to this state of affairs in a jesting manner, but after the Christmas season, he had finally had a serious talk with Engel, who improved himself and the work came around to a form that satisfied Lie – in any case, that is how he characterised the situation for Klein.

During these first years, Lie often stressed that he stood "in good Relations" with his mathematical colleagues, apart from Professor Carl Neumann. Right away during his first year, Lie noted that many of the participants in his seminars got written requests to work instead with Neumann. The approaches came from Neumann's allies. Even after six years in Leipzig, Lie would write: "Neither verbally nor in writing have either Neumann or I ever exchanged a bad Word; but we *ignore* one another." Neumann had been professor at Basel and Tübingen before, at the age of thirty-six, had come to Leipzig in 1868, the same year that the great mathematician Möbius died, following more than half a century of work at the University of Leipzig. Neumann too would remain at Leipzig until his death at the venerable age of ninety-three, in 1925. He had been the youth and student friend of Clebsch and had been among the first to demonstrate the epoch-making nature of Riemann's mathematical works, and together with Clebsch, he had founded *Mathematische Annalen*. But after Clebsch died in 1872, Neumann had gradually withdrawn from the editorial work – it had then been Klein and Mayer who seriously took it over and made *Mathematische Annalen* into one of the leading journals of the era. Neumann's own special field was potential theory – a theory of analytic functions, not far from the theory of differential equations, and a branch of mathematics that received, and still receives, broad application in electrostatics and magnetism.

It was Neumann who, according to rumours, had been able to endorse Lie as a potential colleague, not for his actual abilities and knowledge, but only because Lie was not a fellow function theoretician. According to Lie, Neumann had in his time also been "Klein's Enemy", and that he was one of the reasons that Klein sought to leave Leipzig – something that Klein too in time would acknowledge. When Lie was later to comment upon his first seven semesters as professor in Leipzig, he wrote from Berga, on 28 July, 1890, to Motzfeldt:

> Neumann is a prominent Mathematician, and a rare, outstanding Teacher. He has a rather morbid Tendency to compete with his Colleagues. Whilst the Professors in Norway divide the Fields between themselves, here Each teaches whatever he wants. As long as Klein was here, he [Neumann] taught the same Thing as Klein. Although Klein is for the Time Being Germany's foremost Teacher of Mathematics, it was still Neumann's Chicanery that was the Grounds for his leaving Leipzig.
>
> I am keeping to two Types of Lecturing, elementary, on geometry and higher aspects of my own Theories. Neumann now teaches Geometry every Semester, and always that which I plan to teach the following Semester (I follow a definite Plan in my elementary Lectures). I have been irritated more than I can say by this Petty-mindedness of his. But still my Teaching is going well. The Students are eager to come to me.

And when a couple of years later Lie commented again on the competition with Neumann, he wrote:

> I certainly think that I have now rather effectively beaten him [Neumann]; at least I have Pupils who are rather more different and outstanding than his. All of Leipzig's younger Mathematicians (of whom in my Time a Couple have been called to foreign Universities) group themselves around me and work on my Things.

But it was not first and foremost for his professional substance that Lie gained respect – in any case, not at the beginning, and at least not among the broad majority of the teachers at the university. When, at the end of October 1886, the university rector formally bid all the new professors welcome, Lie attracted attention. The rector – the Dean of Theology and famous Assyrologist, Dr. F. Delitsch – found himself saying in his welcome speech to Sophus Lie that he knew how shockingly the peasant representatives in the Norwegian Parliament had treated the noble King Oscar, so he could well understand that Lie had looked forward to leaving Christiania and accepting his call to Leipzig. But the moment the rector said this, Lie had stood up and protested vehemently before leaving the hall to everyone's consternation. Rector Delitsch later sent a letter to Lie in which, in almost humble terms, he offered his unreserved apology. How Lie responded is not recorded. Yet it was a known fact that Lie felt stifled by having to submit to what he considered were stiff formalities and excessive etiquette. For example, in a letter to history professor Gustav Storm in Christiania, Lie commented on his inaugural lecture as follows: "Every single Affair is accompanied by a Multitude of Events that take a Mass of Time." By way of providing a frightful example of the formalistic and fussy hair-splitting with which he was surrounded, Lie explained that on five occasions he had had to appear at a public office in connection with the "Serving Girl we took with us from Norway [. . .] and so, into Infinity", and he concluded: "For such an informal Man as I, it is consequently all a Bother, especially since here they demand a damnable Punctuality." But Lie also consoled himself with the thought that all such formalistic adversities would surely abate with time – if not, it would come to be a constant vexation, he maintained. However, after living in Germany for a couple of years, he wrote to Professor Storm again:

> Both my family and I look forward to some point in the Future when once again we are able to stretch out on the Grass at Ladegaardsøen [Bygdøy][35] or Frognersæteren[36] without the risk of having a Constable down our Necks. One constantly has the feeling here of having to behave and be careful. And what lovely (certainly too great) Freedom did we not have in Norway.

Meanwhile the conditions at the university continued by and large to make "a good Impression" – the sciences were accorded honour, scientific considerations were always taken seriously, and faculty discussions were "infinitely more elevated than those in Christiania". In addition, the scientific life of Leipzig gave one "rich Opportunities to hear Accounts of interesting Themes presented by thought-provoking Men", and Lie concluded (to Professor Storm):

> Here, however, as in Berlin or Paris, one is taken away from one's scientific works for enormous periods of Time by the Multitude of Inconveniences that accrue when one finds oneself in a Centre of Science. To Date, I have not exactly had an excessive Amount of Time for Work; but it will improve. In Christiania my Work was much too one-sidedly mathematical. I consider it not impossible that little by little I shall find that it is beneficial to all my Activities, that Life is making more diverse Demands upon me.

Lie regularly read Norwegian newspapers while he was in Leipzig – particularly *Aftenposten* and *Verdens Gang*, but also *Dagbladet* and *Morgenbladet*, and he usually commented upon conditions in Norway in his letters to friends in the homeland – above all, to Ernst Motzfeldt, but also to Gustav Storm, Elling Holst, Amund Helland, Axel Lund, and others. In Leipzig Lie seems gradually to have gravitated to the newpaper café, "Café Mercur" when his lecturing was finished, about twelve noon or one o'clock. He then lost himself in these newspapers until the old waiter, Fritz, who was on familiar terms with many professors, interrupted him with the reminder that the clock was going on two. Then, in his letters home, Lie usually gave vent to his critical observations on the political situation in Norway, and when the subject was teaching and conditions at the university, he often became extremely specific in his evaluations and characterizing of personalities.

During the summer of 1888 he admitted to Motzfeldt that he felt it was "funny" that there was so much criticism of the Sverdrup government in Norway, and that he was in agreement with Motzfeldt that it was "humbling for [every] Norwegian" that a man like the peasant, Lars K. Liestøl, had become a member of the government. And when, a year or so later, Liestøl left the government with "3000 Kr. in Service Money", Lie felt this was only to be expected when one advanced "that Kind of a Second-Rate." According to Lie, Law Professor Fredrik P. Brandt, who had been chosen to head the Norwegian University in 1887, was "the most impossible out of

[35] Bygdøy, on the Christiania/Oslo waterfront where embassies, museums and the royal farm are located today, was formerly a complete island farm, and called Ladegaardsøen.

[36] The forested mountain park behind Christiania/Oslo.

all the Impossibles". To illustrate the condition in which Norway found itself, Lie pointed to his old chief at the Observatory, C. F. Fearnley, who had been professor of astronomy for over twenty years, and had a brilliant social position, good income, and a monumental residence, yet, according to Lie, he was "a Ninny" who lacked "sound Sense" and had therefore become "nothing other than a brainless Goat." As "a perfect Example of the Kind of People who enjoy Employment at Christiania University" Lie also pointed to Professor of Physics Oscar E. Schiøtz, who according to Lie, did not have "a single Iota of scientific Merit, run-of-the-mill Teaching Ability, no Initiative. But correct: that will say, always subordinating himself to the so-called Authorities whether or not they were familiar with the Issue at Hand."

That Norwegian science candidates knew much less than did the German, was in and for itself, not so important, nor was it so bad that in most fields the Norwegian University was at a lower level than the German ones: "But that a Person in a technical field in Industry was so infinitely behind in standing, that indeed was an enormous Misfortune for the Country." It was also terrible that Norwegian military officers received a significantly weaker education than did the officers of other countries. Lie also criticised the attitude toward school policy that he felt was dominant in Norway, namely that technical skills could be learned through "the Establishment of Polytechnical Institutes". Quite the opposite, Lie maintained. If one established such technical institutes "weakly equipped, with badly paid *Norwegian* Teachers" decades would go by before the country got any skilled technicians. Lie's recipe was for the Norwegian government to award good pay to "Norwegian Technicians, who have graduated with distinction from a German technical College" – and in this way one would quickly get many skilled people.

And when Lie found out about the hard times into which Norwegian shipping had fallen, he thought the solution was clear: "Norway's seamen should stick to Fishing on a grand Scale. They could satisfy the English and German Markets for Fish. Furthermore, it is incomprehensible that up there one does not grasp that first and foremost, what is needed – it must be emphasized – is technical Education which to a large Degree relies upon well-paid Foreigners as Teachers."

Again and again his conclusion came down to the same thing after these tirades about conditions in Norway: "No, what draw me back [to Norway] are my personal Friends, Considerations of the Future for my Wife and Children, and last but not least, Norway's Nature and Climate." He hoped that Anna and the children will "profit for Life" by their stay in Leipzig. The children were having the opportunity to learn "numerous Things, at least German and Music", and, he added, "Now I can only provide that their physical Development does not suffer." As for himself, he had come to form a school and publish his works, he would otherwise utilize his time to "study the Swiss Tyrol, etc."

Other than this, the reports home always gave the information that the children were thriving, that they made "quick Progress in the Language". But during the first winter the children had had "a remarkable Appetite", particularly Marie, who right after dinner was often hungry again, and did not feel completely well. According to Sophus, Anna was for a time "was convinced that Mai [Marie] was suffering from Heartburn." Sophus himself felt that the child did not get enough food, and against

Anna's better judgement, had decided that for four or five days they should do whatever they could to make sure Marie really felt full. "The Result was surprising. The Child became full and now radiates Energy and Health," he wrote to Motzfeldt. Then he turned to the winter:

> Here the beginning of Winter was hideous, dark and depressing. After Christmas we had a good Winter with Snow and Ice. My small Girls rushed about expertly on Skates. This Winter for the first Time I learned to skate on polished Blades. I had to do so, for there are so few other Opportunities here for other Kinds of Motion. When I began the same, I fell over backwards and injured my Hand, which is still sore. But fortunately I did not allow myself to become disconcerted, and since then I have managed well. Anna is also beginning, after an Interruption of many Years.

He also reported to Motzfeldt: "The Truth, which I think I ought to admit, is that I have had a Couple of beautiful Hiking Tours in the Woods with glistening Snow. The only Problem being that it was not a Coniferous Forest" – he longed to "see the Snow shining through Spruce Trees onto the pure Snow."

During the summer holidays of 1887, the Family Lie once again stayed at their summer house at Berga – that is, Anna, the children and the serving girl probably left before Sophus was finished with his teaching obligations. And when he came to the countryside that summer, he also continued on further, touring on foot through the Swiss Alps, then returning to Berga, and travelling back to Leipzig again with his wife and family. He commented to Darboux in Paris: "The time in the mountains, when I have no rational thoughts in my head, for me have always been the best part of the year."

In the course of that autumn Anna and the children had, unwittingly, to return yet again to the summer house at Berga. Their Norwegian serving girl, Andrine, had fallen seriously ill, and when she had been taken to the hospital they found out that she suffered from "virulent Gynaecological Typhus." Due to the great risk of infection, they all had to move out and the house was thoroughly disinfected – "with smoking Sulphur and Carbolic Acid". It was the first typhus attack in Leipzig that year. Following a week in a hotel, Anna and the children went "to the country". Sophus left for a fourteen-day foot tour in the Tyrol, before he collected the family and they moved back into the house on Seeburgstrasse. Andrine died in the hospital. No one else was infected with the disease, but their new German "Girl" refused to touch anything with which Andrine had been in contact, and consequently Anna had to do most things herself, as Lie confided to his friend Motzfeldt, and added that he himself had been busy for weeks "meeting with Safety Precautions": "The most unsavoury Thing in this Affair is that there possibly are Grounds to fear that there is Trouble with the Water Pipes into my Residence, that has brought the Sickness, and on the other Hand, such that the Contagium is permanently lodged in the House." The typhus episode also threw his "financial calculations" into disarray.

During the autumn of 1887 the printing also began on the first volume of *Theorie der Transformationsgruppen*, whose size was about forty arks, or printer's sheets

(640 pages). Lie had entered into a contract with Teubner Verlag in which an honorarium of 40 marks per ark, and after a sale of 400 copies, a further 20 marks – an arrangement that Engel too – who would receive half of the honorarium – was pleased with. At first, Lie had expected 60 marks per ark. The book was published in June 1888, and immediately thereafter, the second volume was begun. Lie commented:

I am herewith however far from ready. I have so very much more, or more correctly, much more still to do. It grows in my Hands and it fails to go as quickly as one would think. Under these Conditions it will take Time before I will feel that I have carried out my Mission thoroughly. But with Time I will cut my Way through. Although I was under the Assumption that I would be offered decent Conditions, the Prospects thereof are not so good.

Lie was satisfied with his income, he was "very pleased on the Whole", and "Anna and the Children are thriving": "Here, indeed, for those who have the Time and Money, there are rich conditions for amusing oneself." And, he pointed out in a letter to Professor Storm, that along with income, consumption also increased. To Motzfeldt, he wrote (spring 1888):

When my Wife and Children must want for many Things in Leipzig, I always feel that it is my Duty to do all in my Power to make Life pleasant for them; and this costs Money. Thank God we are all doing well here, particularly the Children. Yet it is always my Desire to make my Way back, although I know full well that I shall always feel that the Conditions in Christiania suit me badly. The Issue is this: when one lays out his Life as I have, then no place is completely satisfying. Here it is always uncomfortable not to be capable of talking, thinking and feeling like a German. Given my Nature there is never Talk about assimilating oneself even a little to the local Circumstances. To my Eyes, the Germans have the same Fatal Flaw as the Conservative Party in Norway, namely that they are some of the greatest of Pharisees. I say this even though I willingly admit that Norway's Right stands considerably higher than the Left; nevertheless, I have always had the greatest Admiration for Everything German.

The steadily increasing militarism that found its place under Kaiser Wilhelm II alarmed Lie. During the spring of 1888, conscription was expanded, and expanded so far that Lie, along with all the German professors, barely managed to avoid being called into service. But the possibility of being called up still remained, if war were to be declared and it went badly for Germany: "which I, for my part, cannot imagine so long as Bismarck is at the helm." Anyway, it would be "a highly unpleasant Consequence of my scientific Interests having drawn me to Germany," he commented to Motzfeldt, and recalled the mobilization among the students of Christiania in 1864: "I had little Desire to die for Denmark in '64, and I have in Truth even less Desire to fall on behalf of the Germans."

About the "home front" he had this to say: The World is turned remarkably around backwards up there [in Norway]. It is impossible for me to sympathise with any of the battling Trends. So much is going on up there that seems purely idiotic. [. . .] It is certainly a thoroughgoing Mistake in Norway that Comprehension is lacking, that these Days, Knowledge and Competence are demanded as never before. Here in Germany one, to be sure, overdoes the Acquisition of Knowledge. But in Norway!

He then went into a discussion about how badly both the lower and higher schools kept their standards in the Norwegian school system: "Is not Schooling in Norway thoroughly uneconomical?" – but he usually concluded his comments by saying: "I allow my Pen to run and do not weigh my Words," and pointed out that his strong interest in Norway and Norwegian social and political conditions certainly meant that he judged "in a rush, and all too often a bit askew."

The greatest Danger for Norway is that Competence is not valued. In this respect the Right is not appreciably better than the Left. See, f. ex. how they have not utilised Broch at all.

After the conclusion of the summer semester in 1888, and the family was safely tucked away in Berga, Lie travelled to Norway for the first time since he had left the country.

After his visit to Norway, which had probably lasted for a month, Lie wrote to Motzfeldt: "The whole of my Stay in Norway seems to me like a Dream." The stay abroad had meant that he, so to speak, had obtained other values and "another Measuring Stick for many things":

What for instance made the strongest Impression on me was the Contrast between Christiania's beautiful Setting and Leipzig's boring Environment. This is not to say that I deny that I simultaneously feel how small Things are there at home in comparison with here; but the Large also has its Shady Sides too.

We continue to get on well. I sometimes curse at all the Troubles the Students give me. What is purely comic is the Contrast between a Norwegian Professor's Indifference to his Audience and the sly Attention that a German Professor (in his own Interest) shows in his Students.

Anna and the children had had a good time while he was in Norway – "The Children thrive remarkably well" and: "Anna is doing well" and gets "Gewandthaus concert tickets from our rich Friends".

First among those rich friends stood Adolf Mayer and his wife. Mayer came from a well-off family of merchants, who were close to the banking House of Frege, and therefore in relation to the university, Mayer could quite easily work within the financial confines of an honorary professorship in order to make way for Lie. The Mayers ran a hospitable house, and their parties were said to be a great success, thanks to their capacity for inviting people who got along well together.

There was much social life in Leipzig during the months of January, and February, and the forty-seven-year-old Lie wrote to Motzfeldt:

I am beginning again in my old Age to dance. It is incomprehensible how the German professorial Wives well into an advanced Age (to about 40) set such a Price on Dancing. Thus I find that down here one ought to utilise every Opportunity for healthy Motion, so I have thrown myself into dancing (3 or 4 Times a year) as an easy Means of improving my Health and increasing my Stocks.

It was, moreover, this winter of 1889 that Lie had skis sent from Norway, and radiantly wrote to Motzfeldt:

The Children have been out a lot on Skates, and even – yesterday and today – on *skis*. Today we had a really brilliant Ski Tour; of course with truly hopeless Slopes. But for the Children it was for the time being, adequate.

It was this event that their daughter, Marie, more than fifty years later, brought up as characteristic of her father. She recalled: "We were to be healthy in both body and soul. The whole family had to go out on a long tour every Sunday. We were forced to go out on skates, and at the end of the 80s, he had some small skis sent from Norway. It was certainly something that had never been seen before in Leipzig. We carried them to the nearest park, and we children were far from happy whenever people pointed at us, and called out loud, 'Seht da sind Eskimos!'" [Look at the Eskimos!]

During the spring of 1889 Lie got to know that he had been made a member of the Academy of Sciences of Copenhagen, and in a letter to Zeuthen, whom he thought had been the man behind this appointment, Lie gave his thanks for the honour bestowed upon him, and to this old friend he described a bit about both his life and his views on this and that:

Things are going well for me, even though as a good Norwegian I have heartfelt longings for home in Norway. What particularly pulls me toward the North are Nature and the Climate. Had not the political Conditions been so thoroughly boring and dull in Norway, I would hardly have expatriated myself. If at some Point a little Order were to come into Things there at home, then I would once again move back up. Of course I would first have to have published several major Works, for which I need both linguistic and stylistic Help.

Lie reflected on misfortunes and opportunities and placed great emphasis upon the lecturing and work with students that one did in Leipzig. This obviously hindered one's scientific production, but on the other side, one was forced to "follow Developments in broad Outline":

On the other hand, the narrow Norwegian Conditions are not favourable for original Production. One lives peacefully and follows one's own Thoughts. In Germany and France most the Academics have extraordinary Learning; but Originality one finds only in a few. Of course one attempts to train the Students toward original Production, but the Results do not correspond to the enormous Labour that one expends. The same Thing that happens in Science goes on in Social Life in Germany as well. At first Glance, one is impressed by the Foreign, but one quickly notices that one meets the same Thoughts and the same Forms everywhere. The Germans allow themselves to be conquered by Fashion.

As for the rest, Lie thought that Germany and France had begun, to a certain degree, "to change roles":

The scientific Decadence in France [around 1850] dated from the fact that one ignored Everything that was neither French nor from the Trend-setters in Paris. There is a great Danger that Berlin [now] will exercise a similarly unhealthy Influence. In any case it is striking how few of the younger German Mathematicians are prominent. Where does one find – I will not ask for a Poincaré – but a Picard or an Appell.

This situation was the more remarkable according to Lie if one knew that the number of mathematical students in Germany between 1873-83 was "quite extraordinarily high". In Lie's opinion, the only one who was really running "a school" in Germany was Klein in Göttingen, but to Zeuthen he added:

But one cannot dispute the fact that however many brilliant Abilities Klein possesses, there are still considerable shortcomings in his Work; certainly they could in part serve to counterbalance Berlin's one-sidedness.

Lie expressed similar thoughts in a letter about the same period to Mittag-Leffler:

The Parisians are distinguishing themselves in Mathematics again now as they did in the old Days. On the other hand, as to what bearing this has on Germany, I have for a long time had the Impression that it is going downhill.

The printing of the second volume of *Theorie der Transformationsgruppen* was started at the beginning of June 1889. Lie worked on the proof corrections over the course of the summer. Judging from the dates on the corrected proofs that he sent back to the publisher, it seems he spent the greater part of that August at the summer house in Berga. But according to Engel, Lie himself did not feel himself to be completely "in top form" – the lack of interest that he felt he had noticed on the part of the world of science made him depressed and irritable: "Just wait until the work on transformation groups is ready, then I shall be a completely different human being," he said to Engel.

Whether Lie set aside time for a trip to the Swiss Alps that summer seems uncertain. He certainly looked forward enormously to taking his family and going to Norway the following summer – after four years in Leipzig, he, together with Anna and the children, would spend at least a month in Norway, beginning in August 1890.

But things did not turn out that way. During the autumn of 1889 he began to suffer from insomnia; he became more and more anxious, until in the end he went to a doctor, and in November 1889, he was admitted to a psychiatric clinic in the vicinity of Hanover.

Breakdown

Right at the beginning of his eighth semester as professor in Leipzig, Lie had a complete breakdown. He felt he had been "seized by a boundless despair" and allowed himself to be taken away to a psychiatric clinic at Ilten, near Hanover. Here he remained for a full seven months.

Shortly after he had been released from Ilten in the summer of 1890 and reunited with his family in Leipzig, together with his wife and children, he left for his summer place at Berga. From there he wrote to Motzfeldt to explain what had happened. Fifteen months had gone by since he had last written to his friend in Christiania:

Dear Ernst!
I finally have a chance to give You a little Information on the Reason for my serious Disease of the Nerves, from which I still suffer a little from time to time.

My Nerves have certainly never stood up in Comparison to my Muscle Power and robust Physique. I have felt for many Years that the Power of my Nerves was relatively little; but I never expected that I had anything to fear, since I never felt anything.

I think it has been unfortunate that I pursued this tempting Position in Leipzig, although during these Years I have achieved an extraordinary Amount.

Then, after having listed his scientific victories and advances, and new important works that were being readied to release to the world, Lie candidly explained the difficulties and disappointments he had suffered in his academic work, among students and colleagues. Then he came to what he, with his own eyes, saw as the prime cause of his illness – he went back in time to autumn 1888 when, after having visited Norway by himself, he returned to Leipzig:

Upon my Homecoming from Norway two Years ago, my dear precious Wife told me she was afraid that she had Cancer of the Breast (*do not speak about this*). A Lump really had been found in her Breast, which certainly gave Cause for Anxiety. I cannot tell You what a terrifying Impression that made. I felt that I did not have the Strength to stand up to such a Misfortune as encircled her and us. I was so cowardly that I prevented her from going to the Doctor. It was not until almost a Year later that we went to the Doctor. He gave us Hope that the Lump would shrink with external Treatment. There now followed a Couple of Months of the most frightful Tension. Anna was very brave. But I was on the go Day and Night, and was in Fear and Dread for her and for us all. In the end, I began to sleep badly, and finally I almost did not sleep at all. Now for the first Time it came to me that I was in greater Danger than she was. (In the Course of the Winter Anna underwent a little Operation and it was determined that her Condition was completely without Danger. Thank God that is the Case.)

> If I had been terror-stricken before, my Despair now became boundless. Unfortunately I have always been an impulsive Person. As a Rule I have always had the greatest Self-Confidence. But now all Self-Confidence failed me, it had completely deserted me.
>
> The Doctors said that I suffered from Melancholy, and that was a weak Expression for the fearful mental Condition in which I found myself.

The doctors who had taken Lie in hand during his illness were, first of all, Dr. F. A. Hoffmann, director of the polyclinic at the University of Leipzig. In his report, dated November 16, 1889, he wrote about a "group of impending symptoms of the nervous variety", and sent Lie on to the university's leading psychiatrist, Paul Flechsig, who, on November 24th in consultation with Dr. Max Sänger, sent Lie to the psychiatric clinic at Ilten.

Professor Max Sänger seems to have been a close friend of the Lie family. He was married to a Norwegian woman, Helga Vaagaard from Ringerike, and their son Hans Erling would in time become one of Herman's best friends. Sänger was a well-known gynaecologist; in 1882 he had improved the technique of the classic caesarian operation, also known as the Sänger operation, and he had an extensive scientific output. Helga Vaagaard was one of Nina Grieg's girl friends from youth, and Nina Grieg, also consulted Dr. Sänger – "one of the younger, brilliant Women's Doctors" is what Edvard Grieg wrote home to his brother in the spring of 1888. Eight years later Nina underwent a breast operation that was performed by Dr. Sänger in Leipzig. It therefore seems rather odd that Lie, who had such close ties to local medical expertise, did not consult Sänger during the autumn of 1888 when the lump in Anna's breast was discovered. That help was to be found so near, and yet not sought out, is perhaps a sign that Lie was correct in his self-diagnosis that the germ of his sickness lay here – for to do something in this situation possibly demanded more strength of mind than he possessed.

At Ilten, Lie was treated by Doctor Ferdinand Wahrendorff in his *Privat- Heil- und Pflege-Anstalt für Gemüths- und Nervenkranke* [Private Health and Caring Facility for Diseases of the Mind and the Nerves]. The journal entries that Dr. Wahrendorff kept on the condition and methods of treatment of his patient Sophus Lie, have been lost. Probably, since Lie was a patient paying his own medical expenses, his records were sent back to the doctor in Leipzig who had admitted him, that is, to Dr. Flechsig, with whom Lie also had contact after his return to Leipzig.

Since 1872 Flechsig had been involved in anatomical studies of the brain. He had been the assistant to the well-known Karl Ludwig, one of the founders of scientific physiology, and later, along with the Russian physiologist Pavlov, he became one of Ludwig's most famous students. In 1877, when he was thirty years of age, Flechsig had been made professor of psychiatry at the University of Leipzig, and here he built up a new university clinic for the treatment of mental illnesses. On two occasions Flechsig also functioned as the university rector, and the speech that he gave when he was so chosen for the first time in 1894, aroused great interest in university circles – particularly among philosophers and theologians – and also among a broader stratum of the population when it was published two years later under the title "Gehirn und Seele" [Brain and Soul]. Flechsig, with his great professional gravity, stood as the spokesman for a thorough-going materialistic point of view. Given, as a point of departure, the following facts – that consciousness resides only in and among the living, and that the form of consciousness (experienced in individuals through their own experience of feelings and events), changes with fluctuations of bodily conditions – Flechsig then interpreted the soul as a function of the body. What were called the revelations or expressions of the soul were really the revelations and expressions of life, expressions of a way of living, that merely distinguished itself from other ways of living, of related systems – particularly of the nervous system – in the precise manner that these presented themselves as consciousness.

By means of his anatomical studies of the brain through the course of a long life, Flechsig described the human brain, and he also mapped the "transmission lines" of the central nervous system – one of the important nerve conduits in the spinal column was subsequently named after him, the Flechsig nerve bundle. But his files too, in Leipzig, which probably also contained the records of Lie's illness, are gone, destroyed in the bombing during the Second World War.

But from other sources – Mayer's letters to Klein – one finds that in November 1889, Flechsig felt that Sophus Lie suffered from an advanced neurasthenia (nervous disorder). And Flechsig, whose theoretical basis always linked the psychic with the physical, felt that the main reasons for Lie's condition was homesickness, over-work, and stress in relation to his colleagues. Flechsig, who certainly was familiar with conditions around the university, had pointed to Lie's difficult relationship – according to Mayer – with Professor Neumann. For his part, Mayer seems to have been of a somewhat different opinion. By all appearances, Lie had earlier told Mayer about his worry over Anna's breast tumour, and at the same time had extracted from Mayer a solemn pledge of silence. In a letter to Klein, Mayer remarked, as early as the beginning of November 1889 (before the admission into care), that Lie was anxious and nervous. Lie certainly conducted his lectures and completed, again with the help of Engel, the second volume of the transformation group theory, but he no longer attended the regular meetings at the university – his nervousness only increased, according to Mayer. Mayer also maintained that Lie did not want to discuss the reasons for his edginess, except perhaps to the doctors. But although Mayer recalled a promise of secrecy he had given when Lie "poured out his heart", he wrote to Klein that he did not believe the cause of Lie's anxiety was related to work, but that it stemmed from conditions in Norway. Could it be perhaps that there were powers in Norway who were working to see that his professorship there was filled, and therefore had he been asked about the possibility of returning home? According to Mayer, Anna had begun to notice the changes in the condition of Lie's mind about seven weeks before he was admitted to care. On November 15th, Lie had written to the Faculty of Philosophy and reported that he, due to the condition of his health (Gesundheitszustand), was obliged, under orders from his doctor, to take a long break, and he asked to be relieved of his teaching responsibilities.

Mayer had been at the railway station on November 24th, to say farewell when Lie left for the psychiatric clinic. According to Mayer, Lie had appeared apathetic and had spoken in such a low voice that it had been almost impossible to hear what he said – but among the little that Mayer had been able to understand was the request that his greetings be sent to Klein. Klein, for his part, replied to Mayer that he was completely unprepared for such a turn of events, for he had always regarded Lie as the "Norwegian primal force" who, with the greatest of ease, would transcend the great strains he placed upon himself.

From Mayer's letter to Klein it is also suggested that one of the reasons that Lie was sent to Ilten was Flechsig's acquaintance with the psychiatrist Ludwig Meyer. Meyer, through his years of practice in Göttingen, had sent several of his patients to Ilten, and warmly recommended this clinic near Hanover. In his letter to Klein, Mayer suggested that Klein could probably, through this psychiatrist in Göttingen, get to know how things were progressing with Lie in Dr. Wahrendorff's clinic at Ilten. Mayer hoped Lie would rapidly and fully recover the way Rudolf Lipschitz had done. In his time (1868) Lipschitz had been Klein's professor at Bonn, and was one of Germany's leading mathematicians in the study of Riemann's work.

At Ilten near Hanover, Wahrendorff had – after working as a general practitioner in the Ilten region, and after short study periods at psychiatric clinics in Vienna and Prague – built up his own facilities at the beginning of the 1860s. Already by his second year, in 1864, he had twenty-four patients from the more well-to-do segments of the population, and each of them paid between fifteen and thirty marks per month. Gradually Wahrendorff also took in the sick who were sent by municipal authorities, and right from the beginning the facility was open to patients of both sexes. The number of patients, as well as doctors, nurses and employees in the kitchen, laundry and garden, increased – but some of the treatment methods involved what was called "family care". The number of people ought not be too high; the patients should feel that they were part of family life with certain duties toward the daily upkeep and events. The sick lived separately, or they could live two or three together in one room, and many of the rooms were rented from respectable families in the village. According to the clinic's specifications, such a bedroom was to be spacious and airy, with bed, wardrobe and chair as the minimal inventory. Bed and wardrobe were delivered by the clinic, and the wardrobe shelves contained clothing for Sundays, and clothing for weekdays: skirt and jacket, trousers and vest, two wool hats, six shirts, six pairs of stockings, two pairs of shoes, six pocket handkerchiefs, two neck scarves, and two smocks of blue linen, or two pairs of blue linen trousers. The patients got help and supervision with regard to their personal maintenance, but apart from common meal times and treatments, which seem to have been considerably medicinal – the patients had the freedom to move around the area and among the permanent residents, either with or without their caregivers. In the 1880s professionals came from other European countries and from the United States to study Wahrendorff's methods and practical organization in rural Ilten.

Sophus wrote regularly to Anna during his hospitalisation, and she and the children answered as well as they were able – three of Sophus' letters are extant. In the first, which was written during the first weeks, Sophus began:

My own precious Anna! [. . .] I can almost say that I am now living completely out of the world. Only Your Letters give me a Link to the Straight and Narrow.

He longed to be home and would have come right back, he wrote, if only it had not been for the illness, which demanded "regular Supervision". Had he left there, then he would only have found himself "in another such Facility, where the Difference was not so great". He wrote:

It is very heavy for me to have had to abandon a fine and loving Wife and delightful Children; it is my Prayer to God that I am quickly able to come back. Were I speedily to recover again, I would have perhaps known no Worry about our Economy. What disheartens me, is that my Nerves are in a terrible Condition. The Moment that they improve, my Mental Power should certainly also return, or so the Doctors have said in any case.

Please pray God that I may quickly regain my Health, and let the Children as before pray to God to protect me. Perhaps He will soon look down on us with Mercy. You can be sure that I shall do whatever I am able to regain my Health as quickly as possible. But I must always remind You that You should be indulgent toward me, when I eventually get back. I shall not plague You with Impetuosity and Arrogance.

Who in Leipzig says that I compel Respect as a Teacher?

What You tell about the Children, and particularly about Mai, I find especially endearing. Let Little Lad begin School at Easter. Kiss the Children every Day from me.

Send me the Picture of You that sits in my Office Cupboard, on the right / the most recent. Upon my Departure I had not the Heart to take anything with me, although I had thought of doing so. I do not want Pictures of the Children just now. I am thinking of them all the Time.

It is infinitely pleasing to me that Everyone is showing You Friendship and Kindness. You certainly deserve it.

I hope You get along well now, my dear precious lovely Wife, who is so very much better than I ever deserved. Bear the Hardships with Courage and *Devotion to God*. I do not have the Ability to do so, but You certainly do. Withstand all Temptation, stand firm at the Beginning, for it is unlikely to be more difficult later. Teach the Children to seek out God, so that they can become good and happy Human Beings. God bless You All. Sophus.

In the margin of this letter he wrote: "Influenza has been raging here, but I have remained in the clear."

In a letter from around New Year 1890, Sophus again thanked Anna for her letter: "You have no Conception of how much I value Your Letters, and how moved I am by their Content," he wrote, and stressed that her "loving Character is reflected" in the letters. Otherwise, Sophus had not found it easy to give advice in response to her question about schooling for the six-year-old Herman: "If we were able to calculate that we would be in Norway next Summer then I would have wished that he not go to a German School, assuming that You and Mai could give him a little Instruction." But the way things were now, it was probably best to enroll him at the City School, and Sophus could only hope that he got "a good Teacher and above all, good Classmates." Otherwise, he said that he had begun to read a little, and he thanked her for the book of Walter Scott that she had sent: "Still, at times when I rip my Letters into little Bits because I am not satisfied with what I have written, I tend to repeat this Information. Be that as it may, I have at least begun to read a certain Amount now."

From all accounts he was thinking the whole time about Anna and the children – how big they must have become, who they were together with, and if people were good to them – and he asked Anna to make sure she ate enough, and he felt sorry that Mai's stomach had not become better. He concluded:

About myself there is not much to report. I am not exactly pleased with myself; but I shall do my Best. At least I beseech You to preserve Your Love for me. With all my Faults, I can still in all Honesty say that I have always loved You and the Children and wished the Best for You.

In the third letter, sent from Ilten on February 10, 1890, he wrote that he was happy it had been a mild winter and "so favourable for Walking", for had snow come, they were not "as quick as in Leipzig to clear the Roads of Snow". Volume II of *Theorie der Transformationsgruppen* had been printed and was being launched in Leipzig now in February – Engel should certainly have his portion of the royalties, and Sophus communicated to Anna that he would soon be writing to Teubner so that she should receive money, from which she was "to pay Engel the 300 Marks he wants". But there was something else close to his heart: "I have now come so far that I feel the Need to see the Children's Pictures." But he did not want to have a group photo of the whole family, only the pictures of the children alone. "Later, I will certainly have the Desire to see the Group, with myself in it."

I am not making as rapid progress as I ought. I am unfortunately as unsuited as it is possible to be, beforehand, to undergo a serious Illness. This great Emergency depresses me so shamefully, and it has taken me an improper Amount of Time once again to assemble the Courage for Living. But I certainly hope nonetheless that it shall come to be.

Forgive me, dearest Wife, for all the Worries and Cares that I have inflicted upon You. It will be my highest Goal, according to my Ability, to make Everything go well again. Keep well, my dearest Wife. Ask God to illuminate my Mind and to give me nothing but good Thoughts. It is extremely difficult for a Person of my Character to bear the Cross that has been dealt to me. I shall do my Best, and with God's Help it shall go well.

I am trying to pray to God to make the Burden, that weighs down on You as light as possible. I certainly do not have the right Candour for Prayer. But I must acquire it. *Have no Fear that I shall develop unnecessarily religious Scruples.* I am unfortunately of so little religion from before; now I see that only Religion can give me the Outlook for Recovery. And You, my dearest One, You must seek Consolation and Help from God.

About this same time Mayer reported to Klein that he had received a letter from Dr. Wahrendorff, who was able to relate that Lie was gradually recovering.

In the course of the coming months, Lie also managed to send off a couple of mathematical studies. One of them, *Über die Grundlagen der Geometrie (Erste Abhandlung)*, was quickly printed in the Proceedings of the Leipzig Academy of Sciences – it was a further development of the work he had presented at the meetings of the natural sciences in Berlin in the autumn of 1886, on the theories of von Helmholtz. The other study he sent to Engel, and Engel, who had become a professor, and now also substituted for Lie at the university, edited Lie's manuscript, and a year and a half later it was published (36 pages) under the title *Die Grundlagen für die Theorie der unendlichen continuirlichen Transformationsgruppen (Erste Abhandlung).*

From Ilten, Lie now began to get back into contact with the outside world. Eduard Study, one of the private docents at Leipzig had received a position at Marburg in 1888, and when the first volume of *Theorie der Transformationsgruppen* came out in June of that year, Study had written an important and extremely laudatory review of Lie's work. And now, following the publication of the second volume, and when he heard that Lie was recovering, Study wrote to him at Ilten and thanked him warmly for the new volume. From the bottom of his heart he wished Lie well and sent greeting from his wife as well, and asked Lie to convey their greetings to his own dear wife – in Leipzig the two families had been in good contact, and in the exchange of correspondence between Lie and Study, which continued for many years, there were always greetings to the respective spouses and families.

On May 2nd, Lie wrote a letter from Ilten to Mayer. Mayer had been accommodating by sending Lie the necessary papers to apply for an extended leave of absence, and Lie thanked him in a heartfelt manner. It was in this letter that Lie wrote, when he had his breakdown the previous autumn, he had been "seized by a boundless despair", and then after that for a time it got even worse. Still he hoped it would now be possible "to come back to life", and he acknowledged that things were going somewhat better for him, and that he had begun to work a little again – that he had had a number of his manuscripts sent to him that he was trying to make accessible to a wider readership – but he was basically unsure of his future and of his own health. If things went as he hoped, and he regained his health, he would then continue the work of publishing his theories – if however things went badly for him, something that unfortunately was still a possibility, then the publication of these theories would at least save his children "from ruin". It was sad to be torn away so suddenly from his scientific work, he wrote – not to speak of his family – but the fact that his dear Anna there at home had been so amazing and had shown such "strength of mind" the whole time – was an unfathomable consolation. Lie asked Mayer to convey his greetings to his colleagues, especially Engel. If his health and his strength did not worsen, he would soon, together with Engel, begin work on the third volume, and Lie added that even if he never regained his health, the third volume of *Theorie der Transformationsgruppen* had to come out. For the time being his main thought was to complete this huge work – unfortunately he certainly had to give up the thought of a major work on integration theory, and then he added: "My French student Vessiot shall be my successor in this field, just as Engel, Schur, Killing, Study, etc. are in the field of transformation groups."

In a letter to Klein on May 17, 1890, Lie reported that he was "very sick" and it was extremely doubtful whether he would ever be well again. He said that the hospital would do "all that was possible" for him, but that unfortunately it was impossible for him to feel right in the role of patient, and that therefore it was not going well. But anyway, he did not want to talk about "these sad circumstances", but wrote that in any case he had managed to get himself together enough to begin taking an interest in science. As long as his powers continued, he would work out his ideas to the greatest extent that this was possible: "Unfortunately vanity has been a leading principle of my life. And this still has not abandoned me." In a new one to Klein two weeks later, he wrote:

When I do not talk about my wife and children, that is because it is too painful. Despite all my faults I have lived in and for my family.

A gloomy future lies before me. I must be happy as long as I can work. Work is always my life-long friend.

I certainly hope to regain my health. But for me there is no longer any happiness to be found. May a merciful God protect my wife and my children.

Mittag-Leffler in Sweden also tried to get in touch with Lie. Anna sent him a telegram on May 28th, which went out with the following wording: "Mein mand lange verreist." [My husband is far away.]

But a little over a month later Anna paid a visit to Ilten. In the hospital's reception book, in all its brevity, one finds that Lie, on July 3, 1890, was fetched by his wife – the diagnosis was "a melancholy not cured".

After more than seven months in the hospital Lie was thus back at home in Leipzig. Here he lived for a while in seclusion. He sent a letter to his student Scheffers with a proposal for further work. Lie had probably come to know that during his own absence due to illness, Scheffers had received a tempting offer from Klein in Göttingen, but Scheffers, however, had waited. He now rapidly thanked Lie for his work proposal, and expressed the hope that Lie would soon be in full good health.

After a few weeks at Seeburgstrasse Lie and his family journeyed to the summer place at Berga, and it was in consequence from here that he wrote his friend Motzfeldt about what "really" had provoked his illness. After having recounted how the lump in Anna's breast had filled him with "Fear and Dread", and made him sleepless and nervous to such a degree that in the end he had to give up his lectures and "go into a Nerve Facility". Lie also wrote a little about his experiences at the facility, and the reasons why Anna had fetched him back to Leipzig:

Unfortunately I made an impossible Patient. The whole Time I believed that the Doctors did not understand my Illness. They treated me with Opium in frightening Doses in order to calm my Nerves; but it did not help. And in addition, Sleeping Potions.

3–4 Weeks [ago] I became tired of my Stay at the Facility. I decided that I would, by my own Hand, regain my Mental Equilibrium and my Sleep.

I have now done what the Doctors said no Man could pull off, that is to say, I have discontinued the Opium. This has really taken it out of me. Whilst before I was absolutely robust in Body, the Want of Opium made me faint, gave me a loose Stomach, almost Diarrhoea, strong Night Sweating and other Bedevilment. But now for a Couple of Days (against the Doctors' Advice) I have applied significant Motion; remarkably enough the Night Sweating is decreasing and the Stomach is once again functioning normally. I hope that now in the course of a Week's Time I shall have won out fully over the deleterious Effects of the Opium Cure. I myself believe that the Doctors may have harmed me with their Opium.

My Nerves are considerably battered; but my Body has now regained its old Horsepower. I will now turn my own Hands to curing myself. I shall walk from Morning to

Evening (which the Doctors have declared so much Folly). When I have driven out the First Part of the Opium Weakness, then, with Time, Sleep will naturally return. Such is my Hope.

Anna has faithfully supported me. She has exhibited a Power of the Soul that exceeds all my Praise. I myself have been completely smashed; for I saw no Hope; but she and the Children have been my saving Angels. I lack the Words to express my Amazement at Anna's Strength through this long Period. She has forgotten everything that she herself has had to bear, and has thought only of me.

He hoped that he could begin to lecture – "certainly only lightly" – in the coming semester; that is, the autumn semester that began in the middle of October.

Hopefully in the Course of the Winter I will once more regain myself. Here in Leipzig I have been shown splendid Goodwill by my Colleagues and Everyone. They do Everything they can to bring me back into Equilibrium again.

I am consequently trying to become well again. It is my constant Hope that I can return to Norway. But I cannot give it much Thought until I have regained my Health.

It seems he did not feel that becoming completely healthy again was a realistic hope; in any case he did not think that he would be able to work in the same way as before:

However things go with me and my Family, it is certain that my Name shall survive. It is obvious that I shall, given that a good Part of my Energy is destroyed, never achieve in my Lifetime the Fame that I so hungered after, and which I certainly would have achieved; but my Life's Work shall stand the Test of Time, and with the Passage of the Years, become more and more appreciated, no Doubt about it. Nature has given me a rich Mind.

He ended the letter by thanking Motzfeldt and his wife for all their thoughtfulness and Love:

I am writing only about myself. Sickness makes for Egotism. I hear with deep Sympathy that You and Yours too have been going through heavy Times. May a merciful God watch over us All. His Hand has been lying very heavily upon me. But perhaps better Times shall come again.

Picard in Paris had spoken about Lie's transformation group theory, and laid great stress on its significance. And during the summer of 1890, Lie thanked Picard for the inspiration he derived from knowing that Picard interested himself in his [Lie's] theories, and that they were studied and used in Paris. But it was also now that Lie referred to his "brilliant but nevertheless untalented friend A. Mayer", who had so often corrected his small mistakes, and about his "conscientious but painstakingly slow assistant Engel", who presumably had caused the late publication of Lie's theories.

Lie gradually resumed his work, and he seems for a time to have had trouble starting to write something new. When the winter semester began in October 1890, Lie gave his lectures on "Theorie der Raumcurven und der krummen Oberflächen", but did not have any seminars. Barely a year later he wrote, in a letter to Elling Holst:

Things are going badly for me as before. Certainly I can sleep and I have no bodily Ailments, but my Mind is destroyed. Despondency and dark Despair invade my Soul. The Lectures cost me no Effort, but any higher Endeavour is beyond me. [. . .]

Keep well. Maintain your Kindness toward me. I believe that as unfortunate as my Life has turned out, there has still been in me a certain ideal Endeavour. If I end up with chronic Egotism, then You should know that for my entire Life I have sacrificed Everything in order to work for Science. Now that my Ability is lost, I regard myself not simply as a Wreck, but as Dead.

Do not answer my melancholy Remarks.

Later on Lie stopped writing about his illness and the reasons for his despondency. After his breakdown, his hospitalisation and a period of convalescence, it seems that Lie did not want to dwell any longer on what had happened to his mind and body. Others, however, would maintain that they now noticed great changes in Lie's manner.

Fame

During the summer semester of 1891, Lie was apparently in full activity again. And when the semester was over in August, he left with the whole family for Norway. Holst wrote later that Lie's period of illness lasted until 1892.

Lie produced and published his works. His mathematical creative power had not died, he had not become "some wreck" – his fame increased, and gradually others took Lie's ideas into new fields.

In the spring of 1892, Lie summed up his scientific work and output. The circumstances were that a biographical dictionary was being published in Norway, and the third volume, which included the letter L, was published in May 1892. Since the rather famous dictionary's editor and publisher, J. B. Halvorsen, had himself written the entry on Sophus Lie, he had, prior to printing, sent the text to Lie for his inspection and a certain degree of endorsement. Lie responded on the spot, in a letter to Elling Holst: "From my Point of View his [Halvorsen's] Biography of me is unbounded Impudence. Were Halvorsen to publish this Biography as it stands, then I would have to count him my Enemy." (In their day both Halvorsen and Lie had been members of "The Green Room".) And to get Holst involved and contribute to a reworking of the text, Lie sent a series of pieces of information that Holst "with careful Utilisation" could make use of. When this lexicon was eventually published, there was a footnote to the article devoted to Lie, which read: "The following Account of Professor Lie's – for most of his Countryman both unknown and incomprehensible – scientific Discoveries and Theories, the Publisher was offered the very kind Advice of Mr. Senior Teacher *Dr. Phil.* Elling Holst."

Before Lie summed up his scientific work he stressed to Holst: "I urge You not to inflate anything in the Biography in vivid Terms. But my 20 Years of mathematical Work deserve serious Characterisation. [...] My Biography should still not suffer because Norwegian Newspapers have consequently ignored me." Nor should his "scientific Biography" be reduced to his "Adventure at Fontainebleau", and Cremona's laudatory description ought to be dropped – because, as he wrote, "Cremona is a brilliant Man but he says many Things that cannot be sustained, and anyway, he abandoned Mathematics in 1872, which until then he carried out with Excellence, in order to take up practical Tasks." Lie also stressed that it was "incorrect" to say that he joined Clebsch's school: "I was a Friend of Clebsch, but on the whole I engaged in polemics with most of those in his School. Klein, on the other hand, who I consider a Student of Plücker, I am very close to. We studied together in Berlin and Paris and influenced one another with mutual strength. Klein is now Germany's most famous Docent in Mathematics."

Lie commented that basically the Norwegian press, which in general was "so willing to communicate laudatory Commentaries about the Work of Norwegian Men of Science," had ignored him. Lie wrote: "It could certainly be that with my Character I have made myself Enemies," but he thought mainly that the reason was the conservative newspapers *Morgenbladet* and *Aftenposten* still "found it appropriate to give the Public the Impression that a Parliamentary Professor was an Insignificance."

Lie wanted to give an impression of how his scientific work had been "appreciated in competent Quarters", and he gave the following outline:

> My scientific Work falls into three major Categories
> 1) Geometry
> 2) Differential Equations
> 3) The Theory of Transformation Groups
>
> And here there can also be included
>
> 4) The Application of the Transformation Group Theory to other Disciplines.

And he added:

> Geometry and Differential Equations are old Disciplines to which I, as Others, have made my Contribution.
>
> On the other hand, the Theory of continuous Tranformation Groups is now commonly recognized as a new and important *fundamental Discipline* – founded by me – which is of greatest Significance for many other mathematical Fields.

His first "significant geometric Work" was his doctoral dissertation, and then followed a number of works based upon it. What was called *Sphere Geometry*, and which created much interest in mathematical circles, had later been developed by others – Lie named Klein, Darboux and Reye. Sphere geometry described a transformation where the relationship between the two most important elements in space – straight lines and spheres – was established. Lie reminded Holst that Darboux had described this work of his as "one of the most beautiful demonstrations in modern geometry" ("une des plus belles decouvertes de la géometrie moderne").

From the stream of his many geometric works, Lie then pointed to the investigations into minimal surfaces that in their turn had provoked and influenced many newer works on such surfaces. But he had also "contributed to the newer Advances" in two other geometric fields – these were his work on geodesic curves and the studies of surfaces with constant curvature.

In the second field (point two): differential equations – Lie maintained that it was above all the studies of partial differential equations of the first order, together with the works on contact transformations, which had awakened enthusiasm and been an inspiration for others. This was material that had already gone "to some extent into nearly all the Textbooks on Partial Difl. [differential equations] of the first Order". Of the better known textbooks, he mentioned that of Mansion, and proudly pointed to the fact that "the newest and greatest and best Work" in this field, namely Goursat's textbook, used 134 of the total 348 pages to explicate theories that he had "founded". He also mentioned that the English mathematician Forsyth had, in his book on what was called the Pfaff problem, used forty pages in the explanation of Lie's treatment of the problem:

My Life's Major Achievements are still my Investigations into continuous Transformation Groups, which have absorbed nearly all my Labour Power. I began the same in 1873, and particularly in the Years 1876–1879, gave detailed Information that was published almost immediately in Norway. These Investigations received little Attention. They were entirely new, and moreover [they] were not easily accessible, as Archivet [the Norwegian periodical *Archiv for Mathematik og Naturvidenskab*] was not widely distributed and I only sent out a few Offprints.

In 1880 I finally published a larger Work on this Subject in Math. Ann. [*Mathematische Annalen*] and now Attention began little by little to grow. Picard, a famous Mathematician and Member of the French Academy, in 1883 made a remarkable Review of my Theories and strongly directed Attention to the same. In 1884 Prof Klein and Mayer occasioned Dr. Engel (later Professor at Leipzig) to travel to Christiania to study my Theories and support me in the Publishing of the same. In 1886 I was called to Leipzig myself, to work on my Theories.

Later, Attention to these Theories' Significance, grew to a high Degree.

In his summary of what had happened in the past eight years, Lie stressed that in addition to his own work in the field, there had come out, on average, five new works annually "that solely deal with my new Discipline": "There is every Reason to consider that this Activity will undergo a stark Increase." Of the approximately forty works that Lie considered as treatments of his "new Displine", most were written by German mathematicians, but Lie was eager that it be known that in France, where eyes had opened even wider to the importance of his theories. And to show this, he mentioned that in the course of the past four years, four students had been sent to him in Leipzig from the famous learning facility, l'École Normale Supérieure, in order to study his theories. Of all the real places of learning in Paris – the Sorbonne University, l'École Polytechnique and l'École Normale Supérieure, it was the last, often referred to simply as l'École Normale, where the most theoretically advanced teaching was given. And to stress what this really implied, Lie explained that annually l'École Normale Supérieure took in about twenty young Frenchmen from among the brightest who had sat the examination for university entrance – that is, who had graduated from the French lycées – and "these brilliant young Men" got free tuition and room and board for three years before they were employed in the better positions in the senior classes at lycées and universities. But for the most brilliant at l'École Normale Supérieure it was possible to obtain certain stipend monies for a year's study abroad, and that it was some of these young men who had come to Leipzig. Lie concluded: "This is an extraordinary Acknowledgement of the Importance of my Theories."

When it came to the fourth field of his scientific activity – "The Application of the Transformation Group Theories to other Disciplines" – Lie wrote that he himself, in a series of articles, had made important applications of his "Group Theory to Geometry and Differential Equations", and he pointed to all the applications that the famous French mathematicians, Picard and Poincaré, had made

– in addition there were the works of "a young extraordinarily promising French Mathematician", *Vessiot*.

Vessiot was one of those who had come from l'École Normale Supérieure in 1888 to study under Lie in Leipzig, and precisely this spring of 1892, when Lie was writing this summary for Halvorsen's biographical dictionary, Vessiot was in Paris in the process of publishing a large and important treatise in which Lie's theories were applied to the integration of differential equations. Some months later, in June 1892, Lie wrote to Mittag-Leffler: "I hope that this Work [Vessiot's treatise] will make clear to all Mathematicians that Group Theory in my Sense is important for Function Theory."

To Holst, who would repair the dictionary entry, Lie could also most certainly show a series of most laudatory descriptions of his "Theory's Significance", and by way of conclusion, he pointed to three such descriptions: there were Darboux's commendatory words in the Proceedings of the French Academy, and there was Study's review of the first volume of *Theorie der Transformationsgruppen*, and as well, what Picard had written after he had seen the second volume of *Theorie der Transformationsgruppen*. The printing of Volume Three of this enormous work began that spring of 1892. The processes of printing and proofreading extended over a period of one and a half years, and not until the end of September 1893, was the third volume of Lie's major work finally published. In total it was a little over 2,000 pages. The last volume was dedicated to l'École Normale Supérieure and the scholars Darboux, Picard and Tannery – Lie expressed his particular happiness that "France's most brilliant young mathematicians in competition with a number of German mathematicians, study my investigations of *continuous groups, geometry*, and *differential equations* and utilise them with spectacular results."

It was also in this third and last volume that Lie wrote the foreword with his strong criticism directed to German colleagues, and above all, toward Klein. This foreword would have earth-shaking consequences for Lie personally and for his mathematical reputation, in any case, for a time in Germany.

Eduard Study's review was written after the publication of the first volume in 1888, and was published in *Zeitschrift für Mathematik und Physik*, the journal put out by Oskar Schlömilch. Private Docent Study had read the second proof manuscript of Lie's book, and in his review he pointed out that the work under consideration gave a summary account of the comprehensive theories which, through the years, Lie had developed and gradually allowed to be published in *Matematische Annalen* and in *Archiv for Mathematik og Naturvidenskab* in Christiania. There were valuable contents to these works but they remained nonetheless unknown among the scientific public – a condition, according to Study, that was due in part to the fact that the Norwegian journal was little known, and partly due to the complicated form in which the works had been presented. But when one read the work in question, Study wrote, one could only cherish the good fortune that had allowed Lie, in peace and quiet, to develop his thinking, that the material had matured such that Lie had now been able to "formulate them in a harmonious manner, and think them through independently, far away from the breathless competition of contemporary life." Study continued: "This is not some textbook in which several authors collaborate to spread their theories into wider circles. The work here has to do with the creative work of one individual man, an *original work*, that from the first page to the last deals with completely new things." Study was almost prophetic in his choice of words: "We do not think that we are saying too much when we claim that there are few fields in the science of mathematics that will not be enriched by the fundamental conceptions of this new discipline. But the fields most closely related, differential equations and geometry, will bring a profusion of new discoveries and approaches to new tasks." He ended by writing: "Mr. Lie deserves to find among us what he was looking for when he left his homeland. May he succeed in gathering together a circle of students who will follow him along these newly-trod and promising paths, and who, in a happy collaboration with him, manage to develop the science further in ever-higher and more attractive forms."

The twenty-eight-year-old Study, who accordingly in two years as a private docent, had worked together with Lie in Leipzig, here gave a precise expression of Lie's own hopes and expectations. Parallel with the publication of his works, Lie wanted to found a school at Leipzig, a school for studying and disseminating the theory of continuous transformation groups. Through teaching and above all, with the publication of *Theorie der Transformationsgruppen*, Lie succeeded in creating a mathematical power centre.

Right from the first, Engel was central to the milieu around Lie in Leipzig, who indeed was the collaborator in all three volumes of *Theorie der Transformationsgruppen*. Another important person in the milieu was certainly Lie's old friend Mayer, even if he no longer meant much to Lie in a professional sense. To the students, Mayer was an inspiring lecturer in classic themes concerning the calculus of variation – and for Lie and his family, it seems that Mayer and his well-appointed home were an important gateway to the city's social life.

In addition to Engel and Mayer, of course Private Docent Study too was at the University of Leipzig when Lie came in 1886. Two years earlier, influenced by Hermann Grassmann's *Ausdehnungslehre*, which he considered an adequate foundation for geometry, Study had taken his doctorate in Munich. The following year in Leipzig, under Klein's leadership, he presented his qualifying dissertation, and was by then already familiar with Lie's works.

Friedrich Schur too, who was an assistant and private docent at the Mathematical Seminar, was one of the mathematicians with whom Lie regularly worked, right from his first arrival in Leipzig. Schur conducted exercises in connection to Lie's lecturing, and the two of them – Lie and Schur – as Lie reported to his friend Klein, understood one another "extremely well."

The friendly correspondence with Klein in Göttingen continued during Lie's first years at Leipzig. They exchanged photographs of their children, and Lie reported on his teaching, on the work with students, and the relations between teachers, and he often asked Klein for advice whenever practical and pedagogical problems cropped up. Among other things, Lie got Klein's lecture notes about Cartesian geometry, which he felt were better than the well-known textbooks, and with the publication of his first book, Lie asked if it was necessary to give copies to the colleagues Scheibner, Neumann and von der Mühl – but to Minister of Culture von Gerber, he should surely send a copy, should he not? Klein, who had just left for Göttingen, knew the people and relationships in Leipzig very well – and this was worth a great deal to Lie as he worked to build upon the teaching resources Klein had established.

After only one semester in Leipzig, Lie was able to tell Klein that he had got Study, during the coming semester, to teach invariant theory, using as his starting point Lie's investigations, and added that he also had "plans" for Schur. With great satisfaction, Lie confided to Klein: "As you see, I am beginning carefully to try to group some people around me."

Among the students there was as well the twenty-year-old Georg Scheffers, who quickly began to write down Lie's lectures. Scheffers would also choose themes for his academic examinations in line with Lie's advice. The stages along the route to a German academic career progressed from the state examination or the senior teacher examination, which qualified one for positions in the higher schools and colleges – and then this could be followed by the doctorate, or what was called the dissertation. Even this level of promotion in the German system was given marks, and then, following the doctorate, one reached the final academic test, what was known as "the habilitation", which qualified one to teach at the university.

In the case of the first semester at Leipzig Lie met for the first and only time with another German mathematician who would come to mean a great deal to him and the development of his theories. In the final days of the month of July 1886, Wilhelm Killing came to Leipzig, on his way from his position as professor at Braunsberg in East Prussia to a substitute position in Heidelberg. Killing was five years younger than Lie, had studied mathematics and physics in Berlin – among others with Kummer and Weierstrass – had published articles, in *Crelle's Journal*, on non-Euclidian geometry, and at the same time he had taught at various schools in Berlin and Brilon in Westfalia. In 1881 he had become professor at the remote city of Braunsberg. The esteemed Weierstrass had taught at Braunsberg about thirty years earlier, but the university was poorly equipped and the library lacked so many items that it was difficult for Killing to keep abreast of mathematical developments. Killing would nevertheless remain at his post in Braunsberg for ten years. He taught differential- and integral calculus, basic chemistry and astronomy, and for a time was also rector of the university. He also worked on his own mathematical ideas about the fundamental concepts of geometry. The point of departure for Killing seems above all to have been von Helmholtz's writing of 1868, entitled *Über die Tatsachen, die der Geometrie zu Grunde liegen* (the same work whose shortcomings would be pointed out, and suggestions for improvement made by Lie, at the meeting of natural sciences in Berlin).

Together with his wife Anna, who was the daughter of the composer and music researcher, Franz Commer, Killing was also engaged in social activities, as well as church and religious questions. He was concerned about the new scholastic direction that sought out an identity common to sacred revelations and natural laws; in 1886 he converted to the Franciscan Order. In his mathematical isolation in Braunberg – perhaps not completely unlike Lie in Christiania during the 1870s – Killing developed his ideas on his own, and advanced a long way in the same direction as Lie. Killing began to operate with "systems of infinitely small movements", and when in 1884 he sent his work "extension of the concept of space" to Klein, he quickly got to know about Lie's research and Engel's participation in it. It was later claimed that Killing had tried to contact Lie without success, and that then, acting on advice from Klein, he contacted Engel. In any case, six months after Engel had returned to Leipzig after his nine months of intense work with Lie in Christiania (in November 1885), Killing and Engel began an exchange of letters that would have great significance to all concerned. When in 1886 Killing came to Leipzig, he met Engel for the first time – the two of them met twice more, but not until after Lie's death.

The meeting between Killing, Engel and Lie in Leipzig during the summer of 1886 seems at first glance to have been a joy and an inspiration for them all. Study and Schur were involved in this meeting as well. Lie reported to Klein: "He [Killing] had some really beautiful and important ideas," but he added: "In other fields however he did not give a very solid impression." During the course of his studies Killing had come to view Lie's theories on transformation groups as decisive for research on the foundations of geometry, and at the urging of Engel, Killing had changed his terminology in accord with what Lie had introduced. Later Killing was to maintain that through this discussion with Lie in Leipzig he had received Lie's consent to use this terminology. The work that Killing published in 1886 also had the title, *Zur Theorie der Lie'schen Transformationsgruppen*. The method that was suggested here was to determine all "systems of infinitely small movements in a space form", that is, it was to study the system of transformations which with composition formed a group – much the same as Lie had developed more than ten years earlier, without any acknowledgement. (The task was to find a way of expressing the structural constants that conserved identity, something that led to the study of invariants, and which in its turn, was called Lie Algebra. By using the concepts of the day, the task was to classify simple Lie Algebras, in order to be able to determine solubility – the Killing form of a simple Lie Algebra is non-degenerate if and only if the Lie Algebra is simple.)

A couple of years later, for the first time, Killing's works would become the source of personal anxiety and strife.

"As you see, I am beginning to try to group some people around me," Lie had written to Klein after one semester's activity in Leipzig.

Lie's "plan" with Schur seems to have been to get him to study the mathematical investigations of Riemann, von Helmholtz and Lipschitz in order to clarify the lines of connection to the works of Klein and Lie. For his part, Schur at first does not seem to have had much desire to carry out such a task – and when the thirty-year-old Schur also began to have thoughts about marriage and plans for weddings, there was little time or energy left over for satisfying Lie's wishes. Schur always gave good lectures, but certainly seems (according to Lie) to have felt at times that it was demeaning that his work, as an assistant, was so scrupulously linked to Lie's work. In March 1888, Schur resigned his position in Leipzig to take up a position in Dorpat. Schur would consequently work more with Lie's group theory, even though from time to time Lie reacted strongly both to Schur's work methods and his results. In a letter to Klein (February 1892), Lie wrote that Schur presented the foundation of transformation theory in a much too simplified manner in the work he carried out, and Lie later gave expression to what he thought of Schur, writing that Schur, in citing his own proof for Lie's propositions, ought also to have added that the propositions in fact belonged to him, Lie.

Study would also remain in Leipzig for only a couple of years after Lie's arrival. On several occasions Lie told Klein that he had worked well with Study, that he personally liked Study very much, that his work was always very thorough. Lie added: "He [Study] has perhaps far too much of the becoming modesty that Engel lacks. Study has an amiable nature. He is also gifted. He needs only to find the adequate energy to work." This fear that perhaps Study did not devote himself to mathematical work to a sufficient degree, probably had its basis in the fact that Lie knew of Study's strong interest in biology, an engagement that through the years would result in, among other things, an impressive collection of butterflies. Study taught invariant theory to the best of Lie's students, and it was Lie's hope that Study would master the extensive work that would clearly couple transformation theory to invariant theory, and so demonstrate how the latter could draw benefit from transformation theory.

"The time is ripe for transformation groups," Lie wrote to Klein, but then added: "Transformation theory is a progressive discipline." Study, moreover, was well-read in terms of mathematical literature, and Lie seems to have trusted Study's assessments. In any case, in the autumn of 1887, he mentioned to Klein that Study felt that part of the contents of Poincaré's last treatise in *Acta Mathematica* had already been proven earlier by Max Noether in Erlangen.

Despite the sympathy between Lie and Study – and despite as well the good contact between their wives and children, Study consequently left Leipzig in 1888 with his family, to continue as a private docent in Marburg. Career possibilities were said to be better in the Prussian education system than in that of Saxony. Following a year as guest lecturer in the USA (1893–94), Study would go on to work as professor at several German universities (Göttingen, Greifswald, Bonn), and he advanced Lie's theories in new fields. However, before Study left Leipzig, he also wrote his very laudatory and important review of the first volume of *Theorie der Transformationsgruppen*.

1888 was a very eventful and significant year for Lie. In point form, here is what happened:

The first volume of his major work, *Theorie der Transformationsgruppen* was published.

Private Docent Study and assistant Schur left Leipzig, but they would both go on to do further work on Lie's theories.

For the summer semester "two promising young men" came from l'École Normale Supérieure in Paris to study with Lie.

Lie visited Norway for the first time since moving to Leipzig, and when he returned to Leipzig, his anxiety began about his wife Anna's life and health, and began to have disturbing effects upon his sleep.

In the course of this year he also began seriously to feel that there were possibly powers and persons around him who did not perhaps wish him well. Lie began to feel "badly treated". These were also anxious times for the country. Kaiser Wilhelm I was gone, and his succcessor, Friedrich III died after only ninety-nine days in office, and Kaiser Wilhelm II took over power, and two years later it would be the Kaiser's divergence with Chancellor Bismarck that would lead to Bismarck withdrawing after having dominated German politics for nineteen years.

It was Killing's works and Killing's correspondence with Engel that in the first instance got Lie feeling that he had been "badly treated". At the beginning of 1888, the first part of Killing's large treatise, *Die Zusammensetzung der stetigen endlichen Transformationsgruppen* was published in *Mathematische Annalen*. Killing led off by acknowledging that he was fully and wholly using Lie's terminology, and this, partly through verbal discussions had also been cleared with Lie – and for the readers who were unfamiliar with Lie's works, Killing referred them to Engel's articles. It was these unclear references and relations of priority that disturbed Lie. Lie wrote to Klein that he was "dissatisfied to an unqualified degree" with the way that Engel had, over recent years, in a manner "completely unconstrained", handed out ideas that belonged to him, Lie, and only to him. That Engel had neglected to be "expressly considerate" in this priority, showed that in this instance, Killing referred to Engel's works instead of to Lie's. According to Lie, Killing's work included a number of gross errors, and, Lie stated firmly to Klein: "Mr. Killing's work in Math. Ann. is a gross outrage against me, and I hold Engel responsible. He has certainly also worked on the proof corrections." Lie vented his anger on Engel, even though Klein, who was the editor of *Mathematische Annalen* certainly also was responsible for the printing of Killing's work. "Mr. Engel seems to have come to Norway with the purpose of appropriating all my ideas," Lie pointed out to Klein, and explained that he had spoken seriously to Engel about it, who then admitted that he had behaved "incorrectly". However, Lie did not want to let the matter drop: "Over a period of time I have been treated badly by several mathematicians, but such a remarkable misuse of my confidentiality I had never thought possible. [...] One cannot say about me that I have appropriated other people's ideas, or that I have not acknowledged the merits of others. As far as Engel is concerned, indeed before his arrival in Christiania I had received the distinct impression of a not-uncommon obliging *madness* in his family."

Later in this letter to Klein Lie tried in his characteristic way to dissipate some of the discord. He wrote: "After the passage of some days, I am writing further. I want to stress that my words about Engel should not be taken literally or to the letter. When I become outraged, I express myself much too strongly. But he [Engel] is incomprehensible to me in many ways, even though I also readily admit his good sides. In Christiania I was able to isolate myself completely; it had its bad sides, but for my character, some good ones too." And in the next paragraph of the letter he went on to tell about the great event of two bright Frenchmen having come to Leipzig to study his theories.

The two French students who came to Leipzig for the summer semester of 1888 were twenty-year-old Arthur Tresse and the three-year-older Ernest Vessiot. They had come on the advice and recommendation of their teachers at l'École Normale Supérieure, above all, Darboux, Picard and Tannery. After finishing his examinations, Vessiot had taught for a year in Lyons, while for his part, Tresse was still a student at l'École. Lie reported on his French students to Klein: "They are making good progress and I am expecting something from them, even if they have only been here for one semester. I am working very much with both of them." Vessiot and Tresse would also become important "messengers" for Lie's theories.

Lie's student, Scheffers, had finished his doctoral thesis in the spring of 1889, and to Mittag-Leffler, Lie wrote:

> One of my best Pupils (Scheffers) is sending You a Work, which he has prepared before my Eyes, and who has taken his Doctorate here in Leipzig with a Dissertation that got the best Mark (1). I believe that You can safely accept the same for Acta. It is suitable for making Acta's Readers familiar with my Investigations into Contact Transformations and Groups of such Transformations. S[cheffers] possesses an unusually evident Talent and his Calculations are worked through with great Precision, and bring new Results.
>
> On the Chance that You will accept S's Work, it was my own Thought to send along a short Note of a Couple of Pages, which I hope will be accepted in the same Number.

Lie indicated to Mittag-Leffler that it would be best for Scheffers if his essay was printed as early as some time in the autumn (1889). The paper came to be published in *Acta* a year later, but without Lie's note. On this occasion it had been his illness and his stay at Ilten that hindered him from delivering his contribution to *Acta Mathematica*.

Scheffers "habilitation" dissertation had to do with transformation groups, and it was ready the following year, while Lie was in hospital. By virtue of completing the habilitation thesis, Scheffers was now qualified to teach at German universities, and at this time, while Lie was at Ilten, Klein in Göttingen tried to get Scheffers to go there. But Scheffers, who was considered to have great pedagogical gifts, remained in Leipzig, and he gladly embraced the new work and tasks that Lie gave him.

Around Christmas 1890, the printing began, of Scheffers' written version of Lie's lectures, and six months later the book, 568 pages in length, was published by Teubner, with the title, *Vorlesungen über Differentialgleichungen mit bekanten infinitesmalen Transformationen*, prepared and worked over by Georg Scheffers. The following spring the printing began of a second collection of Lie's lectures. On this occasion the production and printing time lasted for a year and a half, and the 810-page book came out in September 1893, entitled, *Vorlesungen über continuirliche Gruppen mit geometrischen und andern Anwendungen*, prepared and edited by Georg Scheffers.

The contact with l'École Normale Supérieure in Paris became more and more important to Lie. Vessiot and Tresse advanced Lie's studies in different ways. Vessiot worked with the utilisation of continuous groups for solving differential equations and integration problems, and also published a study that Lie hoped would make it clear to everyone that group theory in "his sense" was important for the theory of functions. In these investigations what played an important role was the study of continuous groups that were called solvable or capable of integration. The solution was dependent upon certain aspects (subgroups) of these. (In the terminology of today, the task was to find, by means of structural constants, the Killing form of a simple Lie algebra.)

Vessiot proceeded the furthest with what, for a long time, had been some of the most important problems posed by Lie, namely to work out a respectable Galois Theory for differential equations. Another of Lie's French pupils, Jules Drach, was also working in this direction.

Tresse corresponded eagerly with Lie – four letters still exist from 1892 and six from the following year. Tresse worked with invariants, and during the autumn of 1892 he informed Lie that he could demonstrate that even for infinite groups, the number of invariants was finite – and for the first time it seems, the designation "Lie Groups" had been launched. Why not call them "Lie Groups"? asked Tresse – these groups in which transformations were defined by partial differential equations. And "Groupes de Lie" and "groupe Lie" were designations that Tresse thereafter used in his letters in the autumn of 1893.

One of Tresse's classmates at l'École Normale Supérieure was Élie Cartan. Cartan took his graduating examination in 1891 as the best student, before spending the following year in the military. Then, while Cartan was fulfilling the duties of a sergeant, Tresse was back at Lie's in Leipzig. Upon returning to Paris in the spring of 1892 following his second stay with Lie in Leipzig, he told Élie Cartan about Killing's work – *Die Zusammensetzung der stetigen endlichen Transformationsgruppen* – which had been published in three parts in *Mathematische Annalen* between 1888 and 1890. (It had been the first part of this work that had caused Lie to react so powerfully to the Killing-Engel correspondence.) In these works, Killing had achieved important results concerning continuous groups and their subgroups.

Tresse told Élie Cartan that following the publication of Killing's works, incorrect assertions had been found regarding certain subgroups (the so-called null groups, which in the terminology of today are said to correspond to degenerate Killing forms), and that Engel had taken on the task of correcting Killing on this point. But Tresse encouraged Élie Cartan to look for other inaccuracies in Killing's works, and this indeed became something that Cartan worked on and made part of his doctoral work. A doctorate in France stood much higher than that of Germany. According to Lie, the accepted dissertations in France had "an independence, indeed even significant scientific Value". Élie Cartan's study from 1894 – *Sur la structure des groupes finis et continues* – was 150 pages in length and became a real milestone. But already the year before, Élie Cartan, at Lie's recommendation, had already published a twenty-five-page article "on simple transformation groups" in *Leipziger Berichte*. It was this article that really marked Élie Cartan's debut as a writer.

In September 1893 the third volume of Lie's major work, *Theorie der Transformationsgruppen* came out: 830 pages, counting the indexes and reference lists, plus a twenty-page foreword. And it was in this foreword that Lie both unexpectedly and crassly distanced himself from his former best friend, Klein. The relationship between them had certainly cooled over recent years, although they continued to exchange letters the same way, although not as frequently as earlier. But it was above all professional divergencies that were central to the fact that Lie now broke off the relationship. Following Lie's publication of the first volume of *Theorie der Transformationsgruppen*, Klein judged that there was sufficient interest to have republished his Erlangen Programme. But before Klein's text from 1872 was printed anew, Klein had contacted Lie to find out how the working relationship and exchange of ideas between them twenty years earlier should be presented. Lie had made violent objections to the way in which Klein had planned to portray the ideas and the work. But Klein's Erlangen Programme was printed, and it came out in four different journals, in German, Italian, English and French – without taking into account Lie's commentary on his assistance in formulating this twenty-year-old programme. More and more in mathematical circles, Klein's Erlangen Programme was spoken of as central to the paradigm shift in geometry that occurred in the previous generation. A large part of the third volume of Lie's great work on transformation groups was devoted to a deepening discussion on the hypotheses or axioms that ought to be set down as fundamental to a geometry, that – whether or not it accepted Euclid's postulates – satisfactorily clarified classical geometry as well as the non-Euclidean geometry of Gauss, Lobachevsky, Bolyai, and Riemann.

The information that spread regarding the relationship between Klein's and Lie's respective work, was, according to Lie, both wrong and misleading. Lie considered he had been side-lined but was eager to "set things right", and grasped the first and best opportunity. In front of the professional substance of his work he placed his twenty-page foreword. The power-charge that liquidated their friendship and sent shock-waves through the mathematical milieu was short, if not sweet:

I am no pupil of Klein's. Nor is the reverse the case, even though perhaps this would have come nearer to the truth. [Ich bin kein Schüler von Klein, das Umgekehrte ist auch nicht der Fall, wenn es auch villeicht der Wahrheit näher käme.]

Lie rushed to add that he was not writing this to criticise "Klein's original production within the theory of algebraic equations and function theory:

I highly value Klein's talent and would never forget the participatory interest with which he has always followed my scientific endeavours, but I do not feel that he has satisfactorily discerned the difference between induction and proof, between a concept and its application.

Lie's assertion was that Klein did not clearly distinguish between the type of groups which were presented in the Erlangen Programme – for example, Cremonian transformations and the group of rotations, which in Lie's terminology were neither continuous nor discontinuous – and the groups Lie had later defined with the help of differential equations:

One finds almost no sign of the important concept of differential invariants in Klein's programme. This concept, which first of all a common invariant theory could be build upon, is something Klein has no part of, and he has learned from me that every group that is defined by means of differential equations, determines differential invariants, which can be found through the integration of integrable systems.

According to Lie, Klein's advances in work with non-Euclidean geometry first and foremost consisted of the fact that he had popularised his predecessors. But, Lie continued, in their investigations of geometry's foundation, Klein, as well as von Helmhotz, de Tilly, Lindemann and Killing, all committed gross errors, and this could largely be put down to their lack of knowledge of group theory. Toward the end of this foreword, perhaps as a reaction to having possibly gone a little too far in criticism of his German colleagues, Lie came out with profuse praise of Engel:

Since 1884, with the working out of my ideas, both with the aim of simplicity, and systematic presentation, I have had the good use of the tireless support of Professor *Friedrich Engel*, my excellent colleague at the University in *Leipzig*. [...] Professor Engel holds a special position in my eyes. At the initiative of F. Klein and A. Mayer, he journeyed to Christiania in 1884 to assist me in the working out of an integrated presentation of my theories. He has thrown himself into this task, the size of which at the time we did not comprehend, with a perserverance and an aptitude that has no equal. During this time he has also developed a number of independent ideas, but in a highly unselfish manner, he has renounced presenting them in a detailed and comprehensive manner, and has contented himself with short reports in Mathematische Annalen and particularly in Leipziger Berichte, because he had devoted the talent and the time that his lecturing allows, to work on the very detailed presentation of my theories, so fully and so systematically, but above all, in as *exact* a manner as possible. For these unselfish activities, which have stretched over a period of nine years, I, and in my opinion the whole of the world of science, owe him the greatest possible gratitude.

Engel was also thanked for the help he had given to Lie's teaching at the University of Leipzig. But the book that Lie then mentioned as a further collaboration with Engel – a work on differential invariants and infinite continuous groups, and which would also include the application of group theory to the integration of differential equations – this book was never written, and Engel never seems to have commented on the plan.

But the collaboration with Scheffers continued, and they now began to work up all of Lie's geometric lines of research with a thought to a volume of collected writings. The printing of this extensive material began in December of 1894 and was published a year and a half later – 694 pages, including the index and bibliographical lists – by Teubner, with the title *Geometrie der Berührungstransformationen*. However, on the title page it was announced that the book, written by Sophus Lie and Georg Scheffers – that what was here being brought to the fore was, above all, a more expanded presentation of Lie's works from the period between 1869-72. There would, however, never be a second volume. Scheffers received a professorship in 1896 at the technical college at Darmstadt, and moved there.

In a five-page foreword to *Geometrie der Berührungstransformationen* that Lie wrote in Leipzig in February 1896, he tried to link together the various mathematical disciplines: geometry, analysis, and applied mathematics. He traced historical lines from Greek geometry, via Renaissance algebra, to Descartes' introduction of coordinates, which created analytic geometry, and in turn led to the concept of functions, and through geometric reflections, to integration and derivation, to infinitesimal calculations and differential equations. But more important, and particularly to modern mathematics, was the concept of *transformation*. Through the study of curves and curvature by geometric intuition, transformation gave rise to new important concepts. Lie located his own work clearly as an extension of the work of Abel and Galois, even though these two had scarcely engaged themselves with geometry, it was Abel's ideas which had enriched geometry with a wealth of new points of view, so too, Galois' ideas now had fruitful effects not only on analysis and geometry, but also on mathematical physics. It was important to prevent mathematicians in their own specialized fields – which mathematics had involuntarily given rise to during the nineteenth century – from taking a dismissive or indifferent attitude toward the fields of work of other mathematicians. Lie wrote that he always took care in his own scientific endeavours to stress, as his starting assumption, that analysis and geometry should mutually support and enrich one another, as they had done in earlier times, and he pointed out that in his inaugural lecture ten years earlier, he had also spoken about this. Lie lamented that the major weight in mathematical research in Germany had, in recent times, been placed on analysis, and he added: "I believe in this that I am standing at Klein's side; he understood so well, following Riemann's example, how geometric intuition could stimulate analysis in a fruitful manner."

In addition to these works – three volumes together with Engel (altogether 2,008 pages) and three books together with Scheffers (altogether 2,072 pages) – Lie wrote almost forty studies during his years in Leipzig (altogether 900 octavo and 136 quarto pages). For the most part, these treatises were given out through the sciences association of Leipzig, that is, the publication called *Berichte über Verhandlungen der Kgl. Sächsischen Gesellschaft der Wissenschaften zu Leipzig*. Some of these published papers were expansions, corrections and elucidations of theories he earlier and in concise and suggestive forms had developed and published in the journal *Archivet* in Christiania – but much was also completely new, and often were initiated by the development of his theories by others.

In a sense the production was in three phases: the first were a bit like unpolished drafts published in the Norwegian journal, the second were polished adaptations published in *Mathematische Annalen*, and the third were the serious works undertaken with the assistance of Engel and Scheffers.

In 1892 Lie had been chosen a corresponding member of l'Académie des Sciences in Paris, the very highest acknowledgement that France could award to a man of the sciences. The following year, at the invitation of Darboux and Tannery, Lie visited the French capital. In the Cartan family – Élie Cartan's son is the prominent mathematician Henri Cartan – it was said that Lie's main reason for visiting Paris was to meet Élie Cartan. In any case, Lie had his fling at le Café de la Source on Boulevard Saint-Michel: there he sat during the April days of 1893, often together with the French mathematicians who admiringly followed his explanations as he habitually covered the white marble table top with mathematical signs. When Élie Cartan recalled this more than fifty years later, he said that Lie had stayed six months in Paris – while in reality it was scarcely three weeks. (Lie's concern seems to have been that before being able to find a complete form for treating differential equations with group theory, it was first necessary to specify "the structure" in all individual, finite dimensional transformation groups, and this was before the significance of "structure" for finite dimensional groups had been recognised.)

In 1895 Lie was again honourably invited to Paris, this time for the centennial celebrations of l'École Normale Supérieure. Both this institution and l'École Polytechnique had been founded during the regime of Napoleon, while the Sorbonne could trace its history all the way back to 1257. At the celebratory gathering for l'École Normale Supérieure in April 1895, Lie had been asked to give a speech about the school's most famous pupil, Évariste Galois. Even though Picard was in the process of writing a biography of Galois, it was Sophus Lie who had been invited to Paris to speak about the great Frenchman. This was certainly an acknowledgement that Lie was the man who to the strongest degree had directed Galois' ideas into new and fruitful fields. In his lecture about Galois' influence on the development of mathematics, Lie also had a rich opportunity to draw lines of correspondence to his great countryman, Abel, the "unattainable Ideal" of his youth. In a letter to Picard a year later, Lie elaborated:

> Looked at from the outside, Galois' life was as unfortunate as it is possible to be.
>
> We can only hope that the consciousness of the genius of his powers gave him a sort of compensation. And it was lucky both for him and for science, that he was clear about his abilities, and used them in a field where the results threw light in all directions. His genius also included the instinct which is a necessary component of a ground-breaking mathematical mind.

Lie's original task had indeed been precisely to find a Galois theory for differential equations, a theory that one, by seeing certain symmetrical properties of equations, could determine whether an integration existed. Inspired by the algebraic equation works of Abel and Galois, Lie had started in on differential equations. In the study of these equations, Lie had added a geometric dimension, which had been developed into a general integration procedure; this, in turn, was based on the invariance of a differential equation, or a system of differential equations with respect to a continuous symmetry group.

But Lie's theories concerning differential equations and continuous groups, such that they now appeared in his published writings, had much greater potential that simply to treat Galois groups with respect to differential equations. And it was others who now developed Lie's theories further. Picard proposed another route than Lie's in the development of the Galois theory. Study conceived of using Lie's theories in another field. (In two articles on linear algebra in 1888 and 1894, Ludvig Maurer carried the theories further, and several others could be mentioned.) But first and foremost it was Élie Cartan who carried Lie's theories into mathematical fields that would dominate development, also within the discipline of physics. In the course of the next twenty or thirty years it was Élie Cartan who became an important partner in discussion with Albert Einstein, and an inspiration to the mathematician Hermann Weyl.

Despite the fact that already as early as 1892 Tresse had begun to use the term "Lie group" – and the designation was quite certainly from then on used in mathematical circles – it was really not until about 1930 that Élie Cartan seriously attached the name "Lie groups" to those groups which for a long time had been central to the work of theoretical mathematics and physics. When published again in 1933, Élie Cartan's great treatise from 1894 on the algebraic aspects of Lie's theories, it was unnecessary to make changes. Two of the four volumes of Élie Cartan's collected works have as their overall title "Groupes de Lie". The concept of "Lie algebra" was used for the first time by Hermann Weyl in 1934. Lie groups and Lie algebras have become the modern superstructure of Lie's continuous and infinitesimal groups. Gradually as it has become charted that the laws of physics build on continuous symmetries, and Lie theory has become an ideological foundation, one that has to a great degree distinguished the development of modern mathematics and mathematical model-building. Of all the mathematicians in the world, only Riemann – according to the index in the *Encyclopedic Dictionary of Mathematics* – has more entries than Sophus Lie. Lie theory continues to play an on-going role in new fields in the natural sciences and high technology – a physical systems law of conservation is linked to Lie symmetry; the practical use of the methods of Lie theory give an overview to liquid and gas currents. Wave theory, meteorology, explosions, mechanics, electronics and optics are all fields in which Lie theory is central. It has also become evident that Lie theory is closely related to control theory, the science that studies stability in a system of, for example, items such as automobiles, boats, oil platforms, aircraft, robots and satellites. And this control theory also shows itself to be of utility in the study of biological and physical systems and processes.

Early on in modern physics, Lie theory played a central role. Hermann Weyl was among the first to link the concept of symmetry to the development of quantum mechanics. The symmetry concept has been developed to include as well infinite-dimensional, continuous groups, and such modern theories as gauge theory and string theory cannot be expressed without Lie theory. It is Lie theory that has made it possible to formulate those concepts about what the inside of atoms might look like. Mathematicians and physicists seem to agree that Lie theory represents a paradigm shift in modern physics, that Lie's ground-breaking work has had consequences that can be compared to Einstein's discovery of the Theory of Relativity.

When at the end of the 1920s Einstein was working to find a theory of unity in which both the phenomena of gravitation and those of electro-magnetism could be understood according to the same basic principles, Élie Cartan was one of his most important partners in discussion. In a three-month period, from December 1929 until February 1930, they exchanged twenty-six letters. Einstein was seeking mathematical expertise from Élie Cartan: not the least fascinating aspects were Cartan's theoretical interpretations of general presentations of space based on representations of Lie groups. In the midst of this intense exchange of mathematical letters, a postcard arrived on January 11, 1930, for Élie Cartan in Paris from Einstein in Berlin: "Pater Peccavi! [Father, I have sinned!] One can misuse even the most beautiful theory! Forgive me! Yours warmly, A. E." And he added in parentheses:

"On one occasion I have determined the coordinate system with the help of differential equations, and on the other occasion by maintaining four variables."

At the large international congress of mathematics in Oslo in 1936, Élie Cartan gave a speech on Sophus Lie and the role the theory of Lie groups played in the development of modern geometry. Before he went into the mathematical aspects, he drew lines back to Lie's speech about Galois at centenary celebrations of l'École Normale Supérieure in Paris in 1895, and to the genius, Abel, who had given the country of Norway a standing among the foremost in this science. About Lie, Élie Cartan pointed out:

> The greatest honour that one can show toward a man of science whose genius has opened new roads, is to point out the influence his works have had – that is, the power of the work itself, as well as its expansion into the further development of the science. Lie's theories concerning continuous groups is of interest to such a great number of mathematical disciplines that it is difficult to approach them all [...]

Thereupon, Élie Cartan confined himself to "an attempt to present" how in the development of geometry, Lie's group theory had obviously achieved "a deep unity" of what were until that time divergent points of view.

Today, natural scientists from most disciplines would subscribe to the description made by the French mathematician, Jean Dieudonné in 1974:

> Lie theory is in the process of becoming the most important field in modern mathematics. It had gradually become apparent that the most unexpected theories, from arithmetic to quantum physics, all circle around this field, as around a gigantic axis.

Engel would devote much of his life to producing an edited and annotated edition of Lie's treatises, as had been published over the years in various journals – an effort and a contribution that resulted in *Sophus Lie. Gesammelte Abhandlungen* in seven volumes. (Lie's studies and treatises fill more than 4,000 octavo pages, and Engel's commentaries and notes, more than 1,400 pages.) The first six volumes came out between 1923 and 1937, and the final volume not until 1960, nineteen years after Engel's death. Engel had the help of Professor Paul Heegaard in Oslo to help edit Volume I and Volume II. The publication of these *Gesammelte Abhandlungen* was made possible by a grant in 1919 from Norway's National Research Fund and with support from the academies of science in both Oslo and Leipzig. The economic responsibility for this project was taken up by the newly formed Norwegian Mathematical Association, and the books were printed cooperatively by Teubner Verlag in Leipzig and Aschehoug Forlag in Oslo. (Plans to publish Lie's papers were made right away at the time of his death. In 1912 Teubner sent out a subscription invitation, but this first effort probably raised a subscription only in Norway, and some funds in Germany, but did not generate enough such that production could begin. Engel then wrote an article in a Norwegian newspaper and Parliament granted money for a subscription amounting to forty copies, which secured the publication, and the work thus began in 1919.)

Engel was in Norway on many occasions, both planning and carrying out the work . When he held a lecture at the Norwegian Mathematical Association in 1922, he summed up much of what he knew about Lie. Engel told about their first time together in Christiania in 1884, about the editorial work and the tasks carried out in Leipzig, about Lie's breakdown, and the period that followed. Through the whole thing, Engel enthusiastically stressed Lie's greatness:

> If the power of invention is the true yardstick of mathematical greatness, then *Sophus Lie* must be calculated as the first among all the mathematicians of the time. The new fields that he has opened for mathematical research are so extensive, the methods he has created, so fruitful and far-reaching, that only extremely few can, in this respect, stand to be measured together with him.

Engel pointed out that what is demanded of mathematicians – in addition to their power to invent – is above all acumen, and that Lie certainly was an acute mathematician.

But this perspicacity meant that he did not allow himself to go wild over problems known to be difficult, that is to say, the type of problems that many mathematicians in the past had tried without success to solve. And according to Engel, Lie did not put himself up against "the difficulties that lie hidden in old, supposedly immaculate theories". Lie found his own problems, and his efforts were directed toward choosing such problems as were both intriguing and at the same time capable of solution. It certainly happened as well that along the way he solved tasks that other prominent mathematicians had been stumped by. And when this happened, he was certainly proud, but – according to Engel – only in the sense that he experienced the achievement as irrefutable evidence of the fecundity and appropriateness that lay in the choice of problems, and in their subsequent formulation. He himself did not need such proofs, for he was absolutely certain about his cause, and this unshakeable conviction that his methods and theories at some point in the future would find their warranted recognition and application, never left him, Engel maintained – not once in the darkest times where participation and acknowledgement seems to be lacking everywhere.

In looking back, Engel did not want to dwell upon "the painful memories": "In my memory there exists only a very benevolent and joyful picture of him," he said. But Engel could not avoid mentioning that much was changed when, following his breakdown, Lie took up his work again in the autumn of 1890: "Once he had recovered his earlier suppleness of mind, at least as a mathematician, he became as he had been of old. But not as a human being. His mistrust and his irritability did not dissipate, rather they grew more and more with the years, such that he made life difficult for himself and for all his friends."

Conflicts

As a professor in Christiania, indeed, right back to the time of his engagement to marry Anna in 1873, Lie had worked with continuous groups, which would come to reveal themselves to be of such fundamental significance in formulating the basis of nature's own laws. His work was noticed around Europe, although to a far lesser degree than he would have liked. Throughout the 1870s he had ongoing contact with Klein and Mayer in Germany, and to a certain degree as well, with mathematicians in France. In 1882, while Lie was giving lectures in Paris, a great change occurred in his relationship with the French. And it was Picard, more than anyone else, who had his eyes opened about what Lie's theoretical interpretations could be used for.

In 1886 it had been Lie's German friends and contacts who obtained the professorship for him in Leipzig, in the Kingdom of Saxony. Yet even as he was on his way down to Leipzig he learned – from Professor Schwarz in Göttingen – that his appointment to Leipzig perhaps had also been motivated by rivalries within the mathematical milieu of Germany, and that an important factor in his advancement, it was said, would be his lack of involvement in function theory. In any case, from trend-setting Berlin, Lie's theories – as they now gradually became known with the publication of his monumental writings – were strongly criticised from two directions. On one hand, it was argued that the bases for Lie's theories were so insufficiently developed that they had to be done again from the beginning – Weierstrass would certainly have argued along these lines, and in his eyes what disqualified Lie was his lack of understanding of the theory of Abelian integrals, a subject that was taken up by Weierstrass and his students in Berlin. On the other hand, it was said that for the most part, Lie's work involved an excellent theoretical method which did no more than show how certain problems – above all, differential equations which had already been solved by Euler and Lagrange – could be solved by an alternative and more elaborate method. The mathematician F. G. Frobenius gave expression to such a point of view, but seems later to have changed his views as to the significance of Lie's contributions to science.

In Paris however, Lie was seen as one of the most exciting of mathematicians. This positive assessment, which had been created following his Paris visit in 1882 only increased in scope as his writing gradually became known and studied. Lie's fondness for French mathematics, to which he gave expression on several occasions, was no doubt coloured by the praise he received from that quarter. In consequence, his mathematical ideas and endeavours seem to have corresponded to what in France was considered the correct balance between the utilitarian aspects of mathematics, and the artistic – the artistic was something that tended to go in the direction of "mathematics for mathematics sake". The history of French mathematics is replete with examples of how "utilitarian" and "artistic" mathematicians were evaluated differently at different times. Picard pointed to Cauchy as a mathematician who had managed to unite the two tendencies: Galois was a shining example of "the artistic", while Fourier was considered an example of "the utilitarian". But now in the latter half of the nineteenth century – above all through the work of Picard and Poincaré – both these aspects of mathematical activity were considered to be valuable and necessary. The practical and applicable aspect was abundantly taken care of by mathematical analysis – to a very high degree, the study of natural phenomena had become a study of differential equations. Even during the First International Congress of Mathematicians in Zurich in 1897, Poincaré stressed this relationship between mathematical analysis and the external world: "Whole numbers are the only natural object of mathematical thinking. It is the external world that has inflicted upon us the continuous [. . .] Without it there would be no infinitesimal analysis; all mathematical science would have been reduced to arithmetic [ie., number theory] and to the theory of substitutions [the study of finite groups in the sense of Jordan's *Traité*]. But we have devoted almost all our time and all our energies to the opposite, to the study of the continuous. [. . .] Mathematical analysis opens infinite vistas of whose existence arithmetic would never even suspect."

Following the death of Carl Friedrich Gauss in 1855, one of the greatest luminaries of German natural science was Hermann von Helmholtz. He received his first professorship at the age of twenty-eight in 1849, in pathology and physiology at Königsberg. As the years passed, he captured a series of leading positions and made breakthrough discoveries in several fields. He had been a military doctor in the Prussian army and teacher of anatomy at the Berlin Academy of Art before he was appointed professor in Königsberg. He later became professor of anatomy in Bonn and professor of physiology in Heidelberg, before becoming professor of physics in Berlin in 1877. He then became director of the new *Physikalisch-Technische Reichsanstalt*, a form of schooling that later came to be imitated in other countries.

In 1851, von Helmholtz determined the curvature of corneas and found a method that made it possible to see the interior of the eye sharply and clearly. This led not only to great progress in the diagnosis of diseases of the eyes, but also to a clarification of human spatial presentation, and beyond that, to the question of the character of the basic principles of geometry. He put forward the body of mathematical theory on the conservation of energy and applied this to a series of relations in nature. He worked with light and theory of colour refraction, and in his study of the movement of liquids – in particular, the study of the motion of vortices – he made new mathematical findings. Von Helmholtz was a great lover of music – something he knew a great deal about – and among other things, he investigated sound tonality, tonal combinations, and the significance of overtones in an overall sound picture – and drawing upon his far-ranging fields of interest, he gave a series of popular lectures.

From the list of his scientific achievements it should be mentioned that von Helmholtz was an inspiration to Louis Pasteur and his bacteriological research; he helped Werner von Siemens in the direction of discovering the electrical dynamo; and he laid down the theoretical principles that allowed Heinrich Rudolph Hertz to find experimental proof for James Clerk Maxwell's electro-magnetic waves – what were called radio waves, and again gave support to cordless communication by means of the telegraph, telephone, radio, etc. The point of departure for many scientists was von Helmholtz's 1868 writing on the foundation of geometry: *Über die Tatsachen, die der Geometrie zu Grunde liegen*, which in turn had its beginnings in Riemann's habilitation lecture on the hypotheses of geometry. This was the von Helmholtz treatise that had inspired Killing in the early 1880s to begin the writings, which in his further studies, led him close to Lie's theories. Von Helmholtz had wanted to develop geometric concepts from facts demonstrated through experimentation. Like Riemann's, his point of departure had been to ask: what must be the structure of a multi-dimensional manifold, if a rigid body (that is, a body in which the distance between any pair of randomly chosen points is constant) should be capable of moving around so continuously and monodromically (as such bodies do in real, ordinary, visible, three-dimensional space)?

Very early on, Lie was certainly clear that the transformation theory he was developing, was related to non-Euclidean geometry, and in a letter to Mayer as early as 1875, Lie had pointed out that von Helmholtz's work on the axioms of geometry from 1868, were basically and fundamentally an investigation of a class of transformation groups: "I have long assumed this, and finally had it verified by reading his [von Helmholtz's] work."

Klein too, in 1883, had asked Lie what he thought of von Helmholtz's geometric work. Lie replied immediately that he found the results correct, but that von Helmholtz operated with a division between the real and the imaginary that was hardly appropriate. And a little later, after having studied the treatise more thoroughly, he communicated to Klein that von Helmholtz's work contained "substantial shortcomings", and that he thought it positively impossible to overcome these shortcomings by means of the elementary methods that von Helmholtz had applied. Lie went on to complete and simplify von Helmholtz's spatial theory "without abandoning his line of thought", as he expressed it to Klein. Then in September 1886, a few months after arriving in Leipzig, Lie opened his lecture at the meeting of the natural sciences in Berlin, by saying: "Von Helmholtz's famous work 'Über die Tatsachen, die der Geometrie zu Grunde liegen' deals with a problem that stands in a very close relationship to the new theory of transformation groups." He further asserted that he, by way of Klein, had been asked to consider dealing with this "important, but also special problem," by applying his own transformation theory. Klein was also in attendance at these natural science meetings in Berlin – Lie had only signed up for the meetings when he learned that Klein had promised to be there. Lie presented his results: he had thought to further work out von Helmholtz's "epoch-making investigations", but to go into detailed proofs was impossible.

Lie did further work with von Helmholtz's space problem, and confided to Klein in April 1887, that the earlier works on the problem had now come to a satisfying conclusion – at least when one was addressing finite dimensional transformation groups, and therein, a limited number of parameters. What remained was to deduce some that extended across the board such that infinite dimensional groups could be included. During his stay at the psychiatric clinic at Ilten in the spring of 1890, Lie once again took up his work with von Helmholtz's theory, and both of the studies that he had printed that spring in the Proceedings of the Leipzig Academy of Sciences dealt with this theme.

From his physiological point of departure in the curvature of the eye, von Helmholtz had made a contribution to the basic principles of geometry – to many mathematicians this would be referred to as the Riemann-von Helmholtz space problem.

Only some months after Klein had asked Lie in 1883 what he thought about von Helmholtz's work, Lie asked Klein if he would not consider bringing out his "programme text", and Lie added on that occasion: "This is certainly your most important work from the 1872 period. *Now it will be better understood than it was at that time*." When it was finally republished again in 1892, it was Klein's *Erlangen Programme* that everybody talked about, and it was also the cause of the collapse of the relationship between the two of them. The reason that this had been discussed in 1884 was that Klein was in the process of gathering together some of his earlier works for publication in *Mathematische Annalen*, where indeed he was now the editor. Among these works was also a nine-page paper on the Kummer surface with sixteen singularities that he and Lie had had published in Berlin in 1870. When this work consequently was published in *Mathematische Annalen* (Vol 23, 1884), Lie reacted to the fact that the work they had done together was included, and linked closely to Klein's doctoral dissertation, as well as to some of his other voluminous treatises. Lie seems to have regarded this as a sign that Klein considered their common work to have been a further development of Klein's doctoral work, as though Lie had derived the ideas from Klein.

Be that as it may, Lie found cause, during the spring of 1884, to take up his relationship with Klein. He pointed out that the relationship had been "of the greatest significance", but also, that he, already before making his first journey to Germany, had, through reading Plücker's writings, been steered in the direction of thinking about contact transformations although "of course in an indistinct form." And in any case Lie had definitely been disturbed by the view held by certain mathematicians, that his works (on differential equations and the associated transformation theory) was an elaboration of Klein's ideas.

In the years that followed, Lie alluded frequently to the fact that other mathematicians used his discoveries without acknowledging their sources. For a long time he seems to have felt that the Frenchman, Georges Halphèn's reckless use of the facts was the worst form of dishonesty. But apart from this, he seems to have been pleased that Darboux always cited from his [Lie's] work, and moreover that Picard so strongly stressed the great significance of Lie's investigations for work with differential equations.

When Engel arrived in Christiania in the autumn of 1884 to assist Lie, Lie reacted the moment Engel had begun to talk about "our theory" – and Lie had then, "with all bluntness", made it clear that the theory before them belonged to him, Lie. This was something that Lie talked about for the first time at New Year 1888, when he discovered that Engel had been in contact by letter with Killing for several years, and through this correspondence disclosed Lie's ideas, and even did so in a way that had led Killing, in his studies, to make reference to Engel's works. Lie then stressed to Klein that it was clear that Engel had learned the theories systematically from him; that admittedly Killing had had certain ideas, but the systematic knowledge that he lacked he had obtained through correspondence with Engel. In the foreword to the first volume of *Theorie der Transformationsgruppen*, which came out in June 1888, Lie nevertheless thanked Engel for his contributions. Lie commented upon this in a letter to Klein, by saying that Engel, due to his diligence, certainly deserved these words of praise, but he added: "even though he [Engel] has not for a long while been as bright as he [Klein] and Mayer think." Lie wrote that Engel often made "mistakes of reasoning" – and that Engel in his "remarkable self-confidence, often without informing Lie, had introduced assumptions that limited the scope of the ideas. Lie was also displeased that Klein had given Engel the task of reviewing *Theorie der Transformationsgruppen* in the *Göttinger Anzeiger*: "It will be a good opportunity for him [Engel] to dress himself in borrowed feathers," Lie wrote to Klein, whom he had still not criticised openly, even though indeed it was Klein, in his journal *Mathematische Annalen*, who had publicised Killing's treatises. But Lie was uncertain of Klein's attitude, and wrote to him: "It is a well-known fact that you often have described my works as unreadable. So perhaps you find it quite justified not to acknowledge my theories until they have been reproduced by someone else."

Klein seems not to have answered this, but Lie had continued: "I have always thought too highly of many people, and now I am being punished." He confided to Klein that he certainly ought to be more strict in how he kept his priorities – he had been too naïve: "I am basically good-natured. When I have confidence in a person, then to the greatest possible degree I take into consideration that person's intentions. But now that I feel my life's greatest interests are in danger, then I am thinking to fight for my own interests. Except that toward a friend of so many years, like you, I become deaf and blind." This, Lie wrote to Klein at New Year's of 1889, when the second part of Killing's great work had come out. Lie felt that all his scientific interests were threatened – he felt that the priority of all the fundamental contributions of continuous group theory was in the process of being taken away from him – just as much by Killing as by Engel.

Lie now referred to a well-known example from mathematical history, namely the relationship between Abel and Jacobi – where Lie was playing Abel's role, while Jacobi represented the German mathematicians. When one read Abel's and Jacobi's first works, then it was abundantly clear that Jacobi published his results without proof, Lie maintained, and reminded Klein how Abel and Jacobi had both consequently been given priority. The relationship between Abel and Jacobi had thus been posed in an erroneous fashion. Lie admitted that in the beginning he had been critical of the way Bjerknes had, in his book on Abel, posed this battle over priority, but now, writing to Klein, Lie maintained firmly: "In general, Bjerknes was right, of that there is no doubt, although he was reckless enough to publish his assertions without presenting proof."

"Killing's second piece of work is a disgrace," Lie wrote, continuing: "I have relied upon Engel. He certainly had no right to convey my theories to another mathematician. To me, this is an elementary truth." Lie wanted to send a note to *Mathematische Annalen* about these relations, but when he got to know that the third part of Killing's work would contain a footnote that would set things right and declare where the priority lay, Lie pulled back from the situation and communicated to Klein that if everything was put to right, he for his part was willing to put the whole thing behind him.

And the footnote came out: Lie was satisfied. Killing attributed to Lie the honour of important discoveries and descriptions, and he apologised that Engel had been put forward at Lie's expense. And when the next part of Killing's work came out, Lie praised the treatise in a letter to Klein, saying that he had found it "significantly better edited, than the earlier ones" – the content, to the contrary, he seems to have felt was still unimportant. But Killing was a "bright man", Lie wrote to Klein, adding: "A work can be useful, even if it only brings out matters that seem to me to be self-evident and superficial." In August 1889, Lie repeated this in a letter to Klein. Even though he had far from enough time to check everything in Killing's works, he wrote, they were "of noticeable value" ["beachtenswerth"] – and Lie expressed the hope that in the course of the autumn he would get the time to study Killing's results properly. But this was the autumn that Lie broke down and was sent off to Ilten.

When he began to recover in the spring of 1890, and while still at Ilten, he wrote a letter to Klein that praised Killing: "Killing carries out very good investigations. If his results are correct, which is something I believe, then he is of conspicuous merit." A year later, in another letter to Klein, he had the following to say: "More and more I have had the greatest respect for Killing, even though his early works lack much that could be desired. Among my various successors, Killing is the one who has developed the greatest originality." Despite all the praise for Killing's work, Lie could not refrain from pointing out some important mistakes and omissions in the presentation. In a paper from December 1890, which Scheffers edited, Lie took issue with Killing.

At Easter 1891, Klein came to Leipzig for a few days' visit. In a letter to his old friend and colleague Mayer, Klein had written that he looked forward to meeting old acquaintances, and otherwise see what had happened in the five years since his departure from the city – but above all, Klein had expressed the desire once again to approach "Lie's conceptual circle" and wrote that he calculated on having Engel's help to do so. For his part, by letter, Lie expressed how much he looked forward to meeting and talking again with Klein – Lie certainly had made the reservation that, though for the time being, he was getting "satisfactory natural sleep", he never knew for sure if the situation would get worse, but he reckoned that he would be able to answer everything Klein wanted to know about his work.

Lie also communicated in this letter that he did not agree with the way Klein had formulated the von Helmholtz space problem in an article published in *Mathematische Annalen*. Lie was now able to relate that he was certain of his view on this matter: von Helmholtz proved absolutely nothing fundamental, since he rather simply ignored "the most difficult possibilities" – von Helmholtz's study certainly remained a most beautiful work, but was nevertheless, according to Lie, imperfect.

Of what happened in and around this meeting between Lie and Klein in Leipzig in the spring of 1891, there are only sporadic hints in later letters. Shortly after his visit, Klein reported to Mayer that during his stay he had had detailed discussions with Engel – and that Lie had maintained that he ought to have had Volume Three of his great work ready sooner... (The printing of the third volume did not begin until a year later.) Apart from this, Klein explained that he had become alarmed by how categorically Lie had assessed certain people (Forsyth and "Father Salomon" were mentioned). Lie worked too exclusively "with the perspective of a Leipziger," Klein felt, and had encouraged Lie to compare and comment upon some of his earlier works for more substantial studies to be published by *Mathematische Annalen* in Göttingen – and a collection of their joint work was also discussed. According to Klein, such a work project would "be a good way out of his dark thoughts".

Klein's desire to have Lie edit his old works was certainly well intended, but seems also to have been motivated by the need to create more material for the journal. Nevertheless, Lie did not follow up on Klein's wishes – the work of readying the third volume of *Theorie der Transformationsgruppen* took a great deal of time, and as well, he had to prepare and deliver his six lectures every week. During the summer semester of 1891, Lie gave a series of lectures on "Die Elemente der analytischen Geometrie" – two hours a week for natural science students – and "Einleitung in die Gruppetheorie und Andwendung derselber" for the more advanced, plus practical exercises in the seminar. But in this semester at Leipzig there was a total of only twenty-three students of mathematics. The following semester there were thirty, of whom nine came to Lie's sessions. Among these was the Frenchman, Arthur Tresse, who indeed had studied with Lie three years earlier along with Ernest Vessiot, and who also was one of Élie Cartan's good friends.

In the course of this term, the winter semester of 1891–92, which of course ran from mid-October 1891, to mid-March 1892, much seems to have changed for Lie. The links and points of contact with Paris became closer and stronger, as his friendships with German colleagues cooled and fell apart. The high points in this process occurred with the appointment of Lie as corresponding member of the French Academy of Sciences in June 1892, and Lie's turning so starkly against Klein in the foreword to the third volume of his large work on transformation groups, which appeared in autumn 1893. A whole series of episodes occurred between these two points of extremity, which to a greater or less degree constituted the same trajectory – and this seemed to lie in the direction of both personally and professionally draining his attention away from the German milieu, and flowing instead toward the French.

When in January 1892 Lie had sent a short note to Picard and l'Académie des Sciences in Paris, in which he showed that a new translation of one of Abel's theorems gave a solution to a problem in function theory, he knew that this was a tidbit that would inspire a certain degree of enthusiasm. Picard subsequently presented this information to the Academy in a four-page memorandum. Then only a week later, on February 15th, Picard had received a new note from Lie for presentation to the Academy, this too, in connection to the Abel theorem. And two weeks later there was still another note from Lie, presented to the Academy of Sciences, which this time dealt with a critical commentary on the well-known von Helmholtz work from 1868. These three notes were quickly published in *Comptes Rendus* – Lie had not sent anything to l'Académie des Sciences for twenty-two years.

In his last note, on the von Helmholtz space problem, Lie also recounted Killing's treatment of the problem, and maintained that Killing had more or less made the same mistake as von Helmholtz – and in order to dispel all doubt, Lie would clarify more simply. This he did, among other places, in a study printed in the Proceedings of the Leipzig Academy of Sciences, *Leipziger Berichte* – and here both Klein's and Schur's works were criticised (and also de Tilly and the Italian, Veronese, were criticised, although Lie would later come to regard de Tilly in milder tones). For his part, Killing published a work on problems of the basic principles of geometry – and on a postcard to Lie enclosed with the offprint, Killing wrote that the work had been finished for quite some time. In addition, Killing gave the news that if he now, in March 1892, were to put the work into a final form, he would find it necessary to change certain parts, on the basis of, among other things, what Lie had recently published. Killing concluded: "For many reasons I have preferred not to make any changes, or include any additions. With highest esteem, Yours cordially, W. Killing."

Angry and disturbed, Lie then wrote several letters to Killing, letters that no doubt contained sharp criticism. Killing did not reply right away. In any case it was not until Lie had sent in a communication to *Crelle's Journal*, that Killing replied, and this in a way put a definite end to all further scientific contact. Killing wrote that as far as it went, he agreed with Lie that he might have, in an appendix, referred to Lie's works on the principles of geometry. Killing also mentioned how in the essay he had pointed out that the presentation was indebted to the first volume of Lie's "Transformation Groups". But now that the criticism had reached such a pitch, Killing considered it his duty not to be indulgent in such a manner as to give the impression of confirming everything that Lie had come out with in his various writings. Lie's accusations in his letter to the editors of *Crelle's Journal*, were slanderous. Killing thereupon asked that these accusations be withdrawn, and he backed this up with a powerful counterattack.

Killing stated categorically that the contribution that Lie had made to von Helmholtz's theorem at the meeting of the natural sciences in Berlin in autumn 1886 actually belonged to Weierstrass, and that the first work in which transformation groups were applied to the study of the principles of geometry were his own – Killing's – work from the summer of 1884.

This "outbreak of war" between Killing and Lie happened in March 1892, but their future battles were less conspicuous. At first glance, Lie seems to have planned to send a sharp note to *Crelle's Journal*, and next, a more detailed article to the same periodical, but he let matters lie. Relations with Paris, and the forthcoming election for the corresponding member of the Academy, were more important that Killing's allegations. Perhaps Killing also had ulterior motives for declaring his merits and distancing himself from Lie. In any case, on May 1, 1892, he was appointed professor at Münster.

At the beginning of April, Poincaré had sent the information that the geometrical section of the Academy had nominated Lie to the position left vacant by the death of Kronecker, and requested an overview of Lie's works be submitted. Lie then wrote hastily and from memory – and apologised that there could thus be certain inaccuracies, and in terms of dates, major errors – a list which he felt was unreservedly voluminous since he had not had time to do any editing, and wrote: "I assume that it is not necessary to give reference to my greater works. These are, as well, the execution of my old ideas."

Lie was now also working on a study for the French Academy on the principles of geometry, but he was doubtful as to whether he should send it in. It was already the month of May, the election at the Academy in Paris was approaching, and Lie seems to have been apprehensive of doing anything that might weaken his chances of being elected. He also wondered if Klein was an equally strong candidate, but seems not to have completely put this into words even though his doubts and anxieties grew. In Lie's surviving papers there is, from this period of May 1892, a draft of a letter to Picard in which he tries to get clarification on these matters – and more importantly, there is also the draft of a letter to Klein in which Lie accuses Klein in the strongest terms of having assumed proprietorship over Lie's ideas.

It now seems that Klein's activities of recent years appeared to Lie in a new light. As far back as two years earlier, Klein had given a series of lectures on non-Euclidean geometry, and here he presented what, according to Lie's interpretation, was an attempt to secure for himself the credit for their joint work from the 1870–72 period. Lie had then pointed this out to Klein and they had spoken about it when Klein was in Leipzig at Easter 1891.

Lie's anxiety and indignation toward Klein that spring of 1892 probably came from two quarters. First of all, Lie was afraid that in one way or another, Klein would come to have an influence that would thwart Lie's prospects in Paris, and second, Lie now got to see the written form of the series of lectures Klein had given in Göttingen on non-Euclidean geometry. Lie began a litany of accusations and allegations of injustice and thievery, in letter after letter – but it does not seem that he actually completed and sent these letters – in any case, there were none of this nature to be found among Klein's surviving papers. In the letter Klein received from Lie on May 6th, the tone was relatively mild, although the content came in fits and starts. Lie wrote:

> Your lectures are an on-going attack on me.
>
> You still challenge me to destroy the Helmholtz theory. I shall eventually do so. I also show (see also Compt. Rend. 1892) that H's theory is basically false, and reconstruct it and give [...] two solutions to the problem.

You are publishing an edition of your lectures in Math. Ann. where you treat me quite contemptuously.

I am politely bringing to your attention your colossal mistake.

You do not take back your Insult [. . .] and do not neglect to profit, in this lecture here, and in other places, from my presentations.

You come out with a new *erroneous observation* about me, which places me in a preposterous light.

You have not obviously read my note from 1886.

I am going to take the necessary steps. You are wrong if you believe that you can look for fame in this manner.

For the number of errors in your lecture are greater than you imagine.

S. Lie

Klein tried to reply by making light of the matter, and he stressed, using various approaches, that the whole thing ought to be put to rest as a series of misunderstandings – the lectures were not published (and in addition, the lectures from 1890 existed only in various students' versions), and that if, in his reference to von Helmholtz, he had failed to mention Lie's work from 1886, this was because, on that occasion he had merely referred to what Lie had said at the natural sciences meeting in Berlin in the autumn of 1886, and at the same time he had certainly written that he would return to Lie's group theory on a subsequent occasion.

Then came the month of June and the decision in Paris. Lie was chosen corresponding member of l'Académie des Sciences by a vote of 30 out of 33. (Two had voted for Cremona and one for H. A. Schwarz.) The joy of this seems to have given Lie both peace of mind and a sense of security, or in any case in the coming period it did not lead to the same hectic presentation of his points of view. Mittag-Leffler sent his congratulatory greetings from Stockholm, and Lie replied: "Since already for many Years I have more highly valued the French, over what, in my Opinion, is the more one-sided German Mathematics, I therefore set great Store in the fact that in France they are approaching my Theories."

When the summer semester ended in Leipzig, Lie took his family back to Norway and the town of Moss. Anna and the children probably also stayed for a while at a pension in Åsgårdstrand, while he, for his part, undertook a long and arduous tour through Jotunheimen.

For his part, Klein, from Göttingen, seems to have moved swiftly to make ready an account of their joint works. As early as August, Klein had finished "Über unsere Arbeiten aus den Jahren 1870–72. Von F. Klein und S. Lie" – and in order to make sure that Lie agreed with the presentation and could thereby stand as co-editor of the account, Klein proposed that they meet.

Lie replied at the beginning of October: "I would like it very much if you could come to Leipzig, as you have indicated to Scheffers." And a week later: "Hearty thanks for your friendly letter. I do not doubt that we shall succeed in putting to rights our priority claims with respect to our common publications." But, Lie added, it would certainly be more difficult to put into commission "old unsatisfactory works" for republication. Lie gave the information that his task in the coming years was precisely to make accessible the various ideas that in earlier works which had not had their due, or had not been taken account of in an accessible form. And this preparatory work he expected to occur mainly with the help of his students – but at the same time he informed Klein that as far as it went, he was prepared to give out some of his older works without changing them.

Klein was in Leipzig on October 19, and had detailed discussions with Lie about this publication project of which in consequence, they would both stand as the authors. Scheffers also took part in these discussions – what else happened that day is not known. In any case they did not come to an agreement. Back in Göttingen, Klein wrote a new draft of his account of their joint work – this time with only his name on it. Klein sent this text to Lie on November 1st, and Lie replied six days later:

> It gives me genuine pain to realise that we have so completely misunderstood one another. [...] I read through your manuscript painstakingly. And unfortunately I get the impression that you will not succeed in making a presentation that I can acknowledge as correct.
>
> Even several points that I have already sharply criticised, are incorrect in your current presentation, or at least misleading.

Lie proposed that if they could not manage to agree, then they each could, for himself, give his account of the relationship, and the mathematical public could then form its own conclusions. Klein had also asked if Lie would join the editorial board of *Mathematische Annalen*, something that Lie now refused in the following manner: "I have already tried once to figure as a second-rank editor with *Acta* [*Acta Mathematica*]. I am not planning to do so once again."

There was also another deep-seated relationship that had come to the surface during their meeting in Leipzig, namely the correspondence of their friendship and scientific interest down through the years. As far back as four years earlier, right after the publication of the first volume of *Theorie der Transformationsgruppen*, Lie had written to Klein: "I have again recently and with pleasure read through all your letters (1870–88). Do you have my letters from 1870–74? I would very much like to see them again on some occasion. It will be interesting to see how certain ideas have developed from their first embryonic form. Well, there is no rush."

Now during the meeting and the discussions in Leipzig, naturally enough those letters were talked about which stemmed from the years they were to map out. But Lie's letters to Klein from the years up to 1876 were no longer to be found – it appears that Klein, while tidying up, had burned the letters. After he had now proposed that each for himself should write about how they comprehended the development of their ideas at that time, he commented and continued:

> For the time being I must only lament that you saw fit to burn my letters with their rich contents. In my view, this is vandalism; I had an express promise from your side that you would take care of the letters.
>
> I have already told you that my naïve period has ended. Even though I shall always cherish the good memories of 1870–72, I will nevertheless try to preserve for myself what, according to my perception, belongs to me. It happens that from time to time you think you have had some of my ideas by having made use of them.

The first point that Klein addressed in his letter of reply, was to express his great disappointment: "I thought we really had understood one another, and that now again we could work toward a common goal," he wrote, but continued that now he was also of the opinion that it was best that they publish their separate accounts. Klein had wanted Lie to indicate his points of disagreement, and he had wanted to be able to go through his own letters to Lie from the years 1870–72, so that he could reconstruct the course of events in more detail – and then he would have come to Leipzig again at the Christmas period to thrash out the whole issue: "Instead, you have come out with common accusations and reproaches, which do not lead anywhere," Klein wrote, adding that if Lie did not manage to master his mistrust, then they ought to let the whole matter drop, the sooner the better. Klein concluded that he thus in any case did not want to publish, and comforted himself with the old adage that with the passing of the years, everyone becomes more lonesome.

And with these words and this letter of November 10, 1892, the connection between the two old friends ended.

Lie seems to have been left sitting with the impression that Klein would nonetheless publish his more or less complete manuscript about their common work. Lie thus wanted to give his version of this struggle over priority, and he would subsequently give it in the foreword to the great work in which he was now immersed. A letter written by Klein to Elling Holst following Lie's death indicates that Klein never got his manuscript back – in any case, Holst was asked to look for it among Lie's papers concerning the compilation for the autumn of 1892.

The printing of the last volume of the *Theorie der Transformationsgruppen*, where the development of the theory of transformation groups was examined, where the group theory work of others was commented upon, and where the Riemann-Helmholtz space problem was gone through and criticised – the printing of this volume had been underway since March 1892, and when the work was submitted in September 1893, it consequently contained among other things, the bald assertion that would shock the world of German mathematics: "I am no pupil of Klein's, nor is the reverse the case, although perhaps this would be nearer to the truth."

But Klein never published his manuscript. Lie's attack therefore appeared as a kind of unmotivated bolt out of the blue. It was the thirty-year-old David Hilbert who put into words what many in Germany were probably thinking about Lie: "With the third volume, his megalomania burst into the open."

Otherwise, the period of ten months between the last letters to Klein and the famous foreword, apart from the labour for the great work, was a time when Lie turned his back on Germany. In a letter to Darboux in Paris, Lie gave strong expression to his "love of French science", which had only grown after his move to Germany. He spoke about the year 1870 (the war) – although it had been good for German politics, it had been "fatal" to German mathematics. And perhaps as a hint that he was readying himself for departure, Lie spoke openly about his dream of founding a school in Leipzig before he returned to Christiania. At New Year's 1893, Lie reported to Elling Holst:

Otherwise, things are going remarkably well for me. Along with Sleep, my Joy for Life and Work has returned. For the time being my Theories have advanced tremendously, not only in Europe, but also in America. For the Moment there are, so to say, no bright young Mathematicians in France who are not occupying themselves with my Issues.

Back in Christiania several prominent men – with the leading poet, Bjørnson, and polar hero, Nansen, in the lead – were talking about Lie's discontent down "on the Leipzig plains" and wanted to get him back to Norway's university.

In April 1893 Lie spent eighteen days in Paris. He was received and treated in the very best manner. A laudatory account of his works was published in *Comptes Rendus*; he visited the Academy; he met Pasteur and all the French mathematicians – he visited Picard and Hermite at home, and lunched with Appell and Goursat, went to dinner together with Poincaré at the home of Jordan; he audited teaching sessions at l'École Normale Supérieure; he met his old pupil Vessiot; he met the twenty-six-year-old Norwegian physicist, Kristian Birkeland; and one evening he drank "at 12 o'clock at Night a Glass of Beer with a Swedish Mathematician". He reported home to Anna in Leipzig about all these meetings, and promised to be home with gifts before their son Herman's ninth birthday on April 24th.

In the month of August, after the end of the summer semester, Lie travelled back to Norway with his family, and it seems that this year the whole family joined him on a walking tour in the mountains – but Lie would also set out on a strenuous tour on his own – one evening when he came to a river, it had been so swollen that it had taken out the bridge, and had been unable either to cross over, or to venture back since it had become so dark. As a result, all night long he had walked and run up and down along the river to keep warm, passing the time by whistling all the tunes he knew.

In Christiania they certainly visited his sister Laura at the Eugenia Institution – Lie's other sister, Mathilde, in Tvedestrand, had lost her husband in March of that year and moved to Christiania, and had taken an apartment right across the way (Dalsbergstien 4) – while her son, Professor Johan H. L. Vogt and his family lived at Ullevålsveien 12. Only two months after Sophus, together with Anna and the children, had returned to Leipzig, his sister Mathilde died of pneumonia. Sophus returned to Norway for the sole purpose of attending her funeral.

On the way through Copenhagen, Lie generally visited his friend Zeuthen, and Zeuthen seems to have found it refreshing to have visits from Lie. It was as though the whole of scientific Europe forced its way in the rooms – Lie knew everybody and everything, and brought all the latest news. Zeuthen explained that it was like "new oil for the lamp."

For Lie, the publication of the third and concluding volume of the great work on transformation groups in Leipzig in September 1893 was the fulfilment of his life's work. The next tasks that Lie saw for himself was to make refinements and applications of what had now been completely formulated. But this foreword with its sharp accusations against Klein, caused hindrances to the further work. Because Lie in the same foreword had praised Engel to the skies for his "exact" and "unselfish activity", it now became difficult for Engel to continue to collaborate with Lie –

consequently as well, nothing came of the announced work on, among other things, differential invariants and infinite dimensional continuous groups. As for Engel, his career outlook certainly now lay in other directions than Lie's. According to Lie's German student, Gerhard Kowalewski, relations between Lie and Engel gradually became so cool that they were seldom to be seen in the same place.

As for Klein, he was not in Germany when Lie's accusations against him were published – he was on a lecture tour in the United States. On August 6th, Klein had boarded an America-bound ship at Bremerhaven, and by a twist of fate, met the venerable seventy-two-year-old von Helmholtz and his wife, who had been invited to the USA and the great World's Fair in Chicago. Klein and von Helmholtz had accompanied one another not only to America, but were also on the same ship home again two months later. Klein later told the mathematician, Leo Königsberger, about this trip – among other things that he, Klein, had wanted to discuss science with von Helmholtz, but in the beginning he had been prevented by bad weather. But one day in the middle of the Atlantic, they had discussed the von Helmholtz axioms for space geometry, and von Helmholtz had pronounced himself very satisfied with Klein's explanations, such as they had appeared in *Mathematische Annalen* (Vol. 37, p. 565), and they had also gone into Lie's positions and theories. The following day, Klein explained, Mrs. von Helmholtz had come to Klein and asked him not to discuss such difficult matters with her husband: "He becomes too exhausted by it." About the tragic conclusion to the tour, Klein could also relate that on the way home they had been sitting together one evening in the smoking saloon – the captain had also been with them, and von Helmholtz also had a young doctor from Boston in attendance – the sea was completely calm, but it was hot and humid due to the fact that they were certainly on the outskirts of a hurricane. At ten o'clock, von Helmholtz had got to his feet, said it was bedtime, and disappeared down a very steep stairway. Klein had heard a heavy crash but did not think any more about it before the young doctor shouted, "Something has happened to *der Geheimrath*!" Von Helmholtz had had a blackout and fallen from the bottom-most section of the stairs, had not been able to protect himself, and was wounded and bleeding. With the best medical care on board ship, and in the following weeks ashore, he recovered and took up his work once again. Three years earlier, when he had turned seventy, his birthday had been commemorated in Berlin with a bang-up international celebration, and he was awarded the highest civilian honour by Kaiser Wilhelm II, but a year after this trip to America the life of this great scientist came to its definitive end.

Klein now was increasingly in a more central position in Göttingen. When Professor Schwarz left for Berlin in 1892 to take over the position that had been held by the now-retired Weierstrass, this allowed Klein more freedom of operation. Above all, he began to develop his scientific and teaching reforms, in which the pedagogical aspects were assigned greater weight. Klein's Erlangen Programme became one of the most read of mathematical studies, and his fame as a lecturer and an inspiration to higher mathematics and technical development continued to grow. While Klein was on his lecture tour to America in 1893, and following the Chicago World's Fair while speaking at Northwestern University, at Evanston, Illinois outside of Chicago, he addressed, among other subjects, Lie and Lie's work. On that occasion, Klein emphasised that to understand Lie's mathematical genius, one ought above all to turn to his earlier treatises – where Lie revealed himself "as the true geometer he is."

Then in 1895 David Hilbert arrived in Göttingen – he would become a new star in the mathematical firmament and consolidate Göttingen as a mathematical centre.

In Leipzig, Lie continued to work together with Scheffers – the printing of *Geometrie der Berührungstransformationen* went on through the whole of 1895, and came out the following year. In April 1895, Lie was given the honorary task of speaking about Galois at the centenary celebrations for l'École Normale Supérieure, and once again he was treated royally while in Paris.

Back in Leipzig, Lie tried to begin to work together with another young promising man of science, that is to say, with Felix Hausdorff. Hausdorff, the son of a merchant, grew up in Leipzig, graduated from the city's Nicolai Gymnasium as the top student, and thereafter had begun to study the natural sciences, experimental physics, and mathematics. Under Lie, Hausdorff had studied analytic and projective geometry, under Engel, algebraic equation theory, and under Mayer, differential- and integral calculus, but he also attended lectures on philosophical themes, on the history of socialism, and the labour questions of the day. He had studied for one semester in Berlin, where what seemed to awaken his interest was applied mathematics. But he had been in Leipzig from the summer of 1889, where he had attended Lie's lectures on transformation groups and their application to geometry and mechanics – and by 1893 Lie looked on Hausdorff as a sort of collaborator. Hausdorff took the qualifying examinations that meant he was able to begin working as a private docent from the autumn of 1895, but at the beginning it did not look promising for him. In a lecture he gave on "The Figure and Rotation of Heavenly Bodies" there was said to have been only one person in attendance. At another lecture on "Map Projection", two. Hausdorff's closest teacher at Leipzig was Lie's colleague, Bruns, and Hausdorff worked in Bruns' fields, with optical instruments and light ray mappings precisely by using as his mathematical tool, the concept of contact transformations. Lie wanted very much to get the promising Hausdorff into his own fields of mathematics. And when Professor Bruns – in a meeting of the Leipzig Academy of Sciences, Mathematics and Physics Section, at New Year's 1896 – presented Hausdorff's work, "Die infinitesimalen Berührungstransformationen der Optik" – Lie was given the floor and he made a little presentation in which he pointed out that as early as 1872 he had, in his lectures, demonstrated that in various fields within mechanics and physics, particularly in optics, the most beautiful and best way of proceeding was linked to, and facilitated by, the study of contact transformations. On the basis of this, Lie planned a collaboration with Hausdorff – in which, among other things, Lie had thought to develop his theory of differential equations of the first order, and more of the tasks that Lie had originally hoped to set Engel working on. But after a short while this cooperation broke off – Hausdorff selected another mathematical direction, and quite possibly also felt himself badly treated. Hausdorff started the summer semester of 1896 with lectures on actuarial mathematics, and the following year, lectures on probability theory, and what he called "political arithmetic" – this included lotteries, state loans, finances, and "mathematical statistics" – and his lectures became so popular that he had to repeat the whole series. In 1898 Hausdorff received a permanent position at the newly founded college of commerce in Leipzig, the first school of its type in Germany. Later, in 1898, Hausdorff gave lectures on continuous transformation groups at Leipzig's Mathematical Institute. He introduced new concepts to group theory, and his *Mengenlehre*, which came out in 1914, marked the introduction to set theory and modern topology.

In 1894 in the Russian city of Kazan, an international prize was instituted to commemorate the mathematician Lobachevsky. The prize was to be awarded to a mathematician who had made prominent contributions to geometric research, particularly in the development of non-Euclidean geometry. In 1897 Klein was asked by the prize committee to provide a description of Lie's work, and Klein's evaluation led to Lie receiving the award in 1897 – the very first recipient of this esteemed prize. In his argument, over and above everything else, Klein pointed to the third volume of *Theorie der Transformationsgruppen*, where the theory was applied to the principle axioms of geometry. Klein's evaluation was printed in *Mathematische Annalen* a year later.

Lie thanked Klein for his account in relation to the Lobachevsky Prize, and this letter of thanks, written in 1898, was the first letter that had passed between them since the autumn of 1892. In it, Lie also informed Klein that he had now resigned his post in Leipzig, and in a short time would be making his way back to Christiania. Lie told Klein that the number of mathematics students was increasing in Leipzig, and that the burden of work was great, such that he was unable to give the support and help to students in the same way as he had done in earlier years. Lie complained that he had never succeeded in having any real influence in the faculty, and he was afraid that geometry as a subject in itself at Leipzig, would be phased out. Professor Neumann was still a colleague upon whom Lie could not rely, but: "Mayer is indeed a good human being. But he has this sense of propriety, always plays the nanny, and never understands that I will not be treated like a child, in any case not since I came to know Leipzig conditions thoroughly." Lie added that there was no need to justify the direction of the subject of geometry that Klein in his day had established at the University of Leipzig, and, he continued:

During these twelve years I have kept the geometry banner flying as high as I possibly could. It would be painful to think that with my departure the professorship in geometry would be lost. But I know of no geometer to suggest. I thought about Study, whom I consider a rather gifted geometer. [...] If Study has improved as a lecturer, one could consider him, in any case this is Mayer's opinion.

Lie also mentioned another matter in this letter: "During the year of 1902 in Christiania, we are celebrating the centenary of Abel's birth. We then would like *with Norwegian money* to raise a bronze statue of Abel at our university." Lie mentioned there were also plans to write a history of Norwegian mathematics – Wessel, Abel, Broch, Bjerknes, Sylow, Holst, and he himself had been mentioned – but Lie's main task, which he now explained to Klein, was to gather an international fund, and with the interest from this fund, to award a prize every five years for a mathematical work in the field of pure mathematics – an international committee would select the winner, and Lie mentioned that he already had the support of Hermite, Picard, Darboux, Beltrami, Cremona, but still had not heard from Berlin – and Lie now hoped to have Klein's support.

On May 12, 1898, Klein answered Lie's letter, saying he was surprised by Lie's decision to return home, even though he indeed was well familiar with the difficult conditions in Leipzig. But when it came to the question of whom to employ, Klein commented that he had several current names, but was strongly opposed to either Study or Engel taking over the professorship. And Lie as well was opposed to Engel as his successor.

Lie sent in his official notice of withdrawal to the Saxon cultural minister, Paul von Seydewitz on May 22nd. Here he wrote that it was very difficult for him to give up his "teaching activity at this brilliant university and leave its remarkable mathematical seminar" and continued "with greatest deference and sincere gratitude":

> Now that nevertheless I shall leave Leipzig, the fundamental reason is that the local physical and climatic conditions are not propitious for me. I require many years still to fulfil my plans for publications, and I consider that it is highly possible that Kristiania will be more propitious than Leipzig for my health and my working powers.

A majority of the Faculty of Philosophy voted to ask either Heinrich Weber in Strasbourg or David Hilbert in Göttingen about succeeding Lie – and if neither of them agreed, then Otto Hölder should be the candidate. And by an extraordinary motion, to which among others, Professor Neumann joined, Engel was proposed. Both Weber and Hilbert rejected the offer. Otto Hölder became Lie's successor at Leipzig.

Lie's reputation in Germany was stamped above all by the opinions of Klein and Engel, such as were expressed in speeches and writings shortly after Lie's death. In Göttingen, Klein made a speech that gave rise to much rumour, not least because here, in addition to all his praise for his old friend, Klein suggested the close relationship between genius and madness, and that Lie had certainly been struck by a mental condition that was tinged with a persecution complex – at least, by assessing the point form notes that Klein made for his speech, it seems that this was the expression he used.

In Leipzig, Engel spoke at the Royal Saxon Academy of Sciences – Der Königlich sächsischen Gesellschaft der Wissenschaften zu Leipzig. Engel had remained in Leipzig. Due to his deep involvement in the work of bringing out Lie's three-volume work, he had been passed over in several promotions – most significantly right after the publication of the third volume with its famous foreword. When there was a professorship to be filled at Königsberg, Hilbert wrote to Klein that he considered

Engel completely out of the running for the position: "Even though there is no evidence of his [Engel's] participation in the foreword, to a certain degree I hold him responsible for the incomprehensible and completely unnecessary personal animosity that the third volume of Lie's work on transformation groups is filled with."

In his obituary speech, Engel praised Lie's "powers of invention" which had given mathematicians "rich working materials for generations", and Engel expressed amazement at Lie's "confidence of success that his methods and theories would push on through." Engel's final words were:

> But what is most characteristic of Lie, and what shall become even more so, is his inventiveness, his original mathematical thinking. He did not follow the accepted paths, but rather, his own. I would compare him to a pathfinder in a primal forest, who always knows how to find the way whereas others thrashing around in the thicket, give up out of desperation, and moreover, his pathway always leads past the best vistas over unknown romantic mountains and valleys.

After moving to Münster, Killing had continued his voluminous work on the basic principles of geometry and the "Killing vector field". He wrote textbooks in analytic geometry and engaged himself in the teaching question, and the education of teachers for the gymnasium school system, as well as his church charity work. In 1900, Killing was the second recipient of the Lobachevsky Prize, and his opening words in the acceptance speech were: "Such an acknowledgement of my works, whose shortcomings I so well know, I had never considered possible."

Lie's old friend and assistant, Study, published a book in 1904 on the realistic perception of space problems, and pointed out in a footnote that Lie's critiques of the work of others, and particularly his critique of "the contribution of von Helmholtz, simply seem to have been incorrect." In any case, with this disturbance that Lie had set in motion with his historical-critical discourse on von Helmholtz, with his monumental transformation group theory, he thereby also encountered certain unnecessary hindrances and restrictions. Study hoped that a solution to the Riemann-Helmholtz problem would be found by simpler means and under less qualified and restricted conditions, and stressed that signs indicated as well, that this was in the process of occurring – here Study pointed to David Hilbert and Luitzen Brouwer.

In an article on Lie's sphere geometry that Study wrote twenty years later he said:

> Sophus Lie had the shortcomings of the autodidact, but he was also one of the most ingenious mathematicians to ever have lived. He possessed something, and this, to the richest degree, was something that is not often encountered, and which is now even more rare: creative fantasy. Coming generations will recognise the value of this far-sighted mind better than does the contemporary generation, which exclusively recognises the value of shrewd clearsightedness, while the interconnectedness of all things that are perceivable, has almost completely disappeared from the line of vision.

And then in a publication in 1924, Study repeated words from his 1888 review of Lie's first volume on the transformation groups: "The coming generations will know how to give transformation groups the place in science that this epic work deserves."

In his written assessment of Lie's work for the Lobachevsky Prize Committee, Klein had also pointed out how Lie had found important errors in von Helmholtz's presentation of proof, how von Helmholtz had implicitly and erroneously conveyed his assumptions from the finite to the infinite dimensional. When Klein republished his treatise on non-Euclidean geometry again in 1921, in a footnote he gave a sort of explanation of the superficial treatment he had given Lie's critique of von Helmholtz in the first edition. The main reason was that Klein at the time only knew about Lie's work on the problem from the 1886 period, and not the detailed treatment that Lie came out with later.

The year after Sophus Lie had been employed by the University of Leipzig, yet another prominent man of science from the fringes of Europe found employment at Leipzig. This was the thirty-four-year-old Wilhelm Ostwald. Together with his wife and children he came from Riga in Latvia, and became professor of physical chemistry. Lie and Ostwald, as outsiders, seem to have had similar experiences in coming to terms with the Saxon way of doing things, and with the German university system. It was however not only common customs that drew them together, for they had children of the same age, and this certainly played a role. During their first years in Leipzig the two families frequently visited one another. Forty years later in his memoirs, Ostwald recalled these domestic visits with pleasure, not least because they were so different from the social and culinary forms that were local customary practice, and which they both found insufferable. The only parties, apart from their own, that Lie and Ostwald and their wives seem to have enjoyed, were the gatherings at Professor Mayer's hospitable house. In the course of his nineteen years at Leipzig as teacher, researcher, (textbook) author, and organiser, Ostwald would, so to speak, come to establish physical chemistry as a science in that city. He was in the forefront of areas such as chemical affinity and equilibrium, reaction speed, catalysis, electro-chemistry, thermo-dynamics – among many others. Ostwald's law of dilution (1889) and the Ostwald method of catalytic combustion of ammonia (1901), among other things, earned him the Nobel Prize for Chemistry in 1909.

In his memoirs, published in 1927, Ostwald also gave a more detailed picture of Lie. Ostwald stressed Lie's position and role as one of the time's foremost mathematicians. He experienced Lie as very zealous in his work of refining and expanding new areas that he had made accessible. But Ostwald maintained that Lie did not look like a man of learning. Lie was broad and rough of physique and had a corresponding facial expression: there was something primal about him, something that led one "to imagine him as a mammoth being". As well, something was lodged in his character that was unalterably Nordic. According to Ostwald, Lie was a stranger to social relations of everyday life – the science in which he involved himself with such overwhelming devotion, filled him so completely that there was scarcely room for anything else. But little by little Lie came to suffer from that "special mathematical sickness", which according to Ostwald was an affliction of the brain, that to certain degree fortunately could be cured, because it did not depend on any changes to the inner organs, but rather was due to over-exertion – a form of fatigue that arose from mathematicians treating the most abstract fields of thought, and wherein the brain was unable to get the necessary rest and relaxation that, for example, physicists and chemists got when they checked their ideas against their practical experiments. Ostwald said only that this illness had led to Lie having to interrupt his work and undergo "a cure, from which he returned, suspicious, mistrustful and irritable." In brief, Ostwald commented: "This also had its retroactive effect on our relationship." Professionally this can be translated to mean that a certain degree of conflict arose between the two professors when, during the early summer of 1894 in a meeting of the Leipzig Academy of Sciences, Lie pointed out great mathematical shortcomings in Ostwald's calculations of natural forms of energy, and then subsequently Ostwald had answered him a month later.

Ostwald concluded his memories of Lie by mentioning his untimely death from a lack of red blood cells: "There was no means of combating the malady, and thus the afflicted always clearly senses the approach of the abyss of eternal night. With the aforementioned knowledge, this gloomy fate must have pressed down especially heavily on him [Lie]."

In retrospect Engel explained that Lie had regained the suppleness of his mind, when he recovered from the breakdown, and as a mathematician he returned completely to what he had been before: "But not as a human being. His mistrust and his irritability did not dissipate, but rather they grew more and more with the years, such that he made life difficult for himself and all his friends. The most painful thing was that he never allowed himself to speak openly about the reasons for his despondency." Engel stressed that even the great distinctions he received did not serve to change this very much. Lie had been chosen a member of every single academy of significance – the only noticeable exception being the Berlin academy, he further remarked. All these distinctions – the awarding of the first Lobachevsky Prize included – Lie experienced, according to Engel, as "a tribute due to him, without, to the slightest degree, mitigating his pessimistic opinion of people in general, and mathematicians in particular."

When in 1930, Lie's nephew, Professor Johan H. L. Vogt, tried to characterise his uncle, he said: "We shall avail ourselves of a popular picture. Every person has within himself some normality and some of what may be called madness [Verrücktheit]. I believe that most of my colleagues possess ninety-eight percent normality and two percent madness. I myself admit to ninety-five percent normality and five percent madness. But Sophus Lie certainly had appreciably more of the latter. The merging of a pronounced scientific gift and an impulsiveness that verges on the uncontrollable, would certainly describe many of the greatest mathematicians. In Sophus Lie this combination was starkly evident."

Norwegian Students

Right back to the time of Niels Henrik Abel, getting a stipend for further studies abroad was something that the most promising young Norwegian men of science could almost count on. And now, when the country's great mathematician, Sophus Lie, was in a prestigious position abroad, Lie and Leipzig were a natural travel destinations for the coming generation of Norwegian mathematicians. At the very least, Lie could certainly give the best advice as to where it was most advantageous to go. The first out was Axel Thue.

Axel Thue took his final public service examination in science in Christiania – and it was still called "the science teacher examination" – in June 1889. But he had already some months earlier received a travel stipend of 1,320 kroner for a period of study in Germany. Thue's application had been supported by both Professor Bjerknes and Docent Holst, and Lie had given his highest recommendations with statements about how he would gladly take on Thue as a student. Lie now counselled that Thue should come quickly once he had completed his examinations, even though the summer semester at Leipzig had begun a couple of months earlier, a fact that had made Thue himself want to wait. Lie argued that the lectures he was giving that semester would suit Thue well – they were introductory lectures on his own theories, which Lie would develop in more detail over the course of the coming two semesters, and the material that Thue missed by arriving a little late, Lie would teach him in private. In order to ensure that he got Thue to Leipzig, Lie wrote a letter to Holst after Pentecost, on June 20th:

> You can tell Thue that it would be clever of him to make use of the Interest I have shown him. Truly it has not been my Custom to make Efforts to get Students into my Special Lectures. Now that I for once have set my Eyes upon Thue, I should feel reluctant should my Offer be spurned.

The subtext to this latter statement was the possibility that Thue would choose to go to Göttingen or Berlin instead. Lie had earlier recommended Göttingen to the Norwegian authorities as a place for other Norwegian science graduates to study. Among others, Vilhelm Bjerknes, son of Professor C. A. Bjerknes, had gone to Göttingen, and young Bjerknes had also gone to Bonn and become an assistant to Heinrich Hertz, who was then doing further work on the study of electro-magnetic waves which he had demonstrated and written up three years earlier. Thinking about Thue, Lie now wrote:

In Göttingen, where I would reluctantly send him [Thue], the Lectures for this Semester are unquestionable simple, and next Semester *presumably* even more simple. For the rest I am reluctant to see Thue go to Göttingen. I sent both Guldberg and Bjerknes there. It would [therefore] be far too one-sided if T. goes there too. In Berlin by all Accounts there is little to do. On many Occasions young Math. have left Berlin and come to me.

Apart from this, Lie now told Holst: "In the summer of 1890 I am coming for a Summer Visit to Norway with *Wife* and *Children*. The Fatherland shines with a double Radiance when one is away."

Thue actually went to Leipzig in the summer of 1889, and he attended Lie's lectures on contact transformations, infinitesimal transformations and transformation groups – and since, naturally enough, he was little familiar with this, Lie also gave him private tutoring. Thue reported home to Holst about the reception he had received from Lie: "In addition to the lectures, we are working together almost every day and I am enjoying myself with his vigilant assistance, even now, in the summer holiday season."

But unfortunately this work would not bear fruit. Lie spent the early holiday period tutoring Thue, but at the end of August, he left for his summer place at Berga, where, among other things, he was busy reading the proofs for the first volume of *Theorie der Transformationsgruppen*. In order to take in a bit of the great world abroad, Thue journeyed to, among other places, Dresden, where he contracted a severe case of jaundice such that he had to stay in bed more or less for the next seven months. And that autumn in November, Lie had his breakdown and was sent to Ilten.

From his sickbed in Leipzig, Thue reported home to the Collegium that in his time with Lie he had only managed to go through the introduction of what Thue called "the Lie discoveries", which required much time and persistent work, and that the programme of studies in this field, by virtue of what had happened, had been broken off really before it got underway. Thus Thue could not announce any independent work in the field of group theory – what he had done had been confined only to a new proof for one of "Lie's most important theorems", and Thue had to inform the Collegium, which had awarded him the stipend, that he had not got "any integrated overview of Lie's magnificent and now world-renowned theories." But on the other hand, Thue reported on a great success. While in Leipzig he had also been following Mayer's lectures on calculus of variation, and been so inspired by these that he had continued on his own. During his illness he was also "rich in ideas", and according to him, even found many of his results in number theory. Thue returned home from Leipzig before he was completely well, and soon received a new stipend in the spring of 1890 – 1,400 kroner – to continue his studies. In Leipzig though, Lie's lectures had been discontinued for the whole summer semester. Thue then decided not to continue with his "interrupted Studies" with Lie, and instead, travelled to Berlin. As a continuation of one of Mayer's lectures, Thue had found a theorem on line surfaces and he gave a lecture on this before the Mathematical Association of Berlin. In Berlin he attended lectures on algebraic equations by Kronecker, on rational mechanics by Fuchs and von Helmholtz, and he reported home how he was inspired both by Kronecker's infectious enthusiasm, and how Kronecker praised Abel to the skies: "one of the greatest thinkers who ever lived". Thue remained in Berlin until his return home to Norway in July 1891. Here, three years later he became senior teacher at the Trondheim Technical Institute, and in 1903 became professor of applied mathematics in the capital city, which was now spelled "Kristiania". During Thue's twenty-year stint as professor, he initiated a Norwegian tradition in number theory that is internationally recognized – with names like Viggo Brun, Ernst Selmer, and Atle Selberg. Thue also had followers in the field of logic, particularly Thoralf Skolem.

For a time during the autumn of 1894 there were three Norwegian graduate students in Leipzig, and two of them at least tried to follow "Lie's magnificent and now world-renowned theories". These were Alf Guldberg and Anton Alexander, who both had received their degree in science two years earlier. The third was the physicist, Kristian Birkeland, who had already become the holder of a university research stipend with a permanent working place at the Physics Institute in Kristiania, and was now on a six-month study leave in Germany. Birkeland's northern lights theory, and the discovery of the method for removing nitrogen from the air, which became the basis for Norsk Hydro's fertilizer production, still lay some years in the future.

Alf Guldberg was the son of A. S. Guldberg, and nephew of Professor C. M. Guldberg, and following his graduation he travelled to Leipzig in October of that year, 1892. He studied with Lie and would remain in the city until the following summer, despite the fact that he did not in any ordinary manner appear in the protocols as a foreign student of mathematics. Lie thought highly of Guldberg and spoke on several occasions of how he appreciated Guldberg's work. Guldberg's first works were linked to Lie's studies of group theory, and were close to Picard and Vessiot's works on linear differential equations.

In 1894 Alf Guldberg came back to Leipzig with a travel stipend for foreign study. After a short time he went on to Paris, where he attended Poincaré's lectures on electrostatics, and Hermite's and Picard's on elliptic functions. A couple of years later, Alf Guldberg studied both in Göttingen and Berlin, and attended lectures by all the well-known mathematicians: Klein, Schwarz, Hilbert, and Fuchs. Alf Guldberg also spend a period of time in Copenhagen where he studied the field of insurance and life insurance (actuarial calculations). These stays abroad also occurred, in part, after he became a university grant holder in 1895 and took his doctorate in Kristiania. Guldberg taught mathematics to students at the Military Academy and the military college, and mechanics to those studying geology. His mathematics went more and more in the direction of the applied. When the state economic examination was instituted in 1905, Guldberg taught actuarial calculation, and later his professional fields included both statistics and probability theory. In some of the university's annual reports, Guldberg was described as professor of actuarial mathematics, even though such a professorship did not formally exist.

Science Candidate Anton Alexander received 2,000 kroner for a foreign study tour in 1894. At the beginning of Alexander's application form, it says that he had been – so involved in school work and coaching" that he had had little time for scientific studies, but that he now wanted to study under Professor Lie in Leipzig, and later, function theory either in Paris or Heidelberg. The application was recommended by Professor Bjerknes and Docent Holst – the two who were mainly responsible for mathematical teaching at the University of Christiania.

During his student days, Alexander had been active in Realistforeningen (the Science Association), where at the beginning of the 1890s he went under the name "Philosopher Alexander". The Science Association, to the melody of the song "Behold Norway's vales abloom" (Se Norges blomsterdal), had written verses to all its most prominent members, and about Anton Alexander, who was extremely tall and thin, they wrote, alluding to his height and his good marks: "To an upland spruce so spare / We Alexander can compare / High as a mighty peak / He looms so tall, so sleek / Tra lalala la la / At him we now but stare / A man of one-point-two so rare / A genius he is trala!"

Alexander had met Lie in Christiania in August 1894. According to Alexander's own words, Lie had on that occasion received him in a manner "particularly amiable", and urged him to journey right away to Leipzig. Alexander later explained in a letter to Holst that Lie had also told him about his many students in Leipzig, and that he would probably remain another year in Germany, but otherwise had not wanted to talk more specifically about it. Lie had then given Alexander one of his visiting cards with a note on it, referring to him [Alexander] as "Doctor" and said that now in the summer holidays, this card would open the Mathematics Institute to him, with its well-appointed library.

Alexander went to Leipzig, probably a good while before the winter semester began in the middle of October, and before Lie had returned from his summer sojourn in Norway.

The professional conjuncture between science candidate Alexander, who was "so busy with school work and coaching", and Professor Lie with his mathematical weight, did not prove to be particularly fortunate. In a series of letters home to Elling Holst, his friend and former teacher, Alexander described the conjunction of his own expectations and the great mathematician. A month and a half into the winter semester – where Lie, besides introducing projective geometry, lectured on the application of contact transformation theory to various fields of applied and pure mathematics – Alexander wrote:

> It is this desperate fantasy – that I shall amount to something, something special. It is now worst of all with mathematics. And I *am* only temperament, only humour, bad humour. I have nothing much to talk about with Lie, he belongs to the type of people that I cannot stand. Although he has never shown anything but goodwill toward *me*, I must say, that what I have seen of him up to this point, has not contributed to him standing particularly high in my eyes. I wonder if he does not have the same feeling about me.

For twenty-four-year-old Alexander it was painfully self-evident that he would not come to "presage anything" in "the world of the mathematical sciences", and he added bitterly: "I do not aspire to continue the suffering." For Alexander now understood school-work and where he felt he really fitted, more and more as a life task – he was merciless in his bitterness and repudiation, and he wrote to Holst:

It [mathematical research] is certainly nothing to aspire to, nor to become a colleague of, or allow oneself to become a mental monstrosity like Lie. Yes, I know, I must be cautious: Birkeland has told me so. But at least I can write about it to you.

Alexander lived in the same house as Birkeland in Leipzig, and the two friends – Birkeland was three years older than Alexander – seem to have spent much time together. The defeated and depressed Alexander, wrote about Birkeland in the following manner: "His knowledge is not especially extensive, and his opinion is not always sound, but he has faith. He has an indifference to everything, what I could call an aesthetic comprehension of life, and he has a more sure and instantaneous comprehension than what I have."

Birkeland also knew Lie from earlier. When Lie, in April 1893, had been in Paris and was received with such honours by the French men of science, Birkeland was also in the city, and on several occasions had chatted with him. Birkeland had then been in Paris for some months, working with energy transfers and wave diffusion in connection to Maxwell's equations, and he had begun to develop a fruitful collaboration with Poincaré, who had high expectations of Birkeland. Birkeland, who at the time was involved with applying for an adjunct stipend (the position of a university research grant-holder) at home in Norway, then asked Lie to support his application by sending a recommendation based on a work he had shown Lie on "enumerative geometry". For Birkeland too, it was the university lecturer, Elling Holst in Christiania, to whom he reported from Paris in the spring of 1893:

We [Birkeland and Lie] spent a considerable time together so I got a rather complete impression of him [Lie]. Above all, the first thing that struck me, having heard of his dejection, was that he was in a very good mood, and he looked extremely well. It might well be that the circumstances (for here he is being eulogised to a considerable degree) have brought about a temporary change of condition; I could not say for sure; but it still seems to me rather probable that it all could stem from this.

He has an ugly habit does this great man Lie: he is always belittling the scientific works of others. And the good man is not gracious, but when it comes down to this, he criticises himself almost good-naturedly, indeed, one has almost to forgive him, when toward the end of his diatribe he chuckles and says, "Yes, indeed, I am so mean as to speak this badly of others!"

Birkeland added that Lie was not very interested in the work of others "unless they were occupied with groups," and he almost gave the impression that mathematicians who were not interested in group theory, suffered from "one or another mental deficiency". In the letter's margin Birkeland commented that perhaps this was a bit too strongly worded: "He [Lie] is clearly a kind man."

That autumn of 1894 in Leipzig, Birkeland seems to have worked mostly on his own: it is likely that he had been given admission to all the laboratories of the university. How much he saw of Lie is uncertain; perhaps he sought Lie's advice whenever unwieldy mathematical problems raised their heads and required an approach that would lead to their solution. Birkeland wrote to Holst that on one occasion they had spoken about him [Holst], and Lie had burst out, "Yes, we must find a better appointment for this man; it should not be difficult as he is indeed considerably popular at home."

Lie, in the draft of a letter in connection to the applications for travel scholarships for Guldberg and Alexander, presumably addressed to Professor Storm in Christiania, Lie had commented ironically about Holst's students and his teaching. Were it the case that Alexander could only "invoke a good Examination and Merits in Holst's Seminar", then he ought "unequivocably to rank behind Alf Guldberg", Lie wrote, continuing: "*Just as Everything that King Midas touched turned to Gold, what is quite certainly similar is that whoever sits himself down at Holst's Feet, becomes in his Eyes a Genius.* This is said in full Recognition of E. Holst's Amiability and many-sided Interests."

Otherwise, Birkeland in Leipzig reported to Holst (autumn 1894) that he had the impression that Lie wanted very much to return home to Norway, and he thought that Lie would come as quickly as he was able to do so. In any case, he stated that he "without doubt would thrive much better at home". But it was said that for a time the relationship between Lie and Birkeland was not a completely happy one, and this was certainly because Birkeland had told him he considered Lie had been too sharp in his criticism of German mathematicians in his foreword to the third volume of the transformation group theory. And Lie did not reinstate his friendship until he got to hear about what Birkeland had done one evening at the Neues Theater in Leipzig. The drama which was performed was Henrik Ibsen's *Lille Eyolf*, and when the curtain went up on the third act and Engineer Borheim should raise the Norwegian flag, he raised instead a Swedish flag with the insignia of the Swedish-Norwegian union on it. Birkeland had been so irritated that he had jumped to his feet from his seat in the dress circle, and shouted, "Es ist eine Skandal! Ein Schwedische Flagg! Und Sie sollen ein Norweger sein! [This is a Scandal! A Swedish Flag! And You are supposed to be a Norwegian!]" The audience certainly reacted, and Birkeland immediately left the hall, stormed in to the editor of one of the city's largest newspapers and had printed in the next day's edition a caustic notice about the flag episode. In the performances that followed at the Neues Theater a proper Norwegian flag was used. "Indeed, that was well done!" Lie was said to have declared, chuckling, when he heard of the episode, and immediately invited Birkeland back again. (*Lille Eyolf* was published on December 11, 1894, and a large printing was soon sold out, and it went on to play in many locations across Europe.)

The stay in Leipzig gradually improved for Alexander, though he continued to report his homesickness, even though Birkeland was "a splendid Friend, constantly happy and constantly courageous." It was Alf Guldberg who gave him "courage again" and made sure that Alexander would nonetheless decide to remain in Leipzig until the end of the semester. After having given up on his attempt to follow Lie and mathematical research, he spent more time at the institute library where he studied "a considerable amount of pedagogical mathematics", and he attended a "pedagogical seminar" which had been underway for a year in Leipzig, and which had practicum sessions every two weeks. Alexander remained in the city to the end of the semester, studying the German school system and specialised instruction in professional fields at the university level (which he did not find to be much better than at home). On the other hand, he was very impressed by how much more advanced the French were in this field. He gained knowledge of the French system from one of the students from l'École Normale Supérieure (probably Drach), who was studying with Lie this semester, and who, according to Alexander, was "completely knowledgeable" when it came to the subject of specialised teaching. Alexander also told Holst what this Frenchman "in all confidentiality" had confided to him: "Diese Deutsche haben viel gelesen, sie sind sehr gelehrt, aber sie haben kein mesure" ["These Germans have read a great deal, they are very knowledgeable, but they have no finesse."]

At the end of the semester in Leipzig, Alexander went to Paris, and he was still in that city when Lie arrived in April 1895, to give his lecture on Galois on the occasion of the centenary of l'École Normale Supérieure. At that time Alexander met Lie on several occasions, and was not able to restrain himself from writing home to Holst about his offended feelings: "This man revolts me to the very deepest fibres of my heart. He does nothing but go around speaking ill of me to everybody whom I esteem." It is uncertain whether this referred to Norwegian scientists they both knew, or whether it had to do with the Norwegians among whom he passed the time in Paris. In any case, Alexander spent a great deal of time with the Norwegian artists who at the time would gather at Café de la Régence. He spent the most time with the poet Vilhelm Krag – they were old school mates from the Kristiansand Cathedral School – but also with the Bohemian writers, Hans Jæger and Knut Hamsun, and about the latter (the future Nobel Prize winner), Alexander reported: "Hamsun, who has just finished a new book [*Pan*], is a very interesting fellow. Big and handsome, as he is, he dazzles all the ladies, and to the highest degree makes the most of his appearance and his powers, being fond of plastic arts of himself, and small personal idiosyncracies. He always goes around in black gloves, indoors as well as out, never – out of principle – visits the theatres, is passionately fond of playing whist. For a Parisian he is not out much on drinking sprees, and according to the opinion of intimate friends, is extremely warm-hearted. In many ways makes a naïve impression."

Alexander returned home and in many ways became a man of high profile in the field of Norwegian school policy. He worked as a senior teacher of science, both in Bodø and Skien[37], before becoming senior teacher at the Kristiania Cathedral School in 1903. He was later rector of the colleges in Lillehammer, Drammen and Skien, before he was made inspector by the Pedagogical Seminar in the capital – in order to increase the pedagogical competence of scientists and philologists, the Pedagogical Seminar took up regular activities from 1908. Alexander made astronomical observations for several polar expeditions, among others, for Nansen. For a number of years he was member of the executive and chairman of the "National Association of Philologists and Scientists". In the 1930s, Alexander was a Member of Parliament for the Liberal or Left Party, and he took the initiative to restructure the Teaching Council. Through all his schools activity, Alexander was

[37] Bodø is a city on the Norwegian coast north of the Arctic Circle; Skien, the hometown of Henrik Ibsen, is located on a river near the southwest shore of Oslofjord.

a zealous spokesman for pedagogical competence and specialised professional knowledge, and his own mathematical knowledge was given expression in several textbooks.[38]

[38] A. Alexander's mathematics books were in use until the late 1960s, with the editions of the last twenty-five years being edited and revised by Th. Skjulstad.

Back to Norway

It was Fridtjof Nansen who broached the subject of Lie's discontent "down on the Leipzig plains". Nansen had crossed Greenland on skis in 1888 – a bold and exhausting expedition that brought him much glory, and about which he wrote in a very successful book. Nansen was back from Greenland in the spring of 1889, and in the autumn he married Eva Sars, sister to Ernst and Georg Ossian Sars. Eva Sars was then a well-known concert singer. She had studied in Berlin, and her repertoire included opera, German lieder and Norwegian folksongs. Nansen received 7,000 kroner from the publishers, Aschehoug, for his voluminous manuscript on the tour across Greenland on skis, and in the spring of 1890 the book began to come out in instalments. It was simultaneously translated into English and German. In the autumn of 1890, Nansen left Norway together with his wife Eva, for a lecture tour of Germany. He had an audience with Kaiser Wilhelm II in Berlin. The Kaiser was presented with a copy of the German edition of the Greenland book, *Auf Schneeschuhen durch Grönland*, a book whose popularity did much to spread the sport of skiing, which in Germany and the continent was still considered something rather strange that belonged to the far north. In the course of his German lecture tour, Nansen and his wife Eva arrived in Leipzig, and perhaps Lie was among the many who listened to Nansen lecture about his Greenland expedition. In any case, Lie and Nansen met in Leipzig, and Lie invited Nansen and his wife Eva to visit him and Anna on Seeburgstrasse. The invitation was for November 16, 1890, but Nansen had declined – with a short letter in which, on behalf of his wife and himself, he thanked Lie and his wife warmly for the friendship they showed by inviting the Nansens to their home that evening, and he continued: "As I told You however, my wife is completely mad about music, and since we shall only be in Leipzig for one day, she would very much like to attend the Opera here. Thus it is that I hope You will excuse us, as in such matters I am my wife's obedient servant."

But Lie's position, well-being, and work situation in Leipzig made an impression on Nansen, and when he returned home, he talked about Lie's malcontent in such a way that others began also to take up the idea of getting Lie back to Norway, and when Nansen left for his next great expedition at midsummer 1893 – this time to the high arctic, the "Fram" Expedition, which would extend for over a period of three years – it was others who took the lead in getting the famous mathematician home to Norway. One of them was Nansen's good friend, the geology professor, Waldemar Christopher Brøgger, who on his tour abroad to Italy in 1891–92, took on the errand of travelling via Leipzig so as to talk to Lie about eventually being invited back to take up a position in Norway. The plans were expedited when at Christmas 1892, Elling Holst managed to get the country's leading poet, Bjørnstjerne Bjørnson, to join him in a campaign whose goal was to get Parliament to establish a professorship with a salary so high that Lie would quickly accept the offer and come home. Lie replied, in a letter to Holst:

I am deeply moved by Bjørnson's and Your Idea. I suppose You are the Father of this Plan. Whatever the Designation – Professor of "Transformation Group Theory" refers to in particular, I of course greet this Idea with double Joy.

Bjørnson now cast himself into the thick of things that had to do with the project of getting Sophus Lie back to the fatherland. Bjørnson himself had returned to Norway in the late autumn of 1887 after his stay of many years in Paris, and was involved in countless matters and relations in Norwegian public life – above all, with the question of Norwegian independence in relation to the union with Sweden. And in the union struggle that was approaching, Bjørnson and many others with him, felt that it was decisively important to gather all of Norway's great men back into the fatherland. On February 26, 1893, Bjørnson wrote:

Dear Sophus Lie!

Now, when our people are refused the right of independence, it is for me the people's intellectual power that is the greatest evidence for the renunciation of this outrage. In general, no nation in terms of numbers and time of growth, has won a greater right to lead itself than our own, and this right we have won through fame.

Yours is so great that we have no means of dispensing with it right now. But it is necessary that we continue to do so, so long as You continue to live abroad as a professor at a foreign university.

What if You now, precisely now were called back? Your professorship indeed stands open – if right now a greater salary offer were made to You, what would You say?

The sum You are getting in Leipzig could hardly be offered; but answer me; would You act on this if it were 8 to 10,000 kr.?

For the sake of the fatherland in its struggle? Time reminds us that You are ours, that this is the best sentiment for us, and indeed, it is the strongest means of invoking our parliamentarians' patriotism.

Yours respectfully,
Bjørnst. Bjørnson

And Lie replied:

Most Highly Esteemed Mr. Bjørnstjerne Bjørnson!

Thank you, and again thank you, for your warm-hearted Letter, which I shall always cherish. It is so seldom that a Mathematician finds Understanding and Sympathy. Therefore it makes it doubly good to receive a Glimmer of Sunshine from another Direction.

I am and will always be Norwegian in body and soul. Therein sits my greatest Honour. My young Students quickly learn that I set much greater store in being cited as a Norwegian than as *famous*. Therefore, I would very much like to try to cut myself loose from Leipzig, if my Fatherland will offer me a Position that I can accept, with Respect to maintaining the Well-being of my scientific Work, and at the same time, my Family's Future.

Lie made it clear that he would not demand "full Compensation" for what he earned in Leipzig, or what he "with Certainty" could expect in the future in Leipzig – nor did he want to find himself – as had been his lot earlier at the University of Christiania – "at the lowest Place at Table". He went on to comment to Bjørnson: "I too have my Vanity, and in any Case I must secure for myself such a Position that I can work with full force, free and unhindered." Lie tried to give Bjørnson a description of his position as it was now, in the spring of 1893 – first of all, he had a "significantly larger salary" than a professor in Christiania had ever had, besides which there was a considerable, though variable income to be derived from lectures, examinations, dissertations, together with part of the faculty's administrative income – "Here in general one is paid separately for every professional activity", and:

I am the Manager of a first Rank mathematical Library, a remarkable Collection of mathematical Models, and generally of a mathematical Institute, which scarcely has its Like in the whole World. I have two state-paid scientific assistants, Prof. Engel and Private Docent Scheffers, together with two lower Assistants.

Moreover, Lie explained that among the students in Leipzig he had a rich opportunity to find "useful Forces for the Execution of particular Investigations, which I consider extremely helpful." By means of an extra granting of 3,000 marks per annum he was also in a position to obtain for himself "all the mathematical Equipment" that he desired. But what was still more important was that he had "got three extremely remarkable young Collaborators (Engel, Scheffers, Hausdorff) who are at my own scientific Predisposition." In particular, Lie pointed out Engel for Bjørnson's attention, Engel, who had now supported him for eight or nine years "with remarkable Ability", and understood how to give "broad and comprehensive Form to the Theories, which is as necessary to Mathematics as it is to Literature and Art, whenever new Thoughts are to find full Comprehension." Lie added that evidently the main motivation for these young men was to be able to work together with him, but he also brought attention to the fact that he had to give these co-workers a little recompense in the form of money – about 1,000 marks – "ridiculously Little in Relation to the Amount of Work carried out". Lie now reiterated, in this letter to Bjørnson, that if he now returned to Norway, then working together with these German collaborators would become much more difficult, and "the Opportunity to find new Helpers, much more difficult," and he would have to "pay more for the Assistance". Indeed, he would never be able to find such collaborators in Christiania, nor such who "in the verbal Sense were adequate." This would obviously be a serious loss, but he added:

I can nevertheless overlook all of this, for I long for Relatives and Friends, and perhaps even more, the Land and the People. I imagine that my physical and mental Powers would be much better sustained among Norway's Mountains than on the Plains of Leipzig.

Lie also explained to Bjørnson that never before had he seriously reflected upon the fact that others among his compatriots were requesting his return home: "But the Matter becomes rather different, *when it is You who propose the Motion*, for there is Resonance in Your Words." Lie went on:

You understand that one who has spent his Life straining his Mental Abilities to the utmost, and who, on the other Hand, cannot count on any supplementary Income, which the majority of Norwegian Professors in fact have, then he cannot give up an economically and scientifically good Position without the Certainty of getting an approximately equivalent Position again.

Lie then examined Bjørnson's proposal in concrete terms. He would come if the Norwegian Government could guarantee him 9,000 kroner per annum, and in addition undertake the same obligations as the Saxon state now had toward his wife and children who would otherwise be unprovided for if he were to die – and he added, "I was recently somewhat worn out, but am now again strong and vital, so the Risk to the Norwegian Exchequer is hardly great."

Whereupon Lie set down "as an Old Master of Calculations" a financial estimate that he admitted was perhaps a bit defective, but – when in 1872 Parliament had appointed him an extraordinary professor, it had been "without Expectation other than scientific Work" – and he had applied himself equally to the highest degree, to this scientific work over the past seven years in Leipzig, without costing the Norwegian exchequer any of the salary that Parliament had given him the right to. (Elsewhere Lie compared himself to Broch, who was also an extraordinary professor, and had certainly had his Norwegian salary paid to him, even after receiving his directorship in Paris.) This sum, together with interest, and the interest on the interest, for these seven years, had been a saving to government, and could be calculated as more than 50,000 kroner to the good, Lie maintained, a sum that would more than cover the salary increase he now demanded. Lie also gave the assurance that he would certainly be worthy of the money. There was so much in the homeland that he thought he "would manage to get onto new and fruitful Tracks" – not only at the university – "even about our Officers and Technicians I have words to say." Lie concluded this letter to Bjørnson with the following words:

In the old Days, Norway sent out the Vikings: they were the Terror of the World, but nevertheless, they brought back new Ways, indeed, they found new Countries. In our Days, it is our Poets who impress the greater World with the Audacity of their Thinking and the Brilliance of its Form. Do not think ill of me that I make a Claim upon a little of the same Blood. Without Phantasy one would never become a Mathematician, and what gave me a Place among the Mathematicians of our Day, despite my Lack of Knowledge and Form, was the Audacity of my Thinking. If I veered in too close to the Reefs and Skerries, and even suffered a Shipwreck, I still managed to pull myself together again and carry on to a new Land, in the Struggle for Truth and Justice.

Yours in Deference and Gratitude, devotedly,
Sophus Lie

In a later letter to Bjørnson, Lie deepened this, saying that the Norwegian state, in the same way as the Saxon state had done, ought to look after his wife and underaged children in the event of his demise. If this were difficult or unacceptable to the Members of Parliament, he would nevertheless appreciate it if Parliament were able to conclude that he should receive an annual salary of 9,000 kroner. But if something like this should be taken up, he would certainly not come immediately – in any case, not this first year (1893), and probably not next year either, but "it still ought to be reasonable that sooner or later I come under these Conditions." One of the reasons that he, sooner or later, would return home, he now wrote, was that his wife Anna felt that he got too little exercise in Leipzig, and that the natural environment around Christiania would lure him out "as in the old Days". In an extant draft of this letter written to Bjørnson, he wrote that in fact Anna felt that if he got 9,000 kroner in salary, he ought to return home, even if no promises were made for her and her remaining under-aged children. Lie continued, "As for me, I am holding rather fast to these Terms, but I can see myself, with Time, coming around to my Wife's Point of View. Indeed, we both consider it more likely that my Ability to Work will be better protected in Christiania which has such beautiful Surroundings, that are always dragging me out into the fresh Air, while here, I only go walking as a tedious Duty." And he added in this draft: "It was certainly the case that when I first came here, I was afraid that my best Times were over."

In a letter that he later sent to Bjørnson, he stated: "But I, who have never been able to think of making a Fortune, thus reluctantly give up the Certainty that my Dear Ones' economic Future is at Least secure" – added that if Parliament should come to give him a salary increase over and above the 9,000 kroner, then he could not say how he might react: "Conditions change from Year to Year in more than one Respect."

The last thing that Lie asked Bjørnson about, was that there should not now be any official discussion about this in Norway – the matter must not get "a false Appearance" that he, Lie, was in direct negotiations with Norwegian state power, rather than with private men. Lie drew attention to the fact that if he had contact with other state powers, he was duty-bound by oath to inform the Saxon Ministry, and added: "One gets to howl with the Wolves one travels with."

Bjørnson and Holst also got others to support the proposal for a distinct and high-salaried professorship of transformation theory for Lie. Amund Helland was one of those who quickly and happily supported the proposal. He cited laudatory reviews of Lie's works in *Comptes Rendus* and in the *Bulletin of the New York Mathematical Society.* At first glance, Professor Bjerknes seems to have avoided placing his name in support of the demand, and at the end of March 1893, Bjørnson wrote to Bjerknes and asked if he perhaps felt that Holst's words – that already Sophus Lie was, in relation to his mathematical work, "of similar significance as Abel" – were too strong, and if this was the reason that Bjerknes had declined. In that case, the formulation had to be changed – it was, above all others, Bjerknes' reputable name that was now required, Bjørnson added. Whether this was really Bjerknes' reason seems somewhat doubtful.

The Faculty of Mathematics and Natural Sciences, in a meeting on April 20, 1893, recommended to Parliament, in written form, "that in the event that Professor Sophus Lie returned to Norway, he would get a salary increase personally commensurate with his great scientific significance". The one faculty member who put himself into the lead on the issue of trying to get Lie home was Professor W. C. Brøgger. In a letter to Holst, Lie asked that they move cautiously. It was important to Lie that the matter be handled such that it not be construed that he were in contact with anyone in government. "Nor does it sit well with my Dignity to allow Messrs. Bjerknes and the Collegium to express themselves as to my Competence. I reject the very Thought of this with Indignation."

Henri Poincaré sketched during a lecture, Paris 1908.

To the right, from the top:
Arthur Tresse, Ernest Vessiot, Élie Cartan.

The photos on this and the next page were taken by Carl Størmer, Professor of Mathematics, who, between 1893–96, took a series of photographs of famous men and women in the capital city. He himself wrote about this (in the magazine St. Hallvard*): "I was a young student of 19 on this occasion and had got my hands on an amusing detective camera. It was a round flat cannister hidden under my vest with a lens that stuck out through a buttonhole.*

Top: *Sophus Lie (on the right) together with Amund Helland.*

Middle: *Sophus Lie at the return of Nansen's Fram Expedition on September 9, 1896.*

Bottom: *Kristian Birkeland.*

Under my clothes I had a string down to a hole in my trouser pocket, and when I pulled the string, the photo was taken. I strolled down Carl Johan, looked for a Victim, greeted them, got a sweet smile, and pulled. Six pictures each time, and so home to change the plate. I lived in the neighbourhood and had lots of time, so in the end I had a collection of over 500 photographs."

Left: *Professor Axel Blytt in a carriage on his way to the East Railway Station, Christiania.*

Below: *the same Blytt in front of the university.*

Below: *The two foreign mathematicians at an outing to Sognsvann together with Kristian Birkeland.*

Above: *In front of the university, after Sophus Lie's lecture at Urbygningen, in the old University complex on Karl Johan gate, Kristiania. Alf Guldberg (on the left), who later became professor; in the middle, the American Edgar Odell Lovett (later Director of the Rice Institute, in Texas), and the Austrian Carl Carda. Størmer gives the information that these three, together with he himself, had attended Lie's lecture in Auditorium No. 2. There is much to indicate that this was in the spring of 1896.*

Elling Holst with his daughter, photographed by Carl Størmer.

Elling Holst gathered together rhymes and jingles for the book he published in 1888 entitled Norwegian Picture Book for Children. *He followed this up with a* new collection *(1890), and a* third collection *(1903) – the first of these books is always reappearing in new editions.*

Lie wrote from Leipzig, thanking Holst for these books, and said that the children loved them. The example here roughly says "Jump," said the goose. "Dance!" said the Fox. So we jump, so we dance, So squat we down in our pants! One two three! Don't fall over!

Hoppe!» sa gaasa.
— «Danse!» sa ræven.
Saa hopper vi, saa danser vi,
saa sætter vi os paa huk!
En to tre!
Dætt ikke for de'!

To Holst, Lie stressed that the conditions he had presented in the letter to Bjørnson had been as modest as possible, "since for many Reasons I desire to come home." How great his "economic loss" would be, was impossible to say, "as my Income could not accurately be determined in Advance," but: "Those who know me know I have never lied in order to gain any Advantage. But my Position forbids me from going into Details." Lie's annual salary was known only to him and the ministry, and were he to mention the sum, and this were to find its way into the Norwegian newpapers, he would thereby "according to German Conceptions be compromised." But to orient Holst to his situation, Lie informed him that there were professors at the University of Leipzig who had an official annual salary of around 40,000 marks – but the mathematicians had much less, however recently one mathematician had refused to accept an appointment to Munich for 12,000 marks permanent annual salary (12,000 marks was then equivalent to 10,800 kroner).

In a letter to his friend Amund Helland, who seems gradually to have become central to the campaign to get Lie back, Lie expressed his desire to cut himself free from Leipzig, and he thought he could work for the mathematical field "with Energy in the Homeland [Norway] as well". He now hoped that Parliament would make a good proposal and come out with an acceptable offer. Under no conditions would he come for 8,000 kroner, 9,000 kr. was "acceptable", "although that is not very tempting to me. 10,000 kr. I would consider a propitious Offer." In summary, Lie told Helland:

> What is pulling me to Norway is particularly two Things. 1) the beneficial Influence that Norwegian Nature has on my Well-being and my Ability to work [and] my Desire to work. 2) The desire that my children should become Norwegian.
>
> What is holding me fast here is 1) I am working on a higher Pedestal here, 2) that the economic Provisions are more propitious, particularly for the Wife and Children once I am no more.

In order to invoke acclaim and support for the request that Lie be brought back to Norway, Holst wrote a sixteen-page article about Lie in *Nyt Tidskrift*, in which he pointed out, in every way possible, the greatness and importance of Lie's mathematical works, and his position in the German and French mathematical milieux. Holst pointed to the completely unique position that Norway now occupied in the mathematical world – by the very fact that with Abel and Lie, Norway was in the very forefront – but to those who thought that this could be understood in the same way as literary or musical fame, Holst had this to say: "It is easier for one who has only seen a bucket of water, to imagine for himself an ocean, that it is for one who only knows mathematics from his school days, to conceive of the nature and scope of *Abel's* or *Lie's* thinking, or their greatness." Holst concluded: "Let us show that we have the means of having our great men in our midst. Perhaps when everything is said and done, we *do not have the means* to allow the great lands of culture to have them."

In *Folkebladet* on April 30, 1894, Holst also wrote an article about Lie and the parliamentary motion that was put forward regarding a particular additional grant to put Lie into a condition whereby he could leave "his honoured and golden Exile."

The final outcome in the matter came when Parliament, on July 9, 1894, by a vote of eighty-nine in favour and twelve against, decided to award Lie 4,000 kroner in addition to the ordinary salary of 6,000 kroner. In the Departmental papers it was written that 2,000 to 4,000 kroner were set aside per annum to assist Sophus Lie "wend his way back again to his Position at our University" – and that Parliament had assented that his title should be "Professor of Transformation Group Theory".

Bjørnson, who was abroad again, wrote enthusiastically from the town of Schwaz in the Austrian Tyrol on July 12, 1894, to his Danish friend Christian Hviid:

> Have you seen that we have voted home from Leipzig our great mathematician, Sophus Lie? That we are giving him a double salary? It is so very pleasant that twice this century we have had the greatest of Contemporary mathematicians! The Paris University has annually expended its stipends to young men to go and study under him. The same thing will now happen in Germany. And so too, we have Fridtjof Nansen, as you know, and our great political struggle, our literature, music, art. Ça ira! [It's moving!]

Several days went by before Lie replied to the offer from Parliament, and during this time rumours were circulating that Lie, despite the offer, would not accept the appointment. And what misunderstanding this was based upon no one can say. In any case, not until a week after the parliamentary offer, did Lie's old friend and doctor in Christiania, Axel Lund, dare to send congratulatory greetings and sincere expectations of seeing Lie in Christiania again "in the old Circle of Friends". Also at this time, C. M. Guldberg wrote to Lie to say that the Ministry would accommodate his wishes.

Lie had accepted. He would come "as quickly as Conditions allow". He wrote to the departmental section chief that this could only happen "after some Time has passed", and qualified this by saying he first had to complete the publication of his large works, which were now in the process of being printed in Leipzig.

In the late summer of 1894, Lie and his family arrived in Norway for their holidays. This was the autumn during which Kristian Birkeland, Alf Guldberg and Anton Alexander were in Leipzig.

It was not until a year later that Lie reported to the Norwegian Department of Church and Education that the situation had changed such that he would try to divide his time between Leipzig and Christiania. In the coming two years he would be able to spend only about eight months in Norway – from the beginning of March to mid-October, and teach in Leipzig only during the winter semester. Under these circumstances he made it clear that he would not demand the full salary of 10,000 kroner, but would half of it be suitable? – a compensation for "the Loss of Income that my Departure from Leipzig, repeated Journeys with Family and Effects, Rental of two Accommodations, Money for Schooling, etc., that are entailed." It seems that this was also confirmed by the Norwegian authorities – in any case, Lie considered that former Prime Minister Johan Sverdrup had made promises that would afford him economic security.

In February 1896, Lie wrote to his friend Ernst Motzfeldt in Christiania, and asked him personally to talk to Cabinet Minister Francis Hagerup, and convey Lie's plans to him, that from the very first days of March he would begin lecturing in

Christiania. Lie wrote about his motives in a letter to Motzfeldt, who had served as Minister of Justice under the Conservative government of Emil Stang between 1893-95, and since Motzfeldt had thus been a ministerial colleague of Hagerup. In this letter, Lie revealed the following:

If I am to return to Norway permanently after myself having opened up a huge and splendid Field of Work in Leipzig, then I must have some guarantee that I can find in Norway a Circle in which I can carry out satisfying Work. If this is to succeed, I can only prove to myself by Experiment. For the Moment, I have precisely the best Opportunity for making such a Test. A Series of Lectures and University Matters are now the Order of the Day [in Norway]. I propose to engage in considerably detailed Discussions about these Things. The Goal, which hereunder is taken, from my Point of View, is to gain some Sign as to whether or not Norway has any Use for me. If, for ex., they want to fossilise higher Mathematics, as some individual Persons now understand the Matter, then I shall have nothing to do with Norway.

The "Series of Lectures and University Matters" which stood on the agenda, was a schools- and university debate in which Lie had already participated, with three contributions to *Morgenbladet*. Lie wanted to calculate the strength of his influence in Norway, while at the same time keep up his position in Leipzig. He explained himself to Motzfeldt:

For many Reasons I would like to secure for myself a Retreat in Germany, the same Way that I now keep open a Retreat in Christiania. That the Saxon Minister of Culture, of *his own Inititative* has opened the Way for me to have such an Arrangement, to be able to refresh my Energies at Leipzig University, is a Distinction that is difficult to misconstrue by those familiar with German University Conditions.

But things were not to go as Lie had planned. Two months later Lie gave up these plans to divide his working time between Leipzig and Christiania. There were probably several reasons for this. Lie himself seems to have considered that the major one was the bureaucratic dilatoriness of the Norwegian authorities. He had received leave of absence from the Saxon ministry, and as soon as the winter semester concluded in Leipzig – at the beginning of March – he thus travelled to Norway to begin lecturing, but without having received an official response from the Norwegian side. But not until the middle of March, a couple of weeks after his arrival, had Lie sent a letter to the Prime Minister apologising for writing so late – this was a "Misconduct", but, Lie noted in his own defence, in Saxony such matters were dealt with in the course of a day or two.

It was not until May 2nd – that is, about two months after Lie's arrival – that the government took a final decision, a decision that was completely in line with Lie's wishes. But by then, Lie had already gone back to Leipzig, and it seems uncertain as to whether he knew about the government's decision before he departed from Norway. On the 9th of May, from Leipzig, Lie sent a letter to Cabinet Minister Hagerup, in which he wrote that since he had not received a reply, in the course of the month of April, to his letter from the middle of March, he had assumed that everything was "back to the Old" – and he had communicated the same to the ministry in Dresden, where he had also partially withdrawn from the leave of absence already granted him. He wrote to Motzfeldt: "Dear Ernst! I have an eternal Anxiety." At the same time he reported that his absence from Leipzig for these two months had led to extra costs for him, in so far as he had had to pay for the travel expenses of two foreigners whose visits to Leipzig had been in vain thanks to him.

It had been the parliamentary decision from July, two years earlier, which had made it difficult for the government to find a rapid procedure for dealing with the matter. In the motion that had been tabled May 2nd, it was agreed that the government would overlook the decision of July two years earlier, to change Lie's title to "Professor of Transformation Group Theory". As before, the title was to be "Professor of Mathematics", and on this basis he was granted "Freedom from Service in his official Post for the Winter Months from the Middle of October to February's Departure in the Years 1896 and 1897", and for this arrangement during these years he would receive only an annual salary of 5,000 kroner.

When this decision was sent out by the government on May 19th, Lie's report with the assumption that everything had gone "back to the Old", had arrived. In written form, Lie was then asked by the Department if his last letter to the minister should be considered as "an Application for Full Freedom from Service" in regard to the position at the University of Christiania, that the amendments of May 2nd implied; that is, that his title should be "Professor of Mathematics". Nor did the Department neglect to point out that Lie had certainly been impatient in this matter – in any case they reminded him that shortly after his first letter in March, he had been privately informed that his proposal to divide his teaching time between Christiania and Leipzig would be confirmed – and that in fact the same thing had been communicated to him following the government's provisional considerations of April 22nd.

Why did Lie nonetheless go back to Leipzig, and this right before the government's final response? Had Lie, with his letter about everything going back "to the Old", pre-empted the government's change of the title to his professorship? Or perhaps Lie, in the intervening two months, had found out already what at the outset he had planned to use two years to determine – whether his views were taken sufficient account of, as to whether or not Norway had "any Use for him?"

During these two months in Christiania, Lie had contact not only with former colleagues, but also students and friends. The Mathematics-Natural Sciences Faculty did nothing to help straighten out Lie's work situation, but Lie gave a number of lectures at the university – and both Alf Guldberg and Kristian Birkeland seem to have been in his audience. Almost nothing is known either, as to whether during these two months anything happened in his private life that might have given Lie reason to change his plans.

Shortly after his return to Norway in March, his sister Thea had died – Thea was the sister who had married Johan Vogt in Moss. Sophus attended the funeral, and reported home to Anna that he had spent much of the time together with his nephews, the Vogt boys. Otherwise, it seems that during this stay in Christiania he stayed with his sister Laura at the Eugenia Institution. He sent several postal cards home to Anna and the children, and recounted outings with Laura and the Vogt boys, into the hills of Frognerseter. Apart from this, he described regular gatherings with friends at the home of Ernst Motzfeldt. He had been to a party at the home of Professor Bjerknes "whose Wife sends Greetings". He described how he had been to the home of A. S. Guldberg "whose Son Alf is going to be with me" – and Lie had had with him his two foreign students and they had had a lovely Time, and Professor Guldberg sent Greetings to the Wife. Another evening he visited Axel and Valborg Lund and "had the most charming Time". His advice to Anna and the children was that they must "do everything reasonable for the Maintenance of Health."

Sophus also received regular reports from Anna, and there was nothing upsetting in these reports from Leipzig – there were certainly some problems that had to do with Dagny's schooling, but Herman was doing splendidly and got good marks. Sophus promised him "a Reward", but "watch Yourself when You play with a Spear like the Indians". And with the approach of Anna's April 22 birthday, he told the children to ask their mother for ten kroner [crowns] with which to buy birthday presents for her, and urged them to celebrate the birthday in the normal way: wine with dinner and cakes with tea – and he would pay an additional five kroner for an outing, or some such. As late as April 26th Lie wrote home to say that he reckoned that everything would go well with his position in Christiania – that the reason why he had not received a formal answer from the government was that "the Matter is in Circulation", and to Anna he said he would soon be back in Leipzig to move the family to Norway. (It appears that Anna did not respond to this immediately.)

The decision to interrupt his stay in Norway seems thus to have come to him suddenly. He did not even give himself time to inform his colleagues at the university of his decision. Lie gave Kristian Birkeland the task of informing everyone about the sudden change of plans. In a letter that Birkeland wrote to Lie in Leipzig on May 18th, he said that he hoped he had carried out the task that Lie had set for him, and had done so in the best manner – in any case, Birkeland reassured Lie that he had acted in good faith when he told Professors Schjøtz and Guldberg, that Lie, "due to happenstance" had not returned to continue his lectures. As for his own opinion, Birkeland wrote:

> For me personally it was very dismal to find out that You were not returning to continue Your interesting Course. I had just begun to hope that with Your help I would have a useful tool in hand for the solution of the tasks that I have set myself to concentrate all my powers upon in the near future. Although I have certainly tried to the best of my poor abilities to weld the necessary metals together, I have still not come very far, and perhaps would be otherwise more successful than I am now, had I been able, at the beginning, to base my work on well-founded mathematical measures.

In consequence, Birkeland asked: "Where in your work do I find the treatment of the reflection of light from mirrored surfaces, or with that, related phenomena? It would be of the greatest interest to me if your method could be developed and applied to more common phenomena in mathematical physics."

The only one of his original plans that Lie carried out firmly was to indicate his position in the Norwegian schools and university debate – and that he would do with a vengeance. Teaching and the school system was something that Sophus Lie had always been interested in, he always kept abreast of conditions at home. The reason that these questions arose now in the spring of 1896, with great force, was the publication of the findings of an officially-instituted school commission whose finds were to be presented before Parliament – where great reforms to secondary schooling stood on the agenda. The one part above all, in this complex issue, that Lie set his sights upon was about how teaching in the higher schools should influence university studies.

Already, in autumn 1895, Lie had written three articles for *Morgenbladet*, and in the course of his spring months in Christiania, he threw himself into the debate with yet even grater zeal. The resulting first round was four articles in *Morgenbladet* and two in *Dagbladet*, and in much of this battle, which not least dealt with levels of professional competence and the role of pedagogy in teaching, his target was his old friend Elling Holst.

When with the school law of 1869, fundamental changes had been instituted in higher education – among other things, a new science programme of studies, in addition to the traditional Latin line, and a middle school that was an intermediary link between the basic compulsory school and the gymnasium. Here the intention had been to create a unified common school system in the country. For some years still however, higher schooling continued to be viewed as a training ground for public servants, and therefore it was not very popular in nationalist and liberal circles. The Latin line was anti-nation and foreign, and nor did the science line satisfy the nation's needs in technical and cultural fields. In Parliament, the Left (Liberal Party), under the guidance of Johan Sverdrup and Søren Jaabæk, had on several occasions sought to eliminate higher education as a public affair. It was proclaimed that all public resources ought to be used to advance the common popular school – private initiatives or local authority efforts ought to provide for higher schooling. But when the Left took over government in 1884, there were many in the party who changed their point of view, and they now understood what the Right had long maintained: to stop giving state contributions to the higher schools would not be very democratic. Therefore the thought and demand arose for a revision of the school law which had been ratified in 1869 – many wanted to reduce the position of Latin still more, at the same time the link between the popular school and the higher schools should be strengthened. A school commission was thereupon set up in 1885, and got down to work under new political guidance, but nonetheless, it concentrated most on the popular school and felt that it ought to be up to the higher schools themselves to find their points of connection. As Parliament dealt with the budget for schooling in 1889, all the themes burst out into a mighty debate: the question of the standard school, and therein, the question of the relationship between the popular school and the middle school, the division between programmes of study at the gymnasia, the position of the classical languages, indeed the very existence of higher schools was put on the agenda, together with physical upbringing, and the place of girls in education, and the debate led to a new schools commission being set up. It was the results of this commission that now in 1896 Parliament had to take a position on – the commission's findings had actually been completed two years before, but Prime Minister Hagerup had waited before releasing its proposals.

As a reaction to the working of the new commission, which indeed had been underway since 1890, a new association was founded in Christiania in 1892 with the goal of looking after both the professional and economic interests of university-educated teachers. This was the "National Association of Philologists and Scientists." Soon after its foundation in 1892, the association had 200 members, and at the first ordinary annual meeting – which was held in Christiania in July of 1895 – Elling Holst made an earth-shaking speech. Together with the philologist, August Western from Fredrikstad, Holst presented and defended a resolution in this annual meeting, about the alteration of education for the teachers of higher schools in the country. The resolution was adopted – by one vote – and Holst got so many offers to publish his speech that he allowed it to be printed by *Morgenbladet* at the end of September, entitled "On a Reform of Teacher Education in the higher School".

Lie's first contribution to *Morgenbladet* in autumn 1895 was motivated by Holst's speech and the resolution that the Philologists' and Scientists' National Association had taken. The main content of the resolution was that a demand be raised – "Teacher education ought to be changed such that, wherever beside a Main Subject, or a Group of Main Subjects, which become the Object of a probing scientific Treatment with the Admittance of independent Work, there be demanded immediately to School Requirements – more than now – the corrective properties of propædeutic Instruction as well as comprehensive Pedagogy."

In his contribution, Holst stressed that the divergence between "the School Requirements" and university teaching had always been "remarkably great". He argued that this hiatus had increased in recent years, reminding his readers that the old "Science Teacher Examination with its three *obligatory* sections, had been instituted by Broch in 1851, (and which subsequently Lie had experienced in his own student days), before it was replaced in the 1870s by a "Teacher Examination" with two *free choice* sections to it.

According to Holst, the old all-sided "Science Teacher Examination" was an excellent basis for further scientific studies, and this had also resulted in well-educated teachers in almost every field of science. Holst took the stand that it was incomprehensible that the university courses of studies in the 1870s were reorganised precisely when there was such a great need for science teachers – something that led to the establishment of the science gymnasium and middle school in 1869. The middle school was thus created as an "Tintermediary link" between the popular school and the gymnasium. The new teacher's examination programme, in which the subjects were freely chosen, thus resulted, according to Holst, in learned teachers in practical schoolwork having to teach in a series of subjects that lay beyond their own area of competence. This had given rise to schools equipped with teachers who had "the same holes in their Education" as in the period prior to the 1850s.

Another of Holst's main points was that even if a teacher were university-trained in the field in which he later taught, the university course of studies had been shortened in such a manner that it did practical school teaching very little good. He took examples from his own field, and argued that all of the mathematical course of studies at university lay within the field of higher mathematics, while all the practical teaching work in school fell within elementary mathematics. Holst wrote that "there from the first [in the higher mathematics] not a single Ray of Light fell upon the other Fields, where he [the teacher] had actually to follow his line of work." Yes, indeed, now and then it could even lead to "pedagogical Blunders", Holst argued, such that when a new teacher without having been given any direction, in his actual teaching, seized on "Means of Explanation or Concepts that belong to higher Mathematics."

Holst felt that the education of science teachers at the university had to be set "definitively in the Direction of the School's Need", such that all the science subjects the higher schools required to be taught, should be taught by one and the same man. (Women had gained entrance to public service exam studies in 1884. From 1896 it was also possible for women to find work in the higher schools, but the first female public service graduate was not appointed until 1906 – and then, as a temporary adjunct or lower level gymnasium teacher). Holst himself, as a research fellow, had already tried to repair this shortcoming from which he felt the university suffered. But Lie's departure for Leipzig in 1886, left Holst with the burden of so much extra work, that this did not allow him to continue with what he called "propædeutic Instruction". But Holst had not given up. A year earlier, in 1894, when it looked like Lie would be coming back, it had been proposed by former and current students of the sciences, that Parliament create "a Docent Post in propædeutic Mathematics" for him, Holst. Holst then called attention to the fact that on the part of the authorities, interest had been manifested in this proposal, and he concluded his article with the hope that in the future there would be a "Parallelism between the simultaneous Movements of the School and Teacher Education".

When Lie read this article, in Leipzig at the beginning of November 1896, he, as he put it himself, felt it was "very close to a punch in the Mouth", and immediately he wrote an article which he sent to *Morgenbladet* – in a covering letter to the newspaper's editor, Nils Vogt, Lie added that since he had almost written nothing in his mother tongue for the past ten years, so he supposed that his orthography was old-fashioned, and: "Of course I would very much like a Proof-reader to undertake the necessary Corrections."

The title to Lie's article was "On Mathematical Lessons for Prospective Science Students". He first underlined that through the years he had always supported the science candidates' applications for foreign travel grants, had stressed that mathematical teaching given at the University of Christiania was certainly adequate for prospective teachers of science, but that it was a "terribly meagre Foundation" for young men of mathematical talent. Lie mocked Holst's efforts at giving the public the impression that "advanced and modern mathematical Theories" were taught at the University of Christiania. The actual condition, Lie wrote, was that "there is scarcely to be found a University in the Lands of Culture, where mathematical Teaching is kept within such strict Boundaries as in Christiania." Lie made a comparison with Germany and France, and stressed that mathematical teaching in Norway was not "particularly different" from what Lagrange and Monge taught in Paris a century earlier. Basically what was taught in Norway was only classical mathematics – "excellently suitable for our aspiring School Teachers", Lie argued, but so limiting that it was difficult for Norwegian students to follow the lectures at foreign universities. Lie probably had his former Norwegian students in Leipzig in mind when he wrote:

> It is no wonder then, that the Science Candidates who receive a travel stipend to study Mathematics abroad, almost without Exception, tire themselves out and rarely rise to anything other than negative Criticism and more or less unfruitful Dilettantism.

Lie put forward a series of examples of areas where higher mathematics was a precondition, even for elementary studies. Even such fundamental concepts as "positive" and "negative" were integral parts of higher mathematics, and "the geometric Imaginary Theory, that we can now attribute to our remarkable Compatriot Caspar Wessel," was to a high degree suitable for clarifying "the arithmetic Operations' and Propositions' Generality".

His having pointed to Caspar Wessel in this context certainly can be attributed to the fact that exactly now – in September 1895 – Lie had written the foreword to a new edition of Wessel's forgotten treatise from 1796. Land-surveyor Caspar Wessel, brother to the poet Johan Herman Wessel, had found a geometric means by which to characterise complex numbers, and thereby give an appreciably better account of the nature of complex numbers. And this had been done before C. F. Gauss (in Germany in 1799) and the French Argand (in 1806) and others who received the honour for the discovery. Wessel's work was printed in the Danish *Videnskabernes Selskabs Skrifter* [Proceedings of the Academy of Sciences] in 1799, but subsequently forgotten. It was first brought back into the light of day by a Danish doctoral student, S. A. Christensen, in 1894. In his foreword to Wessel's work, Lie wrote:

> Until recent Times Norwegian Mathematicians have held the Assumption that higher Mathematics in Norway began with *Niels Henrik Abel*. As everyone knows, his immortal Investigations opened broad new Fields, the Treatment of which, down through the Decades, laid the Grounds for the Century's first great mathematical Powers.
>
> Now, however, it comes to light, most remarkably, that there was a Norwegian Mathematician before *Abel*, who, a quarter century earlier, laid down fundamental Ideas that indubitably would have become epoch-making if they had not gone unheeded. [. . .]
>
> If *Caspar Wessel's* Work had come into its own Right, then he would long ago have been accorded as fully great a Name in the Right of Mathematics as his Brother, Johan Herman Wessel has in Norwegian Literature, and his Uncle *Petter Wessel (Tordenskjold)* won in Battle.

Lie also had other examples in his newspaper article, that exemplified the connection between higher and lower mathematics: "The Concepts of Limit, Derivation, and Integral, which form the Basis for all higher Mathematics, cast the strongest Light over the elementary Mathematics' highest related Concepts, namely, the Concept of irrational Numbers, together with the Concepts of the Length of Curves and Area of Surfaces." Lie concluded by writing: "It is incomprehensible to me that a Teacher of Mathematics could forget well-known Truths of such capital Significance." But on the other hand, when it comes to the axioms of geometry, and comprehension of its first principles and mutual relations – where one really must rely upon higher mathematics – then one ought to "throw one's lot in with Euclid, who was still a far greater thinker than the likes of, for example, Hansteen and Holmboe, Legendre and von Helmholtz, to mention only a Couple of Mathematicians", who, according to Lie, had spent their energies on such questions without having the necessary insights.

Nor would Lie allow Holst's praising of the old Science Teacher Examination to go by without challenge, and he "highly doubted that Lectures on Pedagogy would be of considerable Value" – rather one should introduce a practical Teacher Examination more or less as in Leipzig, Lie concluded, and promised the readers of *Morgenbladet* that he would give his views on "several related Questions", if time allowed.

Holst responded to Lie's article on October 31st. He began by expressing great regret upon reading Lie's article, and stressed that many agreed with him when he said that he was "both shocked and stunned over the Tone in which the Article is presented." Toward the end of his piece, Holst mentioned that he among others had received warm support "from People of Stature in the faculty Collegium of the University." However, the greatest cause for regret was that "Lie, in such a summary and frank manner, had declared himself against the whole question itself, and it seems as though he would place the Authority of his Name and his personal Energies in the way of realising this repeated, and more and more unified Desire coming from the Side of the Teachers and Candidate Teachers."

Holst underlined the fact that neither he nor any other had meant to express anything negative about what went on in the way of mathematical teaching at the University of Christiania – it was the "Law and the Statutes" that they worked under, that had to be changed.

Holst also admitted that Lie had given good examples of how higher mathematics could throw light on the lower. But when he, Holst, had used imagery in relation to higher mathematics, saying: "Not a Ray of Light fell on the lower", it was because he "lacked a short and powerful Catchword" to express – in an assemblage where only half were scientists – "the Quintessence of our Disquiet over the current System." No one in the plenary session of the annual general meeting had misconstrued what he meant. Holst now complained that the published expression was "formed in too absolute a manner."

But, on the other hand, Holst argued, "the real Hole in Teacher Education" still existed and still yawned before them; namely, "the Light that currently fell on the Teaching Vocation from higher Mathematics in the Teacher Candidates' Education Programme." And Holst used a picture, or what he called "a Likeness" from what at the time were the popular arctic regions: anyone could be tempted to say that in winter no ray of sunshine or warmth entered the snowed-in cabin of the Greenlander. But for the learned, this would be fundamentally false. The learned would actually say that the seal oil lamp that burned and illuminated that abode, was indeed "transformed Light and Heat from the Sun." [...] "And these learned Men are right," Holst maintained, "But those of us who spoke candidly, are also right. That is the Issue. Perhaps this is what will proceed here. Light does fall. Certainly it has to do so, but not as directly as it seems to do from Lie's Remarks."

Holst repeated his proposal wherein among other things "a propædeutic Course" be undertaken as a major component, and he concluded that in his eyes, these reforms would bring the Norwegian system closer to conditions in other European universities.

The work of educating good teachers for a national, unified school system where material considerations also had their place, was a task that had been given priority in many countries at this time. Germany, which had even more classically oriented schooling than Norway, was also undergoing sweeping reforms during these years. Radical forces in school policy wanted to answer the demands of the times by increasing the amount of German language, culture and history, as well as science subjects on the school curriculum. A great and trenchant German association had been formed after a 22,000-name petition had been submitted to the Prussian National Diet in 1888, calling for a new school reform. In many circles in Norway it was proudly pointed out that the German school reforms were based upon school ordinances that had been developed in Norway since the adoption of the School Law of 1869.

The general developmental features of Europe to which Holst referred when he put forward his plans and reforms, were a welcome point of departure for Lie. The article he sent to *Morgenbladet* in November 1895 had the title "On Higher Teacher Education in Germany and France". In an accompanying letter to Editor Vogt, Lie wrote:

> From my Experience, it is to be regretted that in Norway when it comes to Schooling and University, one passes far too quickly over from the permanent Schemes to something fundamentally different. In this Way one often breaks away from one Extreme to something diametrically Opposite. [...] One should go forward slowly, then one does not risk going too far, so far that one must take a Step backwards again.

Changes in higher level teaching ought to "be undertaken with cautious Hands," Lie advised, hoping that "several elderly University Teachers" would now "come forward and reply to the violent attacks that were coming from different directions, against "the Norwegian Teacher Examination".

Lie's article, "On Higher Teacher Education in Germany and France" was divided and published over two issues of *Morgenbladet*. Lie maintained that he, from his own experience, was familiar with not only the Norwegian and German Teacher Examinations, but also that of France. Indeed, for the past eight years, l'École Normale Supérieure, "which trained the Elite of France's future Teachers for the Schools and Universities, had every Year sent one of its best students to Leipzig to study my Theories under my Guidance." And he declared that through discussions with "these prominent young Men", and through all his teaching work, he knew both the advantages and the disadvantages of the various countries' systems. At the same time he also asked that his Actions not be considered as attacks on any individual person, he stressed that nor would "sentimental Considerations" deter him from struggling against perceptions he considered "incorrect, harmful, indeed dangerous, with all the Seriousness that the Importance of these Matters, in my Eyes, demand." Otherwise the article was sober and to the point in its tone –

it was the school conditions in Germany and France that were Lie's major concern. However, toward the end of the article, and in his conclusion, he became personal once more.

After showing that the placing of weight on pedagogy, as Holst advocated, was to be found neither in Leipzig nor Paris – "and scarcely in any other places": "But there is certainly nothing stopping him [Holst], as have many Idealists before him, from setting up a serious Experiment in this Direction. This would quickly reveal that his Ideas lack sufficient Vital Necessities."

This was as far as the discourse had progressed when Lie came to Norway in March 1896, to see "whether or not Norway has any Use for me." And it began well: in a meeting of natural scientists in Christiania in April, Lie got everyone to agree – except one – to request that Parliament postpone dealing with the question of a new Schools Act and the effects that such a new law would have on higher schooling. Lie followed this up with a long article entitled "On the Education of Teachers of Science Subjects" – an extremely comprehensive presentation that was divided between four issues when it came out in *Morgenbladet* at the end of April and the beginning of May. In the course of these two months he spent in Norway, Lie also wrote two articles published by *Dagbladet* on the subject of the science gymnasium, and here he put history to work. The title of the first *Dagbladet* article was "Joh. Sverdrup, Steen and A. M. Schweigaard on the Science Gymnasium." The second was called "The Position of the Natural Sciences in the Gymnasia".

With the long article entitled "On the Education of Teachers of Science Subjects", Lie enclosed a note to the editor of *Morgenbladet* in which he said:

> When I, who have really only lived for Science, today throw myself into the Struggle of the Day and its agonising Battle, then it is my Intention first and foremost to claim the Authority of the University and the Science Academy.

And he continued by referring, among other things, to a meeting he had had in "the Christiania Pedagogical Society" where he had put forward his point of view under great protest:

> When I hear (most recently in the Pedagogical Society) the Flatness and Immaturity, not to speak of the Ignorance, wherewith young Beginners, although their own Actions are certainly modest enough, attack the University with the most violent Expressions, then I must say, I do not understand why the University's own Men do not sharply defend the University's Reputation, above all, in the popular Press. As far as my Competence allows me, I shall certainly do so. But the Field is great and the Attacks are many, unfortunately not least from those holding State Power.

Lie sent the article, "Joh. Sverdrup, Steen and A. M. Schweigaard on the Science Gymnasium" immediately to the President of Parliament, Johannes Steen, and in an accompanying letter to the president, Lie stressed that now that he had thrown himself into "the Battle of the Day", it was because he had considered it his duty according to his "best Convictions" to uphold "Interests that in my Eyes are sacred". And in order to make sure that the parliamentary president got the main point, he concluded by saying that the enclosed article argued: "the School Commission of 1865–1869 by and large had it Right, both with Regard to the School and the University's Teacher Examination."

Lie continued with his articles on the old school policy; he praised Broch, who with the support of Schweigaard, had made mathematics a main subject in the first Science Teacher Examination programme statutes of 1851, and he now accentuated the teaching work of Bjerknes and Sylow. The restructuring of the Science Teacher Examination that took place in 1876, and that he otherwise had not contributed to, was only a natural revision. The university teachers had consequently decided that the old statutes had placed too much emphasis on the demand for "many-sided Talents and multiple Forms of Labour Power among the Students."

As far as it went, Lie accepted that Holst, and along with him, many others, wanted once again to reorganise the programme of teaching education. Lie wrote: "There are many Roads that lead to Rome, and there are many Schemes for the Teacher Examination that may lead to the Goal." But what stood almost as a "sacred Duty" for Lie was to struggle against what he considered an incorrect comprehension of the relationship "between elementary and so-called higher Mathematics". He reacted powerfully to what was being said and written from many circles about mathematical teaching, and in *Morgenbladet* on April 28th, he again gave expression to the reasons why he felt he had to "step up to the Bar in order to struggle against Opinions, that according to my Understanding, are incorrect and dangerous." He wrote:

> On the whole I have a strong Impression that the General Public do not fully comprehend that for two Thousand Years Mathematics has been the Cornerstone of all higher Education, and above all, that it has been Mathematics and the Natural Sciences which in our century have brought about great intellectual and material Progress in various Fields.

Lie felt strongly that sufficient account had not been taken of his viewpoint and insights. Presumably he had begun to get answers to the "test" he had set for himself and the country before he had come to Norway in March, and he had advanced the menace of "the Paralysis of higher Mathematics in the University". Indeed, if that were the case, then he "had nothing to do in Norway" – and perhaps Lie did not need more signs – on April 21st, the Science Association (Realistforeningen), asked *Holst* and not Lie to lead a discussion on "a reorganisation of the science programme". Realistforeningen was the association Lie himself had got up on its feet and made so much of – and when he now, in his contributions to the newspapers, had defended the old teaching system, which contained no "Lectures on Pedagogy", he had pointed to Realistforeningen as a forum where those studying science got the most outstanding pedagogical practice, as well as insights into, their forthcoming scientific professions.

Lie was being "tested" in the homeland: the science students wanted to have Holst (and not Lie himself) present an orientation about how the first comprehensive Science Teacher Examination of 1851 had been revised in the 1870s – into what Holst called a smaller but professionally more demanding examination – and by the very fact that the demands were greater, this examination had meant that the programme had become more distant from the school.

Holst now suggested that the science programme be divided into a pedagogical division and a professional scientific division – each two years in length. Instruction in the pedagogical section would be given on teaching all the science subjects, and as background knowledge for these subjects, the students would be expected to grasp one of the fields in detail, and this would be the focus for scientific deepening in the next section.

The Science Association did not manage to complete this discussion on April 21st, and continued it a week later – but with the absence of Lie, who had not been invited. The following week, Lie returned to Leipzig – the idea of spending eight months of the year in Christiania was given up – "Everything now goes back to the Old," he wrote.

It seems that now back in Leipzig, Lie rapidly resumed some of his teaching obligations. But when the Leipzig semester was finished in mid-August, Lie and his family left for their holidays in Norway. Among other things, the plan this year was to go walking in the high mountains of Jotunheimen. Lie had promised the twelve-year-old Herman beforehand that they should climb to the peak of Norway's highest mountain, Galdhøpiggen. And for a certain period during this stay in Norway, which stretched from early August to early October, the whole Lie family went to Jotunheimen. But for the greater part of September, Lie was in the capital – Anna and the children were probably staying at the pension in Åsgårdstrand, where in other summers they had usually stayed.

In any case, on September 9, 1896, Lie was – without Anna and the children – in the midst of the great mass of humanity that had assembled at the quay in Christiania to greet Fridtjof Nansen and his men, who were returning from their great polar expedition with the double-hulled wooden ship, the "Fram". They came sailing up the fjord, escorted by warships, torpedo boats, steamships, sailing schooners and a fleet of small boats. The following day, Lie was in attendance at the huge festive party that the city held for Nansen and his crew, at Logen in the old city – whereupon Lie also contributed to the long list of speeches by reading a telegram of congratulations from the two largest science academies of which he was a member – l'Académie des Sciences in Paris and the Royal Society in London, and he added that people at the German universities had followed Nansen and his polar activities "with the warmest Sympathy and the greatest Admiration." Lie was also in attendance when the Geographical Society celebrated Nansen's polar exploits, and perhaps he was also among the throngs who attended the popular celebration at Akershus Fortress on September 13th, where Bjørnson paid tribute to the Fram Expedition, and proclaimed that it had increased Norway's national honour, revitalised the country's independence, and turned the whole nation into "one Family". In any case, Lie was in strong disagreement with Ibsen, who saw Nansen's endeavour as "an Indian Exploit", but both Ibsen and Lie were among the twenty-seven well-known men, who now in the wake of "Nansen and his Fellows' great Achievement", put out a plea for the collection of money in order to create better conditions for scientific work in Norway. It was this collection, with

Professor W.C. Brøgger as its driving force, which resulted in the Nansen Fund. Shortly thereafter, this fund had assembled capital of almost one million kroner. In order to support this work, Lie gave the main speech at a gathering of the Student Society on September 19th . He pointed out the wretched conditions under which science had suffered in Norway, but hoped that now many gifted Norwegians would stand behind the banner of science with the same heroic energy as Nansen and his men had exhibited.

Lie's speech was referred to in both *Aftenposten* and *Teknisk Ugeblad*, and it seems that Lie had strongly affirmed the close relationship between pure mathematics and scientific work, and their link to practical life with all its new advances. A reference from Lie's speech would also be used seven years later, in an appeal to collect more money for the Nansen Fund, and this was:

> These days science is near the centre of every serious forward step, and not least in *material* fields. *This country, if it wants to enter world competition, must hold high the honour of science, and do this not only for the sake of advertisement, but also by deepening consciousness of science's real and ideal value.*

Lie's enthusiasm for Nansen and Norway can also perhaps have been a contributing factor in his decision earlier that year (in March) to try to divide his time between Leipzig and Christiania. Edvard Grieg, who was in Leipzig in February 1896, wrote home to his friend Nordahl Rolfsen: "I met Sophus Lie yesterday, and we shook one another's Hands fortissimo in great ecstasy over Nansen. He too is one of these Norwegians, as I am myself, who feel that when Something Great happens in Norway, it gives as much intense Joy, as if one were oneself the creative Master of it."

During his holiday stay in Norway, Lie also continued to register his point of view on reforms in higher education. Since he had left the country so suddenly in the month of May, the matter had also taken a turn that threw Lie and Holst, and a series of other science teachers and professors, into a state of unity for a time. Parliament had ratified a law concerning more advanced popular schools, which many felt, on many points, would lead to a sinking of the standing that mathematics and the natural sciences had enjoyed, above all, in the science gymnasia. In the newspaper *Verdens Gang* on July 10th, Holst had an article about these conditions – to weaken the Latin line still further, it had thus been proposed that it be replaced by an historical-philological programme. According to many people's thinking, such a course of studies would make it very easy – in comparison with the old Latin line and the existing science line – for those following the gymnasium course of studies, to take up what Holst called "The Common Man's Road to the University." And this, in its turn, would weaken recruitment to the science line, and with that, certainly also at the professional level. According to Holst, in 1896, four out of seven pupils in the gymnasium studied science, and Holst argued: "That will say, more than Half of Those studying have invested in the detailed and important Knowledge of these Fields, where precisely the Exercise of Thought is carried out best to meet with Life's Demands." This formulation was in accord with Lie's interpretations.

But in an article in *Morgenbladet* on July 12th, Holst had also happened to mention – with a certain teasing undertone – that if Lie's presentation of the statutes concerning French teacher education was correct, then indeed, it was not much different from what he, Holst, was proposing for Norway. Nor could Holst refrain from giving expression to a certain reproach for the manner in which he felt Lie had criticised his "propædeutic Lessons". This would provide fuel for yet another newspaper feud between these two old friends.

Back in Leipzig Lie again took pen in hand – the article that was printed in *Morgenbladet* at the beginning of November had the title "Once Again on French Science Teacher Education". Lie had now been in contact with Jules Tannery in Paris, director for science studies at l'École Normale Supérieure, and had posed Tannery a series of specific questions on how and to what degree theoretical pedagogical instruction was given at l'École Normale Supérieure in Paris. Lie had received supplementary responses, and also permission from Tannery to use the information as he saw fit. Thus it was that Lie, in this *Morgenbladet* article, quoted both his questions and Tannery's answers. The conclusion was clearly laid out for all the readers – as well as fully supporting Lie's description on the need and necessity for theoretical pedagogical education for the budding science teachers, what also emerged with the desired clarity was that even at the gymnasium level in France the teaching included what in Norway was called "higher Theories". Lie maintained to the end that Holst's version of what Lie should have written and meant was "incorrect in its Wording and false in its Contents", and he protested against the "unbridled poetic License" that Holst used time after time "to obfuscate if not to say, falsify" Lie's declarations. Lie could not refrain from poking fun at his "colossal Opponent" with his "pedantic Schoolmaster Tone", and after having juggled with the concepts "Holst's propædeutic Lessons" and "Lectures according to Holst's Recipe", Lie underscored that he had not really criticised the fact that Holst held "propædeutic Lessons to be over and above elementary Mathematics" – but, on the other hand, Holst, for his part, had by accident given his views about lectures from long before his own time. Lie ended with the following bold assertion: "But bless me! There are, indeed, Differences between People."

Holst, making a brief reply in *Morgenbladet*, pointed out that the mass of information on the French teaching situation were interesting – and as far as it went, this information revealed that Lie's initial presentation of the French conditions was incomplete – but to compare Norway with the great and rich country of France, and to use the École Normale Supérieure as a model for the Norwegian situation, could really not be considered "sticking one's Finger in our domestic Earth". As for the personal attacks Lie had mounted, Holst felt that Lie's "abnormal Blunders" were now worse than ever, and he wrote: "I am sorry, that a great and brilliant Genius, like Lie, would stand up for such a Position of Weakness as this."

That autumn of 1896 also saw the publication of Henrik Jæger's large *Illustrated Norwegian Literary History*, where "the literature of science" was described in over 300 pages – and where Holst wrote on Norwegian mathematics, from Hauk Erlendsøn in the 1300s to Sophus Lie and his pupils. Holst reminded the reader of the increased professorial salary that Parliament had made available for getting Lie back to Norway, and he concluded: "We can thus hope, that Kristiania again in the future will become the place, from which this Napoleon of mathematics directs his victorious theories out in the field."

Lie's final article was printed in *Morgenbladet* in June the following year. Under the title "On the Scheme for the University Study Programme", Lie gave a down-to-earth overview of the thoughts that he had presented the previous autumn on the people and politics of Norwegian schooling, and which he now very much wanted printed because the matter was once more to be treated by Parliament. In an accompanying letter to Editor Vogt, Lie gave the information that he had had a university colleague to enquire from *Morgenbladet's* editorial room, if they would print Lie's article in the course of the first week following its receipt – if not, he would approach another editorial room. And the reason for this approach, Lie wrote, was: "I now have considerable Experience with the Fact that *Morgenbladet*, like other Norwegian Newspapers, finds it much easier to make a Place for Articles on Politics, Poetry or Polar Research than centrepieces on School and University."

It is uncertain as to which personal experiences Lie was here referring to. But shortly after the November article, on pedagogical conditions at l'École Normale Supérieure, an article was printed in *Aftenposten* under the title "On Abel, Evariste Galois and Ludvig Sylow". Lie's intention here was to secure for Sylow, who was still working as a senior teacher, a position where he could "devote all his Energies to the Working Out of his Mathematical Ideas." Lie drew lines of correspondence from Abel and Galois, to Sylow, citing laudatory statements by French and German mathematicians. So that the readers could have confidence in what he had written, Lie mentioned the fact that "when France's scientific College, l'École Normale Superieure" celebrated its centenary, he, Lie had been the only foreigner invited to have a place in the *festschrift* to describe the influence that the school's greatest mathematical genius, Galois had "exerted on the Development of Science in our Century." The results of this contribution by Lie was that a year later an extraordinary professorship was established in Christiania for Sylow, at the reduced salary of 3,000 kroner per annum.

Following his last article in *Morgenbladet* in June 1897, almost another year would pass before Lie managed to break himself free from Leipzig.

PART VII

At the End of the Road

Eugenies gate 22 as it looks today

The Final Years

When compared with the earlier years, the final period in Leipzig was no easier for Lie; relations to his colleagues still seem to have been coloured by mistrust and misunderstandings – "Dear Mayer! I find You incomprehensible. In private life you show me the greatest of kindness. As a colleague and scientist, You constantly affront me." These were Lie's words to the old colleague who continued to hold the most successful parties in the city.

The years in Leipzig seems to have become easier and easier for Anna and the children. Anna thrived in the city, and the children had certainly gone a long way toward becoming German – Marie was twenty in 1897, and still had one more year to go at the Mädchen [Girls] Gymnasium of Fraulein Dr. Käthe Windscheid; Dagny was three years younger and attended Fraulein Baur's Höhere Mädchenschule [Higher Girls School], and Herman had been placed in the city's venerable Nikolai Gymnasium. A long time later, the family said that one of the reasons that Lie left Leipzig was because he did not want his son to become a German soldier – Herman was fourteen when they left.

During the summer holidays of 1897 Lie was once again on a foot tour in the Swiss and Italian Alps, along Lakes Lugano and Como, to Milan, then on to Turin, Florence and Rome. Accompanying him on this tour was his sister Laura, director of the Eugenia Institution in Kristiania, who this summer did her "grand tour" of Europe. The two siblings then turned north together, and reached Leipzig at the end of September.

It was a great joy for Lie to have been awarded the prestigious Lobachevsky Prize, and it was also quite certainly encouraging to see that the number of mathematics students in Leipzig had increased (from about twenty in the years prior to 1895, to eighty-five or ninety during his final two semesters there). Lie had an audience of well over twenty for his lectures on analytic geometry and contact transformations, and many of these were American or Russian – and in the winter semester 1897–98, Lie conducted exercises entitled "Über einige neuere Gebiete der Mathematik." One of Lie's "elite students" during this final period was Gerhard Kowalewski, who had come to Leipzig in 1896 from the Universities of Königsberg and Greifswald. In his memoirs, written more than fifty years later, Kowalewski described his impressions of Lie during these years. Kowalewski considered it to have been "a particular Divine Grace" that allowed him to have met and become familiar with Lie, "the greatest mathematician of all time". Lie described Kowalewski as one of his most gifted German students, and Kowalewski took his doctorate under Lie's supervision, completing it in the spring of 1898. He took the habilitation exam the following year, and two years later, was appointed professor of mathematics at Greifswald. According to Kowalewski, Lie never felt completely well: "his face was often completely yellow", and he was frequently sleepless, complained of pains and of feeling cold in the head. Perhaps this feeling of cold was also associated with the fact that his once thick hair had now become terribly thin, Kowalewski added, such that Lie now as a rule wore an old, well-used black hat in the auditorium when he lectured. Bowing to family pressure, he eventually purchased a new hat, and the old one was unceremoniously spirited away by Mrs. Anna. Thereafter, Lie attended the Collegium in his new hat, but when he met the students for his lecture several days later, he was discovered wearing the old hat again. The students had laughed, and according to Kowalewski, Lie too allowed himself a smile of satisfaction: he had located his wife's hiding place, and exchanged the new hat for the old, which was softer and sat better on his head than had the new one.

Lie loved teaching, particularly his own theories, and he had lively contact with his students, whom he, according to Kowalewski, identified by calling out their names and asking them questions – and he had been so sure of his conduct and his subject that he never needed to prepare himself, but rather, opened the sessions by asking a student in the first row to see his notes. He would then nod contentedly, and say, "Yes, now I remember," and then launch himself into brilliant interpretations. Sometimes it happened that, as he was rooting into his deepest integration theory, he might burst out, "Come up here, Kowalewski, and tell me what we have to do." Apart from this, one of Lie's favourite pieces of advice to his "elite students" (who in addition to Kowalewski included one Greek, a Russian, and five Americans): "You must exercise with the concepts in my theory the same way a soldier exercises with his rifle."

During the autumn of 1897 Poincaré had written with the recollection that when he had been in Leipzig some years earlier Lie had said something about surfaces that "doubled themselves by translation" – and Poincaré now wondered if Lie had published anything on this. Lie's reply is unknown – there are few remaining traces of his further contacts with French mathematicians of this period. In the course of 1897 Lie published four treatises in the Proceedings of the Leipzig Academy of Sciences. One of these was on the Abelian theorem and the translation manifolds; in another work Lie showed how he had worked down through the years with what came to be called integral invariants, and that theorems Cartan and Poincaré among others now formulated, directly followed from Lie's theories of differential invariants – and in a third work, Lie followed up on this; and the fourth was a study of line geometry and contact transformation.

Whether Lie himself, or together with Anna, had any clear plan about when he would conclude his work in Leipzig, is unclear. But with the conclusion of the winter semester in 1898, the journey back to Norway began to take shape. In the middle of March, Lie went to Kristiania to survey the situation. He talked with old friends, met the Prime Minister, and must also have had consultations in the Department of Church and Education. Lie received the following memo, dated March 23, from

Church Minister Wexelsen: "There is an opportunity for You again to take up Your professorship in *mathematics* at Kristiania university (anc. 1.7.72) with salary, salary increment and personal increment to the total sum of kr. 10,000, according to the parliamentary decision of 9.8.94" [that is, after the scheme that Bjørnson and Holst had initiated, and which gave him the title "Professor of Transformation Group Theory"], "since, in as much as You have hitherto not protested, one shall consider Your letter to the cabinet as a request that the resolution of 2.5.96 be withdrawn" [that is to say, that the proposal based upon Lie's desire in the spring 1896 to be free from service during the winter months, with an annum salary of 5,000 kr. would be withdrawn]. "Within the boundaries that the appropriate laws allow, the Department will do everything possible to ease your children's transition from German to Norwegian conditions, in respect to schooling and university." Five days later Cabinet Minister Wexelsen reported that Lie could come as soon as he had arranged to bring to an end his career as a public servant of Saxony. The Academic Collegium would be informed, and from this date Lie would be regarded as back in the position of professor at the Norwegian university. In the meantime, on March 25th, Lie had been in a meeting of the Kristiania Academy of Sciences.

He stayed with his sister Laura at the Eugenia Institution in Kristiania, and seems already now to have signed a rental agreement that would come into effect from the coming autumn, for a suitable apartment in the area, on Eugenies gate. This was in a different area of the Norwegian capital than where they had lived before. It also entailed a longer walk to the university on Karl Johans gate, than from the old apartment he had dreamed about in Leipzig, the one on Drammensveien overlooking the fjord.

He sent several cards home to Anna and the children, reporting on his stay in the Norwegian capital. In one card to his son Herman, he wrote: "I think I have to write to You to report that *Fridtjof Nansen* came up to Eugenia today to discuss the Abel Monument with me. You can imagine how the Children filled the Stairway and the Windows and shouted Hurrah! for his benefit. There was great Jubilation. I explained that when Kristiania fêted him 1½ Years [ago] at his Homecoming, Laura and I had attended the glittering Nansen banquet. Nansen appeared in very good Health; he is a little taller than I am. His Son is now well, after a bout of Scarlet Fever, and little Siv, who is now five years old, is doing well." In a brief card to Anna, he wrote: "I am doing very well but have nothing in particular to report. Tomorrow I am going to Ernst's [Motzfeldt]. [...] Prof. J. Vogt is in Berlin. He is coming to visit us, so be warned. He is a Rodent." What it was that made Lie use such an expression to describe his nephew is not known – perhaps once again it had something to do with Amund Helland, who on most issues stood on the opposite side to that taken by Vogt, and vice versa.

Lie was back in Leipzig at the end of March 1898, and he allowed two months to pass before he conveyed his decision to the Saxon authorities, who for their part, tried to convince him to stay on. Culture Minister Dr. von Seydewitz personally came from Dresden to speak to Lie, who was led to expect better terms of employment were he to remain.

Lie allowed himself to be listed as usual in the catalogue of lectures for the approaching summer semester in Leipzig, and would indeed have carried this out. But before the summer semester began on April 18th, Lie had been in Rome and had visited Bjørnson who, being out of sorts with things in Norway, was spending a couple of months that spring in the Eternal City. Lie had taken along the whole family on this visit to Italy, and it must have been a joy and inspiration to them all. Bjørnson wrote home from Rome (April 25, 1898) to his old friend, the professor of literature, Christen Collin: "Sophus Lie has been here with his whole family. *He is someone You ought to get to know!* It is a fundamental and life-affirming experience. What a fellow!"

One of the issues to which Lie was warmly attached to at this time was the Abel monument, which he had also talked about with Nansen. Lie had great plans for the centenary of Abel's birth, which was to be celebrated in 1902. But this was indeed only one of three projects for which he wanted zealously to find support – perhaps also to improve his position in his homeland. The first of these was accordingly the Abel monument, a figure cast in bronze that should be located in front of the University of Kristiana. This was an idea that had been talked about in Norway for over twenty years, and which now would really come into being. The second project Lie wanted to see come to fruition was the writing of a large history of Norwegian mathematics, and in this connection, he mentioned the names of Wessel, Abel, Broch, Bjerknes, Sylow, Lie, Holst – a presentation that he conceived as something much more than the thirty-four-page description of Norwegian mathematics that Holst had given two years earlier in Henrik Jæger's *Illustrated History of Norwegian Literature [Illustreret norsk Literaturhistorie]*. The third project was Lie now putting himself at the head of an international movement to collect an Abel fund. Lie's idea was that every five years, a large international Abel prize should be awarded from this fund for a prominent work in pure mathematics. Perhaps it had been the inspiration of the foregoing and well-endowed Nansen Fund that now made Lie the prime mover for an Abel fund. Perhaps he was also thinking of Alfred Nobel, who in January of the previous year, 1897, had made known that he was leaving the major portion of his enormous fortune for the annual award of five prizes – those subjects which were to be favoured in the sciences were physics, chemistry and medicine – but not mathematics – as well as literature, and a peace prize. The honour of awarding the latter was given to the Norwegian Parliament. The other four prizes were (and are) awarded from Stockholm, by the Royal Academy of the Sciences and the Swedish Academy.

Lie now got in touch with all his European mathematical colleagues. He brought them up to date on the forthcoming Abel celebrations and distinctly expressed that the Abel monument would be mounted solely with Norwegian money, while the fund would be international, and an international committee should both award the prize and assess the submissions, even though the final decision would be made in Norway. Lie sent out a subscription invitation and he received an enthusiastic response.

On his visit to Italy, Lie had wanted to seek out Luigi Cremona in Rome, and Luigi Bianchi in Pisa, but both of them were away. Nonetheless, Cremona quickly reported from Rome (on April 4) that he would gladly support the idea. And from Bianchi in Pisa, he received a letter (April 24) in which Bianchi regretted most sincerely that he had been away when Lie was in Pisa wanting to honour him with a visit – Bianchi assured him that he was one of Lie's numerous admirers, that it would be a great honour for him to get to know Lie personally, and he hoped that they could meet when, hopefully soon, Lie next visited Italy – whether in Pisa or Florence. Lie had only to send a message beforehand so that Bianchi would not "miss out on such a valuable opportunity."

Picard (April 6) reported from Paris that Lie could reckon on both his and Hermite's sympathy toward the Abel centenary celebrations. Picard stressed that the admiration for Abel and for Norway were alive and well in France. Picard would put himself at Lie's "disposition", and he reported that his father-in-law Hermite would give 200 francs in advance for the "Prix abélian" that Lie wanted to establish in Christiania. Picard wrote that he also believed that France would be able to contribute a greater sum to the fund, that both the universities and lycées would contribute, and furthermore, he asked to be kept abreast of the work. Otherwise, the only thing Picard could conceive of differently was a more frequent awarding of the prize – an Abel prize every year would be best.

Darboux (Paris, April 27) followed up with similar positive reactions, and reported that he fully and completely was ready to serve in the honouring of Abel and his centennial commemoration. Darboux also thought that Lie could count on support from all the French mathematicians. Lie also got the warmest statement of support (May 13) from A. R. Forsyth at Trinity College, Cambridge: Lord Kelvin would also quite surely contribute to the fund, Dr. Salmon and several others as well, Forsyth added, and in the meanwhile he thanked Lie for the commentary he had written in a letter concerning Forsyth's large edition of Cayley's writings.

Back from his Italian journey, Lie also found time for a quick trip to Berlin before the summer semester began again in Leipzig. In Berlin, his errand was to assemble support for the Abel fund.

He also wrote to Klein in Göttingen – for the first time since the break in their relations more than five years earlier. Before he asked for Klein's cooperation and support for the international fund, he took up two other matters. He thanked Klein for his written assessment to the committee, in connection with Lie's being awarded the Lobachevsky prize, and he reported to Klein that he was now leaving Leipzig and returning to Kristiania. Lie commented as well on his actual situation – on his relations with the students, as well as with Neumann and Mayer – and expressed great worry that the geometric tradition that Klein and Lie had built up at the University of Leipzig was now perhaps ending – in any case, it was still uncertain who would be Lie's successor, or whether there would indeed by a professorship in geometry.

Klein replied (May 12) and said that he found the announcement of Lie's departure to be very unexpected, even though he was certainly well acquainted with Lie's difficulties at Leipzig and could also understand Lie's desire to go back home. Klein suggested various candidates who might possibly succeed to the position in

Leipzig, and concluded by saying that he would certainly support the work of the Abel fund. Klein also thought that Hilbert would support it.

Four days later (May 16), Lie received a letter from Frobenius in Berlin, who wrote that he, and similarly Schwarz, were sceptical to the idea of an Abel fund award for mathematical works. It was the principle of prizes that they had doubts about – prizes often led younger talents astray, away from the real road forward, or so Frobenius maintained. As well (May 18), Fuchs wrote from Berlin to say that he would support all celebrations and commemorations around the Abel centenary, but that the proposal of a fund for a periodic awarding of an Abel prize was not particularly appropriate to Germany.

People in all circles now knew that Lie was leaving his engagement at Leipzig, and he himself was in the process of tidying up and counting down to the date of departure. He went through his papers and gathered piles of letters, articles and unpublished works – the greater part of the 20,000 folio pages that composed his surviving papers, his *nachlass*, seems to have been put in order by Lie himself during the last spring and summer in Leipzig. On June 8th he received an account from Teubner's of his sales from the past four years of his six books published by the press, and the amount of stock on hand in the warehouse. The first volume of *Theorie der Transformationsgruppen* had sold sixty-eight copies, and the number on hand was 358; the second volume, in the same time span, had sold eighty-three copies, and 397 were still in the warehouse – the third volume had sold eighty-one copies, and 476 remained on hand. The book of his lectures on differential equations had sold 175 copies, but there were still 482 on hand, and the book that contained his lectures on continuous groups had sold 119 copies, with a warehouse holding of 464. Lie's last book on contact transformations, which come out in the spring of 1896, now two years later, sat to the total of 642 copies in the warehouse, while only twenty-six had been sold in the last year.

When his lecturing duties were over at the beginning of August, Lie went to Berga with the family for the last time. They stayed at Rathaus Berga, and after some days there he went on, with fourteen-year-old Herman, to Bad Elster. Herman wrote home: "Dear Mama, Now Papa and I are sitting in the Cure Café and listening to a Concert. We are eating Cakes and drinking Iced Coffee. Wonderful."

Whether Lie felt ill or tired, and sought out medical help at this time, is not known. It does not seem that Lie himself commented on the ailment that was beginning to afflict him, and doctors were never the first in whom he confided. On the other hand, Bad Elster was a cure centre on the border with Bohemia, known for its drinking springs and its mud baths, a treatment centre for the alleviation of all sorts of anaemia-related and metabolic ailments, heart and skin diseases, rheumatism and nervous disorders.

Nor are there any records of the departure from Leipzig, on how the Lie family actually packed up everything at Seeburgstrasse and sent it as a removal load to Kristiania – over the years Lie had gathered a large collection of books.

Kowalewski recalled that Lie took leave of him in Leipzig in a "touching manner", saying he would always "hold his hand out" to him. Otto Hölder, Lie's successor in Leipzig, very much wanted to meet Lie before he left, and Hölder wrote from his home in Oberhof on September 7th that he had got to know from Mayer that Lie would be in Leipzig until the middle of September. Engel also wrote in his later accounts that Lie returned to Norway in September 1898. According to Holst, Lie was installed in Kristiania when the autumn term began in mid-September.

In any case, on his way home, Lie stopped in Göttingen to say goodbye to Klein and his wife. The farewell does not seem to have led to any restoration of the old friendship – contact by letter was never taken up again, but it was said that they had come to a sort of reconciliation. Klein must have noticed that Lie was not completely in good health, and one way or another, Klein kept abreast of what happened to Lie's health in the course of the autumn. "You will have heard that unfortunately things have not gone well with Lie," Klein wrote to Mayer five months later, and said it was certainly difficult to recover from "that awful anaemia", which had also made it difficult for Lie to acclimatise himself in his old homeland. When and how Klein got to know about the particular "anaemia" that Lie was suffering from, is not known.

When the Lie family returned to Kristiania in September, they moved into Eugenies gate 22, in a new and comparatively large four-storey brick building, where they had an apartment on the second floor, with a living room and balcony facing southeast, but otherwise the rooms faced north and were without sunlight. Herman was enrolled in the first gymnasium class at the Aars and Voss School, Dagny in the second gymnasium class, and Marie had just graduated and become a university student. It seems that one of the last things the family did in Leipzig was to celebrate her success in the examinations.

According to the catalogue of lectures, Lie took up his teaching duties right away: "Sophus Lie, Dr. extraordinaire, Professor of Mathematics, intends in the hours, which will later be specified, to lecture on differential equations." Holst described Lie's return thus: "He [Lie] began the autumn semester with a brilliant series of lectures, that more than fulfilled every expectation that one might have held. Two foreign students, an Austrian, and an American scientist, were following his lectures, and for them, and an elite group of young Norwegian mathematicians, he held a private tutorial of particularly great interest. But after only two months had passed he had to move that tutorial to his home." Here, presumably, only the two foreign students would meet – the American, Edgar Odell Lovett from Princeton, New Jersey, later sent greetings to Mrs. Lie and the daughters, and was eager to see Lie's collected papers published, something that subsequently did happen (with six volumes appearing between 1922 and 1937, and the seventh in 1960).

What did the family encounter upon their return to Norway in 1898? What were people doing in Kristiania in 1898, and what thoughts did Sophus himself have about what he saw?

In that autumn's newspapers there was considerable discussion about the awarding of the very first Nobel Peace Prize. Bjørnson wanted the first prize to go to Russian Czar Nicholas II, who for his part was said to have remarked that Bjørnson was his favourite writer. Others wanted to divide the peace prize between four winners. The result was that the prize was not awarded that year, for which many were pleased, as some months later by means of a state coup, the czar overthrew the legally constituted government authority of Finland. There were regular reports in the newspapers about the progress of the polar exploration ship "Fram", which under the leadership of Otto Sverdrup, was out on another arctic expedition. November saw the performance of Bjørnson's new play *Paul Lange and Tora Parsberg* about power relations in Norwegian politics, and beside this, the Dreyfus case was cropping up in all the newspapers.

Lie never gave his views publicly on the Dreyfus affair, but he quite certainly had his point of view. Lie had earlier reacted against German policy in Alsace-Lorraine with its suppression of the free press and of freedom of assembly. Some years after Lie had arrived in Leipzig in 1888, a Catholic factory owner, Charles Appell from Alsace had been charged with high treason by the *Reichsgericht* in Leipzig. The factory owner's brother was the mathematician, Paul Appell, who had come to the city to follow the trial, and he had been well received in Leipzig by Lie. Lie even wanted to honour the French mathematician by holding a dinner party in his honour, but his German professorial colleagues did not dare to accept the invitation. Later in his memoirs, *Souvenirs d'un Alsacien*, Appell wrote about Lie's readiness to help him, and praised Lie's courage in that trying situation. As for factory owner Charles Appell, he was sentenced to one year in prison and nine years of hard labour. Six years later the Jewish Alfred Dreyfus, the son of another factory owner in Alsace, was sentenced to life in prison by the French courts. It had been in order to get this case reopened that the writers Zola and Bjørnson, and with them, many others, filled many columns of newsprint with petitions, opinions and new pieces of information on this unjust sentence.

It does not seem that Lie really got down to things in Kristiania – he held some brilliant lectures for the few who attended, but there was no mention of new studies. His last work, on contact transformations and differential equations, was submitted for printing in the Proceedings of the Leipzig Academy of Sciences long before he left Germany. The Kristiania newspapers continued to stress the significance of science in practical everyday life – and there were calls for support for the Nansen Fund, and a committee was established at the New Year, to collect more money. Lie's Abel fund receded more and more into the background and seems to have been a cause that was too difficult and heavy-going to inspire the public.

According to what information is available, Lie was feeling quite unwell as early as November 1898, and he was unable to manage walking from his home down to the university to give his lectures. But he must still have reckoned on recovering his health, and had plans for the coming spring semester – in addition to his private tutorials: "4 hours weekly to lecture from 12 to 1 on differential equations for the more advanced, and 2 other hours weekly at the same time slot on analytic geometry for beginners" in Auditorium No. 2.

At some point or other Lie must have talked about the condition of his health with his friend and doctor, Axel Lund. Lund had had over thirty years' practice as a doctor, the last fifteen of which he had been appointed the city doctor of Kristiania.

He had researched bathing and health spa locations in Germany and Switzerland, and was also interested in such for Norway, but he was in no way a specialist in the disease that he would find Lie was suffering from.

There is no record of when the diagnosis of pernicious anaemia was made; later it was said that Lie first came to know of the diagnosis on Christmas Eve, 1898 – a diagnosis that he immediately, or at least short thereafter, must have experienced as synonymous to a death sentence.

At this time in Kristiania there was a specialist of international dimensions on pernicious anaemia, Professor Søren Bloch Laache. As early as 1881, Laache had received a prize for his study "On the Red Bloodcells' Relationship to Secondary Anaemia", and two years later he published, in German, the book *Die Anämie*, which had given him renown as one of the world's foremost blood researchers. Moreover, Laache's teacher and Lie's former colleague on the editorial board of the journal *Archiv for Mathematik og Naturvidenskab*, Jacob Worm-Müller had studied and written on various types of anaemia. In his book, Laache had sketched the fever curves, the number and forms of red blood cells, and presented a detailed progress report on eleven patients with pernicious anaemia – 130 out of the 300 pages of this book were devoted to this most dangerous form of anaemia. For almost all the patients that Laache had researched, the disease was very progressive – over half of them died in a period between two and six months after Laache had made contact with them, some lived a little longer, and only one made a full recovery.

It might seem strange that Doctor Lund, when he had his suspicions about what Lie was suffering from, did not consult Laache. Laache worked at Kristiania's Rikshospital, and also had his own practice. But perhaps again it was Lie's own unwillingness to seek medical help that affected the turn of events. Once earlier – on the occasion of the lump in Anna's breast – Lie had demonstrated his anguish at facing up to a precise diagnosis. But he had also experienced what such procrastination had cost him. On that occasion he suffered from insomnia and after a time as well, in his own words, was in such boundless despair that he cracked – while Anna's affliction proved to be quite benign.

Pernicious anaemia was a disease of the blood that at the time had no cure – nor in Laache's work was there a discussion of any real treatment besides various types of fever-reducing medication. It was known that the disease was accompanied by a lack of red blood cells, but how was one to create a regeneration? Before the correlation was definitively found, patients were advised to eat iron-rich foods that were thought to increase the quantity of red blood cells. But in most cases the disease was too far advanced before the diagnosis was made, such that a change of diet alone could not help.

Pernicious anaemia is caused by the lack of vitamin B_{12} and the illness develops when the patient is unable to absorb enough of this vitamin, which is completely essential for a healthy division of cells in the body. Therefore a lack of vitamin B_{12} affects all the cells in the body, and in this way can also give rise to many different symptoms – all cells will in principle be vulnerable: the eye problems mentioned earlier, could certainly be a symptom. In many cases such a late cell division will also affect the spinal column and nerves – there is even a condition known as the lack of vitamin B_{12} psychosis.

In all cases the illness will affect the balance, cause a numbness in the arms and legs, will affect the gait, cause the skin to become sallow and pale, reduce sensitivity and affect tongue and chewing motions; toward the end, it is difficult to get out of bed, although death occurs calmly and without pain.

One can be predisposed to pernicious anaemia genetically, and it breaks out in adults and those of advanced age, but because the disease is always so progressive and acute, it seems unrealistic to think that Lie could have been afflicted earlier, and then regained his health, for example, when he had his breakdown ten years earlier.

Sophus Lie died on February 18, 1899, at half-past one in the afternoon at his home in Eugenies gate. Mrs. Anna was certainly with him, perhaps also Axel Lund, at least he quickly wrote out the death certificate: "Cause of Death: Anaemia (pernicious)" – "Length of Illness: ½ to 1 Year." The latter was probably information that Lie himself had given to Lund.

In his final two months Lie must have been clear that he was dying. It is difficult to picture him accepting his death sentence peacefully – he who always wanted so much to be solidly planted where important things were happening. No one in the family ever commented on his reactions to the diagnosis, but whether or not this indicates the presence of exhaustion and indifference – that come with the advance of anaemia – had dampened his reactions, is hard to say.

In any case, Lie's death seems to have come suddenly and as a great surprise to everyone, and reactions streamed in from all quarters. The first was Holst, who the day after Lie's passing, wrote a long article of reminiscence in *Dagbladet*. After conveying the tragic message, Holst described Lie's life, and his greatness in the realm of mathematics – he was "one of the most significant Men our Land has fostered during this Century" and then he went on to describe "his next greatest Passion to his Field", namely, his love of Norwegian nature, Holst concluded, "And nobody has followed more warmly and with more interest the course of events in the Homeland, than he from his honourable Exile."

Lie died on a Saturday, and the following Tuesday the Academic Collegium held a meeting in which it was decided that Lie's funeral should take place at Holy Trinity Church at "the University's Arrangement". The Collegium also reported the death to the Church Department, and urged that Mrs. Lie be given the same economic consideration she would have received in Leipzig. It seems that those in power did their utmost in pursuit of this wish. Parliament decided later, to agree to an offer from Mrs. Lie to sell Lie's book collection for 3,000 kroner, and the books were donated to the university where they constituted the backbone of the collection in the science students' reading room. However, a rather odd incident in connection to the book collection occurred before the sale. Three months after Lie's death, Engel wrote (on May 29th, 1899) to Holst, and appealed to him to find out whether some of the books in Lie's collection might include a number missing from the Leipzig collections – among others, the first edition of Riemann's works,

The family: Anna and Sophus with the children, Herman (b. 1884), Marie (b. 1877) and Dagny (b. 1880).

Two photos of the Lie family. In the lower photo, Sophus Lie's sister Laura is seated on the step with the Lund couple beside and behind her.

The photo above was probably taken when Lie was on a visit to Norway in the autumn of 1888.

To the right:
This sketch by Gustav Lærum accompanied Lie's obituary in Verdens Gang *on February 21, 1899.*

Erik Werenskiold made this sketch the day after Lie's death.

Galois' works, together with some offprints from *Liouville Journal* in blue covers with the stamp of the institute in them. Engel was not sure that Lie had them and therefore did not feel right about asking Mrs. Lie, but he hoped that Holst could find the occasion to go through Lie's library. Whether Holst actually found these items among Lie's books is unknown.

Lie was given a monumental funeral on February 24th. With the President of Parliament, the Prime Minister and Minister of Church and Education leading the country's dignitaries, the Holy Trinity Church was full – and according to newspaper references the church was decorated around the altar, the pulpit and the baptismal font with laurel leaves and palm fronds – in the choir stood the leaders of the Student Society with their banners draped in black crêpe, and the solemn ceremonies began with the students, under their conductor, presenting a song written by Holst:

The stars course deep through space,
Grass-waves shimmer on the breeze,
The sea, when it breaks, foams lace,
In summer's warm, and winter's freeze –
We weigh them all, we measure, number
And with pathways, powers, time encumber.

Thereupon there followed two stanzas on numbers and mathematics that described his great creativity – before the priest, Dr. Krogh-Tonning, spoke on the greatness of science, the guidance of the Lord, and Lie's sacrifice on coming back to the fatherland – which could offer him nothing but the grave. Then the last stanza of Holst's song was presented:

All thanks to the great along our shore,
Who did oft' invite us out to see
The mind's force rise and take the floor.
Lift thine eyes, bend thy knee:
That he who construed from highest thought
Was from our people born and wrought.

The priest's eulogy and Holst's song were reproduced in their entirety in the same day's issue of *Aftenposten*, together with large excerpts of the speeches that were followed by the laying of wreaths on the coffin: on behalf of the university, the foreman of the Collegium, Professor Schønberg, gave thanks for all that Lie had done, and the honour that he, together with his predecessor, Abel, had given the nation and Norway's university, and he also conveyed condolences from the Academies of Science of Rome and Turin. Professor Brøgger then lay a wreath on behalf of the Faculty of Mathematics and the Natural Sciences, and in his speech, which was printed in its entirety in the next day's *Aftenposten*, Brøgger brought to mind Abel, and how surprising it was that a small nation like Norway could, in the course of one century, once again give rise to "a Mathematician who many consider to be the greatest of the Time," and who had taught the Norwegian people to respect and support science, and to understand that science, along with literature and art was "the most important Basis for our Right to live our own Life as our own People."

Then Professor Mohn, who, on behalf of the Academy of Sciences, laid the next wreath, also reiterated the great honour that Lie had given to science and to his fatherland. Parliamentary President Ullmann, in addition to thanking Lie for the contributions he had made to his country, expressed his pride that it had been Parliament which in its day had "elevated him to the University Teacher Position". Then "with deep feelings of sadness", Prime Minister Steen lay a beautiful wreath that was decked with ribbons in Norway's national colours. This was followed by wreaths from the Student Society, placed by its foreman, Wilhelm Aubert, from the Mathematical Association of Copenhagen, from Mittag-Leffler, via Holst, who had laid a large and beautiful wreath decked in the Swedish colours; the representative of the Pictorial Artists Committee, Eilif Peterssen, stepped forward with a wreath. Other wreaths were laid on behalf of the teachers of the Hamar Gymnasium and the Association of Engineers and Architects. According to *Aftenposten* there were several more wreaths, but it concluded its coverage of this portion of the ceremony by saying that Fridtjof Nansen also laid a wreath on Lie's coffin. And to the strains of the organ the coffin was borne out of the church by members of the immediate family, while Professor Helland and former Schools Inspector Lund acted as marshalls, and High Court Judge Motzfeldt carried the medals and orders of the deceased. The funeral cortège then moved off up the hill to the Cemetery of Our Savior, accompanied by the music corps of the military brigade. The priest spoke again at the graveside, and the student choir sang psalms.

The same evening an extraordinary meeting was held by the Academy of Sciences – both Mrs. Lie and her children were in attendance, together with the parliamentary presidential group, the Church minister, professors, and a numerous assemblage of the famed and the learned. Sylow gave the eulogy, and before he went into a detailed description of Lie's life and fame, Sylow pointed out that it was difficult, indeed, impossible, to speak to the memory of this great mathematician: "Namely, it is the Misfortune of the Mathematician, more than of other Men of Science, that his Work cannot be presented to the well-informed and cultured of the general public, and hardly even before a Gathering of Men of Science from all Fields."

Letters of condolence and telegrams streamed in to the Academy of Sciences from all over the learned world, as well as to the university and the family. Anna had a thank you card printed, and sent out her personal thanks until the cards ran out – but in Göttingen however, Mrs. Klein sat and thought it odd that she had received no response to her letter of condolence.

In Paris it was the twenty-five-year-old science graduate, Carl Størmer, who informed many of the French mathematicians of Lie's death. One only understood how great Sophus Lie was when one was abroad, Størmer wrote home to Holst – everyone knew Lie, and everywhere there was talk about how inspirational his ideas had been. According to Størmer, Darboux was particularly wounded by the news of Lie's death – "he [Darboux] had esteemed Lie so highly, he said, that he considered him one of his best friends. He spoke about what great times they had had together, when Lie was down here, and that they had gone to Fountainebleau, and he repeated several times "'je suis tout désolé de ce que vous me dites là' – and said that France would mourn over Sophus Lie – and that it was remarkable that Norway, with such a small population, had such great mathematicians [as Abel and Lie]." Darboux in Paris obtained biographical information about Lie from Holst via Størmer, and then Darboux wrote about Lie for *Comptes Rendus*, and gave a eulogy when the Academy and the French Institute held a commemorative meeting for Lie on February 27th. (The Academy – l'Académie des Sciences – was a section of the French Institute [l'Institut de France]).

Størmer took along the poet, Gunnar Heiberg, to this commemorative meeting at l'Académie des Sciences. Heiberg was living in the city and reported home about various events in the French metropole. At that time, Norwegian literature, music and the visual arts had a high standing in France – Ibsen, Bjørnson, Grieg and Munch were the frontline figures in a whole series of Norwegian artists. But Heiberg also saw clearly that in the natural sciences, Norway also had stars that the French admired. Polar research with Nansen in the lead was duly noted, but still stronger was the radiance emanating from the Norwegian mathematical giants, Abel and Lie. In the March 7th edition of *Verdens Gang*, Heiberg wrote about his experiences "in the French Institute" – after having described the subdued light that entered the windows through yellow curtains high up along the walls and slanted down through the long oaken hall, where about seventy members of the Mathematics and Natural Sciences Section of the French Institute held their meetings every Monday, Heiberg noted that "of all these Principal Personae" there were far fewer bald heads that one otherwise saw in a French theatre, or from the gallery of the Chamber of Deputies: "But they did not seem content and satisfied. One had not the Impression that Science had enlivened and excited them. But it was rare to see a Face that had not, once one had set one's Gaze upon it, become handsome. And it was an Enjoyment to sit for an Hour's Time and watch this Senate of Masters of Matter and Numbers."

Gunnar Heiberg described several of the French scientists, whom he knew about from their honorary political posts, from their participation and points of view in the Dreyfus case, and also from their scientific discoveries – before he went on to describe three of Lie's colleagues: Paul Appell, Henri Poincaré, and Gaston Darboux. He reported that Appell, who was tall and erect, and had a blonde beard, rose and said "it is not at all strange that such an Inventive Genius as Sophus Lie had worn himself out." Heiberg pointed out that Poincaré was becoming a Universal Genius in the great Riches of Mathematics," and even though externally he did not look like Lie, there was something similar about him: "There was a similar Anxiety of Struggle about them both. For over Sophus Lie, for all his Huge Presence, with his big Head and prominent Frontal Lobe, there was an eternal Anxiety." As to Poincaré, Heiberg wrote: "His Back is curved, and almost humped. His Head is dense – filled in a way. Calm, brown melancholy Eyes, which suddenly, either when he speaks to someone, or is sitting alone, begins to search, flit back and forth, to play. He is searching for something. Something dark that the Eyes will search out and illuminate. Then a while later, they are again calm and melancholy." And about Lie, Gunnar Heiberg recalled:

Once when I was at the back of a Tram I came to look at his Head, and that of the others who were standing there. And I had to smile at what I saw there. What was impossible, was that the other Heads, though assembled there by Chance, could have been one and the same Body Part, with the same Function. All we others resembled one another. He was unique in himself.

There was always something about his that was a tremble. Like a Brain Flower quaking on its Stem. As he had dared to go out and penetrate the Dark, and then a sudden shining Light had assailed him in the Eyes with Force, which is not something that we ordinary Mortals comprehend.

In reference to Darboux's speech, Heiberg wrote about the great mathematician's statement that "Lie was one of the world's greatest Mathematicians [. . .] his Discoveries and Theories were epoch-making", about how highly Lie was appreciated in France, and "what deep Sorrow his Death had awakened among all Mathematicians". Darboux had explained how Lie had been arrested at Fountainebleau in 1870 as a spy, how he had been released from prison with the help of the French Institute, and Darboux had concluded with a degree of levity that the French Institute could be proud of having done something so meritorious in relation to such a Man as the great Norwegian Genius, Sophus Lie.

A month after Lie's death, Bjørnson thanked Holst for all the work he had done in relation to the deceased. Bjørnson wrote: "Once again, with the death of Sophus Lie, I admire you. Your poetry, your industriousness at having got all the rest of us to cherish him, the scrupulous truthfulness you have applied throughout, for someone who had taken a slice of your life and your honour with him, indeed, who even tried to remove your honour; but whose excuse was that he was not aware of doing so."

On April 20th, the university held a commemoration for Lie in Kristiania, and again it was Holst who had been asked to give the speech – the poet Theodor Caspari had written memorial verses for recitation both before and after the speech. After a long and thorough description of Lie's life and laudatory accounts of his mathematical achievements, Holst reminded the audience of a proposal that had already been launched, about raising a monument to Lie, "Abel's worthy Successor". And in addition, Holst proposed "another Monument": "a Collected Edition of his numerous and far-reaching Treatises." This was something that Darboux too, had suggested in his eulogy at the Institute in Paris on February 27th – it would be of invaluable utility to have collected in chronological order, and according to their subjects, the whole series of studies, "in which Lie laid out first Hand the complete Originality of his Genius." According to Holst, who had the support of Darboux, it would constitute a completely necessary supplement to the great classic Treatments, where he [Lie], like the great Artists of the Renaissance, has allowed the Hands of his Pupils to fill in the large Surfaces of the Picture."

Appendices

Deichman Library, Oslo. Axel Revold's fresco from 1932 is visible in the background. The fresco depicts modern science and technology (with the figures of P. A. Munch, Sophus Lie and Niels Henrik Abel), and poetry (represented by Henrik Wergeland).

Chronology

1842 December 17th, Marius Sophus Lie is born at the Eid vicarage in Nordfjordeid. His father, the vicar, Johan Herman Lie, 39 years of age, is chief councillor of the village and active in all the affairs of education and schooling. His mother, Mette Maren, née Stabell, 35 years of age is thought highly of, by all and sundry, and not least for her work on the vicarage estate, which is considered a model farm.

1843–51 Childhood spent at the vicarage at Nordfjordeid, together with six siblings – three brothers and three sisters. Sophus is the pentultimate child. The village is marked by its degree of growth and prosperity. A new church is built in the settlement. For some periods of the year, officers and soldiers assemble here from the surrounding districts, to conduct their exercises and drills.

1851 The family moves to Moss (here Father Lie remains the parish vicar until his death 22 years later). Sophus, along with his brother John Herman, who is two years his senior, is placed in the municipal School of Science (Realskole) in Moss.

1852 Mrs. Lie, Sophus' mother, dies in April. The eldest brother, Fredrik, who is 18 takes his school-leaving exam (examen artium) and begins to study science at the university.

1852–57 Following the death of Mrs. Lie, the atmosphere around their home in Moss seems to have become more austere. Domestic affairs are taken over by a housekeeper and Sophus' elder sisters. In 1854, Sophus' little brother dies at the age of ten. Sophus completes the curriculum at the science school, and has private tutoring for a year before being sent on to further schooling in the capital city.

1857 Sophus is enrolled at Nissen's Latin and Grammar School, a school that placed greater emphasis on science subjects and modern languages than what was otherwise normal in the established Latin schools. At Nissen's School he meets the same-aged Ernst Motzfeldt, who becomes a faithful friend and an important source of support throughout Sophus' life.

1858 Sophus receives high marks in his examen artium, and begins to prepare for the obligatory secondary examination, a rather comprehensive preparatory exam at the Royal Fredrik's University of Christiania (now Oslo). He takes this examination in the autumn of 1860 and graduates as one of the two best in his cohort.

1861–65 He studies the sciences, does brilliantly in the examinations, and graduates as a cand. real. in the autumn of 1865. His student days are marked both by social activities and academic absorption (at the Scientists Association [Realistforeningen]), together with long tours on foot into the mountains during the summers. Sophus is notable for both his physical strength and his mood swings.

1865–69 He supports himself by privately tutoring students in the sciences, gives a series of popular lectures in astronomy, and is otherwise in great doubt as to what he will do next – he is worried that he does not have a calling in one or another direction. During the summer/autumn of 1868 he discovers that "there is a mathematician in him".

1869 Lie's first mathematical work, "Repräsentation der Imaginären der Plangeometrie" is published both by the Academy of Sciences in Christiania and in the renowned periodical *Crelle's Journal* in Berlin. Lie receives a travel scholarship and comes to Berlin in September. In the rich mathematical milieu of Berlin he also meets young Felix Klein. Lie and Klein become best friends; they begin to work together and the friendship becomes of vital significance to them both.

1870 In February, he leaves Berlin for Göttingen, where he also seeks out the leading mathematicians – and then on to Paris. In April, Klein arrives in Paris as well. Lie and Klein stay at the same hotel and a very inspiring time ensues as they meet with French mathematicians Darboux and Jordan. On July 19th, the Franco-Prussian War breaks out. Klein must leave the country in the greatest of haste, and Lie decides to walk to Milan to visit the mathematician Cremona. Outside Paris, at Fountainebleau, he is arrested under suspicion of being a German spy. Four weeks later he is released, and takes the train to Milan, then back up through Germany, meeting Klein again in Düsseldorf. They continue to collaborate, and he returns home to Christiania in December.

1871 Lie receives an adjunct stipend; that is, he becomes the holder of a university research grant holder in mathematics. He finds additional employment as a substitute teacher at Nissen's School. In June he defends his dissertation, "Over en Classe geometriske Transformationer" – he receives his doctorate with distinction, but according to Elling Holst, Lie's pupil and biographer, "Not one mother's son had grasped a single word of it." It is said in Paris that Lie's work is "one of the most beautiful discoveries in modern geometry." In the course of the autumn, Lie applies for the vacant professorship at Lund in Sweden. In radical circles it is considered that it would be a defeat, if the country's great mathematical talent had to abandon the fatherland and move to Sweden. (Above all, one must not make the same mistake again as had been done with regard to Abel, the country's former mathematical genius, who never received any permanent position in Norway.) Action is taken in several quarters, and a proposal is made that an extraordinary professorship be created for Lie.

1872 In February, Parliament establishes an extraordinary professorship by a vote of 85 to 16. Lie thus becomes the country's first "Parliamentary Professor" (the next to be appointed in this manner is historian Ernst Sars). In the autumn, Lie travels to Germany and visits Klein, who has become professor at Erlangen, and is working on what later will come to be known as the Erlangen Programme. Lie returns to Christiania, proposes to the eighteen-year-old Anna Birch, and she says "yes" on Christmas Eve.

1873 Father Lie dies at Moss. Sophus engages in a zealous exchange of letters with his fiancée, Anna Birch in Risør, and otherwise carries on an extensive mathematical correspondence with Klein and Mayer in Germany. This is the year Lie begins his life work, what today is called Lie theory.

1874 Sophus and Anna marry in August. For their honeymoon, they journey to Paris, where among other things, Lie also looks for Abel's missing Paris Treatise. Together with Ludvig Sylow, Lie had taken on the task, in 1873, of editing a new collected edition of Abel's works. (This is published in two volumes in 1881.)

1876 Lie works on his geometrical investigations, and together with zoologist G. O. Sars, and the doctor of medicine, J. Worm-Müller, starts the periodical *Archiv for Mathematik og Naturvidenskab*, which becomes the Norwegian voice for "the modern breakthrough" in the sciences. In the coming years, Lie publishes a series of articles in this periodical, often as first drafts of what would, in later versions, be published abroad in the journals of other countries.

1877 Sophus and Anna have their first child, Marie. He does further work on differential invariants and on his group theory.

1880 Sophus and Anna have their second child, Dagny. He is in Sweden and meets Mittag-Leffler in Stockholm; he raises the idea of a Nordic mathematical journal, something that leads to *Acta Mathematica* coming out with its first issue a year later.

1882 During the autumn, Lie spends a couple of months in Paris, where French mathematicians (Picard, Poincaré and Halphèn) seriously look into Lie's theory of transformation groups, and its applicability.

1884 Sophus and Anna have their third child, Herman. Lie publishes substantial articles on differential invariants and differential equations in the journal *Mathematische Annalen*, of which Klein has now become the editor. In the autumn, Friedrich Engel arrives (facilitated by Klein and

Mayer, who are now both professors at Leipzig) in Christiania to help Lie with the elaboration and writing up of the new ideas and theories on transformation groups. Engel remains with Lie for nine months, and the work they now begin will result in the large three volume *Theorie der Transformationsgruppen*, properly speaking, Lie's major work – which will total 2,000 pages.

1886 Lie moves to Leipzig with his family to take over the position of professor that is vacated when Klein moves to Göttingen. Lie also takes over leadership of the Mathematical Seminar. Lie gets three private docents (Engel, Schur and Study), who work on themes linked to his theories, and the student, Georg Scheffers, begins to write down Lie's lectures – something that later will result in the publication of three extensive books. Otherwise Lie is publishing papers, one after the other, in various periodicals.

1886–89 Promising young students come from all over Europe and North America to work and study under Lie in Leipzig – what is most prestigious is the fact that students come from the École Normale Supérieure in Paris to be instructed in his theories. In 1888, the first volume of *Theorie der Transformationsgruppen* is published.

1889 Teaching and advising take more time than he would like; he feels he has not mastered the German language well enough, and being unaccustomed to the conditions at the university, together with personal divergencies, he becomes sleepless and depressed. In November he suffers a breakdown and is sent to a psychiatric clinic near Hanover. He remains hospitalised here for seven months.

1890 In June he is released from the hospital, and gradually resumes his mathematical work. The second volume of *Theorie der Transformationsgruppen* is published.

1891–93 Following his breakdown, friends and colleagues in Germany believe they can see great changes in Lie's attitude and behaviour. He attacks others for using, and almost stealing, his ideas. Lie continues his mathematical work, which is increasingly gaining approval. In 1891 his lectures on differential equations are made available – edited by Scheffers – in book form, and two years later, his lectures on continuous groups with geometric applications, are written up and edited by Scheffers. In 1892 Lie is chosen corresponding member of l'Académie des Sciences in Paris, the highest distinction that France can give a foreign scientist. Lie turns more and more toward Paris and its mathematicians, and he is honoured by being invited to Paris to attend various events. In 1893 the third and final volume of his great work, *Theorie der Transformationsgruppen* is published (by Teubner Verlag in Leipzig). In a twenty-page foreword Lie mounts a stiff attack on the German mathematical milieu, and what creates a particular uproar is the following declaration: "I am no pupil of Klein's, nor is the opposite the case, although perhaps this would be nearer to the truth."

1894 Lie wants to go home. There are many as well in Norway (with Fridtjof Nansen, Bjørnstjerne Bjørnson and Elling Holst in the lead) who want to repatriate the country's most famous man of science. Parliament approves changing the title of his professorship to "Professor of Transformation Group Theory" and offers him a salary of 10,000 kroner per annum, almost double the salary of an ordinary professor.

1894–97 Lie postpones the move back home to Norway. It takes time to wind up his works and his obligations – Anna and the children are thriving in Leipzig, but during the summer holidays, the whole family is back in Norway (as had become customary since 1891). Lie engages himself deeply in the Norwegian school and university policy debates. In 1896 he tries to divide his time between Leipzig and what is now called Kristiania (today, Oslo), but abandons this plan a short while later. His final great book project is also published in Leipzig in 1896. This is *Geometrie der Berührungstransformationen*, edited by Scheffers. In 1897 he receives the first awarding of the Lobachevsky Prize for his work in geometry.

1898 In May he gives notice, and resigns his post in Leipzig. In September he returns with his family to Kristiania. Here several people notice that he is not really in good health, and it gradually becomes clear that he is suffering from pernicious anaemia – at that time – an incurable blood disease. He holds his lectures through the autumn, and the last from his sickbed at Eugenies gate No. 22.

1899 Sophus Lie dies on February 18th. Six days later, following funeral ceremonies of great solemnity in a packed Holy Trinity Church, he is laid to rest in Our Saviour Cemetery in Kristiania/Oslo.

Notes and Commentaries

Abbreviations:
The National Library of Oslo (Nasjonalbiblioteket i Oslo) = NBO
The National Archives (Riksarkivet) = RA
The University of Oslo = UiO
Lie's *Gesammelte Abhandlungen* (Collected Works) in 7 vols. = GA

PART I

Tracking Him Down: A Torrent of Stories

Page 3 The story of Sophus falling asleep with a half-eaten pancake in his mouth has been taken from an unpublished manuscript written by Anne Christine Skavlan, the sister of Olav Skavlan (who were children of Vicar Aage Schavland). She was married to the magistrate (and cabinet minister in Sverdrup's government in 1884), Ludvig Daae. Anne Christine Skavlan Daae's manuscript is in the possession of Professor Bjarne Rogan.

Page 7 The letter from Sophus to his father is quoted here from a speech given on the occasion of Sophus Lie's one hundredth birthday, by his son, Herman, on December 17, 1942 [Herman Lie 1942]. At that point in time Herman had several letters from his father, which since then seem to have been lost. Sophus also wrote to his own father that he was training himself to leap up and over a moving horse (see also Herman Lie in *Studentene fra 1901*, Oslo. 1951).

Page 11 Alexander Kielland came to Christiania and took his examen artium in 1867, and his secondary exam the following year. Information about Lie's private tutoring is to be found in Engel's article [Engel 1922]. According to Engel, Lie also expressed his enthusiasm for the first works of Bjørnson and Ibsen, but was sceptical about what he came to consider Ibsen's mysticism, an attitude that in other respects was quite common. "Mystical experiences" beyond the realm of the empiricism of everyday life were phenomena from which most of the informed and forward-looking upright citizens distanced themselves "Mysticism" was considered a scandal in the age of Brandes' progressive modernist fervour.

Page 14 "Norwegian Man of Science Imprisoned as German Spy", *Dagbladet*, October 12, 1870. During his arrest, Lie quite likely had a letter with him from Klein (with their "suspicious" mathematical contents); this letter was stamped "Sedan", which was the Prussian headquarters. Right after his return home, Lie recounted in detail (at Realistforeningen – the Association of Scientists) his experiences, and in later tellings of the story the letter from Klein was said to have been stamped "Metz", where the German expeditionary forces were stationed. When Lie was released after four weeks of imprisonment, it was only a few days before the Prussians occupied Fontainebleau.

Page 17 On April 17, 1875, Parliament ratified the law by which Norway would join the Scandinavian Currency Commission. The crown (krone) would be based on the gold standard and would be legal tender in Denmark, Norway and Sweden. That same autumn gold kroner coins were minted, as were small change coins of silver and bronze. The decimal system was new, and to ease the transition from the old tripartite system (1 speciedaler = 5 ort = 120 shillings) to the two-part system (1 krone = 100 øre), calculation tables were distributed. Currency which had its denominations described in both systems (daler/kroner and shilling/øre) was in use in Norway from 1874. The pure kroner currency first appeared in 1876. Sweden and Denmark had established the kroner system in 1873, and this had been ratified in Stockholm the previous year. But when the kroner system first came before Parliament in Norway it was defeated.

Ole Jacob Broch was one of the main men behind this currency reform, and he worked long and hard to have the common Nordic kroner system become a step on the road to an international currency, a "world currency", as it was called in the years around 1870. Broch's recommendations were a contributing reason why Norway, two years after Denmark and Sweden, adopted the kroner system. But when it came to the establishment of the metric system, thanks to Broch, Norway was the first of the Nordic countries to convert. As early as 1875 the metric system was implemented. The Swedes followed suit in 1878 and the Danes more than ten years later. (Norway's copy of the official 'Parisian meter' was the third one ever made.) As the leader and director of the international bureau for weights and measures, at Sèvre in Paris, Broch would also spend the last ten years of his life (1879–89) striving to metricise Europe and the rest of the world. (In 1875 Broch was chosen corresponding member of the French Academy of Sciences, Mechanics Division).

Page 20 Theodor Caspari [Caspari 1945] tells about "the impulsive professor who had garnered renown for himself not only as the world-famous mathematician, but also for his intractable impulsivity [... and who] horrified Norwegian hikers by pottering around naked when the sun was too hot for him."

PART II

The Family Tree

Page 29 About the Lie family in Copenhagen (1804–07) and later at Gilstad Farm: Lars Lie was copy clerk in the Royal and State Courts, and the eldest son, John, (who subsequently became a city personality and a doctor, who attended to the health of, among others, the poet, Henrik Wergeland), was trained at the Basedow's Institute. But otherwise, the family lived in straitened circumstances. Mrs. Caspara Gill Lie had two more children, one of them dying a short time after birth. And during the great bombardment of Copenhagen in September 1807, Lars Lie was almost killed by falling beams and tumbling walls while large sections of the city were being destroyed by English cannon-fire. The Lie family wandered about in the streets, homeless for a period before they found refuge at Amager. And from there they allowed themselves to be driven in retreat back to Norway, poorer than ever.

Safely home at Christmas time, 1807, Lars Lie got the joyful news that he had been appointed magistrate at Stjør- and Værdalen, north of the city of Trondheim. Given the conditions of the day, Gilstad was a large farm – they pastured four horses, seven milk cows and forty heifers, and a large garden of many hectares, complete with a summer house gazebo. But despite the size of the farm, and the income from his public office, Magistrate Lie was in terrible economic straits. Admittedly, the farm had cost him 8,000 riksdaler, but it was said that the reason for his scanty economic situation was that he had been the victim of thievery or fraud. *Thieves* were a phenomenon that Magistrate Lie was particularly vigilant about – even in his extensive fruit and berry gardens he had rigged out an ingenius system of cords attached to bells that rang whenever uninvited guests entered the garden at night. And with the clang of the bells, the magistrate would come running, shooting in all directions with a shotgun loaded with rocksalt.

Before the children were sent to the cathedral school in Trondheim, they were home-tutored at Skogn by Hans Ulrik Midelfart. Midelfart was both priest and one of the founding fathers of the Norwegian Parliament. He was known to be one of the brightest vicars in the diocese. While pupils at the cathedral school in Trondheim, Johan Herman (Sophus' father) and his brothers received very little money from home, due probably to their father's alleged economic losses. In any case, Johan Herman explained that during his school years he himself had to find the money for clothing and books, and that he had had free room and board at Major Stabell's house. Apart from this, the disciples at the cathedral school underwent intense physical tempering. They reported that they bathed outdoors in December, and often swam out to the island of Munkholmen; they also engaged in sprinting; they were zealous about skiing and skating in winter, and when they reached the senior forms, they went to dancing school, to "Pastry Shops and Restaurations", and frequent balls and dinner parties as well.

Mrs. Caspar Gill Lie died, after a couple of days of grave illness, it is said, after she had visited her ill friend, Mrs. Middelthon, the priest's widow, in the summer of 1826, at the age of 48. Magistrate Lars Lie's second marriage was to Dorothea Heidman, but he died at New Year's 1829, probably of heart failure, at the age of 59. At that time, only the youngest child, the daughter Edle, born in 1815, was still living at home. Following the death of the father, she moved in with her eldest brother, John, who already had a medical practice in the capital city. Later, Edle would also live with her brother, Johan Herman, and while there, would among other things, teach little Sophus. In her old age, Aunt Edle would make jokes about how she had been the first to teach mathematics to Sophus Lie.

Page 31 Regarding the maternal side of the family:

Lieutenant Colonel John Herman Lie, Sophus' brother, worked out a family tree which shows that he and his siblings were 32nd generation descendants of King Harald Hårfagre. In the 19th generation – that is in the 1400s – via Johanne Andersdatter of Asdal – there is a family branch leading back as well to historic persons such as Ragnvald Mørejarl and Gange-Rolf, not to mention the first Swedish and Danish kings (Svein Tjugeskjegg who married Sigrid Storaade), founder of Russian czardom (Rurik), Emperor Leo IV of Byzantium, Charles the Great of France, and Alfred the Great of England. Lieutenant Colonel Lie was probably on much more secure footing when he worked out his mother's last five or six generations of antecedents: Maren Stabell's great grandmother was Mariane Christine Vind, whose close ancestors were Rear Admiral Jørgen Vind and Prefect J. F. Vind. Mariane Christine married Lieutenant Colonel Ditlef Scharffenberg, and their son, Herman Nicolai (colonel in Trondheim) had a daughter, Elisabeth Magdalene, who married Major Mathias Cecelius Stabell.

LITERATURE

Vogt 1914; "Af Bataillionslæge Jo[h]n Lies Optegnelser om sin Slægt og sig selv" by D. Thrap, in *Personalhistorisk Tidsskrift*, Sixth Series, III (2), Copenhagen, 1912; "Doktor John Lie" in *Skillingsmagasinet*, 1890, No. 50 and 51.

The Priestly Family at Nordfjordeid / Father's Home in Moss

Page 34 About the vicarage itself (at the time of the Lie family's arrival):

According to the minute scrutiny of the dean – the vicarage buildings that the public was dutybound to maintain (that is, the main building, the animal barn and the servants' hall and servants' quarters – later the residence of the vicarage farmer – had a number of things lacking. These were items that ought to be repaired forthwith. The houses the priest himself was responsible to maintain (that is, the episcopal salon and living quarters, blacksmith shop, the horse stables, the millhouse, together with a granary and hayloft) had so much missing that the dean counselled Lie to "allow the arrangement of a legal Inventory, for which the former Vicar should be liable for the requisite Reparations."

Vicar Lie's predecessor at the parish farm was Nicolai Nielsen, one of the founders of the Norwegian Parliament, who had been the parson at Eid for fourteen years. Nielsen had also become Dean of Nordfjord, before he applied for the much better calling of Borgund in Sunnmøre County. He had a lordly grand manner, "an extremely fine Organ and an uncommonly fortunate Declamation," and had educated bright "school keepers" [itinerant lay teachers], it was said. But he had nonetheless allowed the buildings on the vicarage to fall into disrepair. Farming had never been one of Vicar Nielsen's great interests, even though he came from a small village outside Odense on the Danish island of Fyn, where his father was a bell founder and blacksmith. In the ecclesiastical record book of the Eid parish, he had written that he had "a particular desire to set his eyes upon the Polar Regions of the cold North". This had led him to apply for a clerical position in Vardø in the northern county of Finnmark in 1807. There he married the daughter of the commander of Vardøhus, the local military detachment, before, a couple of years later, becoming parish priest at Ytre Holmedal in Sunnfjord. Here, in 1814, he gave a speech that inspired the powerful bishop, Johan Nordahl Brun of Bergen to laudatory expression: "The Honourable Brother has spoken like a Man." It was later said that this sentence led to him being chosen the district's representative to attend the constitutional assembly at Eidsvoll, where he was a staunch supporter of independence, but otherwise did little of note.

Page 42 In an exchange of letters between Nils B. Maurseth (of Nordfjordeid) and Adler Vogt [cf. Vogt 1914, 1951], Maurseth recounts that his maternal grandfather, Peder Eriksen Oterdahl (Father Lie's first assistant), explained that he (Oterdahl) was the one who had sent his own gifted hired man, Batolf Nilsen Gausemel to the Eid vicarage in 1847. [Material at NBO, Manuscript Collections.]

Page 46 The merchant, Claus Wiese, who apparently was also an engineer, made the drawings for the church, and the honorarium was six speciedaler. Eid church was restored in 1915, with the interior being decorated with the traditionally-painted rose designs common in rural Norway - the painter was Lars Kinsarvik, and the altar cloth had been made by Cecilie Dahl, at the recommendation of Erik Werenskiold.

LITERATURE

Visitaser i Eid og Hornindal 1800–1990, Part I, collection of Erling Tomasgard, Hornindal, 1990; "Eid kirkje 1849–1949" in *Kyrkjeblad for Eid og Staarheim*, Eid, 1949; *Eid Kyrkje 150 år*, Nordfjordeid, 1999.

Father's Home in Moss

Page 48 About 1850 there were fifteen higher, or secondary schools in Norway, of which nine prepared pupils up to the *examen artium* (there were 90 students who took *examen artium* in 1850), while the remainder were lower "realskoler", that is, the more demotic "grammar schools". In 1848 the resolution was taken to establish state-organised grammar schools, when Parliament granted the means to restructure and expand the learned (Latin) schools into joint Latin-and-Grammar Schools. Here the pupils studied the German language for two years before beginning Latin. Once Latin was underway, the pupils were divided into two streams: the Latin school, which prepared children for university, and the Grammar school, whose public mandate was to prepare children partly for entry to higher schooling (for example, the military college), and partly "to step out into Life and onto a purely practical Way of educating oneself for a practical Vocation in some higher Calling." The civic Grammar School of Moss is today Kirkeparken Videregående Skole.

Page 53 O.P. Nyquist, in the *Moss Tilskuer*, 1906 (no. 138, 141, 150), writes about Vicar Lie of Moss. A letter to the editor (*Moss Tilskuer*, 1862, no. 62, 64) reflects upon the question: "Can one go too far in the true Fear of the Lord?"

The obituary for Vicar Lie is found in the *Moss Tilskuer*, 22.01.1873 and 29.01.1873. (See also p. 179).

LITERATURE

Vogt 1914, 1951; Ringdal 1989; *Indbydelsesskrift til den offentlige examen ved Moss kommunale middelskole*, Kristiania, 1886.

PART III

Nissen's School

Page 59 John Herman Lie was enrolled as a cadet at the Military College from January 1, 1857, to September 11, 1861. [Riksarkivet's roll of officers from 1860 to 1892.]

Older brother Fredrik lived at Akersgaten 49, and after finishing his university examination in the autumn of 1861, moved to Dronningensgate 22.

Page 60 Nissen's School was founded in 1843. It was here that in many ways the reform programme of A. M. Schweigaard was brought to life.

Page 65 Hans Ross – teacher at Nissen's, later parliamentary professor (appointed by Parliament 23.05.1881) and collector of languages – made his debut with a book in English: *Europe as it Ought to be at the End of 1861* (London, 1860). M. J. Monrad's book, *Om de classiske Studiers Betydning for den høiere Almeendannelse [On the Significance of Classical Studies for Secondary General Education]* was translated into Swedish (1863), and a new, updated edition came out in Christiania in 1891.

Page 68 Louis Enault published a book in Paris in 1857 entitled *La Norvège* (400 pages), an excerpt from which was translated into Norwegian by Ludvig Daae and appeared in *Ill. Nyhedsblad*, 1857, p. 160.

Page 69 Olav Skavlan's diary came into the possession of the NBO, Manuscript Collections, in 1999. Here he tells about Sophus Lie's sailing tour: some of his mates wanted to go out sailing, but due to the strong wind, it was difficult to find the manpower. Even though it was right in the middle of the written part of the examen artium, they approached Sophus Lie, and the discussion was said to have gone as follows:

Listen, Sophus. Wouldn't you like to be out sailing today. It would be terrific.
Sure, but it's blowing something awful.
Yes. But that is precisely what is so good. There's a new boat we want to try out, to see what it can take.
To see what it can take – what does that mean?
We'll find out how high the sea is, and whether the boat can take it.
Yes, that is what I would love to see.

And in a flash Sophus had made himself ready "to see" – and subsequently they had gone out and capsized among the islands. Skavlan commented that Sophus always wanted to be along when the wind was strong, but big and strong as he was, he preferred just to sit in a midship position, almost as inert as ballast. (Later [see p. 192] Lie describes himself as a "poor Wretch" at sea.) Skavlan also reports that Lie seldom, or never, spoke about his family "except when he was sick, and one often heard him complain that things were going so badly for him that he was a Disgrace to his Parents as well."

Page 70 There is more about Lie's examen artium in 1859 in Strøm No. 2, 1997.

LITERATURE

Indbydelsesskrift of Nissen's School for the years 1845–59, together with *Efterretninger om Nissens Latin- og Realskole i Aarene 1853–55*, and *1858–61*.

Student Life, 1859–65

Page 71 "Reallærereksamen" [Science teacher examination] was established by law on September 15, 1851, and the exam quickly became a prerequisite for gaining employment as a "senior teacher" of science.

Page 72 The scheme for obligatory and/or elective subjects in the secondary examination elicited the following response from Lie: "The Objective of this Free Choice was, one supposes, to open a certain Room for Manoeuvre for the special Interests and Desires of the various Talents. In Fact, by and large, Things have arrived at such a Situation that prospective Science Teachers and Medical Doctors choose, as a Rule, the Science Subjects, while on the other Hand, Philologists and Theologians prefer to follow the humanistic Subjects. In any Case, the newly-established Free Choice, which is not particularly in Harmony with the public Conception of the Secondary Examination, *brings to Mind the Thought that this Examination not only be regarded as the Closure of General Education, but also that it serves as an Introduction to the various Academic Studies leading to Positions in the public Service*." [*Morgenbladet*, 13 June, 1897: "On the System of University Programmes of Study" – "Om Universitetsstudiernes Ordning" – and also in other articles from 1897–98, Lie commented on the Act of 1824 and the Act of 1845, together with the reorganisation of 1857, among other things.]

When later, with the law of 1875, there was instituted "almost unlimited Free Choice in the Secondary Examination", this provoked reaction and led to its being rescinded in the University Committee in 1883. At that time, Lie was one of the proposers from the Faculty of Mathematics and the Natural Sciences. Here Lie expanded the view "that one ought to go right back to the public Conception of the Secondary Examination, that consequently this Examination should alone serve to fill in the Gaps in the Education of those who have respectively followed the Latin or the Science Lines at School." His proposal, which implied that the secondary examination should include obligatory subjects, received the support of the Medical Faculty; but the other three faculties wanted to keep the elective system to such an extent that the secondary examination could serve as an introduction to these faculties' programmes of academic study toward public service positions.

The discussion on whether the secondary examination should be the conclusion of general studies or the preparation for scientific studies did not end until 1903 when a reorganisational process determined it should be preparatory to studies in the sciences.

Page 79 Regarding teaching books, cf. p. 111.

Page 83 It was a young theologian, Niels Hertzberg who gave a lecture at the Student Society in 1856 (strongly inspired by Søren Kierkegaard), entitled "Is it advisable to become a Priest in the State Church?"

Page 83 Realistforeningen (the Association of Scientists) was founded on April 18, 1859, and its name was "The Scientists' and Mineralogists' Association", and held its meetings on the property of Organ-builder Brantzeg at Akersgaten 61. The Association existed in various forms for five years before, due to a lack of members (when in 1864 the mobilisation for participation on the Danish side in the war with Germany was so strong among the students), it stopped its activities. And four years later when Realistforeningen was resurrected, it was at the initiative of Sophus Lie. At later commemorations of this event, which occurred on October 25, 1868, an issue was made of the fact that Sophus Lie had carried the heavy blackboard on his broad shoulders from the Student Society quarters on Universitetsgaten, up Svartfeierbakken to the new premises on Nedre Vollgate. Among the members of Realistforeningen were Carl C. Berner, Axel Blytt, Jacob Aall Bonnevie, Hans Geelmuyden, Axel Guldberg, Cato M. Guldberg, and Henrik Mohn. A central person during Lie's first membership seems to have been Thorvald Broch. Lie was chosen chairman of the association on January 17, 1871; that will say, right after returning from his first tour abroad.

And *à propos* the education of mathematicians: Thorvald Broch writes (in *Indbydelsesskrift fra Skiens skole*, 1870) about the situation in the 1850s (that is, before the science teacher examination of 1851 had come into actual effect). He writes that mathematics teachers were to be found "for whom the x's and y's of mathematics had been (and remained) in the strictest sense unknown quantities." And Rector A. Brinchmann pointed out (in *Indbydelsesskrift fra Molde skole*, 1867): "It is not long since this important subject [mathematics], like a poor relative of the school system, was pressed first on to one, then another teacher ..." In the parliamentary discussion of 1851 – before the establishment of the science teacher examination – it was also pointed out that "now there are extremely few teachers who can be called brilliant mathematicians". Hartvig Nissen stated [Nissen 1876] that the state of Norway's mathematical teaching in the 1860s scarcely had its counterpart in any other land that could in fairness be compared with Norway.

Apart from this, Thorvald Broch criticised (in *Indbydesskrift fra Aars's og Voss's skole*, 1864) the mathematical teaching book of Ole Jacob Broch. For Thorvald Broch, mathematics was more than practical knowledge and the learning of propositions – instruction ought to be concerned with a holistic understanding of the subject, and give the pupil part of a mathematical mode of thinking, such that one "at a given moment, without the application of his memory of what the individual propositions involved, could reproduce the whole mathematical building." Thorvald Broch himself published textbooks, both in "Elementary Arithmetic" (1864) and in "Numbers and Algebra" (1866).

Jacob Aall Bonnevie also wrote several mathematical textbooks in the 1870s, by way of answer to the new law on the examen artium of 1869. Bonnevie's books had great significance for teaching both in the grammar schools and the gymnasia specialising in the sciences, where they were used for over fifty years. (As a school director in Trondheim, Bonnevie, as early as 1877 also stepped forward as a spokesman for the use of local dialects in school instruction. [See also p.420, regarding the schools debate in which Lie partook].

(More on Realistforeningen is to be found in Strøm, no.2, 1997; Johannessen 1959.)

Page 85 Kristofer Janson subsequently became one of the country's most diligent itinerant lecturers, and at the time apart from the writers Vinje and Garborg, was the most noticeable spokesman for New Norwegian – the national language in the language debates of the nineteenth century. Through Janson, a broad section of the population came in contact with the growing movement for a national language (as distinct from the prevalent Norwegian-Danish used as the official administrative tongue). He would frequently hold several lectures in each place he stopped. He presented the Norse sagas and modern poetic works (both his own and others'): he recounted folktales and narratives from mythology; and in a Grundtvigian manner, he explained the symbolism of these narratives and drew out the everyday wisdom from them. He also held lectures on historical themes, particularly about 1814, the year that Norwegians forged their own constitution.

The question of using New Norwegian in writing and teaching had been debated ever since Vinje began to publish the journal *Dølen [The Dalesman]* in 1858. Janson used both New Norwegian and standard Norwegian (which was evolving from Danish-Norwegian) in his talks and speeches. When, in 1876, Parliament granted Janson a poet's salary (as had been done for Ibsen and Bjørnson), this was in effect a vote for and against New Norwegian: 30 Members of Parliament voted against. In his memoirs, Janson writes as well [Janson 1913]: "I never felt personal sympathy toward Vinje. But I admired his playful wit and his mind, although I admit he allowed himself to be dragged around to all kinds of meetings, as just another dancing bear for the public's amusement." And about Aasen, the founder of New Norwegian – the new Norwegian language - Janson wrote: "I was as well sometimes up to visit Ivar Aasen at his garret on Theatergaten. Never have I met anyone who has defended his peasanthood right in the heart of the most civilised surroundings as has Ivar Aasen. There sat this brilliant and learned researcher of language, as shy and bashful as a virgin and hid himself away, in a cramped and filthy closet where the stove was cracked and a fire hazard."

Page 88 The anecdotes concerning Lie being taken for a murderer at Levanger, and walking so fast along the national highway such that the sheriff with his cariole was unable to catch up with him, quite certainly pertain to the summer of 1863. The murder referred to here, had been committed north of Steinkjer that year.

LITERATURE

About examinations, subjects and teachers, see *Norske Universitets- og Skole-Annaler* for the years 1859–66, together with *Universitets årsberetning* for the years 1859–66, in RA under UiO, registries.)

The Lack of a Calling

Page 97 The episode from Tvedestrandsfjord in the summer of 1866 has been commented upon in several connections. Poul Hegaard mentions it in his article on Sophus Lie in *Norsk biografisk leksikon*, Oslo, 1923–69. According to Erik Krag – in a letter to Viggo Brun 02.04.1967 [Brun's surviving papers, NBO, Manuscript Collections] – who got the story from his father, who had been a schoolmate of J. H. L. Vogt, and had heard the story from Vogt himself, the reason was thus said to be that little Johan had irritated his uncle by maintaining that "since animals could swim, so too should humans be able to; or else it had been Sophus who had stated this, and so, instantly, he proved his point. Also, J. H. L. Vogt himself talks about the episode [Vogt 1930].

Page 99 Members of "The Green Room" – the radical circle of the Student Society, where Scandinavianism and the union with Sweden, the press and literature, theatre, music and painting, theology and science were zealously discussed – were: Frits Hansen, H. E. Berner, Carl Berner, Cathrinus Bang, Christian Ross, Hans Ross, J. J. Jansen, E. E. Schiøtz, Gustav Storm, Einar Skavlan, Olaf Skavlan, Elling Holst, Lars Holst, Nordahl Rolfsen, Hans Geelmuyden, N. R. Heyerdahl, Fredrik Wallem, Amund Helland, Kristoffer Lassen, Jens Brage Halvorsen, August Mohr, Otto Blehr, Theodor Blehr, Hans Brecke, Sofus Arctander, Oscar Nissen, Wollert Konow, Ernst Sars, among others. (cf. photo, p. 78). They described themselves as "the Cream of Society", and they quite likely meant by that, the cream of the Student Society – in addition to the vehement discussions, it was also a kind of entertainment club. In a cantata written by Hans Brecke in 1867, we learn that:

I sing not of Sweden, I sing not of War,
I sing not of the little Pig beaten;
I sing for the sweet calm Days of Yore,
At Abraham's where after dinner's eaten,
We relax in Style at Table, Chair and Sofa.

"The Green Room" was in existence between 1864 and 1873; the association took the name "O. S. V" [Et Cetera], and from 1889, it changed to "Andvake" [the name of the ancient kings' royal horn, the "lur"]. "Andvake" is still in existence and holds regular meetings.

More on "The Green Room" is found in Onsager 1939.

Page 101 Rasmus Nielsen's lectures were on the subject of "faith and knowledge", and to find a truer content to his philosophy, Nielsen used knowledge gleaned from mathematical studies; he had also published *Filosofi og Mathematik* (1857) and *Mathematik og Dialektik* (1859). (When Nielsen became professor of philosophy in 1841, it was a post that Søren Kirkegaard had wanted to get.)

Page 104 "The Road to Mathematics was long and heavy" – in a letter to Klein, January 1884, and to Holst (1893).

PART IV

Into Mathematical History

Page 107 Armauer Hansen, who was in Vienna (and Bonn) from May 1870, to May 1871, and studied microscopic anatomy, reports that in Vienna he (by chance in a bookshop) came across one of Darwin's books for the first time. Neither he nor any of his fellow students had ever heard a word about Darwin in Norway, or so he asserts. Meanwhile Michael Sars held his last lecture at the Academy of Sciences in 1869 (some months before he died), where to a great extent he agreed with Darwin's work. This led to the Academy's members ardently wanting to know more; and the secretary, M. J. Monrad promised to develop the matter further. This led to Monrad's *Tankeretninger i den nyere Tid [Contemporary Schools of Thought]* [Monrad 1874] where he turned against Darwinism and defended the old romantic theories.

Darwin was also discussed in the Scandinavian meeting of the natural sciences in Christiania in the summer of 1868 – in one lecture it was maintained that the species are not constant, that there are no natural boundaries found between them, but that all species are related through transformational forms. [*Forhandlinger ved De Skandinaviske Naturforskeres tiende Møde i Christiania – Proceedings of the Tenth Meeting of the Scandinavian Natural Science Researchers in Christiania* 1869.] G. O. Sars treatise on the starfish Brisinga in 1875 won the praise of Darwin himself.

Page 109 The books Lie was particularly inspired by were, first, Poncelet's *Traité des propriétés projectives des figures* (1822) – the book whose greater part was written while Poncelet was imprisoned at Saratov – second, Poncelet's *Théorie des polaires réciproques* [*Crelle's Journal* 1828], and third, Poncelet and Plücker's "System der Geometrie des Raumes in neuer analytischer Behandlungsweise" [*Crelle's Journal* (1846)].

The lending records of the university library show that Lie had also familiarised himself to the works of Carnot, Hamilton, Cremona, Möbius and Hunyadi among others [Strøm No. 2 and 3, 1997].

Page 113 Holst's treatise was published as a university programme; that will say, a scientific work in which the relevant professor in the candidate's field gave his official imprinteur to the candidate's work by placing his own professorial initials on the front page: eg., "Published by xx [Prof. xx]" . . .

There was a Mathematician in Him

Page 119 Church Department (Christiania 14.07.1869) allotted 6,000 spd. for foreign study tours for men of the arts and sciences – Lie received 400 spd., as did Henrik Ibsen.

A little about Lie's applications:

In November 1867, Lie first applied for what was called the adjunct stipend; this was awarded to a talented scientist who had qualified by passing his public service examinations. In his application he referred to his marks in the student examination, the secondary examination, and the science teacher examination. He commented on the latter exam in the following manner: "My special Marks on this Examination's various Divisions were respectively 1, 2, 2, and I was absolved in the 2nd Division where, in the relevant Protocol, there was added a statement that the Mark of 2 in this Instance should not override the Possibility of Recommendation." Lie went on to explain that since then (following the science teacher examination in the autumn of 1865) he had continued his mathematical studies. This he had done with the exception of the first half-year "wherein the Condition of my Health hindered me from doing so," and that he had only taught to the degree necessary to obtain for himself "a meagre Subsistence" [UiO, Kollegiet 97, Journalsaker, Journal No. 570].

Much of the same information was given in his application for the travel grant of the Hjelmstjerne-Rosenkrons Legacy – likely made in March 1868. Here, Lie concluded his applications as follows: "I propose essentially to pursue the study of the Application of Mathematics to Branches of Physics." Lie announced that he wanted to go to Paris, and from there, to a university in Germany. With its report and recommendations on the 1868 applications for the adjunct stipend, the Faculty of Mathematics and Natural Science stated that Lie had provided too little information about his scientific production for the faculty to make a determination.

In March 1869, to the Collegium, the Faculty recommended giving 100 spd. to Lie from the Rosenkrons Legacy, so that Lie could "exclusively" be able to "employ himself with the Editing of his mathematical Works", and at the same time advised (in a meeting on March 24th) the Collegium to recommend to the government that a foreign travel and study grant (of 150 to 200 spd.) be awarded to Lie. This latter request seems to have been made not least due to Professor Bjerknes' written recommendation. With the Faculty recommendation for the adjunct stipend of 1870 (taken in a meeting 01.12.1869) Lie, by a vote of five to four, was placed third on the list of recipients (after Axel Boeck and G. O. Sars). Other applicants for the adjunct stipend during this period were J. E. Sars, J. W. Müller, J. D. C. Lieblein, O. Skavlan, Fr. Petersen, and J. C. Johanssen.

In his adjunct stipend application of February 1869, Lie commented on the just completed work in his "imaginary theory". He wrote: "In this Work, which in my Opinion has scientific Merit, I presume to realise an Idea, that Wallis put forward in 1865, and that mathematicians of the present century, like Argand, Poncelet, Grassmann and Hamilton are known to have busied themselves without reaching their Goal." [RA, Church and Education Department, Schools Office D, state stipends.]

The difficulty of fully understanding Lie's "imaginary theory" is perhaps indicated by the fact that the eight-page article (together with his elaborated versions) in *GA*, Vol. I, are accompanied by Engel's commentary, an 'explanation' that is well over 100 pages long.

Page 120 An expression of attitudes in the capital city of Norway at this time is, to some extent, found in a piece written by Kristian Elster in *Aftenbladet* (March 8, 1870, and reissued in Kristian Elster, *Fra det moderne gjennombrudds tid.* Willy Dahl, Bergen. 1981) in which he writes about a performance of Shakespeare's *Hamlet*: "For our Contemporaries, it must seem most strange that at any Time one could disagree about what Hamlet is. That is to say, in our Day, the Hamlet Character is the one most often critically assessed, ethically weighed, and poetically presented. Anyone who has read the aesthetic Literature of our Day, will have plenty of Hamlet Physiognomies in his Mind. In any Epoch, the Hamlet Nature certainly has its common Manifestations in Life, but particularly in a Time so afflicted with Reflection, like our own, it is not noticed: for most possess the Hamlet Nature to some Extent, and their Tragedy is played out quietly in simple everyday Life, unrecognised except by those who know them the most intimately, and these few perhaps do not understand."

Page 120 Lie thanks Broch on behalf of the students of the sciences because "his hospitable House" had always been open to them [*Morgenbladet* 26.03.1869].

The First Tour Abroad

Page 123 Felix Klein, born 25.04.1849, was said to be a child prodigy, and it was pointed out that all the numbers in his date of birth are the square of other numbers: 5^2, 2^2, 43^2. Klein's letter of 31.10.1869 to his mother is reprinted in Lie's *GA*, Vol. 1, p. 636.

Page 125 "The Academy of Sciences of Christiania" was founded in 1857, in 1925 changed its name to "The Norwegian Academy of Science in Oslo", and today is known simply as the Norwegian Academy of Sciences. (Kjerulf was one of the founders in 1857.) Another such organisation, the Royal Norwegian Academy of Sciences was the country's first science association, and was founded in Trondheim in 1760.

Page 136 In a letter to the Collegium dated 25.02.1870, Motzfeldt reported on Lie's stay in Berlin ("since medio September"). He wrote that Lie now, during "these Days" was travelling via Göttingen to Paris, and would later (from the beginning of September) likely stay for three months with Professor Cremona in Milan, and had now requested, through Motzfeldt, about 150 to 200 spd., from the Hjelmstjerne-Rosenkrons Legacy. In this request, Motzfeldt pointed out that if Lie did not get the money he would have to return home directly following his stay in Paris – the 400 spd. that he had already received was used up. Lie himself wrote from

Berlin in February (1870) to the Collegium, explaining his plan to travel either to Milan or to Cambridge. With this letter he enclosed the study printed in *Göttinger Nachrichten* and brought to the Collegium's attention as well the work he had published in *Crelle-Borchhardt's Journal* (i.e., what is commonly referred to simply as *Crelle's Journal*). He also listed his two studies published in the proceedings of the Christiania Academy of Sciences – and finally he informed the Collegium that he had given Professor Bjerknes "a Number of Documents which hopefully will support this, my Petition."

In a letter from Lie (Munich 14.11.1870) to Bjerknes (concerning the adjunct stipend application): "A Year ago You wrote that You supported the *diligent* Stipend-Holder. Going strictly by the Book, I reply, and You must admit with Justice, that Examinations and Stipends have nothing to do with moral Properties. As appreciative as I am to the University for the two Travel Stipends I have received, were I to be passed over I would be filled with infinite Bitterness."

Page 137 Theodor Reye later gave Lie his book, *Geometrie der Lage*, with a dedication, in which Reye had come out with the following laudatory description: "Sie haben mit Ihrer Geometrie des Imaginären einen sehr glücklichen Fund gethan." (cf. p. 126).

Page 141 Lie studied differential equations with great zeal, by, among other things, reading Imschenetsky's teaching book on partial differential equations, and he soon came to know both what was known and unknown in the field. Jordan's presentation of Galois theory gave Lie insights into prospective means for solving differential equations.

Page 143 From the middle of June 1870, and for about a month, Ernst Motzfeldt was in Paris together with Axel Bruun. They were together with Lie "immediately prior to the French-German War," Motzfeldt writes in his diary [RA, Motzfeldt family private papers].

Page 143 Broch had (for health-related reasons) stayed at Ems in 1870, and there he saw both Minister Gambetta and Wilhelm I of Prussia. Broch had left Ems only a few days before the ill-fated meeting between the French ambassador and the Prussian king (July 13th) – and war was declared six days later. Broch also stayed at the bathing spa of Ems in the spring of 1871. He suffered from a throat infection that made it difficult for him to speak [Seip 1971, p. 424].

Page 145 Translated from the German: "Denn der Zweck einer mathematischen Arbeit kan vernünftiger Weise nur der sein, verstanden zu werden, nicht der, Bewunderung für den Autor zu erregen."

Page 145 In Norway there was great harmony of viewpoint concerning the Franco-Prussian War. In 1864 there had been great antagonism toward Germany due to its attack on Denmark, and in 1870, the blame for the war was unilaterally laid on Bismarck and Germany.

Page 145 Only a week before Lie and Klein parted in Paris in July 1870, Lie had made his famous "line-sphere discovery" – in which lines are mapped to spheres. These transformations (contact transformations) had nice properties relating asymptotic curves on one surface, to curves on another surface.

Page 151 The work that they had done in Berlin and which would be published in *Monatsberichte*, was a commission Klein now described as something that had "hovered" between them for a long time ("unserer schon so lange schwebender Arbeiten").

Page 151 Kummer's expression in German was: "Frankreich hat sich in diesem Kriege als eine sittlich sehr tief gesunkene Nation gezeigt und darum geht es, wie ich überzeugt bin, seinem weiteren Verfalle immer mehr entgegen."

À propos mathematicians in prison:

In his memoirs (*Souvenirs d'apprentisage*, Basel 1991, English translation, 1992), the French mathematician André Weil tells about how in 1939 in Helsinki he thought about Sophus Lie at Fontainebleau in 1870. André Weil found himself in a comparable situation. Weil, together with his Eveline, had been staying in Finland since the summer of 1939. They had visited the mathematicians Ahlfors and Nevanlinna, and in the course of the summer, they had spent some fine weeks at a hotel on Lake Salla. Eveline was a stenographer, and to keep up her shorthand skills, Weil liked to dictate to her from Balzac's novel *La Cousine Bette* while they sat on the shore and enjoyed the beautiful landscape. In the politically-charged situation between Finland and the Soviet Union, this literary activity was construed as suspicious – especially since what was written down seemed to have to do with an intense scrutiny of the surrounding landscape.

With the greatest of secrecy, Weil's activities were reported to the police, and on November 30th, the day the first Russian bombs fell on Helsinki, Weil, due to his foreign appearance, was stopped by the police, and they already had a file on him. (Earlier that autumn Eveline had returned to France.) André Weil was imprisoned, and when he had been forced to show the police the apartment in which he had been living, they discovered a lot of stenographic writing in shorthand, his reassurances that this was only the text of Balzac's novel, were not particularly convincing. And when, in addition, they found a letter from the Russian mathematician, Pontryagin – about a possible visit to Leningrad – the issue became terribly obvious in the eyes of the police: the man that they had in front of them was a Russian spy!

It was said that what prevented him from being taken out and shot was the fact that Nevanlinna (who was also a brigadier), had happened to meet another officer at dinner, who in the course of their conversation, mentioned that the next day they had to shoot a person who claimed to know Nevanlinna. After having learned the name, Nevanlinna was said to have asked if it was really necessary to shoot the man, and would it not be better simply to escort him out of the country. And this is what in fact happened: Weil was placed on a northbound train to Haparanda; he crossed into Sweden and arrived in Stockholm, where he was given assistance by the mathematician Cramér. Subsequently, from Viggo Brun in Norway – with whom Weil had corresponded – he received money to get himself to Bergen, and over to Newcastle.

Parliamentary Professor

Page 154 The common father-in-law of Guldberg and Waage was Cabinet Minister Hans Riddervold. And when Waage's wife died, Waage married Guldberg's sister.

Page 156 The fact that Lie wrote his doctoral dissertation in Norwegian, rather than in German was not from his own choice. Professor Gustav Storm, as General Secretary of the Academy of Sciences, told Engel (in a letter dated 27.01.1900) that Lie had first written the treatise in German. He had been forced to translate it to Norwegian due to the rigorous rules that ordered the only languages acceptable for doctoral theses were Latin and Norwegian. Storm also pointed out that the title "Over en Classe ..." is quite atypical of Norwegian, and is rather a literal translation of the German "Über eine Classe ...". No German version of this dissertation, however, is to be found among Lie's surviving papers.

More on Lie's doctorate [RA, UiO, Collegium, jnr. 97]: Lie reported to the faculty on March 25, 1871, that he wanted to defend a doctorate. He submitted "a hand-written Thesis", and wrote: "In the two first Years (1866–68) following my Public Service Examination, I occupied myself almost exclusively with the Philosophy of Mathematics. In the Year 1868 I applied myself to the Dedication of my Time as far as possible to expand my Knowledge of Mathematics in general, and in particular of what is called 'modern Geometry'." At the same time, Lie described his works from 1869: "The Principles of an Imaginary Theory which is based upon one of the Ideas put forward by Plücker: to introduce the Line as a Space-Element". He brought to their attention the study "by my Friend Klein and myself" in the Berlin Academy's *Monats-Berichte* (15.12.1870). His first lecture as part of the doctoral examination, was on May 19th, under the self-chosen title "On Plückerian Line Geometry". Subsequently – on May 26th and 31st – he had to lecture on given themes, and after approval, the oral defense, the disputas, was held on June 12th in the university's main hall.

Page 157 Olav Skavlan took his doctorate on "Holberg as a Comedy Writer" 23.03.1871. (Since then there have indeed been no doctorates taken on Holberg in Norway.) In his diary [NBO, Manuscript Collections] Skavlan reports that Lie had great respect for the verses that Skavlan made, and that on one occasion, Lie was to have asked, "Hey, how do you make these verses? I want to do this too." Skavlan answered that it was not so easy to do – one had to have a "natural talent, but you know, it's not something that you possess!" "No, but can't you teach yourself with Practice?" Lie asked, and after receiving an affirmative reply, said, "OK, teach me. Let's get down to it right away!" According to Skavlan the two of them decided to improvise with a string of given rhyming words, but they did not achieve particularly good results. On one occasion Lie had seen a poem with the syllables "-ater" rhyming 12 times – a task that Skavlan seems to have given to someone else who had also wanted to learn to write verses – and Lie had said, "I must have such an exercise too." Skavlan had then given him the ending "elle", and he said about "elle" that he could easily "make 15 Rhymes, instead of 12. Lie took great care of the verses Skavlan wrote for him: "This I can use as a model,"

Lie replied on another occasion when Skavlan had written a verse for him. But the third time, Skavlan reports, "he [Lie] became rather emotional, when he came right up to me and placed his Arm around my Neck and said, 'I also want to write such a verse for you . . . '."

The verses written to honour Lie's doctoral defense were set to an old melody from the rural folk tradition called "stev", and printed ahead of the verses were the following nonsense forms (with some catch phrases from Lie's studies): "When the given Curve of the Length of Zero has a Point, thus the corresponding Curve in the line Complex has a stationary Tangent $f(xyz\,dx\,dy\,dz) = 0$."

Page 158 Lie applied to the Collegium (17.02.1872) for 200 spd. from the Hjelmstjerne-Rosenkrons Legacy for a scientific study tour abroad. With this application he enclosed his work from *Göttinger Nachrichten 1870*, and gave notice that he would renounce all claims to a stipend if he were "in the course of the Autumn to receive Employment as Prof. in Christiania or Lund." Other applicants for foreign study grants were Kristofer Janson, Professor Guldberg, Professor Kjerulf, Professor Schübeler, Professor Bugge and Professor Esmark.

Page 158 Carl Johan Danielsson Hill was the father of Carl Fredrik Hill, who today is considered one of Sweden's great painters. In particular, his expressive drawings of illness were trend-setting in their influence. (C. F. Hill was diagnosed in 1876 as suffering from schizophrenia.)

Page 161 Daniel C. Danielsen seems to have been the prime mover behind the motion to set up a special professorship for Sophus Lie. Chief of medical staff at the St. Jørgen leprosy clinic in Bergen, Danielsen was highly respected internationally as a man of medical science, and he had published a series of works on leprosy and other diseases. He joined G. O. Sars' Arctic Ocean Expedition in 1876–78 (with, among others, G. Armauer Hansen) and took over the editorship of *Nyt Magazin for Naturvidenskaberne* after Sars resigned – when Sars, together with Lie and Worm-Müller established *Archivet*. Danielsen was also a firm supporter when in 1874 Ernst Sars was also made a parliamentary professor. It was Danielsen who got Ernst Sars to give out a declaration that he was not a positivist, a declaration that was later used in a vehement struggle over basic philosophies of life.

Page 161 Lie's behaviour in the parliamentary gallery in 1872 is described in Ludvig Daae's political diaries [Daae 1930].

Page 162 King Karl XV died 18.09.1872, and some of the political suspense abated when now, King Oscar II took over.

Page 182 Lie commented on the parliamentary professorship in a letter to Klein: "Es ärgert mich aber immer die Absurde Geschichte zu hören: Ich wäre nie Professor geworden, wenn nicht meine politischen Sympathien dem Storthing bekannt gewesen wären." Lie used the same expression in his letter to Mittag-Leffler (in July 1882) – that the allegation that he was "among the reddest of the red" was completely groundless.

Page 163 Motzfeldt wrote many letters to his wife Else from his hiking tours in the mountains. Lie reported from Bjølstad in Hedalen (the Valdres area), 31.07.1872, saying that he and Lie, plus marine painter Boll, had set off handsomely: Boll first in a cariole, while Lie jumped up on the wagon, before continuing on horseback. Motzfeldt, writing to Else in the summer of 1875: "Lie rowed Gjendin." (Gjendin, a Jotunheimen lake the length of a day's hike, was made famous by Ibsen in the lyrical boastful daydream by Peer Gynt, where he explains why he forgot to watch his mother's herd). And on that occasion they also had been to Sikkelsdalsgården.

Page 164 There is an interesting observation on the unique character of the geometric discipline in a letter Klein writes to Lie on February 16, 1872: "In der Geometrie erstrebe ich – sofern ich wirklich Geometrie, und nicht nur, wie gewöhnlich Analysis treibe – eine volle sinnliche Anschauung von den räumlichen Dingen und den Gesetzen, die zwischen ihnen stattfinden."

Page 165 Translated from German – Klein (August 1872): "Ich bin ein meist sehr lustiger Mensch geworden, der allerdings Mathematik treibt, auch mit ernst Mathematik treibt, aber doch nicht mit diesem finsteren Ernste, wie damals." In a letter written a couple of months earlier, Klein had complained that he lacked intensity and endurance in his work: "Ich habe gar keine Intensität und Stetigkeit im Denken mehr" (29.06.1872) – and he stressed that Lie's visit in terms of science was essential to him: "Dein Hierherkommen ist für mich wissenschaftliche Lebensfrage."

Page 166 There is more about the Erlangen Programme in Hawkins 1984 and Rowe 1989.

PART V

"My Inner Life Has Been Most Mighty"

Page 179 Vicar Lie felt himself pressured by his "excessively pious" parishioners. Great parts of the lay movement were however gradually channelled into the state Church, thanks, not least, to Professor Gisle Johnson. Johnson accommodated this religious revival by stressing that there is forgiveness to be found for all sins, and by underlining that God's creation work was complete, and in no way lacked any improvement or perfecting (in the form of the Darwinian theory of evolution or positivist points of view). (In the 1880s a new revival movement led to the formation of a series of new denominations coming into being – Baptists and Methodists – outside the state Church, and the Pentacostals and Independentists on the fringes of the state Church – see also pp. 54–55.)

Hans Riddervold stepped down as Minister of Church Affairs on July 1, 1872, after 24 years of service, and was replaced by Dean C. P. P. Essendrop, who was expected to keep a lower Church profile.

Page 183 Niels Henrik Abel's mother's father, Niels Henrik Saxild Simonsen had a son, Daniel Barth Simonsen, who became the bailiff of Namsdalen and married Sophie Margarethe Hildrum. Their daughter, Marie Elisabeth Simonsen, subsequently married Gottfried Jørgen Stenersen Birch and became the mother of Anna.

Page 185 Dikken Zwilgmeyer's letters, currently in an unorganised collection, are at NBO. (The politician Ludvig Daae was, otherwise, mother's brother to the Zwilgmeyer sisters.)

The Birch-Reichenwalds felt Anna could at least tell the Horn family in Risør about the engagement, and Sophus commented: "Think it over, my Dearest, and do as You think best."

The Birch family traced their genealogy back to the German-born Johan Gottfried Reichenwald, who had settled down as a rural merchant at Fåberg, in the Guldbrandsdalen area, and died there in 1806. He was married to Marie Elisabeth Birch, the daughter of a merchant from Læsø in Denmark. Their son, Paul Hansen Birch, rose to the top in the military – after having served under Christian August in 1806, and been commander of the guards during the deliberations at Eidsvoll in the spring of 1814, when Norway's constitution was laid down. Thereafter he became captain, major, lieutenant colonel (and King Karl Johan's accompanying officer at the coronation at Trondheim Cathedral in 1818), brigadier, adjutant general, Chief of the Trondheim Brigade, commandant and general commissioner. In 1813 Paul Hansen Birch had married Cathrine Hoffmann Stenersen (the daughter of the bailiff of St. Thomas, in the Danish colony in the West Indies, where she was also born) – and among their children was Anna's father, as well as Christian (the Prime Minister who subsequently took his paternal grandfather's name), and Mrs. Motzfeldt.

Page 193 Ernst Sars was in Fredrikshald in 1872 [Fulsås 1999, p. 134], and moreover, seems to have been there the following year as well. Here he gave twelve historical lectures that were a further refinement of the lectures he had given the previous year while he was a research fellow at the university. When Professor Lochman wanted to drive Sars out of the university (with his complaints that Sars was a positivist, and so on), Sars decided to publish these lectures, which became the first volume of *Udsigt over den norske Historie [An Overview of Norwegian History]*, which came out in 1873 and became an influential historical work.

Page 203 Regarding Helland and the Academy of Sciences:

Helland had heard from Lie that the General Secretary (Professor Monrad) had sent Lie's works directly to the printer – perhaps not so strange considering Monrad's experience with Lie's first work, which indeed none of the country's mathematicians had understood. But this method of operation, which was not actually in accord with the Academy's statutes, was certainly not something that Monrad applied when it came to contributions from non-members like Helland. (Among other things, this applied to the lecture Helland gave to the Academy of Sciences' thematic meeting on March 22, 1872, on the formation of the fjords and mountain lakes.) In addition to the fact that the content of Helland's lecture was highly controversial, many also felt that the text of the lecture showed no respect, and its mocking tone would have been satisfactory for a *Dagblad* article, but not for a scientific study. In a vote on October 24, 1873, the so-called "Helland attack" – where Lie spoke powerfully in favour of Helland – something happened that had never occurred before: a proposal for membership was turned down when the motion came to the vote.

The issue of Helland's membership was taken up in a new meeting on December 5th (where again Lie spoke), and again on February 20th, 1874. However, the issue was postponed, and through the whole of 1874 the matter was moved forward to the Academy of Sciences' general assembly, where attendance was high in the expectation of renewed conflict. Not until the meeting of February 19th, 1875, was the conflict settled, and then because Helland himself announced that he would not enter the Academy even if he were elected.

The Helland Affair sharpened the contradictions in the milieu of the sciences – one consequence was the G. O. Sars resigned from the editorship of *Nyt Magazin for Naturvidenskaberne*, and together with Lie and Worm-Müller, started the new periodical *Archiv for Mathematik og Naturvidenskab*. When the application for support for the new publication came before Parliament, Ludvig Daae noted in his diary: "A Number of our younger Nature Experts – Worm-Müller, O. Sars, and S. Lie – who do not quite see Eye to Eye with the Old Guard, want to give out a new natural sciences Journal and hereto petition for a Contribution." Politican Daae went on to explain with alarm that the budget committee had even received an anonymous communication countering this allocation [Amundsen Vol. 1, p. 125ff].

The Department of Church and Education supported *Archivet* by subscribing for thirty copies (for scientific institutes and journals) and the editorial board stated that in return the periodicals that they received would belong to the university library.

Lie wrote to Bjerknes (undated, but probably in connection with Helland's application to the Faculty of Mathematics-Natural Sciences in 1876 for a travel stipend, in which among others he was competing with W. C. Brøgger): "However I can only protest that You, who like several other Members of the Faculty, have Themselves failed to form a personal Position on this Matter. I myself am convinced that it is a Human Fact that a Case such as the Helland Affair is a remarkable Example of Partisan Passions that can blind Men of Learning, men who I consider to be both gifted and honest." Helland did not get the travel stipend [Hestmark 1999, p. 78].

Page 204 Helland's prose essay, "On the Mountains [Om Fjellene]" was published in the Tourist Association's "Yearbook 1872". Helland opposed the conserving of the old forms of the language – a point of view attested to by the announcement in this issue that the writers have expanded "the retention of their own orthography".

"My Life's Good Fortune" / Marriage at Last

Page 205 Moreover, Head Teacher Fredrik Lie (Sophus' eldest brother) from Kristiansand, was also appointed to the "artium deputation" of 1873.

Page 212 In *Gesammelte Abhandlungen*, Vol. 7, pp. 175–219 (an article dated 1895), Lie gives an account of the historical development of the theory of partial differential equations of the first order. He writes in detail about the relation between his works and those of Mayer – and here Lie presents critical remarks about a number of other writers. Lie also stated that during the years between 1873 and 1876 he, in the mathematical sense, lived only for transformation groups and integration problems [Engel 1822] – and that he vainly hoped to awaken interest in his investigations.

Page 215 Elling Holst wrote in *Nyt Tidsskrift* (Dec. 1893) about strolling with Sophus Lie during the autumn of 1873 as Lie explained his new studies. The publication of the transformation group theory began three years after this, in 1876. The groups that Lie used were closely connected to the geometric interpretation of differential equations – "they constituted a bridge that joined together the geometric and analytic sides of this same problem. He [Lie] called them transformation groups. Today we call them Lie Groups in his honour" [Lindstrøm 1995, p. 531].

Page 216 Lie's letter of response (translated from the German) to Mayer, written one day in November 1873, has been published in *G.A.*, Vol. 5, p. 584 – all of Lie's letters to Mayer are now located in the NBO, Manuscript Collections.

Page 218 The newspaper serial that Sophus comments upon with such commitment seems to have been "A Golden Man" [Et Guldmenneske], a novel by the Hungarian, Maurus Jokai, translated into Danish. (The book was published in Denmark in 1874, translated by A. Damkier.) This series came out in *Aftenbladet*, between September 10th and December 18th, 1873. Jokai was an extremely popular writer. In March 1874 his story "Paa Flugten [In Flight]" was serialised, as was Jokai's novel, *Other Times, Other Places*, between March and the end of July 1874. Otherwise it was at this period of Sophus and Anna's engagement that a piece of Mark Twain's, "The Amazing Frog-Jump" was presented. Other series in the spring of 1873 were Mrs. Edwards' "To Visit or Not to Visit?" and the novel, "Willing to Die", by the writers of "The Rose and the Key". "Ranch" by Rhoda Broughton came later in 1873, and during the summer of 1874, "Without Purpose" by Florence Marryat.

Page 218 A brief overview of Norwegian newspapers: *Morgenbladet* (Conservative, founded in 1819), *Dagbladet* and *Aftenbladet* (Liberal), *Verdens Gang* (the most radical, founded in 1868). *Dagbladet* was founded by H. E. Berner in 1869, *Aftenposten* was founded by Christian Schibsted in 1860, and for the first half-year it was called *Christiania Adresseblad*). It had made great progress during the Franco-German war of 1870–71 (with lead writers Yngvar Nielsen and Fredrik Bætzmann; while *Morgenbladet* had Christian Friele). *Vort Arbeide*, founded in 1884, came to be called *Sosial-Demokraten* two years later, and was published from 1894 as a daily; it was taken over by the Labour Party (that is, it was the forerunner to *Arbeiderbladet*).

Page 225 Klein, in a letter to Lie of 26.04.1874: "Ich hatte jetzt zu Dir kommen wollen, um ganz wieder in den Geist Deiner Arbeiten einzudringen."

Page 226 Many thought that Sophus Lie and the writer, Jonas Lie, were relatives, and perhaps Sophus also thought so (in any case, it was Lie's novels that he gave to Anna). Jonas Lie made his debut with *Den Fremsynte [The Visionary]* in 1870, and quickly became one of the nation's great poets. He published two books in 1872: *Fortællinger og Skildringer fra Norge [Stories and Sketches from Norway]* (which among other things included the short story "Nordfjordhesten" ["The Nordfjord Horse"] and *Tremasteren Fremtiden [The Three-Masted Future]*. Lie received "the poet's pay" in 1874; the same year he published *Lodsen og hans Hustru [The Sea Pilot and his Wife]*. When Sophus wrote to Anna "that things shall go as splendidly for us as in any Novel", this was probably with reference to Jonas Lie. Jonas Lie's permanent address would later be Paris for a period of 24 years. However, there is no information as to whether, while visiting the French capital, Sophus Lie ever made contact with the writer, but at the gathering place, Café de la Régence they must certainly have met.

A little more from Sophus' letters to Anna:

It was also important for Sophus to present himself to his intended as young and active, as well as a man who kept his promises. He reported that, as agreed, he had had his poor Locks clipped, and was convinced that many years would still go by before he needed to apply a Wig – the growth of his hair had, in fact, become stronger in recent Years. And as small Attempts at being Worthy of Love, he had enclosed with his letters a *daler*, which she could use for her Pleasure, but he was eager to add, "I also do this with the Calculation that I will get a little Praise for being kind and loving." The first Requirement in any Marriage was, according to Sophus, that both partners behaved with proper sincerity toward one another, and since indeed they could both do this, he maintained: "Thus can we be completely convinced that the other Issues will take care of themselves." He expressed the view that he felt the time (the spring and summer of 1874) crept forward: "so horribly slowly, that we shall never reach August" (i.e., the wedding). Perhaps she saw Things differently than he did, he commented, but went on: "You too are looking forward to August, are you not? Say so, say that you are my sweet Girl. Am I bad to demand all Sorts of Things from You? But still, at that, it is not a preposterous Desire." And he wanted to know: "Have You, at any Time since Christmas, wanted the Wedding to be postponed again? Tell me *the full Truth*."

Herr Professor mit seiner Gemahlin

Page 232 Sylow had a leave of absence (from his head teacher position at Halden) from August 1873 to August 1877 in order to devote his time to Abel's works. Weierstrass had sent Lie information from Berlin about the divergence between Abel's original letters to Legendre and the published versions in *Crelle's Journal*.

Page 233 In a letter to Mayer in May 1875 ("Haben Sie Puiseaux Bericht über das Problem dreier Körper gelesen?" etc., published in *GA*, Vol. 4, p. 482 – original in the NBO, Manuscript Collections), Lie also commented that the German mathematician, Jean Charles Radau had reduced the three-bodies problem, and Lie mentions that Paul Gordon had been Klein's colleague in Erlangen.

Page 234 In one of his first letters (dated Erlangen 22.02.1875) after their meeting in Düsseldorf, Klein wrote to Lie: "When I think of you, I so often have the Feeling of a long, indeed perhaps hopeless Separation from my better Self." ("Wenn ich an Dich denke, habe ich so oft die Empfindung einer langen, ja vielleicht hoffnungslosen Trennung von meinem besseren Selbst.") The romantic conceptions of an alter ego or "twin souls" seems to have been alive and well in Klein – and Klein's statement after Lie's death, about the close kinship between genius and madness is another such romantic concept. From his own mental breakdown Klein too was familiar with "the romantic madness."

Page 235 The letter to Mayer a short time following his homecoming in December 1874, that the stay in Düsseldorf would be crucial to his coming works, is published in *G.A.*, Vol. 4, pp. 523–24. All of Lie's letter to Mayer are found in NBO, Manuscript Collections, Collection 52.

In the Lee of "the Modern Breakthrough"

Page 240 In relation to Holst's application (April 28, 1874) for a travel study grant, Lie writes to the Collegium [RA, KUD, j.nr. 523 D 74]: "Among all those studying the Sciences, with whom I have come in Touch since my own Student Days, Elling Holst is the only one in whom I have observed an innovative mathematical, and especially geometric Gift. [...] So far as I know, in addition, Holst is the only Norwegian Student who in the past 6 or 8 Years has submitted Works to the mathematical Periodicals of our Neighbouring Lands." (Holst became a student in 1868, took the secondary examination the following year, and began to study science in 1870.) Holst's report on his study tour is found in RA, UiO, Collegium reports, Box 2, 1872–79, and also KUD, schools office D, state stipend. 43.

Page 256 About this related series of communiqués in *Archiv for Mathematik og Naturvidenskab*, Lie later gave the following description (in a letter to Holst in 1892). In relation to Holst's article for Halvorsen's dictionary of writers, he wrote that because the ideas were new and *Archivet* had a very limited circulation, thus these studies of continuous transformation groups, particularly in the years between 1873 and 1876, went virtually unrecognised. Today it has been pointed out – by among others the mathematician, Eldar Straume [Straume 1983, 1992] – that Lie's works in the 1870s are somewhat cryptic in form, and therefore difficult to read. In other words, it is somewhat difficult to reconstruct the logic of Lie's reasoning and the leaps in his thinking. Straume stresses how Lie was continually "seduced" into researching new ideas before he had elaborated and written down the old ones in a readable form. Not until Lie had prepared a major work for *Mathematische Annalen* in 1880, was serious attention paid to his theory.

Page 257 In October 1876, Lie corrected the proofs of his work on integration methods of what was called the Monge-Ampère Equation – a work that was published in *Archivet* at New Year's 1877. Lie, in a letter to Mayer (November 1876), wrote about his research, saying he was working "most acutely", that he was applying his ideas about transformation groups and was full of great expectations. "Ich habe zu viele Pläne!" ["I have too many Plans!"] he wrote in a letter to Mayer. In his studies of partial differential equations, Lie now also wanted to deal with the Pfaff problem, the problem of homogeneous and non-homogeneous differential equations. He writes [in a letter that Mayer received 03.04.1877]: "Ludicrous as it is, it continues to pain me to write, due to my shoulder ... " [letter to Mayer, received spring 1877]. "Wenn Sie und Klein mir helfen werden, hoffe ich, dass es einigermassen gelingen wird, obgleich ich zuweilen halb verzweifele" [letter to Mayer, 27.02.1877]. Lie wanted to go to Germany in the autumn of 1877 and hoped that Mayer and Klein would assist him. He expressed this desire in a letter to Mayer (received 12.07.1876). [NBO, Manuscript Collections, Collection 52.]

A Steady Stream of Works

Page 270 In a letter to Klein (May/June 1878) Lie summed up his tour abroad in the autumn of 1877: "Im grossen Ganzen, ich war ärgerlich und unzufrieden. Und ich konnte nur mein Gleichgewicht dadurch erhalten, dass ich etwas Mathematisches leistete, was mir meine volle Selbstachtung wiedergab. [...] Da ich anfänglich keinen grossen Erfolg mit den Minimalflächen hatte, wandte ich mich zu meinen Transformations Gruppen, die wahrscheinlicherweise die wichtigste Leistung meines Lebens werden wird. [...] Mein Haupt-Resultat die sich auf alle Dimensionen ausdehnt, ist dass die lineare Gruppe die einzige ist, die im Unendlichkleinen die grösste mögliche Transitivität besitzt. Hierbei ist also meine alte Vermuthung glücklich realisiert." In his treatise "Sätze über Minimalflächen" [*Archiv for Mathematik og Naturvidenskab* 1878, pp. 166–76], Lie refers to Henneberg and Herzog – and Lie had written to Henneberg from Munich, right after the meeting there of the natural sciences in September 1877 [*GA*, Vol. 1, pp. 786–89].

Lie did not report to the Collegium about his stay abroad during the autumn of 1877, until January 28, 1880 [RA, UiO, Collegium, Reports, Box 2, 1880]:

"About the 20th of August, 1877, I travelled to Munich, where I took part part in the general Meetings of the Natural Sciences, where on that Occasion I presented a Lecture on Minimal Surfaces." Otherwise, Lie explained that he had met German and other foreign mathematicians, and after: "a shorter Stay in Leipzig and Berlin, I returned to Christiania at the End of October 1877." And by way of documenting the profitability of the university's investment of the 175 spd. he had used, he referred to his publications from this period.

Page 271 Inserted between practical knowledge and commentaries, Klein writes to Lie: "Gerade, was Du über Stringenz schreibst, interessirt mich sehr." (Exactly what you write about Stringency is of particular interest to me.") From the context in which it occurs, it implies that by "stringency" Lie was referring to something in the direction of "proper editing".

Page 272 In a letter to Sylow (26.10.1878), Lie is astonished by the fact that Broch, who is in Paris, would have taken action to start a subscription campaign about Abel's works. He had come to hear from Holst that Bjerknes had written to Hermite, who reported back from Paris that he did not know about any such campaign. Lie wrote to Sylow that he could certainly ask his Paris acquaintances what had happened, "but then because many Years have passed since I activated my French Friendships, I do not have much Desire to do so." Lie asked Sylow if he, instead, could not write to ask Jordan whether or not there really was such a subscription campaign, and if so, by whom.

Page 274 Johan H. L. Vogt described the mountain tour with his Uncle Sophus during the summer of 1879 in a speech given in 1930 [Vogt 1930]. Johan H. L. Vogt had become a (science) student in 1876, and had returned to Christiania after spending a year at the Dresden Polytechnikum. (Another Norwegian also attended the technical college in Dresden that same year of 1877 – Nils Johan Schjander, who returned two years later. Schjander frequented the Bohemian circles around Hans Jæger, and quite probably attended Lie's lectures between 1883–86. A couple of years later he journeyed to Argentina, found work as an engineer, and was the only member of the Christiania Bohemia who actually reached Patagonia [Fløgstad 1999, pp. 92–97, Fosli 1994, p. 435]. There is no evidence that Lie engaged in the debate about Jæger and the Christiania Bohemia – but as for Holst, he was strong in his criticism of Jæger, and in a letter to E. Sars (05.02.1882), he made an urgent request not to support the Jæger Affair (by supporting "an Address" if the action against Jæger should collapse) and thereby give "Your and the Liberal Party's Enemies such a Triumph."

Johan H. L. Vogt took the mineralogy examination at the University of Christiania in the autumn of 1880 – then, after various travels, both within the country and abroad, and a study period in Stockholm (where he also became a university lecturer at Stockholm College), and to Freiberg, Clausthal and Leipzig, he returned home in the autumn of 1885 and the following year became professor of metallurgy.

Page 275 The discussion with Lieblein: Lie's letter of 29.10.1879 is published in the Norwegian University's and School's Annals, 3rd Series, Vol. XVII, pp. 8–10, where as well the whole matter is presented. Lieblein gave reassurance that he did not believe what, from time to time, was maintained, namely: "that certain Individuals absolutely lack the Ability to learn Mathematics", but he had many examples of many gifted people having great difficulties acquiring mathematical knowledge. And indeed, since everything pointed to the fact that one could well be "a cultured Person and even become a brilliant Scientist in many fields without having deeper, more thorough-going Knowledge of Mathematics," Lieblein felt that the door of the university (due to a too strict written examination in mathematics) ought not be closed to these persons.

Page 277 Holst's friend who told about Lie being given notice in 1879-80 and having to move from the Tandberg property to the Schønheyden's estate (later Drammensveien 84) due to the sword blows to the ceiling, was State Prosecutor

Arne Lommerud, who told it to Johannes Arneson, who wrote it down [Arneson 1964, p. 115].

The Mathematical Milieu at Home and Abroad

Page 280 Lie wrote about Bjerknes' Abel biography in a letter to Sylow (17.02.1881, and 11.08.1882): "It seems absurd to me [that] Bjerknes seems to have denied Jacobi a Part of the Honour for his greatest Discovery: the Conversion of algebraic Integrals," and according to Lie, there were several unfortunate expressions in the book – like Jacobi's "Agressiveness" and that "Jacobi trod on his [Abel's] Heels" – an expression that otherwise Bjerknes had found in Hansteen's letter to Schumacher in 1828).

Page 282 About Bjørnson and Motzfeldt and the sale of Aulestad, see Keel 1999, p. 32.

Page 282 About Bjerknes' "hydrodynamic Institute at our University", in a letter from Lie to Mittag-Leffler 06.09.1881.

Page 282 Bjerknes' hydrodynamic investigations were established in the university in 1876, with, at the beginning, private support from the Lettersted Association. In 1880 it gained its own posting in the university budget, with its own premises and employees [Haugland 1999]. A protest against grants to Bjerknes is to be found in Lie's surviving papers, Nachlass-pakke nr. 27.4 (dated 05.10.1881).

Page 284 About their work on Abel's collected writings, Lie writes in a letter to Mayer: "It is amazing how much time a work of editing can consume. In the last years my own original work has been considerably hindered thereby."

Abel's collected works were presented at the Christiania Academy of Sciences on December 9, 1881 – at which time Lie made the following remark: "I want to make it expressly clear that I regard Head Teacher Sylow the main editor of Abel's works. That is to say, in the 8 years that this publication has required, it has been Sylow, who long before our collaboration began, had published studies, that testify to the deep and comprehensive study of Abel, left all his own independent work aside, to devote himself exclusively to Abel's works." [See also Lie's article "Om Abel, Évariste Galois and Ludvig Sylow" in *Aftenposten* on 25.11.1896.]

Page 288 Lie to Klein (ca. 1882): "Bei mir würdest Du immer eine *unbedingte* Sympathie und Freundschaft finden" – "Für mich ist es eine capitale Geschichte Deine Gesellschaft zu gewinnen."

Page 290 The exchange of letters between Klein and Poincaré began on June 12, 1881 – it was a friendly correspondence and a "competition" in which they both tried to formulate a great theorem – but in any case resulted in Klein's illness [Rowe 1992].

Page 293 Bjørnson on "France's moral conduct" is found in a letter to Ernst Sars 02.03.1884.

Page 297 After Mittag-Leffler had brought out the first issue of *Acta Mathematica* in Stockholm, Lie wrote to Klein: "Die entdeckungen der späteren Jahren sind ganz sicher epochemachend in der Geschichte der Mathematik."

Page 298 Regarding Klein's reaction to his breakdown, see Klein's *Gesammelte Abhandlungen*, Vol. 1, p. 380, and also Renate Tobies 1981.

"It is Lonely, Terribly Lonely"

Page 299 Lie, in a letter to Klein written in September 1883: "Es ist einsam, schrecklich einsam hier in Chra., wo kein Mensch meine Arbeiten und Interessen versteht."

Page 299 In the autumn semester of 1882, the students C. J. F. Brostrøm, H. N. Drejer and Iver Hesselberg chose Lie as their private tutor (but none of them seems to have developed any special relationship with Lie). Brostrøm got a position in Sweden, Drejer became a lieutenant in Trondheim, and Hesselberg became a teacher at Anderssen's School in Christiania.

Page 299 In the Norwegian election of 1882 – known as the impeachment election – 72,000 people voted. In the course of the 1880s, 250,000 people emigrated from Norway [Sørensen 1984].

Page 300 The first issue of *Acta Mathematica* was reviewed in Sars'/Skavlan's journal, *Nyt Tidsskrift* (February/March 1883), and by Bjerknes in *Morgenbladet*. In addition to Poincaré, contributions to the first year's issues of *Acta* were made by Hermite, Appell, Picard, Goursat, Bourguet, Reye, Fuchs, Schering and Zeuthen, and moreover, by the Swedish Malmsten and Gyldén. The only Norwegian contribution to the first number was a large portrait of Abel. The first real contributions came from Holst (1886) and Sylow (1887–88).

Page 302 According to Lie, he and Sylow had decided that in the course of 1883 they would publish "*all* Abel's letters, fully and in *Norwegian*". It is uncertain what prevented this; in any case, the letters were not published until the Abel jubilee in 1902, and at that time with commentaries by Elling Holst.

Page 303 Lie was extremely familiar with the mountain landscape of Nordmøre, and was among those who made the area known to wider circles.

Page 304 Translated from the German: "Wenn ich einmal die Kunst mir aneignen könnte mich geltend zu machen, so wäre es schön. Dazu scheint es indess nie zu kommen." (Lie, in a letter to Klein).

Page 309 In the letter to Sylow (01.02.1884) it appears that Lie by chance had come to know that Poincaré and Stephanos would translate "an old Programme of Klein's into French for *Acta*, and that Klein would consequently add new notes, a submission that Mittag-Leffler had refused, and after that the work was put aside. Lie comments that he too had encouraged Klein to publish this work – this work "which stands in intimate relationship with my old Themes in a very effective Manner." Lie continues: "As for the rest, I have taken no Part in Poincaré's Proposal, apart from the fact that P. and I talked about the Work in Question, which in my Eyes is the most significant mathematical-philosophical Work in my Time." Lie felt that Mittag-Leffler's refusal was a "Blunder", and that accordingly this was due to his desire to follow "in Berlin's Footsteps" – and Lie concluded: "But what irritates me is that this Man could, as it were, pass over a Writer and a Work, both of whom I am so close to, without me knowing anything about it."

Page 309 Sonja Kovalevsky [Sofia Kovalevskaya], 33 years of age, came to Stockholm in the autumn of 1883 at Mittag- Leffler's initiative, and it was Mittag-Leffler who found her employment at the Stockholm College, first as a docent, and later as professor. In Berlin, Sonja Kowalevsky had been Weierstrass' private pupil (1870–74). She had written an important mathematical work and taken her doctorate at Göttingen. She attracted attention in Stockholm and elsewhere as woman mathematician, professor and author.

Page 310 Lie's first letter to Engel (received in Leipzig on June 30, 1884) is published in Purkert 1984. Engel's work was already commented upon in the correspondence with Mayer in 1882.

Page 310 What Lie felt in relation to this lack of "Redaktionsfähigkeit" is indicated by the following, from a letter to Klein in August 1884 (also published in *GA*, Vol. 6, p. 793): "Dein Buch über das Ikosaeder habe ich empfangen, wie ich Dir schrieb. Ich fühle mich sehr geschmeichelt, dass Du mich in so ehrenvoller Weise in der Vorrede genannt hast. Allerdings fühle ich recht gut, dass ich es nur halb verdient habe. Ich schäme mich, dass ich Dich in meinen letzten Arbeiten nur mit einer gewissen Reservation citire. Ich habe indess gelernt, dass ich vorsichtig sein muss. Denn, wenn ich jemand unbedingt citire, so glaubt man, dass der Andere *Alles* gemacht hat. Ich verstehe nicht, woran es liegt. Wahrscheinlich daran, dass man ohne weiteres voraussetzt, dass meine Ideen mit meiner Redaktionsfähigkeit proportional sind."

Summoned to Leipzig

Page 312 Engel described his relationship to Lie several times [Engel 1899, 1900, 1922]. Engel wrote good Norwegian and exchanged letters with a number of Norwegians, among others, Bjørnson. Engel even wrote about Norway's struggle about its languages – "Sprachenkampf in Norwegen" in *Nachrichten der Giessener Hochschulgesellschaft* 1938. "Ich bin so gewiss, absolut gewiss, dass diese Theorien einmal in der Zukunft als fundamental anerkennt werden" from [Engel 1899, p. XLIX, regarding Lie's letter to Mayer, January 1884].

Page 315 Axel Thue – later professor of mathematics – took very precise notes of Lie's lectures, and his clear writings are found in NBO, Manuscript Collections. Another who wrote down Lie's lectures during these years was Vilhelm Bjerknes. But even Thue had his difficulties in clearly writing out his lecture notes – in any case, Thue reports (from Tønsberg, June 20, 1885) to Elling Holst that he had written off some of the less interesting lectures in higher geometry."

Page 316 See [Alfsen 1929], who describes that Lie's lectures in Christiania could both become congenial chats, and yet could be broken off with him throwing the chalk at the blackboard and storming out.

Page 316 Helland had again been proposed for membership in the Academy of Sciences – together with, among others, W. C. Brøgger and Head Doctor Armauer Hansen – but Helland sent in a message that he did not desire to become a member [Hjortdahl in a letter to Brøgger, 14.03.1885]. (Brøgger took over the professorship in 1888, after Kjerulf, despite the fact that Parliament had decided that upon Kjerulf's death, Helland should advance to his position.)

Page 316 The Mathematical Association that was founded 02.03.1885, had as its first paragraph: "The Mathematical Association's aim is by means of lectures and talks to address mathematical questions, which as a rule, ought to fall under the categories demanded of the public service examination in the sciences." Realistforeningen, which Lie had revived in the autumn of 1868, existed for ten years. The Mathematical Association was now a new "revival"; and the following year (spring 1886) it once again took the name Realistforeningen. Elling Holst founded the Mathematics Seminar on September 29, 1886, to acquaint students with "such theories as were not subjects of their regular lectures." This seminar was a forerunner to the Norsk Matematisk Forening.

Page 316 To a certain degree, Lie seems to have had no eye for personal animosities: Mrs. Lie reported that Sophus had invited home Peter Birch-Reichenwald, Ernst and Axel Motzfeldt, Bailiff Aubert and Dr. Axel Lund – all on the political Right - but along with them, Amund Helland and Sofus Arctander, who were vehemently on the Left. The guests had not exchanged a word all evening [Herman Lie 1942].

Page 317 From early on, Lie seems to have had a positive attitude toward an appointment to Germany, and "through such a move" he would thus "have the possibility of overcoming the isolation" in which he lived in Norway [*Leipziger mathematische Antrittsvorlesungen. Auswahl aus den Jahren 1862–1922.* Teubner Arch. Z. Math., 8, Leipzig 1987].

Page 318 Whilst the "poet's pay" that Ibsen and Bjørnson, Lie and Janson received was 1,600 kr., a starting professor salary, even for a parliamentary professor, was 4,500 kr. In an undated letter (about Christmas time 1885), Lie wrote to Bjerknes that he hoped Bjerknes would support the ordinance concerning leave of absence, and he expressed "a Fear that the Newspapers will get their Finger in the Affair before everything is settled. [...] With the Incorrection whereby they frequently refer to things, they could to a crucial Degree cause me Unpleasantness." Lie also announced that should Bjerknes want an alternative to Holst as lecturer in geometry upon Lie's departure, or should Bjerknes desire a reorganisation of the subjects, then Lie would not be insistent regarding his proposal.

Page 319 Lie's sister-in-law Petra, in Bergen, comments in a letter to Sophus' sister Laura (01.02.1886). "I think that Sophus leaving the country with his family has indeed been a blow to You. We too have not yet gotten over the surprise, which we read in the newspapers. Does Anna really think well of it? Were I in her shoes I should not think I would be very pleased about such a change of residence, especially abroad, where in consequence, one comes face to face with so much that is new and so many unaccustomed conditions, and what seems to me the most lamentable, to be so far away from family and friends. In one's maturity it is difficult to make new friendships."

Page 320 At first Schwarz misunderstood Klein's handwriting and instead of "Lie", read "Sie" [Ullrich 1999]. A draft of a letter from Klein to Schwarz seems to indicate that in fact it was Klein who first launched this "function theory" argument. Klein's draft is to be found in Niedersächsische Staats- und Universitätsbibliothek, Göttingen, Abteilung für Handschriften und seltene Drucke, Cod. Ms. F. Klein 8:941/Anl.

Page 320 The celebratory dinner prior to Lie's departure was held at the premises of the Sisters Larsen in Kristian Augusts gate No. 6, second floor, on Friday, April 10th, at 6 o'clock. The price for each participant was 8 kroner, and Professor Gustav Storm was in charge of the arrangements on behalf of the faculty and the scientific milieu. Bjerknes gave the main speech [NBO, Letters, Nr. 469A]. And Ludvig Daae, who in the Academy of Sciences was most frequently on the opposite side, would certainly have expressed how rare and remarkable the appointment to Leipzig (and the great land of culture) was.

Page 321 (Volapyk was thus a new international auxiliary language devised by the German, J. M. Schleyer – and consequently an earlier form of esperanto). The verses that have been omitted (half of the first and all of the second stanza), are:

The Sciences, 'tis often said,
Stride down Republic's Path,
And at their trusted lifelong Head
Who could they place but Math?
For nothing else so pure is known;
The Matter's, with economy, concluded –
All pared away to Skin and Bone –
No Whiff of Ambiguity exuded.

No Thought there is of Application, when
He conceives his Theory, and it tests,
Like a pious Hermit in his lonely Glen,
We find our Solitary out on Quests.
He's driven on by Truth's stern Thirst,
Knowing that, perhaps with secret Laughter,
One comes to see its Application first
But a thousand Years thereafter.

Page 322 Before Lie left Norway, he purchased shares in Hypotekbanken with what he "in the Course of Time" had laid by – "The Gratuity from Abel included" [Lie to Motzfeldt, Christmas 1886].

Page 322 With Lie's departure for Leipzig, the sum of 1,600 kroner was granted to Holst so that he could teach and give examinations in geometry. The Mathematics-Natural Sciences Faculty [R. A. Mat. Nat. faculty records] show that at the same time Lie had given five teaching sessions "each Semester and even 6 weekly Lectures of which however by no means all were required by the Course of Studies for the public Service Examination." On average, during the same period, Bjerknes had lectured somewhat more than six hours a week – earlier, between 7 and 8 – and consequently, in no way could Lie's lectures have been taken over by Bjerknes.

PART VI

The Coming Period: A More Difficult Balancing Act

Page 326 During his first year at Leipzig, Lie had little or no contact with his French colleagues. The news that Lie had taken up a position in Leipzig seems to have reached mathematicians in Paris via Ole Jacob Broch, who since 1879 sat as leader and director of the "international bureau of weights and measures" at Sèvre in Paris. During the spring of 1887, Lie was informed by Darboux that he, Lie, had been placed on the esteemed list of potential corresponding members of the Academy of Sciences of France. Lie immediately communicated this news to Klein, saying that indeed he was one of the many proposed, and stressed that no corresponding member had died since 1884. But nevertheless, Lie admitted that the news had given him both courage and energy. For his part, Klein had been in Paris in the summer of 1887, and could certainly also consider himself as German candidate for the prestigious appointment to the French Academy of Sciences, when his colleague Kronecker died. Leopold Kronecker passed away in December 1891, and six months later it was consequently Sophus Lie who was chosen as Kronecker's successor in Paris. See also pp. 367 and 435.

Page 326 Citation from Kowalewski, 1950.

Page 328 From a letter to Motzfeldt dated 28.07.1890.

In the Big City of Leipzig

Page 331 Engel's view of the reasons for Lie's choice of residence was (in Engel 1922) "highly unfortunate". Attendance records for Lie's lectures are found in Engel's Nachlass, Giessen.

Page 335 The gravestone monument to Niels Henrik Abel is in Froland churchyard.

Page 335 In his "Antrittsvorlesung", Lie gives an overview of the historical development of mathematics. He dwells particularly on the interaction between geometry and analysis, and he describes the role that concepts like group and invariant play in mathematics. He ends by saying that he will lead geometric developments further in the same spirit as his predecessor Felix Klein. [His inaugural lecture is published in *GA*, Vol. 7, pp. 467–76].

In a letter to Motzfeldt (24.02.1889) Lie thanked him for "the really excellent Photograph from that remarkable Sunday out at your fine Cottage". This photo is from the autumn of 1888, and was taken during his first return to the homeland after moving to Leipzig. In this letter Lie writes, "My entire Stay in Norway was to me as a Dream." See also p. 347.

Sigurd Lie – Sophus' nephew – was a well-known composer in the 1890s. His romances were often played, and his melodies for Norwegian (and Danish) poetry were frequently sung. Following his studies in Leipzig and Berlin, he spent three years as the concert master at the Bergen Harmonium, and later as orchestra conductor at Christiania's Central Theatre. In 1900 there were great plans to set Ibsen's Terje Vigen as an opera in Berlin, with music by Sigurd Lie. Before music seriously "took him" he studied the sciences for a period, and while a music student in Leipzig, he visited Sophus and Anna frequently at their home on Seeburgstrasse. Sigurd Lie had contracted tuberculosis at an early age, probably before he was twenty, and died in 1904 at the age of 33. He had a close relationship to his Uncle Sophus – in the spring of 1896 he wrote from Copenhagen: "Dear Uncle! I have heretofor contracted a debt and do not want – in any case not at present – to ask my father to bail me out. [...] Would you send me 100⁰⁰ kroner as soon as possible?"

Felix Klein

Friedrich Engel

Eduard Study

Hermann Weyl

Bust of Sophus Lie, created by Elisabeth Steen – the bust stands in Church Park, Eid, and was unveiled during Sophus Lie Memorial Week, August 17–19, 1992.

Below:

The grave of Sophus Lie and Anna (née Birch) at the Cemetery of Our Saviour in Oslo.

Aftenposten

pris: Innenbys Kr. 9.50 pr. kvartal » 3.50 » måned | Oslo, onsdag 15. juli 1936 | Abonn.pris: Utenbys Kr.

Matematikerkongressen — hyldest til Sophus Lie.

Den franske matematiker professor Cartan holder sitt foredrag ved avsløringen av Dyre Vaas Sophus Lie-byste.

Matematikere fra 32 land var samlet i Aulaen, da Sophus Lies byste blev avslørt. Efter programmet skulde direktør Sejersted Bødtker overlevert bysten på givernes

ker, professor Cartan nu vil holde et foredrag for oss som vil illustrere Lies betydning for geometriens utvikling. Det er en særlig glede for mig ved denne anledning

De russiske matematikere som uteblev.

Var man redd for at de ikke vilde vende tilbake til paradiset?

Vi har tidligere bragt en meddelelse om at de russiske matematikere som var anmeldt til kongressen ikke er kommet. Ved henvendelse til den russiske legasjon får man bare det svar at man overhodet intet kjennskap har til saken. Kongressens sekretariat har fra et par av de anmeldte fått korte meddelelser om at de ikke kommer, fra andre har man overhodet ingen oplysninger fått. Dekanus ved det naturvidenskapelige fakultet, professor Heegaard, kunde heller ikke gi nogen oplysninger. — Men for et år siden, forteller han, var det en russisk videnskapsmann som fikk lov til å

nnene vil
se.
uttalelse
ngen. –
sd, sier
es

itikken banlyst
ndsen, pre-
allforbund:
sonlig for min-
og finner det
denne ikke blev
get strengt på
e upolitisk, ut-
veriges idretts-
o Lindman,
ør i moderne
fremragende
under en te-
ning av gårs-
utning.
å bli idrettens
er to forbund,
være ett. At

Både
Indust
på Håndv
Underhand
Filipstad.
allerede til

Fra Håndver
utstillingen i Osl
en tid siden sen
om at det allered
blitt trangt om
utstillingen.
Vi har snakk
sekretær Mow
om dette forhold
at situasjonen er
så godt som sp
— Vi har mo
anmeldelser fra
i Oslo, uttaler h
disse storbedrifte
egne paviljonger.
vil bli utformet a
arkitekter for at
sikker på at de
anleggets ramme
vil ikke medføre
ensformighet, ide
vil sørge for at
den ytre og in
individuelle

At the International Mathematics Congress in Oslo, 1936, Élie Cartan gave the keynote speech – "Le Rôle de la Théorie des Groups de Lie dans l'Évolution de la Géometrie Moderne". Dyre Vaa's bust of Sophus Lie was presented to the University by Lie's family and friends. The bust is now located in the Sophus Lie auditorium, Blindern Campus, University of Oslo; and a bronze cast is in the Library of Mathematics at Blindern.

In his first winter semester (1886–87), Lie lectured on "Projective Geometry", and "Einleitung in die analytische Geometrie der Ebene und des Raumes", and in the seminar, he held exercises in infinitesmal transformations. [These lectures are found in Lie's Nachlass, Packages 37 and 51.]

Page 341 The expression "société thuriféraire" was used by Paul Du Bois-Reymond [Ullrich 1999].

Page 341 Despair over Engel's dilatoriness, among other things, is found in a letter to Klein dated 02.11.1886.

Page 342 Commentary on the "competition" with Neumann is found in a letter to Motzfeldt dated 27.02.1892.

Page 343 Farmer Lars Liestøl had, in addition, worked both as sheriff (in Setesdalen – he lived at Frøysnes) and as language assistant to Hans Ross, among other things. (Liestøl was also a polished creator and performer of songs with spontaneous rhyming verses, or "kvedar", and an acomplished fiddler. He entered Parliament for the first time in 1874, and at the beginning had difficulty taking sides in the growing division between left and right in the political landscape. But when it finally dawned on him what the struggle was actually about – the power of royalty or the power of the people (Løvland 1913) – he whole-heartedly took a stand, but was nevertheless in doubt when in 1888 Sverdrup asked him to become a member of his moderate regime. Liestøl felt like a foreigner in the government and resigned after eleven months. Perhaps Lie's description of Liestøl is coloured by the fact that he was writing to the conservative, Ernst Motzfeldt.

Page 346 Sylow visited Lie in Leipzig during the spring of 1888, and in a letter dated March 3, 1888, Sylow wrote to Mittag-Leffler: "I talked with Lie in Leipzig. He mentioned that Bjerknes' Abel Biography had caused an unpleasant Sensation in Germany, and for that Reason, felt that it was better to let matters lie, namely that I should not talk to Kronecker about the Abel statue." (According to Sylow, Kronecker felt that Abel had been accorded too much honour in the commentaries that Sylow and Lie had written for the publication of Abel's collected works.)

Page 347 Dance activity in Leipzig is found in a letter to Motzfeldt 24.02.1889.

Page 347 Some more on Lie's comments about conditions in Norway [in letters to Motzfeldt]:

"Conditions in Norway are laughable" – in that the government could appoint as Professor of Latinate Philology, the Norwegian L. B. Stenersen over the Danish A. B. Drachmann, was once again a crowning example of "personal Considerations" being set above scientific and scholarly ones, and it happened "for the good Reason that they lacked any Insight." (This passing over of A. B. Drachmann in 1888 was otherwise something to which many reacted, the protests were particularly strong from Danish university people, as well as scientists and scholars.)

Lie commented as well on "the Norwegian Administration of Justice" which was "far from being perfect", and he felt that it was easy to ascertain that the decisions of the High Court were often "unfortunate".

He felt that it was "absurd that our old Friend Carl Berner" had appeared as a head of the Left (Liberals) – and he seemed to find it disappointing that (in the summer of 1888) his friend, P.O. Schjøtt "was weak enough to enter government", but Lie did not believe that Schjøtt's political track record would be very long, and he asked Motzfeldt (who in political terms was a conservative) to bear with him.

On October 20, 1888, Lie wrote to his nephew Johan H.L. Vogt – following the death of Kjerulf: "I set Helland much higher than You, although I still fear that Conditions have forced him to occupy himself very much with thankless Politics." Johan H.L. Vogt, who was now a member of the Collegium, asked his uncle for a statement on Elling Holst's qualifications – and in an undated letter, Lie replied: "Holst could have become a very good Mathematician if he had had the Energy. As a docent, he is sincerely gifted. He has a lively Interest in his Science, and understands how to impart a Love of Mathematics to his Audience."

Meanwhile during the summer of 1887, Johan H.L. Vogt had married Martha Johanne Abigael Kinck (a niece of the poet, Hans E. Kinck), and on their honeymoon tour they had visited Leipzig and stayed some days with Sophus and Anna on Seeburgstrasse.

Another who perhaps visited Lie at this period was Andreas Aubert. (Aubert was a good friend of Engel's – in Engel's Nachlass there are 28 letters from Aubert.) In 1888 Aubert was in Germany and particularly in Berlin and Dresden looking for material for his important book on the Norwegian painter J.C. Dahl – and it was Aubert who now discovered the German painter, Caspar David Friedrich (who had died in 1840 and had been forgotten in his homeland), and this discovery created great European renown for Aubert as an art critic. Aubert wrote about Friedrich's emotional intensity and sincerity, and the harmony between the human soul and the life of nature – in *Kunstchronik* in 1895 – and became an interpreter of neo-romanticism. The Germans became very enthusiastic about the newly discovered Friedrich in the 1890s. [Walter Baumgartner: "Long long gone ... sight or vision?" BASAR 1/78, pp. 68–72.]

The alienated bourgeoisie of the imperial Wilhelmian realm, with its intense technological and economic expansion was unable to reconcile/solve its great social problems and ideological conflicts. Disasters and problems were not only naturally-given (like subjugation and exploitation, crop failure and epidemics) but also were socially created. "The social" had become a concept in science – and a growing working class began as well to make itself heard politically.

Breakdown

Page 351 Dr. F.A. Hoffmann's report, dated November 16, 1889 [Universitätsarchiv Leipzig, Personalakte Nr. 693 – Sophus Lie, Blatt 41]. How close the contact was between the Lies and Max Sänger and his Norwegian-born wife, Helga Vaagaard, is not known for certain. (Helga and Max married in 1884 – she was the sister of Hanchen Alme, who again was a close friend of Nina and Edvard Grieg.) In Lie's correspondence there is no mention of a relationship between the families in any sense other than the fact that their sons, Herman and Hans, were close friends

– these two, Herman and Hans, kept in touch by letter, even after Hans Sänger came to Norway in 1934. In connection with Hans Sänger's application to visit Norway in 1934, the editor Thommesen wrote that Sophus Lie had considered Max Sänger "as his only intimate friend" in Leipzig. In his speech [Herman Lie 1942] of recollections in connection with the one-hundredth anniversary of his father's birth, Herman Lie introduced him to the audience and said, among other things: "My friend, Hans Sänger, son of my father's friend of many years in Leipzig, the great gynaecologist, Max Sänger."

Could the fact that Lie did not consult Sänger in the autumn of 1888, perhaps be an underlying cause, a sign of a great mental dislocation? (The sudden loss of his mother when he was ten years old, and his little brother two years later, must have left their marks.) Lie's need for planning and security could also be seen as having been driven by an anxiety about the unexpected. At the same time for him, the unexpected has, in a parallel sense, also gained greater power – the uncertainty inherent in everything, the fact that the whole equilibrium could collapse at a single blow.

Page 355 Sister Thea in Moss commented about Sophus' illness in a letter to Anna in December 1889: "It is a very heavy weight, when the Cross and the Ordeals are placed on us, but indeed we know, there is a cherished God who is directing Everything, and He has a Divine Intention for Everything."

Page 359 In the autumn 1892, Lie wrote to Holst, "I am sleeping almost as before, and every day better. By this means I am gaining new lust for life and life-force. I am now once again living in a rational manner, something that for the past two years I have not done."

Page 359 The illness that Lie succumbed to ten years later – pernicious anaemia – is a disease of the blood that can also affect the distribution of matter in the spinal chord in such a way that the proteins in the nerve cells become damaged. Thus, it is said that in 50 percent of pernicious anaemia cases are also accompanied by psychic changes, by virtue of which the patient may become depressive and irritable, often in relation to the loss of sleep – something that can also lead to paranoid symptoms. In the medical literature, one finds descriptions of a form of what is called pernicious psychosis. Changes in the retinas and optic nerves can lead, more or less, to a marked sensitivity to light, as well as blinking and flickering of the eyes. How long, and to what degree this blood disease advances before it becomes fatal is to some degree uncertain. In Lie's time, the causes of, and reasons for the disease were unknown, and there was no treatment.

The mathematician Hilbert contracted the same illness – pernicious anaemia – in 1925. Earlier that year the doctors, Whipple and Robscheit-Robbins had discovered that the consumption of raw liver had a favourable effect of the disease – methods of treatment were further developed by G. R. Minot. Hilbert received this treatment and regained his health.

A safe treatment for this disease of the blood was developed in 1926 by the Americans, G. R. Minot and W. P. Murphy – the disease was caused by the blood's inability to absorb Vitamin B_{12} – and in 1934 Minot and Murphy received the Nobel Prize for Medicine for their work in this field.

Fame

Page 362 As yet another example of his "Theories' Importance", Lie mentions (in a letter to Holst in 1892) that the winner of the Jablonsky Science Association prize dealt with his, Lie's theory – applied to a particular example.

Page 366 It seems that contact with Killing had possibly led to a dilemma for Engel. Engel knew that Lie watched vigilantly over his priority, and what were the limits to that which Engel himself could vouch for, and that he had learned from, in relation to Lie's theories?

Page 369 Zorawski, another of Lie's students, published his doctorate on invariant curves in *Acta Mathematica* in 1892 – again, an application of Lie's theories. Zorowski later became a central man in the development of mathematical research in Poland. He also took part in the reconstruction after 1945.

Page 370 In a letter to Poincaré (in October 1892, also showing his gratitude for his membership in l'Académie des Sciences) Lie commented on Riemann and von Holmholtz, and presented the following point of view: "Ich glaube die einfachsten Axiome für die Geometrie einer *drei*fach ausgedehnten Zahlen Mannigfaltigkeit gefunden zu haben, nämlich:

1) alle Bewegungen bilden eine Gruppe bei denen zwei Punkte eine und nur eine Invariante haben
2) getrennte Punkte bleiben immer getrennt."

Page 372 In his foreword to *Geometrie der Berührungstransformationen* (1896), Lie reflects upon his work against the background of the development phases of mathematics. He writes (translated from the German): "Among the ancient Greeks, geometry was, so to say, the only mathematical discipline. Abstract analysis still had not been found. Even astronomy and mechanics were subordinated to geometry. Following the noteworthy discoveries of Diophantes and the Indian mathematicians, it was not until the Renaissance that an algebra was created, and a formal apparatus, which was lacking among the earlier, and otherwise so discerning Greeks. With Descartes' introduction of the coordinate system, together with his creation of analytic geometry, there followed an epoch-making correlation between geometry and analysis. This quickly led to the fundamental mathematical concept of *function*. As early as Archimedes' geometric investigations lay the germ of the concepts integral and derivative." And after going briefly through how these concepts (integral and derivative) – through applications to geometric, kinetic and mechanical problems in the works of Kepler, Cavalieri, Descartes, Wallis and Fermat – had led inexorably to *infinitesimal calculation*, Lie continued as follows. "No less characteristic, but something which receives less attention, is *transformation*, which in the newer mathematics marches forward ever more strongly as a fundamental concept in its own right, and which stands beside the function, has its origins in the most ancient geometers' treatment of certain particularly lucid and naked facts [that will say, in geometric intuition]. Also the closely related concept of *differential invariants* first appeared in geometry, namely in curve theory."

Page 373 Élie Cartan says in "Un Centenaire: Sophus Lie" (1942) that in the spring of 1893 Lie had stayed in Paris for six months. (Lie's concern seems to have been that before one could find an independent group theory method of treating differential equations, it would be necessary to specify "the structure" in all actual finite dimensional transformation groups [see also the 1895 study in *GA*, Vol. 6, p. 601]. This was before the significance of "structure" to finite dimensional transformation groups was known.)

Before Élie Cartan published his 25-page article "On Simple Transformation Groups" in *Leipziger Berichte*, he had only published two short notes in *Comptes Rendus*.

Page 374 Lie worked most with differential equations, and he was perhaps not so concerned with the most fruitful directions that his ideas created the foundation for. With Picard, Vessiot, among others, and above all, with Élie Cartan, Lie's theories took form and developed into what are now called Lie Theory. As well, Appell further elaborated some of Lie's ideas of transforming a graph of one function to another – a very general point of view because not only the function, but also the argument is transformed.

Emmy Noether took up some of these ideas in 1918; Hermann Weyl and Élie Cartan in the 1930s (Cartan worked with Lie algebra for twenty-five years.) C. Chevalley worked on Lie material in the 1940s, and with "Seminaire Sophus Lie" in the 1950s in Paris (most of this was written down by P. Cartier), and see the book *The Theory of Lie Groups*; E. Kolchin (1948). Later, Armand Borel, Sigurdur Helgason, Shiing-Shen Chern, Wu-yi Hsiang, H. C. Wang, A. A. Kirillov, and others, worked on these ideas.

Page 374 The description of Galois is from a letter Lie wrote to Picard in December 1896, also in gratitude for Picard's biography of Galois.

Page 376 The index of the great works of reference – like *The Encyclopedia of Mathematics* (Dordrecht 1994) – show that Bernhard Riemann is the mathematician with the most references. The second is Sophus Lie, and the third, Niels Henrik Abel.

Page 376 In relation to the publication of *G. A.*, there was a report in *Aftenposten* (22.03.1921), under the headline "A Mathematical Major Work". This article reports that what Teubner had already decided by 1912 – that is, to publish Lie's Collected Works in seven volumes under the editorship of Engel – had been postponed due to the war. This was now in process again, as the Norwegian Mathematical Association had applied to the state research fund to obtain financial support for the printing. Moreover, because "at the moment it is impossible to set up a budget for the future", 5,000 kroner should be granted as an initial contribution from the research fund. To assess the significance of such a disbursement (and indeed to lay the basis for its own application for support) the Norwegian Mathematical Association had communicated with several mathematicians abroad – and *Aftenposten* published extracts from some of the replies. From Vessiot at the Sorbonne: "Everything that can make the dissemination of Sophus Lie's ideas easier, will be doing a great service to the future of mathematical science. Therefore I can only applaud the plan of the Norwegian mathematicians, to publish, with notes, the treatises of their famous countryman, and I vitally hope that their plan will turn into reality." From Klein in Göttingen: "I would be most pleased, if this [the grant] should occur, and I particularly recommend not to save on the commentaries, for the geometrical thinking, of which Lie solely and originally made use, and which also was the secret of his productivity, now today still meets with only slight understanding." Bäcklund at Lund sent the following: "To a most unusual degree this work is of a ground-breaking character, because at his period there was nothing else with which to compare in terms of outlook on a multitude of geometric contributions and methods, and as well achieved this not only for geometry in the ordinary sense, but also for mathematics in general, new research fields."

In May 1937, the German Mathematical Association (DMV) held a four-day conference on Lie's mathematics (at the University of Hamburg, with a great number of young Germans and a whole section of French mathematicians participating). Élie Cartan had come from the Sorbonne, and gave the keynote address – "Jede Lie'sche Gruppe ist isomorph einer linearen Gruppe" – Professor Witt from Göttingen lectured on "Lie'sche Vectorfiguren", and Professor Landherr from Rostock on "Lie'sche Ringe" – besides which, there were lectures on soluble groups and groups of topological transformations.

Engel's manuscript (with his editing finished) for Volume Seven of *Gesammelte Abhandlungen* (Engel died in 1941) came through World War II undamaged, but nevertheless was not published until 1960, and then without the foreword that Engel had written about the "priority relations" between Lie and Klein.

Page 377 Lie had a large mathematical correspondence, and most of the letters are to be found in his surviving papers. The correspondence dealt with everything from specific mathematical problems to general commentaries and points of view, and the development of treatises. Some of the letter-writers have not been mentioned in the main text, particularly from the period after Lie arrived in Leipzig in 1886. These are: Emil Waelsch and Georg Pick in Prague (Waelsch wanted to visit Lie in Leipzig in 1892); J. E. Manchester, who would come to Lie in 1896; the American, James Morris Page, who was Lie's student in Leipzig, and considered himself, according to later letters (from the University of Arkansas, 1895), as the main representative of Lie's "brand of mathematics" in his part of America. Page also published articles on transformation groups in the American *Annals*, both in 1895 and 1898, and became a mathematician of considerable profile in the new group theory milieu of America.) According to a letter dated 10.12.1889, from Frederick Carlos Ferry (who earlier had also been Lie's student in Leipzig, and was now at Clark University, Worcester, Massachusetts), Lie had been invited to Clark University's tenth anniversary celebrations in June 1899 – Lie was offered $500, or more if he so required. (Otherwise at Clark University, Ernest W. Rettger had taken his doctorate the previous year on Lie's theory of continuous groups.)

G. Bohlmann wrote from Berlin in the autumn of 1892 that he wanted very much to come to Leipzig to study Lie's theories. Lie and Bohlmann kept in contact for the next three- to four years (the two of them also met, and it seems as though Lie, through the intermediary Schwarz in Göttingen, was one of those who initiated the publication of the collected works of Weierstrass). One finds four letters exchanged between Lie and Friedrich Wilhelm Franz Meyer (each letter composed of four pages on differential invariants and on diverse publications) from the period 1892–93.

There are letters from Hermann Werner (one of Lie's German students), Walther Dyck (Munich), Juels Drach, K. Zorawski, H. Grassmann (Halle), H. A. Schwarz, H. Weiner (Halle), P. Stäckel (Halle), E. Lemoine (Paris), Maurice Lévy, Georg Cantor (who in 1893 recommended himself to Lie with a plan for international mathematical research), R. LeVavasseur (Moulins), Arnold Peter (Paris), J. Dantschoff, E. Borel (Lille), Alfred Achermann (Leipzig), G. von Escherich (Vienna), Carl Heumann (Uppsala), Dr. Domsch, A. Duraud, W. de Tannenberg (Lyons), and Gabriel Koenigs.

From Max Noether there is a letter from 1873, as well as four letters and a postcard from 1892. These letters deal with interpretations of Abel's theorem, which had been presented by Picard at the French Academy of Sciences on 08.02.1892 – and several letters, from 1892, from Heinrich Martin Weber concerning Abelian integrals. (At this time Weber was preparing the second edition of Bernhard Riemann's *Gesammelte Mathematische Werke und Wissenschaftlicher Nachlass.*)

Conflicts

Page 378 For descriptions of how Lie was referred to in Berlin, see Hawkins 2000.

Page 382 Lie admitted that he had been critical to the way Bjerknes had presented the struggle over priority – Lie then added, now we know better thanks to the indiscretion of Bertrand.

Page 385 About Lie's nomination to the position of corresponding member of the French Academy of Sciences, see *Comptes Rendus* 1892, Vol. 114, p. 1329. On 12.04.1894, Lie received one of the highest Saxon honours – Ritterkreuz 1. Classe des Verdienstordens.

Page 388 About the Erlangen Programme: In Klein's first publication of the programme [Erlangen 1872] ("Vergleichende Betrachtungen über neuere geometrische Forschungen"), one finds, at §9, pp. 32–36: "Wesentlich mündlichen Mitteilungen von Lie entnommen." (This has been substantially taken from Lie's verbal communications).

The fact that Klein, in his lectures at Göttingen 1889–90 seems only to have referred to Lie in connection with the attack on von Helmholtz's discovery of the foundations of geometry, was also a factor in the conflict. Klein seems to have considered the Erlangen Programme as also a condition for Lie's works. For

Lie on the other hand, the Erlangen Programme seems more to have stood out as an obvious result of their cooperation in 1869–72, and therefore also had no epoch-making significance for his later investigations. The only thing Klein lacked (before the Erlangen Programme) [Hawkins 2000] "was an explicit statement that the transformations defining a geometry, always form a group, and the declaration that any specification of a (continuous) group of transformations acting on a manifold defines a geometry."

Page 388 A letter from Klein (24.04.1899) to Elling Holst indicates that Klein never got his manuscript back. (This manuscript is now at NBO, Manuscript Collections.)

Page 391 About von Helmholtz: Von Helmholtz was presented to Norway in an article by Sophus Torup in *Nyt Tidsskrift*, 1895, pp. 95–114. (Von Helmholtz's autobiography, entitled *Vorträge und Reden*, came out in two volumes in 1903.)

Page 392 In relation to the Chicago World's Fair of 1893 (to commemorate that the previous year had been the 400th anniversary of Columbus' discovery of America) a large mathematical congress was held in that city – a congress that was to have the greatest significance for American mathematics [Smith 1934]. Felix Klein was the honorary president of the congress, and several European mathematicians had been invited (Study also attended and gave a lecture), and many of them had sent in their papers: Dyck, Hermite, Hilbert, Hurwitz, Meyer, Minkowski, Noether, and Weber, among others. Following the Chicago congress, Klein, together with a number of other participants, went to Northwestern University, Evanston, Illinois, and here Klein gave a series of lectures between August 28th and September 9th, 1893, which were edited by Professor Alexander Ziwet and published the same year (2nd edition, 1911).

Page 393 About Hausdorff:

Apart from his epoch-making mathematical work and his social engagement, he also made himself known as a poet and philosopher. Using the pseudonym Paul Mongré, he produced a Nietzsche-inspired work of prose, *Das Chaos in kosmischer Auslese* (1898), that was reviewed by, among others, Rudolf Steiner. He published a collection of poems, *Ekstasen* (1900) and a one-act play, *Der Arzt seiner Ehre* (1904), performed in both Berlin and Hamburg. When in 1913 Hausdorff became professor at Greifswald, he succeeded Friedrich Engel, who had become professor at Giessen. Hausdorff's epoch-making *Mengenlehre [Set Theory]* came out in 1914 and several prominent mathematical works in set theory, mass theory and fractal geometry followed. In 1921, Hausdorff became professor in Bonn; he was by now a looming presence in international mathematical circles. During the autumn of 1941 when the vast majority of Jews from Hamburg were interned – and after the Wannsee Conference on January 20, 1942 – Hausdorff was completely without hope of a future in Nazi Germany, and together with his wife and sister-in-law, committed suicide on January 29, 1942 [Eichhorn 1994, Brieskorn 1996]. Hausdorff wrote two letters to Lie, dated Leipzig 23.03.1892 and 28.03.1894, and these are printed in Eichhorn 1994, pp. 62–65.

Page 393 The Lobachevsky Prize:
Lie did not travel to Kazan to receive the prize. See also his letter of thanks, Item No. 19, below, in "Chronological Bibliography, Other Printed Material". Killing received the prize in 1900 (nominated by Engel), and Hilbert (nominated by Poincaré) received it in 1904.

Page 394 Klein's draft of his obituary speech for Lie – "Zum Vortrag über Lie von der Societät am 29. April 1899" is found in Klein's Nachlass. Niedersächsische Staats- und Universitätsbibliothek, Göttingen. See note to page 234.

Page 395 Killing's words in 1900 upon receiving the Lobachevsky Prize are cited from Killing 1997 [letter 80]. Following the Russian Revolution in 1917, the Lobachevsky Foundation, which had been founded by A. Wassilyev, was liquidated. Later on, it was established again, and today is still a prestigious prize.

Page 395 Regarding Study's thoughts about the space problem: the definitive formulation was not arrived at until 1953, and then by Jacques Tits.

Page 396 About Ostwald:
Immediately upon arriving in Leipzig, Ostwald got down to establishing a periodical on physical chemistry. In 1888 he began to publish a series of writings that would come to be very significant to the informed public's knowledge of important scientific discoveries. In "Ostwalds Klassiker der exakten Naturwissenschaften" he presented classical works of physics, chemistry, astronomy and mathematics. Professor Bruns supported and collaborated in the astronomical section of this work, and Ostwald asked Lie if he could, in a similar manner, contribute to the mathematical section. Lie described this in a letter to Klein.

After at first characterising Ostwald as "a bold and outstanding man", and about these plans for a "classical series" as being "audacious", Lie wrote: "I told him [Ostwald] naturally enough that I was not the man for such an undertaking (his inquiry was perhaps not seriously meant)." But since Lie had "sympathy" for Ostwald's "audacious" plans, he asked Klein if he had anyone to propose.

Engel would certainly have been useful as an assistant, particularly with translations and editing, but Engel would be, according to Lie, far too one-sided in his choice of current works, and besides, Ostwald would like to have a mathematician with a degree of renown.

"He [Ostwald] must certainly begin with something by Gauss," Lie wrote to Klein, and requested a speedy reply if Klein were interested. It is not known whether Lie received any proposals from Klein, but Ostwald's classics series became a great and enduring success. (Klein's Erlangen Programme was also included in this classic series.)

Like Lie, Ostwald did not always feel well-adjusted to life in Leipzig, and in 1906 he resigned his professorship over divergencies with the university – but he remained active within his field for another 25 years. He lectured as well on natural philosophy, and published a journal of modern natural philosophy. He worked to provide energetics as a basis for a science of culture, and took his standpoint from scientific materialism – matter and spirit were two internal forms of the same phenomenon, he argued. He enthusiastically supported the notion of uniting the states of Europe, and his great interest in painting and music led him to develop his own theory of colour in the 1920s. His home, "Landshus Energie" was converted into a museum – and his son Wolfgang (a year older than Sophus Lie's son Herman) also became a professor of chemistry at the University of Leipzig.

Lie was consequently very critical of Ostwald's energetics – a theory that (simply put) should make it possible to feedback all phenomena in nature to a definite property, namely, *energy*. And thus it was, after a lecture on this subject that Ostwald gave (to the Academy of Sciences of Leipzig) during the autumn of 1893, that Lie (in the Academy of Sciences on June 4, 1894) attacked the mathematical foundations of the theory. Ostwald delivered a response to Lie's attack at the end of July, again at the Academy of Sciences.

At Realistforeningen in Christiania on December 10, 1902, Kristian Birkeland spoke vehemently against modern "energy theory" – and particularly against the form in which Ostwald had developed it. Otherwise, Ostwald's theories were presented in Norway through the book entitled *Energi og Kultur*, Oslo, 1911 (translated by K. Visted); see also Marthinsen 1954.

Page 397 Citation from Vogt 1930.

Page 397 The American mathematician, E. O. Lovett (who was one of those who in 1898 followed Lie from Leipzig to Kristiania) wrote in 1902 [Scriba 1982] to Engel: "You who were so well acquainted with his abnormal vanity and sensibility would certainly enjoy the following anecdote: when I arrived in Leipzig as a student, I asked Lie for advice about which lectures I should attend, and then asked of course which of your lectures I ought to go to. 'To Engel? No, that is forbidden!' answered Lie."

Apart from working on Lie's treatises, Engel was also one of the editors for the publication of Hermann Grassmann's works – three volumes, each divided into two sections, that was brought out by Teubner between 1894 and 1911. In 1903, at the initiative of Bjørnson, Engel was made Knight of the Order of St. Olav, First Class, and at the University's celebrations commemorating the one-hundredth anniversary of the death of Abel, on April 6, 1929, Engel was awarded an honourary doctorate by the University of Oslo.

There was an international mathematics congress held in Oslo in 1936, at which Erhard Schmidt laid wreaths (decorated with swastikas) both at the Abel monument in the Royal Palace Park, and at Lie's grave in the Cemetery of Our Saviour. Engel was disappointed that he could not (for political reasons) travel to Norway. (The newspaper accounts of the event indicate that Professor Lietzmann was the head of the German delegation.) Schmidt and Lietzmann had both been Hilbert's students in Göttingen.

Norwegian Students

Page 398 Lie wrote to Holst from Leipzig in February 1889 in connection with Thue's eventual journey: "You know that I always counsel young Norwegian Mathematicians to travel abroad as soon as possible. That is to say, certainly they have a remarkable Opportunity in Christiania to study *classical* Mathematics; yet on the other hand, as the *newer* Mathematics is lectured on only to a very small Degree, it is difficult there at home to dedicate oneself to the Multitudes of different Forms of Knowledge that are necessary for anyone who wants to become a mathematical Writer."

Page 399 Axel Thue's report to the Collegium [RA., UiO, Collegium, Reports, Box 5].

Page 400 Alf Guldberg's report [RA, UiO, Collegium, Reports, Box 6].

Page 401 There are many references to Anton Alexander in the minutes of Realistforeningen (1890–92). In a discussion with Magnus Alfsen, among others, about how, in order for mathematics teachers to be able to give good lessons, they had to know mathematical history, Alexander maintained that all who studied the sciences had an adequate amount of such knowledge. Therefore, he said, it was unnecessary to make mathematical history part of the curriculum, and a topic to be tested in the examination. (There are letters from Alexander to Holst from Leipzig 29.11.1894, 01.12.1894, 21.12.1894, similarly from Paris, 22.04.1895, and from Gjendeboden 31.07.1895.)

Page 402 Birkeland reported that Lie was in Paris in 1893 from ca. April 1st –20th. Birkeland reported to Holst (07.07.1893 and 26.04.1893) that he had been "considerably together" with Lie, and that he had shown Lie the work on enumerative geometry, a wide-ranging field in which it has been demonstrated that mathematicians and physicists have many identical interests.) The research fellowship that Birkeland applied for in April 1893 was, indeed, the position that had been left vacant when Holst became a docent. Otherwise, between 1893–95, Birkeland worked together with Poincaré in Paris and Heinrich Hertz in Bonn.

Back to Norway

Page 410 In a letter to the Collegium, dated April 20th, 1893, Holst makes it clear that Lie wanted to come home to his old position with a "personal supplement" in terms of salary. Holst gave the Collegium arguments for advancing this proposal before Parliament and the government. Lie's theories had been taken up by textbooks, and they were being studied by leading universities in Europe. Holst pointed to Lie's "brilliant innovation" (transformation group theory), and brought to mind Lie's membership in the French Academy, and in closing, mentioned the similarity with Abel and the fame the country had received through him, and also cited a letter received from Lie in Leipzig: "I long for Relatives and Friends, and perhaps even more, for the Land and the People. I imagine that my physical and mental Powers would be much better sustained among the Mountains of Norway than on the Plains of Leipzig." But Holst wrote to the Collegium four days later, April 24, 1893: "Due to recent occurrences, I would like leave to request that the honourable Collegium keep in abeyance the question of Sophus Lie's summons home until further notice." [NBO, Letters, 234].

The reason for this postponement was probably that Holst and Bjørnson and others wanted to gather the broadest possible support before the proposal was put forward. Bjørnson, for example was waiting for support from Bjerknes. Otherwise, Bjørnson was at this time involved in the most stormy period of his relationship with Erika Nissen, while he also busied himself with plans to build a house on Munkedamsveien in the capital. Three years earlier, Bjørnson had transferred Aulestad to his youngest son for 50,000 kroner.

W. C. Brøgger was upset that he had not been asked to assist in the campaign. In a letter, dated Høvik, 06.04.1894, Holst admitted that the reason for this was that Helland did not want to have Brøgger involved – similarly, a whole series of Holst's friends were cast aside by Helland, who also wanted to have some representatives from the Conservative Party, and got Holst to ask Gustav Storm and Bernhard Getz. But Holst felt he was being "dethroned in the affair by Helland". He complained to Brøgger, and wrote that he was afraid there would be "little joy in the sacrifice, that I for my part am bringing on behalf of my hero Lie" [NBO, Letters, 298].

Page 415 From Holst, in *Nyt Tidsskrift* (December 1893).

Page 416 Holst writes to Mittag-Leffler on 24.04.1895: "When he [Lie] was here in the autumn [1894] he assured one and all, and said so in large gatherings, that he was predisposed to the offer from home – but when the move could be undertaken, on this he was more undecided."

Page 419 More on Sophus Lie in Norway during the spring of 1896: In Bergen, his brother, John Herman and John Herman's wife Petra, were concerned about the nephew, Sigurd Lie (the composer – see page 492), and Sophus reported home to Anna that Sigurd (in Bergen) had just suffered an attack similar to that which he had had while in Leipzig.

The "Test" and the "Experiment" Lie had spoken of going through, was thus to go to Norway and teach from the beginning of March until the end of June when the semester ended in Christiania, and then, following the Norwegian summer holidays, to teach in September and up until the middle of October, when the winter semester began again in Leipzig.

Page 421 A little more about the schools-struggle that Lie participated in:
"The Philologists and Scientists Association", founded in 1892, became the "Norwegian Secondary Teachers Company" [Norsk Lektorlag] in 1939, and the "Norwegian Teaching Federation [Norsk Undervisningsforbund] in 1984.

August Western (Dr. Philos. 1893) was also involved in the question of the relationship between oral and written forms of the language, and in 1888 he wrote on diphthongs – which in Christiania were understood to be very vulgar, and the double forms of spelling, such as *kjød* and *kjøtt*. Western determined, perhaps somewhat arbitrarily, that *kjøtt* was a masculine variant, and was found "in general only among men – "ladies say kjød." [Vinje 1999]. Western was later known for his large *Norsk Riksmålsgrammatikk* [Standard Norwegian Grammar] (1921). Western was also the man behind the Standard Norwegian Dictionary.

Contact between the pedagogues of schooling in Germany and Norway was also advanced by an article in *Samtiden* in 1893, which was entitled "Natural Science Teaching in the Schools", and was translated into German by Vilhelm Bölsche in *Die deutsche Gesellschaft für ethische Kultur.*

About Lie's involvement in the schools debate, see also below, "Chronological Bibliography". Holst's contributions and replies are in *Morgenbladet* 1895 (nr. 539 and 607), *Morgenbladet* 1896 (nr. 408 and nr. 722), and in *Den høiere skole*, nr. 2 and 3, 1909. Sylow also went into print (on Lie's side) in *Morgenbladet* 1896 (nr. 651).

Page 423 Lie in a letter to Sylow (ca. Dec. 1896): "I have become considerably vexed over the fact that the Danes have wanted to monopolise C. Wessel. It seems to me quite miserable. In fact, it was a great misfortune for Wessel and Norwegian Science, that this gifted Mathematician was going around in an Environment that lacked all Understanding of him. And nor does the Manner whereby Wessel's Treatise is finally taken up truly do any Credit to the Honour of Danish Mathematics. You know though, Christensen's and Juel's Part in this Matter. The Issue is the old One: the Danes suffer a Sense of Injustice through Time toward the Germans, and at the same Time permit themselves the Same Injustice toward Norway, as they blame the Germans for."

Page 426 The editor and founder of the popular culture periodical, *Samtiden* [Current Times], Gerhard Gran, addressed himself to Holst (in a letter 20.02.1896) with the following questions: "Does there never occur anything in Your field that is of general public interest? Could You not consider dangling something entertainingly mathematical in the pages of *Samtiden* from time to time? What do You say?"

Page 429 Lie's speech about Nansen, in the Student Society on 19.09.1896, is referred to in *Aftenposten* on 20.09.1896, and *Teknisk Ugeblad* nr. 39, of 24.09.1896, p. 302ff. [Hestmark 1999, p. 328].

Page 429 In an undated letter from Sophus to his sister Laura, he asks her, out of the 150 kroner that have come from their brother John Herman in Bergen (that is, from the shipping shares that John Herman was administrating for himself and his siblings), 100 kroner should be sent to the Nansen Fund; 30 kroner to the Geographical Society, which had paid for Lie's participation in the Society's festivities of Nansen in 1896; and 20 kroner to his mountaineering *alma mater*, the Tourist Association, as a partial payment on his debt "which is possibly somewhat greater."

Page 431 Lie, in a letter to Sylow (Leipzig, 26.11.1896): "Right this moment I see that a Poet [Randers] has come out in Aftenposten blaming Mathematics for having got such a great Place in Jæger's history of literature. These howling Mongrels! They go into a make-believe World, believing that a Country can live materially and intellectually from Poetry of doubtful Quality, and ditto for Politics. When will they ever succeed in giving these *Poseurs* any Idea about what Norwegian Mathematics has accomplished!"

PART VII

The Final Years

Page 436 Kowalewski (1950) tells about Lie's lectures and his time at the newspaper café ,"Café Mercur". On his walking tours during these years, Lie was also frequently accompanied by the Danish theologian, Frants Peter William Buhl, who in 1890 had become Professor of Old Testament Theology in Leipzig, following the death of Professor Delitzsch. (Delitzsch was the rector who held the speech that Lie, upon his arrival in 1886, marched so demonstratively out of.) The Swedish mathematician Anders Wiman was also in Leipzig in the summer of 1898. According to Kowalewski, during Lie's final couple of years in Leipzig, the following were his (elite) students: the Greek, Papazachariu, the Russian, Sintzov, and the Americans, Blichfeldt, Bouton, Rohtrock, and two brothers by the name of Arnold, as well as van Etten-Westfall.

Page 437 À propos Lie's comments on Vogt: disagreement between Professor Vogt and Professor Helland seems to have been of an almost chronic nature. For example, in the Collegium affair of October 6, 1892 – which concerned management, leadership and future employment at the botanical museum – Helland was alone in voting against the establishment of a professorship in systematic botany, while Vogt was the only one who voted against Helland's subsidiary proposal to grant 1,600 kroner annually for extra lectures in botany, and 800 kroner in salary increase to the conservator at the museum.

Page 439 In a letter to Klein (April 1898), see also page 393, Lie gave expression to the fact that he felt that teaching work had become increasingly more satisfying in the past 7–8 years – that the students' dissertations were by and large rather insignificant in quality, arose from the fact that he was no longer helping them in the same way as during the first years. But otherwise, conditions were the same: Neumann had only become more and more impossible, and now, in his 66th year, he was also considerably weakened by age – nor did he hear so well, and he misunderstood everything, or so Lie maintained.

Lie also complained of Mayer's paternalistic behaviour, his continual inclination to act like a babysitter. In addition, Lie mentioned that he had been in Rome and Berlin, and that both the Italians, and above all, certainly all the mathematicians of l'Académie des Sciences in Paris would sign a subscription invitation to an Abel Fund – but that he had still not heard anything from Berlin – "but they will certainly go along with it," Lie wrote, and listed all the German mathematicians he had approached: Fuchs, Frobenius, Hilbert, Klein, Noether, Mayer, Schwarz, and Weber.

Klein replied on May 12th. He said that despite understanding well Lie's difficulties in Leipzig – and therefore understanding Lie's desire to go home – he found the report of Lie's departure unexpected. He could scarcely believe it. Klein quickly shifted to suggesting various candidates who might be possible successors to Lie at Leipzig. He mentioned Hilbert, Hurwitz, Minkowski, Wirtinger, Study,

Brill, Voss, Schur, Staude, and Kneser. Klein concluded by saying that he would certainly support the work with an Abel Fund, something he also thought Hilbert would support.

Page 439 About the Abel work: In 1902 the 100th anniversary of Abel's birth was celebrated in a grandiose manner [Stubhaug 1996, Hestmark 1999]. But an Abel monument did not come into being before 1908 (Vigeland's Abel sculpture, in the Abel Garden of the Royal Palace Park, Oslo).

Nansen wrote to Holst from London on December 7, 1906, recalling that the precondition for the Abel monument was that it should be raised with only Norwegian money. This was, he wrote – "in the manner that Sophus Lie so often stressed [...] but beyond this that an *Abel's Fund* should be collected from contributions made by mathematicians and mathematically interested people in all countries". But due to the monument, it had been inadvisable to gather money for such a fund: "which we still ought to have, especially since the blessed King Oscar's promised Abel prize went to heaven along with the union [between Norway and Sweden]." At the same time, Nansen explained that he had made contributions to many foreign monuments. There was nothing unusual about that, he stressed, with the exception that then, one would have to be clear that it would not be possible to inscribe on the monument: "Your Countrymen have raised this monument to You".

An Abel prize was in fact established during the jubilee in 1902 – a gold medallion with the value of 1,000 kroner (contributed by King Oscar II) which was to be awarded every five years by the king. But the political events of June 7, 1905, with Norway's independence from Sweden, meant that the Abel prize "went to heaven". But as late as February 1905, L. Sylow and C. Størmer were working on the protocol for the medallion.

Page 441 The letter from Klein to Mayer is dated 27.01.1899.

Page 441 The two students who followed Lie from Leipzig so as to be able to attend his lectures in Christiania, were the Austrian, Karl Carda, and the American, Edgar Odell Lovett (Princeton, N. J.). Lovett returned to the USA, and on 09.12.1898, he sent Christmas and New Year greetings to Lie, "Mrs. Lie and the Misses Lie" with the following inscription: "Ich komme bald wieder nach Kristiania zurück." (Lovett wrote later as well, in 1909 from Houston, Texas, to Mrs. Lie, with reassurances that Lie's collected works would be published, even if not in Norway. See also Lovett's greeting to Engel, note on page 505.)

Page 443 Neither in the archives of Rikshospital or Laache's private hospital are there any records indicating that Lie consulted Laache.

Page 444 Death notice – City Archives, Municipality of Oslo.

Page 444 On the printed notice of death that Mrs. Lie sent out to friends and acquaintances, there was written: "Mein inniggeliebter Mann, unser treusorgender Vater *Professor Dr. Sophus Lie* entschlief heute Mittag sanft und ruhig 56 Jahre alt. Kristiania 18.2.1899. Anna Lie, geb. Birch. Marie Lie. Dagny Lie. Herman Lie."

Page 449 The funeral at Holy Trinity Church: The vicar, Dr. Krogh-Tonning, officiated. Holst's song was sung to the melody of 'Bedre kan jeg ikke fare' ['A better way I cannot find to go'], by Reissiger. The text had been printed and distributed to all in attendance, and took as its theme, a verse from the Bible that describes the giving to all things "size and number and weight". The wreath decked in the Swedish national colours, that Holst laid on behalf of Mittag-Leffler, purportedly cost 20 kroner. Holst also laid a wreath on behalf of the Danish Mathematical Association – and consequently the newspapers wrote: "the colours of the three nations embellished the bier".

Both before and after the ceremony of sprinkling the earth upon the coffin, two verses of the hymn, "Med sorgen og klagen hold maade/Lad Guds ord dig trøste og raade" [Hold off thy sorrow and lament/May God's word console thy torment]. This was the same hymn that twenty-six years earlier had been sung at the funeral of Vicar Lie. At the funeral of Sophus Lie this hymn was printed on the same sheet as Holst's song.

Page 450 After Sylow's speech at the Academy of the Sciences (24.02.1899) there followed a commemorative silence. This in turn was followed by a lecture on sun spots by Professor Kristian Birkeland.

Page 450 Judging from its records, Realistforeningen does not seem to have marked Lie's passing with speeches and so forth – there was talk about him in the meeting on 02.03.1899, and on March 23rd there was a speech at Realistforeningen in connection with Axel Blytt's unexpected death.

Page 450 Klein in a letter to Holst, 24.04.1899 – on the lack of an answer from Mrs. Lie to Mrs. Klein's letter of condolence.

Page 450 Carl Størmer reported home in a letter to Holst on February 20, 1899.

Page 452 Scarcely a month after Lie's death – and before the university's memorial gathering – Holst had lost his wife. Bjørnson had this in mind as well when he sent a letter of consolation to Holst (19.03.1899) and commented on all the work Holst had done in connection with Lie's death, right in the midst of his own sorrow.

In the university's records for 1899 (published in 1900) one finds about Holst that due to illness, his lectures did not begin until 25.01.1899, and that they were postponed with the approval of the Collegium due to various work "that was imposed upon him partly by Professor Lie's death and funeral, and partly by the docent's responsibilities to the examination deputation committee with regard to teaching in the science teachers' programme. In this manner and at the request of the Collegium, he wrote the song for Lie's funeral at Holy Trinity Church, and gave the eulogy at the later memorial ceremony held subsequently by the university."

Page 452 The idea of a statue of Lie was quickly taken up under the leadership of Nansen, Brøgger and others, in the Academy of Sciences, and approved on May 16, 1899. The Abel Committee, which was working on the Abel monument, was requested to take charge of collecting money that would ensure that Lie too got his statue. However it soon became clear that it would be difficult to work on two monuments simultaneously, and the plans for the Lie monument were put to aside.

Sophus Lie's Descendants

Precisely when the family returned to Norway and Christiania in the autumn of 1898 is somewhat uncertain. Marie (21 years of age) graduated from her Leipzig gymnasium (Mädchengymnasium of Fraulein Dr. Käthe Windscheid) – and a large party was held in her honour in that city on September 17th. (The invitation card with the menu and the programme of music are found in Herman Lie's Nachlass.)

The apartment they moved into, at Eugenies gate 22, was in a new, handsome brick building, built in 1892 (and designed by architect Johan Grøstad), in the neo-renaissance style – the façade of dressed masonry with plaster columnar embellishments, the main cornices grandly sculpted, and a terrace that faced southeast.

This final year of life must have been a continuous disintegration of all Lie's plans and desires for a good life. He who so much wanted to be in control, he who liked so much to be admired, he who loved and enjoyed the great outdoors, sport and rigorous activity – he now experienced everything crumbling around him. As early as that autumn he must have found that his body could not tolerate the recreation he had been looking forward to in the capital city's "beautiful Surroundings". He did not manage to walk the sickness out of himself, as, in his own words, he had done ten years earlier. He turned fifty-six on December 17th, and if he now actually did receive the diagnosis of pernicious anaemia on Christmas Eve, 1898 (as has been said), so it must have seemed to him a strange and contrary sign, opposite of the communication he received on Christmas Eve twenty-six years earlier – when Anna answered his letter of proposal, and in his own words, "settled [his] Fate".

Pernicious anaemia was a death sentence, and Lie must have been fully informed about it. Clearly, and with all his senses, he must have been in full consciousness of the proximity of death. For a period of more than two months he must have known that everything would soon be over. He who in all his deliberations and earlier choices, had thought above all of his family, had now put his family in a situation that was worse than if they had remained in Germany. Maybe it was Lie himself in his despair, and from his deathbed, raised the question that perhaps Anna could move back to Leipzig after he had passed away?

In any case, the fact was that only two months after her father's death, the eldest daughter, Marie, was back in Leipzig, and one of her errands seems to have been to look into the possibilities of a return there, or at least that is how Mayer in Leipzig understood matters. However, the Saxon state no longer had any obligations toward the Lie family, and thus were unable to produce any sort of contribution. According to Mayer, Mrs. Lie had made remarks to the effect that she expressed the hope that a financial collection could be set in motion for Lie's surviving family – such as had been done for, among others, the surviving relatives of the mathematician Clebsch, and as had been done, and such as had been bequeathed earlier to the family of Michael Sars.

Mayer commented on this in a letter to Klein (26.04.1899): "My relationship with Lie, even though he once ended his friendship with me in writing, was nonetheless so cordial that I would not like to let something remain untried when it comes to assisting his surviving family." However, Mayer stated that he was in some degree of doubt as to whether anything could be accomplished – most in the milieu seemed to think it was difficult to defend such a collection of funds. Teubner, however, had declared itself willing to contribute, and the appeal might eventually go a long way out into scientific circles. Mayer wrote that Klein's recommendations would now be decisive to the attitudes taken by both Engel and Hölder. What Klein wrote in reply is not known. But nothing came of the collection of money for Lie's family.

At first this seems to have been a great blow to Mrs. Lie – later in the year she wrote to Mittag-Leffler: "Now that I have seen there is nothing to expect from the Saxon Side, I believe that I still have a Duty to my Children to investigate one way or another the possibility of improving the Position. [...] Now that the Germans have shown themselves so heartless toward us, perhaps it is possible that someone from the French Side would do something. I believe that the Science that Sophus Lie advanced to such a high Degree, in return ought to allow his 3 Children to greet the Future in an untroubled Manner." And as the Scandinavian colleague who had the greatest knowledge of European conditions, she urged Mittag-Leffler to look into the possibilities. How Mittag-Leffler answered her is unknown – but no economic support was ever talked about.

Marie remained in Leipzig (where she had her friends and acquaintances). She studied languages at the university and married an optometrist, Friedrich (Fritz) Leskien, who was the son of August Leskien, the famous professor of Slavic languages. Together with her husband, Marie began to translate Norwegian literature, first and foremost, Alexander Kielland, and in connection with this, on later visits to Norway, she got to know Stavanger and Jæren – "Kielland's landscape", as she called it. Marie also translated Hans E. Kinck and tried to work on Hans Aanrud, but felt it was more difficult "to find the tone", as she wrote to her brother Herman. She also worked on the texts of Nina Roll Anker and O. Thommesen.

Marie had two children, Hans Peter and Ragna. Hans Peter Leskien, who trained as a medical doctor, opposed the Nazi regime by immigrating to Palestine in 1936, and after World War II, continued to live under there when the Palestine Protectorate had become the State of Israel, even though he was not Jewish. Marie's daughter, Ragna, married Ernst Hölder, professor of mathematics at Leipzig, and had two children, Peter and Birgit. The Hölders lived at Albertstrasse 37.

In the first correspondence (1899–1902) between Marie in Leipzig and Dagny in Christiania [Kristiania], it emerges that Dagny criticised her sister for having abandoned them. Dagny wrote: "You [Marie] are without doubt the family's cleverest member' you are the only one who can make things jolly and lively for others" – and Dagny tells that she and her mother often sit in their apartment on Eugenies gate, lost in their thoughts: "We speak very little to one another, and most frequently, we are bored when we are at home." And when the summer of 1901 approached, she stated that she and her mother did not know what to do – Dagny wanted to go to Romsdalen, but her mother thought it was too expensive there, and from her mother's point of view, the area around Galdhøpiggen [Norway's highest mountain] was "not attractive" – so there would be a tour to Rondane again this year, as there was last year, Dagny concluded.

That the eldest daughter was well-liked is also perhaps indicated in a letter of condolence (in connection with her father's death) that Marie received from the theologian, Anathon Aall (who was ten years Marie's senior, and was staying at Oxford in February 1899): "Your splendid father could not other than win over all those who came near him personally. He had a remarkable joy in seeing others happy." Marie was also close friends with Emma Lund, the daughter of Axel and Valborg Lund (in Christiania), and the close relationship between the Lie and Lund families seems to have persisted (a letter to Herman of 19.05.1902, is signed "Aunt Valborg"). Contact with the Motzfeldts on the other hand, seems to have been very sporadic. Sophus Lie's sister, Laura, died in 1911, and was then (and later) honoured and paid tribute by former students for the fact that they "in the days of youth at the Eugenie Institution learned from her noble personality" [*Eugenia Stiftelse gjennom 100 år*, Oslo 1927].

Dagny also moved back to Leipzig (ca. 1902), after having graduated from Aars og Voss Skole (Latin line), ready to become a university student. In Leipzig she became engaged to the six-year older Walther Straub – against, at first, the strong disapproval of her elder sister, Marie – and she married, in Christiania, in 1904. Straub became a professor of pharmacology and they set up house in Freiburg im Breisgau. They had two sons – Harald (born in 1905) and Peter (born in 1909). At first they lived in Leipzig, and later, due to Straub's university career, in Marburg, Würtzburg, Freiburg and finally in Munich. Walther Straub died during the bombing of Munich in 1944; Dagny was seriously wounded, but escaped that bomb-ravaged city by moving to a little village in northwest Bayern, where she lived temporarily at the home of her son Harald and her daughter-in-law. She died the following year (28.12.1945) from the wounds she had suffered. (The Karl-Straub Institute of Pharmacology, which Walther Straub had been building up, was also destroyed during the Second World War.) Their son Harald was a physicist, and established himself in the USA after the war, returning to live in Germany during his pension years. Peter, a jurist, worked in diplomacy in Rome, became an officer in Rommel's army, was imprisoned in May of 1943 in Tunis, and sat in an American prisoner of war camp. He later became mayor of a small city in Schwarzwald, married a teacher, and had no children.

Herman too, upon the family's return home in the autumn of 1898, was placed in Aars og Voss Skole, and was confirmed at Uranienborg Church on April 9, 1899. He chose the Latin line at school and was a hard-working pupil. In a letter to his sister Marie he explained that he had dropped out of dancing school because it took up too much time, and because "the ladies have been so off-putting. They were disgusting." Herman became a student at the university in 1901, and in a letter to Marie, wrote that he was proud (09.06.1901) that he, somewhat astonished, had achieved the full top marks in his *examen artium*. He continued, writing that there had been fewer with top marks among those studying the sciences, and he added (in contrast to the views of his father): "It is funny that those classically educated thus show their superiority in the intellectual area."

Herman began to study law, was on the executive of the Student Society in 1903–04, executive member of the Student Choral Society in 1906, and took part in the great choral tours to America (1905), Nordland in northern Norway (1907), Denmark (1908); in addition to this, he was active in the Norwegian Students Rowing Club. He took his public service examination in law in 1907, and then worked in banking and finance.

A letter to Engel, dated 15.01.1907, indicates that Herman now tried to get money from the Nansen Fund to publish his father's works. From Leipzig on December 21, 1912, Marie reported to Herman that in *Neueste Nachrichten* had just (on the occasion of their father's seventieth birthday, December 17th) announced that "Papa's collected works will be coming out at Teubner's." Marie observed that it was certainly not by chance that the announcement was made exactly on that date, and moreover she reported that it was thought that all seven volumes would be published in a ten-year period. In a letter written four months later, Marie wrote that she had very much wanted to subscribe to her father's works, but could not afford to do so due to having to move house, among other things. In the same letter (20.04.1913), Marie gave the news that all the siblings (Marie, Dagny and Herman) received money from Teubner – and Marie expressed her joy that there was so much money for each of them.

Both Herman and his mother wrote regularly to Marie in Leipzig (less often to Dagny, it seems), and in June 1906, they both mention that Herman was invited to the St. Hans Aften (mid-summer celebration) "at Werenskiold's out at Lysaker". Herman added: "I have come quite regularly into the Lysaker clique." (In addition to Werenskiold, the artistic Lysaker circle included, among others, Fridtjof Nansen, the artists Gerhard Munthe and Eilif Peterssen, and the art critic Andreas Aubert; moreover, the wealthy Axel Heiberg lived in the vicinity.)

Herman Lie became engaged to Karen Inga (Basken) Werenskiold (the daughter of Erik Werenskiold) in the summer of 1913, and they married at New Year's 1914. (It was said that banker Lie now administrated his father-in-law's finances.) On February 2nd, 1922, Basken gave birth to a son, who was named Per. Three years later Herman and Basken were divorced. Herman moved to Jevnaker in 1940, and his surviving papers, among other things, were left with his friend Ole B. Styri at Hauger Farm.

A little more on Herman Lie's working career: following his juridical public service examination, he worked for a year at the banking firm, N. A. Andresen & Co. in Christiania, followed by two years in New York and two years in Paris. In both places he was occupied with practical banking activities and theoretical banking study. In 1913–14 he started Lie & Co. Bankierforening [Banker Association], which was taken over by Andresen's Bank in 1920, and where he worked for three years before he went in as a partner in the firm Hieronymus Heyerdahl Jr.

In the 1930s he worked together with the director, J. Sejersted Bødtker, in the field of currency exchange and stocks and bonds, conducting business with Paris, London and New York – and about his activities after the interruption of the war (which for Norway extended from 1940 to 1945), he writes in *Studentene fra 1901* (Oslo 1951): "Unfortunately, after the war this kind of business was unable, to a great extent, to get back on its feet due to currency restrictions and all the state intervention we were then blessed with. But even in the bleak post-war conditions there were fields in which a finance firm could prevail."

And in his short summing up of his life, Herman Lie makes place for the following: "I have not taken part in public life and unfortunately have never made myself deserving of any decorations or honourary offices. In any case, I would like to mention that if my father, Sophus Lie had lived, he would have been 100 years old during the occupation – that is, on 17.12.1942. He was commemorated on that day with a celebration to which I had invited my closest friends, at the 'Valkyrie' in Majorstuen [an area of shops and restaurants on the west side of Oslo], which at that time was one of the city's most aristocratic gathering places. I opened the party by telling a little bit about my father, particularly about the period of his youth, and many other, among them Professors Vilhelm Bjerknes and Olaf Broch, also had many things to recount about him."

This party at the restaurant "Valkyrie" seems to have had great significance for Herman – he describes it in detail in letters to his sisters, and preserved his speech and accounts of the event [Herman Lie 1942]. Among the thirty guests, in addition to Bjerknes and Broch, were also the mathematics professors Carl Størmer and Ingebrigt Johannesson, together with a number of relatives – "My father was extremely fond of relatives," Herman said – and the gathering also included the sons of several of his father's friends: Ketil and Peter Motzfeldt (sons of Ernst and Axel Motzfeldt), Hans Blehr (son of Theodor Blehr), Herman Reimers (son of Herman Reimers, Sr.), Jon Kielland Peterssen (son of Eilif Peterssen), Hans Sänger (see also note to p. 351), and others.

Mrs. Anna Lie continued to live in the apartment at Eugenies gate 22. The special widow's pension that Lie had spoken so eagerly about having included in the parliamentary motion (in relation to the extraordinary professorship in "transformation group theory") in 1894 had subsequently been given up. But after his death, Parliament rapidly granted 2,000 kroner to Mrs. Lie, and, in ordinary widow's pension she received about 1,000 kroner. In a letter to Mittag-Leffler (in relation to enquiries about eventual French support for Lie's family), Mrs. Lie commented: "Indeed, Parliament has done for us what one could dare hope for. But nonetheless, it is not just that such a great Man's Family should have to live in such tight Circumstances." (About Mrs. Lie's pension in the Pension Protocol of the Widow

Fund, see New Department 2 [nye afdeling nr. 2], together with *Morgenbladet* 124, 1899.)

Mrs. Lie tried to sell Lie's collection of books on the open market. When she discovered that the collection lacked a couple of issues in the 12th and 13th volumes of *Acta Mathematica*, she wrote to Mittag-Leffler to get to know how she could come by these issues. She added: "Indeed, this reduces the Value significantly, when individual Parts are missing, and it is most important to me to get the Books sold in the mathematical World." An antiquarian bookdealer in Paris offered the most for Lie's collection, "but still, it is not what I hoped for," Mrs. Lie commented to Mittag-Leffler at the New Year 1901, and at the same time reported that it was no longer so important to find the missing issues of *Acta Mathematica*. This was because Fridtjof Nansen had proposed, namely, that the University of Kristiania should buy the book collection for 4,000 kroner – she hoped this proposal would go through, and in any case, Sylow and Guldberg knew what was missing. Already in the autumn of 1900, Parliament had decided to purchase Lie's book collection [Royal Resolution of 11th October, 1900: "Purchase of Professor Dr. Lie's book collection"]. The intermediary between the family and the state was the old family friend, Sofus Arctander, and the price was settled at 3,000 kroner.

Otherwise, it seems Mrs. Lie lived in peace and quiet, and maintained a broad correspondence with her children, and as time went by, she seems to have travelled occasionally to visit her daughters in Germany. The tumour in her breast grew once again, and was removed without subsequent repercussions.

When Adler Vogt was working on a book in 1912 about the Lie family [Vogt 1914], it seems that he wrote to all the relatives to obtain both information and money for the printing costs. In a letter from Marie (in Leipzig, 17.02.1912) to Herman, we find that their mother was opposed to the writing of such a book. For her part, Marie would gladly give the 100 kroner that was mentioned, but declined to write anything biographical about her father. First of all, she thought that Herman now wrote better Norwegian that she did, "and second, You [Herman] are much more into everything, that concerns Papa's life. You are familiar with his correspondence, You once organised the library," but Marie wanted to contribute in any way she could – and thus she recounted that some days earlier she had received a package from Mrs. Professor Adolf Mayer, containing all Lie's letters to Mayer (Mayer had died in 1908). Marie comments on these letters by saying that they contain much mathematics, but also a certain amount of personal observation, and an extremely funny letter about "the carnal prizes in Leipzig" in 1886 – and "it also contains a couple of matters from the doctors at Ilten from 1890, and a letter from Papa himself from Ilten," (These letters are now in NBO, Manuscript Collections – the medical explanations however, are not included).

Mrs. Lie died in Leipzig on June 12, 1920 [*Morgenbladet* 12.06.1920] – Herman gives the date as June 10th [in *Studentene fra 1901*] – and on June 17th she was buried beside her husband in the Cemetery of Our Saviour in Oslo.

Chronological Bibliography of Sophus Lie's Published Works

Christ. Forh. = *Forhandlinger i Videnskabs-Selskabet i Christiania*, Christiania. (Until 1875 these *Proceedings* were paginated for every year; after this date, individual studies were paginated separately.)

Arch. for Math. = *Archiv for Mathematik og Naturvidenskab*, Kristiania. (The periodical that was founded in 1876 and edited by Sophus Lie, G. O. Sars og J. Worm-Müller, appeared in four issues per annum – at the beginning, regularly every three months, but after 1881, often very much delayed.)

Math. Ann. = *Mathematische Annalen.*

Leipz. Ber. = *Berichte über die Verhandlungen der Königlich Sächsischen Gesellschaft der Wissenschaften zu Leipzig: Mathematisch-Physische Classe.* Leipzig. (Published in issues shortly after works were submitted.)

Comptes Rendus = the journal published by l'Académie des Sciences in Paris.

Leipz. Abh. = *Abhandlungen der mathematisch-physischen Classe der Königl. Sächsischen Gesellschaft der Wissenschaften zu Leipzig.*

G.A. = *Gesammelte Abhandlungen*, Vols. 1–7. Leipzig, and Oslo 1922–60.

This is the standard work in which all of Lie's articles and papers are located, collected and edited by Friedrich Engel and Poul Heegaard. Volumes 1–6 contain the papers that Lie himself had published – a total of 3,627 pages. Volume 7 contains 35 papers and treatises (476 pages) from Lie's surviving papers. In addition, this *Gesammelte Abhandlungen* contains 1,409 pages of remarks and commentaries, the vast majority written by Engel.

1869 Repräsentation der Imaginären der Plangeometrie. (Jeder plangeometrischer Satz ist ein besonderer Fall eines stereometrischen Doppel-Satzes in der Geometrie der Linien-Congruenzen). Eight pages, quarto format, privately printed in Christiania during the month of February. Later on, published in *Crelles Journal*, Vol. 70, Issue 4, pp. 346–353 under the title: Über eine Darstellung des Imaginären in der Geometrie – and a more elaborated version is found in *Christ. Forh.*, pp. 16–38, (with 12 figures) – and with a continuation: Repräsentation der Imaginären der Plangeometrie, *Christ. Forh., årg. 1869*, pp. 107–146. G.A. 1, pp. 1–11 and pp. 14–66.

1870 Über die Reciprocitäts-Verhältnisse des *Reye*schen Complexes. *Göttinger Nachrichten*, Nr. 4, pp. 53–66. G.A. 1, pp. 68–77.

Sur une certaine famille de courbes et surfaces. Co-authored with Felix Klein, *Comptes Rendus*, Vol. 70, pp. 1222–1226, 1275–1279. G.A. 1, pp. 78–85.

Sur une transformation géométrique. *Comptes Rendus*, Vol. 71, pp. 579–583. G.A. 1, pp. 88–92.

Om en Classe geometriske Transformationer. *Christ. Forh.*, pp. 506–509. G.A. 1, pp. 93–96.

Über die Haupttangenten-Curven der Kummerschen Fläche vierten Grades mit 16 Knotenpunkten. *Monatsberichte der Berliner Akademie*, pp. 891–899. Co-authored with Felix Klein. G.A. 1, pp. 97–104.

1871 Over en Classe geometriske Transformationer. *Christ. Forh.*, pp. 67–109. G.A. 1, pp. 105–152.

Über eine Classe geometrischer Transformationen (Fortsetzung). *Christ. Forh.*, pp. 182–245. G.A. 1, pp. 153–210.

Über diejenige Theorie eines Raumes mit beliebig vielen Dimensionen, die der Krümmungs-Theorie des gewöhnlichen Raumes entspricht. *Gött. Nachr.*, Nr. 7, pp. 191–209. G.A. 1, pp. 215–228.

Über diejenigen ebenen Curven, welche durch ein geschlossenes System von einfach unendlich vielen vertauschbaren linearen Transformationen in sich übergehen. *Math Ann.*, Vol. IV, pp. 50–84. Co-authored with Felix Klein. G.A. 1, pp. 229–266.

Zur Theorie eines Raumes von *n* Dimensionen. *Gött. Nachr.*, 1871, Nr. 22, pp. 535–557. G.A. 1, pp. 271–285.

1872 Über Complexe, insbesondere Linien- und Kugel-Complexe, mit Anwendung auf die Theorie partieller Differential-Gleichungen. *Math. Ann.*, Vol. V, pp. 145–256. G.A. 2, pp. 1–121. (An expanded version of his doctoral dissertation.)

Kurzes Resumé mehrerer neuer Theorien. *Christ. Forh.*, pp. 24–27. G.A. 3, pp. 1–4.

Neue Integrations-Methode partieller Gleichungen erster Ordnung zwischen *n* Variabeln. *Christ. Forh.*, pp. 28-34. G.A. 3, pp. 5–11.

Über eine neue Integrationsmethode partieller Differential-Gleichungen. Erster Ordnung. *Gött. Nachr.*, Nr. 16, pp. 321–326. G.A. 3, pp. 12–15.

Zur Theorie partieller Differentialgleichungen erster Ordnung, insbesondere über eine Klassifikation derselben. *Gött. Nachr.*, Nr. 25, pp. 473–489. G.A. 3, pp. 16–26.

Zur Theorie der Differential-Probleme. *Christ. Forh.*, pp. 132–133. G.A. 3, pp. 27–28.

Zur Invarianten-Theorie der Berührungstransformationen. *Christ. Forh.*, pp. 133–135. G.A. 3, pp. 29–31.

1873 Über partielle Differential-Gleichungen erster Ordnung. *Christ. Forh.*, pp. 16–51. G.A. 3, pp. 32–63.

Partielle Differential-Gleichungen erster Ordnung, in denen die unbekannte Funktion explicite vorkommt. *Christ. Forh.*, pp. 52–85. G.A. 3, pp. 64–95.

Zur analytischen Theorie der Berührungs-Tranfomationen. *Christ. Forh.*, pp. 237–262. G.A. 3, pp. 96–109.

Über eine Verbesserung der *Jacobi-Mayer*schen Integrations-Methode. *Christ. Forh.*, pp. 282–288. G.A. 3, pp.120– 25.

Neue Integrations-Methode eines $2n$-gliedrigen Pfaffschen Problems. *Christ. Forh.*, pp. 320–343. G.A. 3, pp.126–148.

1874 Allgemeine Theorie partieller Differential-Gleichungen erster Ordnung. *Christ. Forh.*, pp. 198–226. G.A. 3, pp. 149–175.

Zur Theorie des Integrabilitätsfaktors. *Christ. Forh.*, pp. 242–254. G.A. 3, pp. 176–187.

Veralgemeinerung und neue Verwerthung der *Jacobi*schen Multiplicator-Theorie. *Christ. Forh.*, pp. 255–274. G.A. 3, pp. 188–206.

Begründung einer Invarianten-Theorie der Berührungs-Transformationen. *Math. Ann.*, Vol. VIII, pp. 215–303. G.A. 4, pp. 1–96.

Über Gruppen von Transformationen. *Gött. Nachr.*, Nr. 22, pp. 529–542. G.A. 5, pp. 1–8.

1875 Allgemeine Theorie partieller Differentialgleichungen erster Ordnung. II. *Christ. Forh.*, pp. 1–15. G.A. 3, pp. 207–220.

Diskussion der Integrationsmethoden der partiellen Differentialgleichungen erster Ordnung. *Christ. Forh.*, pp. 16–48. G.A. 3, pp. 221–251.

Allgemeine Theorie der partiellen Differentialgleichungen erster Ordnung. *Math. Ann.*, Vol. IX, pp. 245–288. G.A. 4, pp. 97–162.

1876 Theorie der Transformationsgruppen (Erste Abhandlung). *Arch. for Math.*, Vol. I, pp. 19–57. G.A. 5, pp. 9–41.

Theorie der Transformationsgruppen (Abhandlung II). *Arch. for Math.*, Vol. I, pp. 152–193. G.A. 5, pp. 42–77.

Vervollständigung der Theorie der Berührungstransformationen. *Arch. for Math.*, Vol. I, pp. 194–202. G.A. 3, pp. 252–259.

Resumé einer neuen Integrationstheorie. *Arch. for Math.*, Vol. I, pp. 335–366. G.A. 3, pp. 260–286.

1877 Neue Integrations-Methode der *Monge-Ampère*schen Gleichung. *Arch. for Math.*, Vol. II, pp. 1–9. G.A. 3, pp. 287–294.

Allgemeine Theorie der partiellen Differentialgleichungen erster Ordnung. Zweite Ahandlung. *Math. Ann.*, Vol. II, pp. 464–557. G.A. 4, pp. 163–264.

Die Störungstheorie und die Berührungs-Transformationen. *Arch. for Math.*, Vol. II, pp. 129–156. G.A. 3, pp. 295–319.

Synthetisch-analytische Untersuchungen über Minimal-Flächen. I. Über reelle algebraische Minimalflächen. *Arch. for Math.*, Vol. II, pp. 157–198. G.A. 1, pp. 286–318.

Berigtelse. (Berichtigung.) [Published with the preceding study, that was Lie's lecture at the Meetings of Natural Science Research in Munich 19.9.1877] *Christ. Forh.*, G.A. 1, pp. 319–320.

Theorie des Pfaffschen Problems. (Erste Abhandlung.) *Arch. for Math.*, Vol. II, pp. 338–379. G.A. 3, pp. 320–354.

1878 Mathematiske Sätninger. (Mathematische Sätze.) *Christ. Forh.*, G.A. 1, p. 322.

Petite contribution à la théorie de la surface Steinerienne. *Arch. for Math.*, Vol. III, pp. 84–92. G.A. 1, pp. 323–328.

Theorie der Transformationsgruppen. III. Bestimmung aller Gruppen einer zweifach ausgedehnten Punktmannigfaltigkeit. *Arch. for Math.*, Vol. III, pp. 93–165. G.A. 5, pp. 78–135.

Sätze über Minimalflächen. *Arch. for Math.*, Vol. III, pp. 166–176. G.A. 1, pp. 331–337.

Sätze über Minimalflächen. II. Bestimmung aller algebraischen Minimalflächen, die sich in einem algebraischen Kegel einschreiben lassen. *Arch. for Math.*, Vol. III, pp. 224–233. G.A. 1, pp. 340–347.

Sätze über Minimalflächen. III. Ueber die in einer algebraischen Developpable eingeschriebenen algebraischen Minimalflächen. *Arch. for Math.*, Vol. III, pp. 340–351. G.A. 1, pp. 349–356.

Theorie der Transformationsgruppen. Abhandlung IV. *Arch. for Math.*, Vol. III, pp. 375–460. G.A. 5, pp. 136–198.

1879 Klassifikation der Flächen nach der Transformationsgruppe ihrer geodätischen Curven. *Universitetsprogram*, Kristiania. G.A. 1, pp. 358–408.

Beiträge zur Theorie der Minimalflächen. I. Projectivische Untersuchungen über algebraische Minimalflächen. *Math. Ann.*, Vol. XIV, pp. 331–416. G.A. 2, pp. 122–215.

Geometriske Meddelelser. (Geometrische Mitteilungen.) *Christ. Forh.*, G.A. 3, pp. 355–356.

Theorie der Transformationsgruppen. V. *Arch. for Math.*, Vol. IV, pp. 232–261. G.A. 5, pp. 199–223.

Beiträge zur Theorie der Minimalflächen. II. Metrische Untersuchungen über algebraische Minimalflächen. *Math. Ann.*, Vol. XV, pp. 465–506. G.A. 2, pp. 219–265.

Bestimmung aller in eine algebraische Developpable eingeschriebenen algebraischen Integralflächen der Differentialgleichung $s = 0$. *Arch. for Math.*, Vol. IV, pp. 334–344. G.A. 3, pp. 357–366.

Zur Theorie der Flächen constanter Krümmung. I. Bestimmung ihrer Haupttangentencurven und Krümmungslinien. *Arch. for Math.*, Vol. IV, pp. 345–354. G.A. 3, pp. 367–374.

Zur Theorie der Flächen constanter Krümmung. II. Das sphärische Bild der Haupttagenten- und Krümmungs-Curven. *Arch. for Math.*, Vol. IV, pp. 355–366. G.A. 3, pp. 375–386.

1880 Weitere Untersuchungen über Minimalflächen. *Arch. for Math.*, Vol. IV, pp. 477–506. G.A. 1, pp. 414–437.

Über Flächen, deren Krümmungsradien durch eine Relation verknüpft sind. *Arch. for Math.*, Vol. IV, pp. 507–512. G.A. 3, pp. 387–393.

Resumé af en Integrationstheori. *Christ. Forh.*, Nr. 1. G.A. 3, pp. 394–397.

Sur les surfaces dont les rayons des courbure ont entre eux une relation. *Bulletin des sciences mathmatiques et astronomiques,* 2nd Series, Vol. IV, pp. 300–304.

Theorie der Transformationsgruppen I. *Math. Ann.*, Vol. XVI, pp. 441–528. G.A. 6, pp. 1–94. (Translated into English by Michael Ackerman, with commentary by Robert Hermann, Brookline, Massachusetts, 1975.)

Meddelelse. (Mitteilung). *Christ. Forh.*, p. 8.

Zur Theorie der Flächen constanter Krümmung. III. *Arch. for Math.*, Vol. V, pp. 282–306. G.A. 3, pp. 398–418.

Geometriske Meddelelser. *Christ. Forh.*, p. 10. G.A. 3, pp. 419–420.

Zur Theorie der Flächen constanter Krümmung. IV. Bestimmung aller Flächen constanter Krümmung durch successive Quadraturen. *Arch. for Math.*, Vol. V, pp. 328–358. G.A. 3, pp. 421–446.

1881 Zur Theorie der Flächen constanter Krümmung. V. *Arch. for Math.*, Vol. V, pp. 518–541. G.A. 3, pp. 447–466.

Mathematiske Sätninger. (Mathematische Sätze.) *Christ. Forh.*, p. 6. G.A. 3, pp. 467–468.

Mathematiske Sätninger. *Christ. Forh.*, p. 7.

Diskussion der Differentialgleichung $d^2 z / dx\, dy\, F(z)$. *Arch. for Math.*, Vol. VI, pp. 112–124. G.A. 3, pp. 469–479.

Transformationstheorie der partiellen Differentialgleichung $s^2 - rt = (1 + p^2 + q^2)\ ^2/a^2$. *Arch. for Math.*, Vol. VI, pp. 153–167. G.A. 3, pp. 480–491.

Mathematiske Sätninger. *Christ. Forh.*, p. 12.

Über die Integration durch bestimmte Integrale von einer Classe linearer partieller Differentialgleichungen. *Arch. for Math.*, Vol. VI, pp. 328–368. G.A. 3, pp. 492–524.

Om algebraiske Differentialligninger, der tilstede infinitesimale Transformationer. *Christ. Forh.*, Nr. 15. G.A. 3, pp. 525–530.

1882 Zur Theorie der geodätischen Curven der Minimalflächen. *Arch. for Math.*, Vol. VI, pp. 490–501. G.A. 1, pp. 440–448.

Bestimmung aller Raumcurven, deren Krümmungsradius, Torsionsradius und Bogenlänge durch eine beliebige Relation verknüpft sind. *Christ. Forh.*, Nr. 10. G.A. 3, pp. 531–536.

Bestimmung aller Flächen, die in mehrfacher Weise durch Translationsbewegung einer Curve erzeugt werden. *Arch. for Math.*, Vol. VII, pp. 155–176. G.A. 1, pp. 450–465.

Über Flächen, die infinitesimale und lineare Transformationen gestatten. *Arch. for Math.*, Vol. VII, pp. 179–193. G.A. 5, pp. 224–235.

Untersuchungen über geodätische Curven. *Math. Ann.*, Vol. XX, pp. 357–454. G.A. 2, pp. 267–373.

Meddelelse. *Christ. Forh.*, p. 13. G.A. 5, pp. 236–237.

Untersuchungen über Differentialgleichungen. I. *Christ. Forh.*, Nr. 12. G.A. 3, pp. 537–547.

Untersuchungen über Differentialgleichungen. II. *Christ. Forh.*, Nr. 22. G.A. 3, pp. 551–555. (Lecture given November 3rd, at the Société mathématique, Paris.)

Meddelelse. *Christ. Forh.*, G.A. 3, pp. 548–550.

Über gewöhnliche Differentialgleichungen, die eine Gruppe von Transformationen gestatten. *Arch. for Math.*, Vol. VII, pp. 443–444. G.A. 5, pp. 238–239.

1883 Klassifikation und Integration von gewöhnlichen Differentialgleichungen zwischen *x, y,* die eine Gruppe von Transformationen gestatten. I. *Arch. for Math.*, Vol. VIII, pp. 187–248. (This study and the one that follows were also reprinted in *Math. Ann.*, 1888, Vol. XXXII, pp. 213–281.) G.A. 5, pp. 240–281.

Klassifikation und Integration von gewöhnlichen Differentialgleichungen zwischen *x, y,* die eine Gruppe von Transformationen gestatten. (Abhandlung 2.) *Arch. for Math.*, Vol. VIII, pp. 249–288. GA 5, pp. 282–310.

Untersuchungen über Differentialgleichungen. III. *Christ. Forh.*, Nr. 10. G.A. 5, pp. 311–313.

Über unendliche continuirliche Gruppen. *Christ. Forh.*, Nr. 12. G.A. 5, pp. 314–361.

Klassifikation und Integration von gewöhnlichen Differentialgleichungen zwischen *x, y,* die eine Gruppe von Transformationen gestatten. III. *Arch. for Math.*, Vol. VIII, pp. 371–458. G.A. 5, pp. 362–427.

Untersuchungen über Differentialgleichungen. IV. *Christ. Forh.*, Nr. 18. G.A. 3, pp. 556–560.

Sätninger (Sätze). *Christ. Forh.*, p. 20. G.A. 3, pp. 561–562.

1884 Bestimmung des Bogenelements aller Flächen, deren geodätische Kreise eine infinitesimale Berührungstransformation gestatten. *Arch. for Math.*, Vol. IX, pp. 40–61.

Über die allgemeinste geodätische Abbildung der geodätischen Kreis einer Fläche. *Arch. for Math.*, Vol. IX, pp. 62–68. G.A. 1, pp. 487–491.

Mathematiske Meddelelser. I. *Christ. Forh.*, Nr. 8. G.A. 5, pp. 428–431.

Über die Haupttangenten-Curven der Kummerschen Fläche vierten Grades mit 16 Knotenpunkten. *Math. Ann.*, Vol. XXIII, pp. 579–586. (Reprint of the work co-authored with Felix Klein from 1870.)

Klassifikation und Integration von gewöhnlichen Differentialgleichungen zwischen *x,y,* die eine Gruppe von Transformationen gestatten. IV. *Arch. for Math.*, Vol. IX, pp. 431–448. G.A. 5, pp. 432–446.

Zur Theorie der Transformationsgruppen. *Arch. for Math.*, Vol. IX, pp. 449–451. G.A. 5, pp. 447–448.

Mathematiske Meddelelser. II. *Christ. Forh.*, Nr. 9. G.A. 5, pp. 449–452.

Untersuchungen über Transformationsgruppen. I. *Arch. for Math.*, Vol. X, pp. 74–128. G.A. 5, pp. 453–498.

Bestemmelse af kontinuerlige Grupper. (Bestimmung kontinuierlicher Gruppen.) *Christ. Forh.*, p. 12.

Mathematiske Meddelser. III. *Christ. Forh.*, Nr. 15. G.A. 5, p. 499–502.

Über Differentialinvarianten. *Math. Ann.*, Vol. XXIV, pp. 537–578. G.A. 6, pp. 95–138.

1885 Allgemeine Untersuchungen über Differentialgleichungen, die eine continuirliche, endliche Gruppe gestatten. *Math. Ann.*, Vol. XXV, pp. 71–151. G.A. 6, pp. 139–223.

Über gewöhnliche lineare Differentialgleichungen. *Christ. Forh.*, Nr. 21. G.A. 5, pp. 503–506.

1886 Untersuchungen über Transformationsgruppen. II. *Arch. for Math.*, Vol. X, pp. 353–413. G.A. 5, pp. 507–552.

Bemerkungen zu v. Helmholtz Arbeit über die Thatsachen, die der Geometrie zu Grunde liegen. *Leipz. Ber.*, Supplementary issue, pp. 337–342. G.A. 2, pp. 374–379. (Lecture presented at the Natural Science Research Meetings in Berlin 21.09.1886.)

1887 Nogle mathematiske Sætninger. *Christ. Forh.*, p. 4. G.A. 1, p. 493.

Die Begriffe Gruppe und Invariante. *Leipz. Ber.*, Issue I & II, pp. 83–88. (Also printed in the *American Journal of Mathematics*, 1889, Vol. XI, pp. 182–186.) G.A. 6, pp. 224–229.

1888 Beiträge zur allgemeinen Transformationestheorie. *Leipz. Ber.*, Issue I & II, pp. 14–21. G.A. 6, pp. 230–236.

Theorie der Transformationsgruppen. Erster Abschnitt. Unter Mitwirklung von Dr. Friedrich Engel. Leipzig (June 1888), 632 pages, with a 10-page foreword. (Reprinted with the addition of *errata*, and a detailed subject index, New York, 1970.)

Zur Theorie der Berührungstransformationen. *Leipz. Abh.*, Vol. XIV, Nr. 12, pp. 535–562. G.A. 4, pp. 265–290.

Zur Theorie der Transformationsgruppen. *Christ. Forh.*, Nr. 13. G.A. 5, pp. 553–557.

1889 Ein Fundamentalsatz in der Theorie der unendlichen Gruppen. *Christ. Forh.*, Nr. 7. G.A. 5, pp. 558–560.

Die infinitesimalen Berührungstransformationen der Mechanik. *Leipz. Ber.*, Issue II, III & IV, pp. 145–156. G.A. 6, pp. 237–247.

Reduction einer Transformationsgruppe auf ihre canonische Form. *Leipz. Ber.*, Issue II, III & IV, pp. 277–289. G.A. 6, pp. 248–259.

Über irreduzible Berührungstransformationsgruppen. *Leipz. Ber.*, Issue II, III & IV, pp. 320–327. G.A. 6, pp. 260–266.

1890 *Theorie der Transformationsgruppen. Zweiter Abschnitt.* Unter Mitwirkung von F. Engel. Leipzig (February 1890). 555 pages, with a 6-page foreword. A supplement of four chapters, taken from Lie's unpublished papers, is printed in G.A. 2, pp. 691–791. (Like Volume 1, Volume 2 was also reprinted in New York, 1970.)

Über die Grundlagen der Geometrie. (Erste Abhandlung.) *Leipz. Ber.*, Issue II, pp. 284–321. G.A. 2, pp. 380–413.

Über die Grundlagen der Geometrie. (Zweite Abhandlung.) *Leipz. Ber.*, Issue III, pp. 355–418. G.A. 2, pp. 414–468.

Neuer Beweis des zweiten Fundamentalsatzes in der Theorie der Transformationsgruppen. *Leipz. Ber.*, Issue IV, pp. 453–477. G.A. 6, pp. 267–287.

Bestimmung aller r-gliedrigen transitiven Transformationsgruppen durch ausführbare Operationen. *Leipz. Ber.*, Issue IV, pp. 478–490. G.A. 6, pp. 288–299.

1891 Die linearen homogenen gewöhnlichen Differentialgleichungen. *Leipz. Ber.*, Issue II, pp. 253–270. G.A. 4, pp. 291–306.

Vorlesungen über Differentialgleichungen mit bekannten infinitesimalen Transformationen. Bearbeitet und herausgegeben von Dr. Georg Scheffers. Leipzig (July 1891). 568 pages (with a 14-page foreword).

Die Grundlagen für die Theorie der unendlichen continuirlichen Transformationsgruppen. (Erste Abhandlung.) *Leipz. Ber.*, Issue III, pp. 316–352. G.A. 6, pp. 300–330.

Die Grundlagen für die Theorie der unendlichen continuirlichen Transformationsgruppen. (Zweite Abhandlung.) *Leipz. Ber.*, Issue III, pp. 353–393. G.A. 6, pp. 331–364.

1892 Bemerkungen zu neueren Untersuchungen über die Grundlagen der Geometrie. *Leipz. Ber.*, Issue I, pp. 106–114. G.A. 2, pp. 469–476.

Sur une interpretation nouvelle du théorème d'Abel. *Comptes Rendus*, Vol. 114, pp. 277–280. G.A. 2, pp. 481–483.

Sur une application de la théorie des groupes continus à la théorie des fonctions. *Comptes Rendus*, Vol. 114, pp. 334–337. G.A. 6, pp. 365–367.

Sur les fondements de la Géometrie. *Comptes Rendus*, Vol. 114, pp. 461–463. G.A. 2, pp. 477–479.

Über einige neuere gruppentheoretische Untersuchungen. *Leipz. Ber.*, Issue III, pp. 297–305. G.A. 6, pp. 368–375.

Untersuchungen über Translationsflächen (Abhandlung I). *Leipz. Ber.*, Issue V, pp. 447–472. G.A. 2, pp. 484–506.

Untersuchungen über Translationsflächen (Abhandlung II). *Leipz. Ber.*, Issue VI, pp. 559–579. G.A. 2, pp. 507–525.

1893 Über Differentialgleichungen, die Fundamentalintegrale besitzen. *Leipz. Ber.*, Issue IV, pp. 341–348. G.A. 4, pp. 307–313.

Sur les équations différentielles ordinaires, qui possèdent des systèmes fondamentaux d'intégrales. *Comptes Rendus,* Vol. 116, pp. 1233–1235. G.A. 4, pp. 314–316.

Über die Gruppe der Bewegungen und ihre Differentialinvarianten. *Leipz. Ber.*, Issue IV, pp. 370–378. G.A. 6, pp. 376–383.

Vorlesungen über continuierliche Gruppen mit geometrischen und anderen Anwendungen. Bearbeitet und herausgegeben von Dr. Georg Scheffers. Leipzig (September 1893). 810 pages, with a 14-page foreword and 54 figures.

Theorie der Transformationsgruppen. Dritter und letzter Abschnitt. Unter Mitwirkung von Prof. Dr. Friedrich Engel. Leipzig (September 1893). 831 pages, with a 27-page foreword. 2nd edition (reprinted like Vol. 1 and Vol. 2, New York, 1970.)

1894 Bemerkungen zu Ostwalds Princip des ausgezeichneten Falles. *Leipz. Ber.*, Issue II, pp. 135-137. G.A. 6, pp. 384–385. (Critique of Ostwald's note in *Leipz. Ber.*, 1893, Issue VII, pp. 599–603, and Ostwald's reply is found in *Leipz. Ber.* 1894, Issue II, pp. 276–278.)

Zur Theorie der Transformationsgruppen. *Leipz. Ber.*, Issue III, pp. 322–333. G.A. 6, pp. 386–395.

1895 Untersuchungen über unendliche continuirliche Gruppen. *Leipz. Abh.*, Vol. XXI, Nr. 3, pp. 43–150. G.A. 6, pp. 396–493.

Zur allgemeinen Theorie der partiellen Differentialgleichungen beliebiger Ordnung. *Leipz. Ber.*, Issue I, pp. 53–128. G.A. 4, pp. 320–384.

Bestimmung aller Flächen, die eine continuirliche Schaar von projectiven Transformationen gestatten. *Leipz. Ber.*, Issue II, pp. 209–160. G.A. 6, pp. 494–538.

Verwerthung des Gruppenbegriffes für Differentialgleichungen. I. *Leipz. Ber.*, Issue III, pp. 261–322. G.A. 6, pp. 539–591.

Influence de Galois sur le développement des Mathématiques. I. *Le Centenaire de l'École Normale 1795–1895.* Paris, 1895, pp. 491–489. G.A. 6, pp. 592–600.

Über die aus dem Jahre 1874 herrührende Integrationstheorie eines vollständigen Systems mit bekannten infinitesimalen Transformationen. *Leipz. Ber.*, Issue IV, p. 400. G.A. 6, p. 601.

Beiträge zur allgemeinen Transformationstheorie. *Leipz. Ber.*, Issue V & VI, pp. 494–508. G.A. 6, pp. 602–614.

1896 Die infinitesimalen Berührungstransformationen der Optik. *Leipz. Ber.*, Issue I, pp. 131–133. G.A. 6, pp. 615–617.

Die theorie der Translationsflächen und das ABELsche Theorem. *Leipz. Ber.*, Issue II & III, pp. 141–198. G.A. 2, pp. 526–579.

Geometrie der Berührungstransformationen. Dargestellt von Sophus Lie und Georg Scheffers. Erster Band. Leipzig (April 1896). 694 pages, with 92 figures and a 12-page foreword. No subsequent volumes were forthcoming.

Zur allgemeinen Transformationstheorie. *Leipz. Ber.*, Issue IV, pp. 390–412. G.A. 6, pp. 618–638.

Zur Invariantentheorie der Gruppe der Bewegungen. *Leipz. Ber.*, Issue IV, pp. 466–477. G.A. 6, pp. 639–648.

1897 Das ABELsche Theorem und die Translationsmannigfaltigkeiten. *Leipz. Ber.*, Issue I & II, pp. 181–248. G.A. 2, pp. 580–639.

Die Theorie der Integralinvarianten ist ein Corollar der Theorie der Differentialinvarianten. *Leipz. Ber.*, Issue III, pp. 342–357. G.A. 6, pp. 649–663.

Über Integralinvarianten und ihre Verwerthung für die Theorie der Differentialgleichungen. *Leipz. Ber.*, Issue IV, pp. 369–410. G.A. 6, pp. 664–701.

Liniengeometrie und Berührungstransformationen. *Leipz. Ber.*, Issue V & VI, pp. 687–740. G.A. 2, pp. 640–688.

1898 Zur Geometrie einer MONGEschen Gleichung. *Leipz. Ber.*, Issue I & II, pp. 1–2. G.A. 4, pp. 385–386.

Über Berührungstransformationen und Differentialgleichungen. *Leipz. Ber.*, Issue III & IV, pp. 113–180. G.A. 4, pp. 387–488.

Studies Published from Lie's Papers Following His Death

[NBO, Manuscript Collections]

Shortly after Lie's death, Carl Størmer, Alf Guldlberg and Elling Holst began to go through Lie's papers – an archive of approximately 20,000 folio pages. Størmer published an overview of more than 7,000 folio pages, arranged in 27 packages in "Verzeichniss über den Wissenschaftlichen Nachlass von Sophus Lie. Erste Mitteilung", *Videnskabs-Selskabets Skrifter I. Math.-naturv. Klasse* 1904, No.7. A 'Zweite Mitteilung' with an overview of the remaining packages, numbers 28 to 56, were published by Alf Guldberg in *Videnskabs-Selskabets Skrifter. Math.-naturv. Klasse* 1913, Nr. 7; however, the dating is imprecise, and a more exact dating of the papers is found in Elin Strøm: "Datering av Sophus Lies Nachlass i Universitetsbiblioteket i Oslo" [The Dating of Sophus Lie's Surviving Papers in the University of Oslo Library], *Historisk institutt*, Oslo, 1998.

Størmer and Guldberg published one work from Lie's Nachlass: Über Integralinvarianten und Differentialgleichungen. *Videnskabs-Selskabets Skrifter*, 1902. G.A. 6, pp. 702–749.

Vol. 7 of Lie's *Gesammelte Abhandlungen* (edited by Friedrich Engel before 1940, but not published until 1960, and at that time completed by Ernst Hölder and Viggo Brun) contains the following works from Lie's surviving papers:

Zur Theorie eines Raumes von *n* Dimensionen II, pp. 1–10.
Über partielle Differentialgleichungen zwischen vier Variabelen, pp. 11–25.
Über partielle Gleichungen erster Ordnung mit bekannten infinitesimalen Transformationen, pp. 26–42.
Partielle Differentialgleichungen erster Ordnung mit bekannten infinitesimalen Berührungtransformationen, pp. 43–53.
Über partielle Gleichungen erster Ordnung, pp. 54–70.
Über partielle Gleichungen erster Ordnung zwischen *n* Variablen, pp. 71–88.
Über Differentialgleichungen, welche bekannte infinitesimale Transformationen gestatten, pp. 89–95.
Partielle Differentialgleichungen und Pfaffsches Problem, pp. 89–95.
Zur Invariantentheorie der Berührungstransformationen, pp. 96–100.
Über das Pfaffsche Problem, pp. 107–111.
Partielle Differentialgleichungen und Berührungstransformationen, pp. 112–114.
Neue Integrationsmethode eines beliebigen Pfaffschen Problems, pp. 115–131.
Semilineare und quasilineare Differentialgleichungen 1. Ordnung und Pfaffsche Systeme, pp. 132–156.
Verallgemeinerung der Cauchyschen Integrationstheorie der partiellen Differentialgleichungen erster Ordnung, pp. 157–174.
Geschichtliche Bemerkungen zur allgemeinen Theorie der partiellen Differentialgleichungen erster Ordnung, pp. 175–216.
Zur Geschichte der partiellen Differentialgleichungen erster Ordnung, pp. 217–219.
Über simultane Systeme, die in den unbekannten Funktionen linear und homogen sind, pp. 220–224.
Schar von ∞^3 oder ∞^4 Kurven des Raumes, die eine Gruppe von Punkttransformationen gestattet, pp. 225–230.
Die Transformationsgruppen einer Gleichung: $s - F(x,y,z) = 0$, pp. 231–241.
Die Transformationsgruppen einer Gleichung: $s - F(x,y,z,p,q) = 0$, pp. 242–250.
Über partielle Differentialgleichungen von der Form: $s = F(x,y,z,q)$, pp. 251–256.
Über partielle Differentialgleichungen, pp. 257–261.
Sur les groupes continus infinis et les équations différentielles, pp. 262–269.
Zur Invariantentheorie der unendlichen Gruppen, pp. 270–271.
Gruppentheorie, angewandt auf Geometrie, pp. 272–300.
Über einen Linienkomplex im R_4, pp. 301–310.
Bestimmung der Haupttangentenkurven einer Flächenfamilie, pp. 311–314.
Zur Flächentheorie, pp. 315–322.
Ausdehnung des Meusnierschen Theorems, pp. 323–325.
Neue geometrische Deutung und Verwertung des Abelschen Theorems, pp. 326–360.
Funktionalgleichungen, welche die Abelschen Integrale erster Gattung definieren, pp. 361–385.
Einzelne Aufzeichnungen über Funktionalgleichungen, welche die Abelschen Integrale erster Gattung definieren, pp. 386–402.

Translations-M_3 zweiter Art im R_4, pp. 403–446.
Einzelne Aufzeichnungen zu den Translations-M_3 zweiter Art im R_4, pp. 447–457.
Über die Plückersche Liniengeometrie, pp. 458–466.
Über den Einfluss der Geometrie auf die Entwicklung der Mathematik, pp. 467–476.

Other Printed Material

1. A letter to the Academic Collegium, dated 29.10.1879, published in the Norske Universitets- og Skole-annaler, 3rd Series, Vol. XVII, pp. 8–10, August, 1881.
2. Remarks on the occasion of the publication of Abel's collected works in 1881. *Christ. Forh.*, 1881, from a meeting held 09.12.1881.
3. On Abel [Om ABEL], in *Morgenbladet*, Nr. 188A, 1883.
4. On Mathematical Teaching in our Schools. [Om Mathematikundervisningen i vore Skoler]. *Dagbladet*, 11.12.1884 (Nr. 441) – given as a lecture at the Academy of Sciences, 28. 11.1884, and published in *Christ. Forh.*, 1884, Nr. 16., Christiania, 1885 (8 pages). Was later printed in *Jahrbuch über die Fortschritte der Mathematik*, Vol. XVI, Year 1884, p. 56. Berlin 1887. (See Package 22, Lie's papers.)
5. On Mathematical Teaching in our Schools, II. [Om Mathematikundervisning i vore Skoler. II]. *Dagbladet*, 26/27.02.1885, (Nr. 67 & 68) – given as a lecture at the Academy of Sciences, 13.02.1885 and printed in *Christ. Forh.*, 1885, Nr. 11, Christiania, 1885 (10 pages). (Provoked by teacher, and later cabinet minister, Bonnevie's protest, in *Dagbladet* 09.01.1885 (Nr. 9) against Lie's first lecture (see above entry). (Both of Lie's lectures were published as separate papers, and sold for the price of kr. 0.25 each.)
6. Comments on the occasion of the appearance of G. Scheffers' work entitled "Zur Theorie der aus *n* Haupteinheiten gebildeten complexen Grössen." *Leipz. Ber.*, Year 1889, Issue II, III & IV.
7. Foreword to E. Goursat's "Vorlesungen über die Integration der partiellen Differentialgleichungen 1. O." Leipzig 1893. G.A. 4, pp. 317–319.
8. Comments on the appearance of the work of L. Maurer "Über die lineare homogene Gruppe." *Leipz. Ber.*, Year 1894, Issue II.
9. About the Mathematical Instruction of our Science Teachers. [Om de mathematiske Forelæsninger for vordende Reallærere]. *Morgenbladet* 31.10.1895 (Nr. 607, morning edition).
10. On Secondary Teacher Education in Germany and France. [Om den høiere Læreruddannelse i Tyskland og Frankrige]. *Morgenbladet*, 29.11.1895 and 04.12.1895 (Nr. 666 & 678).
11. Om Uddannelsen af Lærere i Realfagene. *Morgenbladet*, 28.04.1886, 30.04.1886, 10.05.1886, and 13.5.1886 (Nr. 241, 246, 270, 276).
12. Joh. Sverdrup, Steen and A. M. Schweigaard on the Grammar School. [Joh. Sverdrup, Steen og A. M. Schweigaard om Realgymnasiet]. *Dagbladet*, Nr. 110, 1896.
13. Academic Reading Rooms [Akademiske Læseværelser]. *Dagbladet*, 1896, Nr. 143.

14. The Position of the Sciences in the Gymnasia. [Naturfagenes Stilling i Gymnasierne.] *Dagbladet,* 24.5.1896 (Nr. 146).
15. Once again on French Science Teacher Education. [Atter om den franske Reallærer-Uddannelse.] *Morgenbladet,* 2.11.1896 (Nr.700 morning edition).
16. On Abel, Évariste Galois and Ludvig Sylow. [Om Abel, Évariste Galois og Ludvig Sylow.] *Aftenposten* 25.11.1896 (Nr. 811).
17. Foreword (2 pages) to the new edition of Caspar Wessel "Om Directionens analytiske Betegning, et Forsög, anvendt fornemmelig til plane og sphæriske Polygoners Opløsning." *Arch. for Math.*, Vol. 18, Issue I, Kristiania, 1896. G.A. 2, pp. 689–690.
18. On the Regulations Pertaining to University Studies. [Om Universitetsstudiernes Ordning.] *Morgenbladet,* 13.07.1897 (Nr.357).
19. Comments on the publication of E. Study's work entitled "Über Bewegungsinvarianten und elementare Geometrie. I." *Leipz. Ber.*, Year 1896, Issue V & VI.
20. Letter of thanks for the Lobachevsky Prize (dated Leipzig, 01.05.1898). to A. Wassiliev, President of the Physics-Mathematics Society of Kazan. Lie expresses the desire "dass sich die Verhältnisse so entwickeln werden, dass der Lobatschewsky nach und nach als einen geometrischen Preis ohne irgend welche Beschränkung aufgefasst wird." The letter is printed in *Bulletin de la Société physico-mathématique de Kasan,* II. Series, Vol. VIII, Nr. 3. Kazan, 1898.
21. *Dear Ernst. [Kjære Ernst].* 60 letters from Lie to Ernst Motzfeldt. Edited by Marianne Kern and Elin Strøm. History of Science writing series [Videnskapshistorisk skriftserie], nr. 4, Dept. Mathematics, University of Oslo, 1997.
22. A number of accounts (of his studies given out in Norway) published in *Jahrbuch über die Fortschritte der Mathematik* – here, among other things, Lie reports as well on Elling Holst's doctoral work (1882, Vol. XIV, pp. 599–600). *Jahrbuch über die Fortschritte der Mathematik* was established by Carl Ohrtmann, and was published in Berlin. In Leipzig, Leo Königsberger and Gustav Zeuner brought out *Repertorium der literarischen Arbeiten aus dem Gebiete der reinen und angewandten Mathematik,* where on occasion Lie also sent works he had written.

Acknowledgements

Apart from the Department of Mathematics, University of Oslo, The Norwegian Non-Fiction Writers and Translators Union (NFF), The Johan and Mimi Wesmann Support Fund, and The Nansen Fund – there are several I would like to thank for their help and inspiration during the work of putting this book together.

The desire to have a more comprehensive biography of Sophus Lie was seriously advanced in connection to the commemoration of the 150th anniversary of Lie's birth, in 1992. The Department of Mathematics consequently arranged "Sophus Lie Memorial Week" between August 17–23, 1992. The initiator and leader of the organising committee was Professor Olaf Arnfinn Laudal. The programme of events was divided into three parts: an exhibition on Lie's life and mathematics; a symposium on Lie Theory; and the unveiling, at Nordfjordeid, of a bust of Lie (see p. 494). The proceedings of the symposium were printed as *Proceedings. Sophus Lie Memorial Conference*, edited by Professors O. A. Laudal and Bjørn Jahren (Oslo 1994). The exhibition was made by historian Elin Strøm, who was also responsible for the biographical portion of the exhibition (the text of which also appears in the *Proceedings*), while Professor Eldar Straume was responsible for the mathematical portion of the exhibition. Beyond this, Elin Strøm gathered together comprehensive source material on Sophus Lie, and she has written several rich accounts of parts of Lie's life and work (see List of Literature). Elin Strøm has also developed a "Sophus Lie archive" with a chronological overview of his scientific work and his correspondence. Throughout my work on this project I have had access to all her material, and I thank Elin Strøm for having put this at my disposition – and for the discussions we have had *en route*. And thanks to Strøm, Herman Lie's papers, among other things, were accessible to me (see Unpublished Sources, Nasjonal biblioteket, avd. Oslo) – thanks as well, to Ole B. Styri and Eva Styri for having taken care of this material since Herman Lie's death in 1960.

Jan Erlvang, administrative leader of the Department of Mathematics for over a generation, was the main force behind me, at New Year's, 1997, obtaining office space at the department and beginning work on the book. I thank him warmly both for having supported the project the whole time, and for having found the economic means for carrying out the work. I also want to thank the two principals, Arne Bang Huseby and Arne B. Sletsjøe and the two office managers, Harald Bjar and Yngvar Reichelt of the Department who whole-heartedly supported the project throughout. In the university environment it is extremely seldom that an institute or department invites an outside writer into its internal circles – and it had been an unqualified delight for me to be in this milieu. Thanks – both for the direct help, and for the mathematical understanding which to some extent I have taken in "by osmosis" and which has given inspiration to the work – to Professors Geir Ellingsrud, Tom Lindstrøm, John Rognes, Bjørn Jahren, Ragni Piene, Kristian Ranestad, Erling Strømer, Per Holm, Per Tomter, and Ola Bratteli. Thanks also to Bent Birkeland, Nils Voje Johansen and Torger Holtsmark for their interest and comments. I am indebted to Bernd Fritzsche in Leipzig and Peter Ullrich in Giessen for useful information in the course of things – and also for the willing assistance of Hermut Mettlach at the German Embassy in Oslo, to Dr. W. Becker at Klinikum Wahrendorff in Sehnde, to Ragna Hölder (Sophus Lie's grandchild) in Mainz, and to Professor F. Hirzebruch at the Max-Planck Institute in Bonn.

Special thanks to Professor Laudal who has been the professional support and inspiration to the project – his ongoing reading of the manuscript, his comments and clarifications have been an indispensable sounding board for the work. (However, any lack of professional clarity that remains in the book is the fault only of the author.)

Thanks to Jon Haarberg for the indefatigable reading of the manuscript and his redeeming comments. Many thanks as well to both Professor Laudal and Dr. Henry Crapo (École des Hautes Études des Sciences Sociales) for their suggestions concerning mathematical terminology used in the English language edition. Thanks as well to Librarian Oddvar Vasstveit and Librarian Sigbjørn Grindheim at the National Library's Manuscript Collections; they have even gone so far as to find unknown material for my work on the book. And thanks to Eva Mandt for her ability to interpret difficult handwriting. Thanks to Professor Stein Evensen for the medical-historical consultation, and to Professor Inger Nordal for the enthusiastic feedback. I am grateful to the Mathematical Library's Tina Mannai and Leyla Rezaye-Golkar for all their willing assistance to find and to order the most obscure books and offprints.

And thanks to Harald Engelstad of Aschehoug Publishers for having been aglow with the project throughout, and for his firm grasp at the concluding stage of the manuscript.

List of Literature

Unpublished Source Materials

• *Nasjonalbiblioteket avd. Oslo (NBO), Håndskriftsamlingen*
[National Library, Oslo Division (NBO), Manuscript Collections]:

Letter Collection 289, which contains letters from Lie to Klein during the period 1870–76, together with 425 letters from foreign mathematicians, written to Sophus Lie.

Letter Collection 52, which contains letters from Sophus Lie to Mayer, Engel, Steen, and Zeuthen, among others.

Letter Collection 1 (from the Church and Education Department [Kirke og Undervisn. Dep.], as well as from Teubner, Grøndahl, Cremona).

Letter Collection 7 (Sylow's collection).

Letter Collection 8a (from O. J. Broch).

Letter Collection 12 (from Weierstrass).

Letter Collection 86 (to Storm).

Letter Collection 234 (Holst's letter collection that contains letters from Lie, Bjørnson, Thue, Alexander, Birkeland, and others.

Letter Collection 298 (Brøgger's collection).

Letter Collection 469 (C. A. Bjerknes' collection).

Letter Collection 704 (Helland's collection).

Bjørnstjerne Bjørnson's letter collection.

Ms. 8° nr. 3839, that contains the Sophus Lie-Nachlass. (Unpublished works, drafts of letters, and letters from diverse foreign and Norwegian colleagues and friends. (See also, Chronological Bibliography.)

Ms. 8° 574: Proceedings of Realisternes og Mineralogernes Forening [Association of Scientists and Mineralogists].

Ms. 4° 3239: Realistforeningen's minutes and reports.

Approximately 190 letters from Sophus Lie to his spouse, Anna Lie, together with letters from Sophus Lie to his children (plus various other materials) are currently in a raw, unorganised condition in this collection.

Personal papers left behind by Herman Lie – containing letters between Herman og his sisters, between Mrs. Lie and the children. Photo album belonging to Sophus Lie, together with diverse notebooks from his student days. An exchange of letters between Nils B. Maurseth and Adler Vogt when Vogt wrote his book about the Lie family. This material is not organised, recorded or registered.

Olaf Skavlan's papers, in Ms. 8°. 617: A: 3.
Axel Thue's lecture notes, in Ms 8° 3052 og Ms. 8° 391.
Student Society Files.
University Library lending records, 1859–69.

• *State Archives [Riksarkivet (RA)], Oslo:*

Records of proceedings and copy books for the Collegium, University of Oslo (UiO), 1860–1900, together with journals and reports concerning the UiO. [Catalogue designations: UiO, kollegiet, journalsaker; and: UiO, kollegiet, innberetninger, Box 1–8]. UiO, Matrikler, 65. Oppgaver, Ex. art. 1859.
Church and Education Department Archives [Kirke- og undervisningsdepartementes arkiv]: KUD, skolekontor D (with journal nr.)
The Motzfeldt family archive (private archive 234), which in addition to Lie's letters to Ernst Motzfeldt [Strøm 1997], contains E. Motzfeldt's diaries from the mountain tours together with Sophus Lie.
Records of the Faculty of Mathematics and Natural Sciences, 1860–1900.
Reports on the Post of the Vicar of Moss parish (RA 567).
Ecclesiastical rolls for Eid (RA 135).

• *Nordfjordeid Municipal Archives:*

Registry list and minutes of meetings, 1838–94.

• *Aker sykehus, Psychological Division:*

Journal concerning Gottfried J. S. Birch.

• *Oslo City Archives:*

Registration of Death for Sophus Lie (signed by Axel Lund).

• *Mittag-Leffler Institute, Djursholm:*

25 letters from Lie to Mittag-Leffler, together with letters to Mittag-Leffler from Holst, Sylow, and C. A. Bjerknes.
Four letters and one telegramme from Anna Lie to Mittag-Leffler and drafts of his response.

• *Niedersächsische Staats- und Universitätsbibliothek, Göttingen:*

Klein-Nachlass X and XXII: Letters from Sophus Lie to Klein. (ca. 120), 1876–1899, together with letters from Holst to Klein.

• *Sächsische Akademis arkiv – University Library, Leipzig:*

Documents pertaining to the Matematical Seminar and the Matematical Institute; minutes of meetings of the Sächsischen Akademie, where Lie was a zealous giver of lectures; Mayer's Nachlass (which also includes three letters from Lie to Mayer).

• *Universitätsarchiv Leipzig:*

Personal documents. Faculty archives. Collegium archives. Faculty meeting records. Catalogue of lectures. Annual reports [Jahresberichte]. Letters from Engel, Klein, Hausdorff, Kowalewski, Scheffers, Mayer, and others.

• *Zentrales Staatsarchiv Dresden:*

Ministerium für Volksbildung, Nr. 10281 /212. (Lie's letter to the Minister of Culture, Paul von Seydenwitz, 22.5.1898, together with letters from Mayer, Scheibner, and Neumann in relation to Lie's departure from Leipzig.)

• *Bibliothek des Mathematischen Instituts, Giessen:*

Friedrich Engel's surviving papers – including among other things, an overview of the number of students attending Lie's lectures in Leipzig, diverse letters to Herman Lie and friends in Norway. The papers are being indexed by Peter Ullrich.

Reference Works

Dictionary of Scientific Biography, 14 vols. New York, 1970–76.
Norsk biografisk leksikon, 16 vols. Oslo, 1923–69.
Store norske leksikon. Aschehoug & Gyldendal, Oslo, 1978–81.
Norsk biografisk leksikon, vol. 1. Oslo, 1999.
Salmonsens Konservationsleksikon, 22 vols. Copenhagen, 1915–28.
Norsk Forfatter-Lexikon 1814–1860, J. B. Halvorsen, 6 vols. Kristiania, 1885–1908.
Matematikk-Leksikon, Oslo, 1997.
Encyclopaedia Britannica. Biography.
www-history.mcs.st-andrews.ac.uk/history/BiogIndex.htl

Biographical Materials on Lie

Alfsen, Magnus. Sophus Lie. In *Vore Høvdinger,* Halvdan Koht, ed., Trondhjem, 1929, pp. 162–168.
Arneson, Johannes. *100 år og minnest mangt.* Oslo, 1964.
Baas, Nils A. Sophus Lie. *Det Kgl. Norske Videnskabers Selskabs Forehandlinger,* 1992, pp. 43–48.
Baas, Nils A. Sophus Lie. *The Mathematical Intelligencer,* Vol.16, 1994, pp. 16–19.
Bjørnstjerne Bjørnson og Christen Collins brevveksling 1889–1909. Oslo, 1937.
Brun, Viggo. Universitetsprofiler: Sophus Lie. (Unpubl. Manus., 1961).
Caspari, Theodor. *Fra mine unge Aar.* Oslo, 1929.
Caspari: Theodor. *Fra Friluftsliv på Vinsterflyen.* Oslo, 1945.
Engel, Friedrich. Sophus Lie. *Norsk matematisk tidsskrift.* 1922, pp. 97–114.
Engel, Friedrich. Sophus Lie. Berichte über die Verhandlungen der Königlich-Sächsischen Gesellschaft der Wissenschaften zu Leipzig. Math.-Phys. Klasse, 51. 1899, pp. XI–LXI.

Engel, Friedrich. Sophus Lie. Ausfürliches verzeichnis seiner Schriften. *Bibliotheca Mathematica.* Gustaf Eneström, ed. 41 pages, Leipzig, 1900.

Engel, Friedrich. Jahresberichte der deutschen Mathematiker-Vereinigung. Vol. VIII, Leipzig, 1900.

Engel, Friedrich. Zur Auseinandersetzung mit Felix Klein. (Actually written as the foreword to Volume 7 of Lie's *Gesammelte Abhandlungen,* but is now to be found only in manuscript form.)

Daae, Ludvig. *Politiske dagbøker og minner.* Publication of Den historiske Forening. Oslo, 1971.

Domer, Yngve. On the foundation of Acta Mathematica. *Acta Mathematica.* Vol. 148, 1982.

Freudenthal, H. Marius Sophus Lie. Dictionary of Scientific Biography 8. 1973, pp. 323–327.

Fritzsche. Bernd. Biographische Anmerkungen zu den Beziehungen zwischen Sophus Lie, Friedrich Engel und Edvard Study. I. G. Czichowski, B. Fritzsche (eds.) S. Lie, E. Study, F. Engel, Beiträge zur Theorie der Differentialinvarianten. Teubner-Archiv zur Mathematik, 17. Teubner-Verlag, Stuttg. – Leipzig 1993, pp. 176–217.

Fritzsche. Bernd. Sophus Lie. En skisse av hans liv og verk. Symmetri, Trondheim 1995, pp. 7–20 (Translated from the German by Torger Holtsmark.)

Fritzsche. Bernd. Sophus Lie. A Sketch of his Life and Work. Journal of Lie Theory. Vol. 9, pp. 1–38.

Fritzsche. Bernd. Einige Anmerkungen zu Sophus Lies Krankheit. Historia Mathematica 18. 1991, pp. 247–252.

Grieg, Edvard. *Brev i utvalg 1862–1907.* Vols. I–II. Finn Benestad (ed. and commentary). Oslo 1998.

Günther, P. Sophus Lie. 100 Jahre Mathematisches Seminar der Karl-Marx-Universität Leipzig. Berlin 1981. pp. 111–133.

Hawkins, Thomas. The geometrical origin of Lie's theory of groups. *Symmetri,* Trondheim 1995, pp. 21-29.

Hawkins, Thomas. Line Geometry, Differential Equations and the Birth of Lie's Theory of Groups. *The History of Modern Mathematics.* New York 1989, pp. 275–327.

Hawkins, Thomas. Jacobi and the Birth of Lie's Theory of Groups. *Archive for History of Exact Sciences* 42. 1991, pp. 187–278.

Hawkins, Thomas. The Birth of Lie's Theory of Groups. *The Mathematical Intelligencer* 16, Nr. 2. 1994, pp. 6–17.

Hawkins, Thomas. *Emergence of the Theory of Lie Groups.* New York, 2000.

Heegaard, Poul. Marius Sophus Lie. *Norsk Biografisk Leksikon,* Vol. 8, Oslo, 1938, pp. 353–357.

Helgason, Sigurdur. Sophus Lie, the Mathematician. In *Proceedings, Sophus Lie Memorial Conference. Oslo 1992.* O. A. Laudal and B. Jahren, eds. Oslo, 1994, pp. 3–21.

Helland, Amund. Professor Dr. Sophus Lie. *Ny illustreret Tidende,* 11. 04.1886.

Holst, Elling. Sophus Lie. *Nyt Tidsskrift* 1893, pp. 127–142.

Holst, Elling. Sophus Lie. *Folkebladet* 30.04.1894.
Holst, Elling. Mathematik. *I Illustreret norsk Litteraturhistorie,* Henrik Jæger, ed., in Vol. 1. "Videnskabernes Literatur i Det nittende Aarhundrede", Kristiania, 1896, pp. 68-101
Holst, Elling. Speech to the University's Memorial Ceremony for Sophus Lie. Kristiania, 1899.
Holst, Elling. Træk af Sophus Lies ungdomsliv. *Ringeren.* Kristiania, 1899, pp. 98–102.
Holst, Elling. Tale ved Universitetets Mindefest 20. 4.1899 [Speech to the University's Memorial Ceremony for Sophus Lie 20.04.1899]. (Printed in *Aftenposten.*)
Holst, Elling. Matematikken. In *Det Kongelige Frederiks universitet 1811–1911.* Festschrift, Vol. 2, Kristiania, 1911, pp. 425–472.
Holst, Elling. Sophus Lie. In *Normænd i det 19de Aarhundrede.* Compiled and edited by Gerhard Gran. Kristiania, 1914, pp. 99–143.
Johansson, Ingebrigt. Sophus Lie. *Norsk matematisk tidsskrift.* 1942, pp. 97–106
Killing, Wilhelm. *Briefwechsel mit Friedrich Engel.* Dokumente zur Geschichte der Mathematik. Vol. 9. Edited by Wolfgang Hein. DMV. Braunschweig/Wiesbaden, 1997. (Reviewed by Bernd Fritszche in *Jahresbericht der DMV,* 1999.)
Klein, Felix. *Korrespondenz Felix Klein – Adolph Mayer.* Leipzig, 1990.
Klein, Felix. Zur Ersten Verteilung des Lobatchewsky-Preises. (3.11.1897.) Kazan, 1897.
Kowalewski, Gerhard. *Bestand und Wandel.* München, 1950.
Leipziger Adress-Buch für 1887–1898. Leipzig, 1887–1898.
Lie, Herman. Speech on the occasion of the 100th anniversary of the birth of Sophus Lie, 17.12.1942. Unpublished manuscript.
Lie, Sophus. *Gesammelte Abhandlungen.* Edited by E.Engel and P. Heegaard. 7 Vols. Leipzig–Oslo, 1922–1960.
Lie, Sophus. *Theorie der Transformationsgruppen.* 3 Vols. Leipzig, 1888–1893.
Lie, Sophus. *Kjære Ernst.* 60 letters from Lie to Ernst Motzfeldt. Edited by Marianne Kern and Elin Strøm. Vitenskapshistorisk skriftserie, Nr.4. Mat. inst., University of Oslo, 1997.
Mittag-Leffler, Gösta. Sophus Lie. (Obituary.) *Acta Matematica,* 1899.
Noether, Max. Sophus Lie. *Mathematische Annalen,* 53, 1900, pp. 1–41.
Ostwald, Wilhelm. *Lebenslinien. I–II. Eine Selbstbiographie.* Berlin, 1927.
Proceedings, Sophus Lie Memorial Conference. Oslo 1992. O. A. Laudal and B. Jahren, eds., Oslo, 1994.
Personal-Verzeichnisse der Universität Leipzig. Leipzig.
Purkert, Walter. Zum Verhältnis von Sophus Lie und Friedrich Engel. *Wissenschaftl. Zeitschr. d. Ernst-Moritz-Arndt-Universität Greifswald. Math.-nat.wissensch. Reihe,* 33. 1984, pp. 29–34.
Rowe, David. The Early Geometrical Works of Felix Klein and Sophus Lie. *The History of Modern Mathematics.* New York, 1989, pp. 209–273.
Rowe, David. Der Briefwechsel Sophus Lie – Felix Klein, eine Einsicht in ihre persönlichen und wissenschaftlichen Beziehungen. *NTM-Schriftenr. Gesch. Naturwiss. Leipzig,* 1988, pp. 37–47.

Straume, Eldar. Sophus Lie - eit tversnitt av hans liv og arbeid. *Normat,* 1984.
Straume, Eldar. Lies kontinuerlige og infinitesimale grupper. *Normat,* 1992, pp. 160–170.
Straume, Eldar. Sophus Lies og differensialligninger. *Normat,* 1992, pp. 171–179.
Straume, Eldar: Sophus Lie i historisk perspektiv. University of Tromsø, 1983.
Straume, Eldar: Sophus Lies matematiske idear. University of Tromsø, 1992.
Straume, Eldar: Sophus Lie (1842–1899). *Newsletter of the European Mathematics Society,* Nr. 3, March 1992.
Strøm, Elin. Marius Sophus Lie. In *Proceedings, Sophus Lie Memorial Conference. Oslo, 1992.* O. A. Laudal and B. Jahren, eds., Oslo, 1994.
Strøm, Elin. *Studenten, 1859–1865.* Vitenskapshistorisk skriftserie, Nr. 2, Mat. inst., University of Oslo, 1997.
Strøm, Elin. *Kandidaten, 1865–1869.* Vitenskapshistorisk skriftserie, Nr. 3, Mat. inst., University of Oslo, 1997.
Srøm, Elin. Datering av Sophus Lies Nachlass i Universitetsbiblioteket i Oslo. Hist. inst, University of Oslo, 1998.
Strøm, Elin. *Sophus Lie og oppgjøret med Felix Klein.* Vitenskapshistorisk skriftserie, Nr. 5, Mat. inst., University of Oslo, 2001.
Strøm, Elin. Da Stortinget "ansatte" professor i matematikk. Unpublished manuscript. 1997.
Study, Edvard. Theorie der Transformationsgruppen I (Rezension). *Zeitschrift für Mathematik und Physik* 34, 1889, pp. 171–191.
Study, Edvard. Über Lies Geometrie der Kreise und Kugeln. *Mathematische Annalen* 91, 1924, pp. 82–118.
Study, Edvard. *Die Realistische Weltansicht und die Lehre vom Raume.* Braunschweig, 1914.
Svare, Helge. Sophus Lie. *Normat,* 1992, pp. 148–159.
Sylow, Ludvig. Sophus Lie. *En Mindetale.* Kristiania, 1900.
Sylow, Ludvig. Sophus Lie med Fortegnelse over hans vitenskabelige Arbeider. *Archiv for Mathematik og Naturvidenskab,* 1899, pp. V–XXIII.
Sylow, Ludvig. *Meddelelser af Sophus Lie til Videnskabsselskabet fra Aarene 1869–1871.* Videnskabsselskabets Skrifter. Math.-naturvit. Klasse. No.9, 1899.
Thalberg, Knut. To leilighetsviser angående Sophus Lie. Addendum to *Norsk Matematisk Tidsskrift* 4/1947, pp. 1–6.
Tobies, Renate and David R. Rowe (eds. and commentary). *Korrespondenz Felix Klein - Adoph Mayer.* Leipzig, 1990.
Vogt, Adler. *Slegten Lie.* (76 pages). Drammen, 1914.
Vogt, Adler. *Slegten Vogt i tre generasjoner.* Oslo, 1951.
Vogt, Johan. Herman Lie. Sophus Lie. (Manuscript of the lecture held at NTH, Trondheim, 1930).
Yaglom, Isaak M. *Felix Klein and Sophus Lie: Evolution of the idea of symmetry in the nineteenth century.* Boston, 1998.

General Literature

Aaland, Jacob. *Nordfjord fraa gamle dagar til no.* Eid, 1909.

Aas, Einar. *Kristiansands Katedralskoles historie.* Oslo, 1932.

Akivis, M. A. and B. A. Rosenfeld. *Élie Cartan (1869–1951).* American Mathematical Society. Vol. 123, 1993.

Amdan, Per. *Molde høgre Skole 150 år.* 1982

Amundsen, Leiv. *Det norske vitenskapsakademi i Oslo 1857–1957.* 2 Vols. Oslo, 1957–60.

Andresen, Anton Fredrik. *Opplysningsidéer, nyhumanisme og nasjonalisme i Norge i de første årene etter 1814.* KULT's Skriftserie Nr. 26. Oslo, 1994.

Arneson, Johannes. *100 år og minnest mangt.* Oslo, 1964.

Bedeutende Gelehrte in Leipzig. 2 Vols. Leipzig, 1965.

Biermann, K. R. *Die Mathematik und ihre Dozenten an der Berliner Universität 1810–1933.* Berlin, 1988.

Birkeland, Bent. *Norske matematikere. Litt om deres liv og virke.* Oslo, 1993.

Birkeland, Bent. Ludvig Sylow's Lectures on Algebraic Equations and Substitutions, Christiania, 1862. *Historia Mathematica* 23, 1996, pp. 182–199.

Bomann-Larsen, Tor. *Den evige sne.* Oslo, 1993.

Boyesen, Einar. *Hartvig Nissen 1815–1874 og det norske skolevesens reform.* 2 Vols. Oslo, 1947.

Broch, Hjalmar. *Zoologiens historie i Norge til den annen verdenskrig.* Oslo, 1954.

Budden, F. J. *The Fascination of Groups.* Cambridge, 1972.

Cartan, Élie – Albert Einstein. *Lettres sur le Parallélisme Absolu 1929–1932.* Palais des Académies. Bruxelles, 1979.

Cartan, Élie. *Notice sur les travaux scientifiques.* Paris, 1974

Christensen, Carl. *Livserindringer.* Kristiania, 1920.

Christophersen, H. O. *Marcus Jacob Monrad. Et blad av norsk dannelses historie i det 19. århundre.* Oslo, 1959.

Collett, John Peter. *Historien om Universitetet i Oslo.* Oslo, 1999.

Comptes Rendus du Congrés International des Mathématiciens, Oslo 1936. Oslo, 1937.

Dahl, Helge. *Knud Knudsen og latinskolen.* Oslo, 1962.

Dahl, Helge. *Lærerutdanningen ved Universitetet i Oslo fra 1814 til i dag.* Oslo, 1964.

Dahl, Kristin. *Den fantastiska matematikken.* Stockholm, 1991.

Det Kongelige Fredriks universitet 1811–1911. Festschrift in 2 Vols. Kristiania, 1911.

Dietrichson, Lorentz. *Svundne Tider I.* Christiania, 1913.

Domar, Yngve. Matematisk forskning under Stockholms högskolas första decennier. (Lecture, 29.09.1978, on the occasion of the 100th anniversary of Stockholm University).

Daae, Ludvig. *Politiske Dagbøker.* Tillegg til Historisk tidsskrift, 1930.

Edvard Grieg: Brev i utvalg 1862–1907. Vol. 1., Finn Benestad, ed., Oslo, 1998.

Egeland, Alf. *Kristian Birkeland. Mennesket og forskeren.* Oslo, 1994.

Eichhorn, E. and E.-J. Thiele. *Vorlesungen zum Gedanken an Felix Hausdorff.* Berliner Studienreihe zur Mathematik. Vol. 5, Berlin, 1994.

Eid Kyrkje 150 år. Eid sokneråd. Nordfjordeid, 1999.

Engel, Friedrich. Wilhelm Killing. *Jahresbericht DMV,* 1930.

Engel, Friedrich. Eduard Study. *Jahresbericht DMV,* 1931.

Evensen, Stein, Per Stavem og Lorentz Brinch. *Blodsykdommer.* Oslo, 1993.

Eves, Howard. *Great Moments in Mathematics.* Washington, 1983.

Ewald, William. *From Kant to Hilbert. I–II.* Oxford, 1996.

Farestveit, Åsmund. *Hvordan norske matematikere reagerte på den ikke-euklidske geometrien.* Graduating thesis, Høyskolen i Agder, Kristiansand, 1998.

Frei, Günther, ed. *Der Briefwechsel David Hilbert – Felix Klein.* Göttingen, 1985.

Forhandlinger ved De Skandinaviske Naturforskeres tiende Møde i Christiania. Christiania, 1869.

Fløgstad, Kjartan. *Eld og vatn.* Oslo, 1999.

Fosli, Halvor. *Kristianiabohemen.* Oslo, 1994.

Fulsås, Narve. *Historie og nasjon.* Oslo, 1999.

Furre, Berge. *Soga om Lars Oftedal.* Oslo, 1990.

Gran, Gerhard. To stadier i Universitetets liv. *Samtiden,* 1908, pp. 231–253.

Grey, Jeremy. *Ideas of space.* Oxford, 1979.

Gravir, Olav. *Kyrkje og Folk i Mo i 100 år.* Skien, 1939.

Gårding, Lars. *Matematik och Matematiker.* Lund, 1994.

Hagemann, Gro. *Katalog over vitenskapshistoriske arkiver.* Oslo, 1982.

Haraldsen, Haakon. Cato Maximilian Guldberg and Peter Waage. *The Law of Mass Action. A Centenary Volume 1864–1964.* Oslo, 1964.

Hansen, G. Armauer. *Livserindriger og betragtninger.* Kristiania, 1910.

Hansen, Hans Magne. *Sigurd Lie.* Graduating thesis, UiO, 1961.

Haugland, Karen Mathilde, C. A. Bjerknes. *Norsk Biografisk Leksikon.* Oslo, 1999.

Hawkins, Thomas. The Erlangen Program of Felix Klein. Reflections on its Place in the History of Mathematics. *Historia Mathematica* 1984, pp. 442–489.

Hawkins, Thomas. Non-Euclidian Geometry and Weierstrassian Mathematics. In *Studies in History of Mathematics.* 26 Vols. Esther R. Phillips, ed. New York, 1987.

Heegaard, Paul. Felix Klein. *Norsk Matematisk Tidsskrift,* 1926, pp. 45–51.

Helgerud, Randi. Industriutstillingen i Drammen. *Rundtom,* Nr. 2, 1998.

Helland, Amund. *Norges Land og Folk. Topograisk-statistisk Beskrivelse over Kristiania I–III.* Kristiania, 1917.

Hem, Erlend and Per E. Bördahl. Det første keisersnitt i Norge. *Tidsskrift for Den norske lægeforening 1998,* pp. 4648–4653.

Herresthal, H. and Ladislav Reznicek. *Rhapsodi Norvégienne.* Oslo, 1994.

Hestmark, Geir. *Fridjof Nansen og arktisk geologi.* Det norske Videnskaps-akademis årbok 1992.

Hestmark, Geir. *Vitenskap og nasjon.* Waldemar Chr. Brøgger 1851–1905. Oslo, 1999

Hirzebruch, Friedrich. Centennial of German Mathematical Society. *Miscellanea Mathematica,* 1991, pp. 177 194.

Hjeltnes, Guri, ed. *Vitenskapelige profiler på 1800-tallet.* Forum for universitetshistorie skriftserie 1/1997. Oslo, 1997.

Hoffmann, Karl H. Zur Geschichte des Halbgruppenbegriffs. *Historia Mathematica* 19, 1992, pp. 40–59.

Holmboe, Thorolf. *Fredrik og Mathilde Vogts etterslekt.* Oslo, 1956.
Høigaard, Einar. *Den norske skoles historie: en oversikt.* Oslo, 1947.
Ibragimov, Nail H. Sophus Lie and Harmony in Mathematical Physics. *The Mathematical Intelligencer.* Vol. 16, Nr. 1, pp. 20–28.
Jansen, Jens Jonas. *Oplevet og tænkt.* Kristiania, Aschehoug, 1909.
Janson, Kristofer. *Hvad jeg har oplevet.* Kristiania and København, 1913.
Johannessen, Viggo. *Realistforeningen* 1859–1959. Oslo, 1959.
Keel, Aldo. *Bjørnstjerne Bjørnson.* Oslo, 1999.
Klein, Felix. *Vorlesungen über die Entwicklung der Mathematik im 19. Jahrhundert.* Berlin, 1926
Klein, Felix. *Gesammelte mathematische Abhandlungen.* 3 Vols. Eds. & commentary by R. Fricke and Ostrowski. Berlin, 1921–23.
Klein, Felix. *Das Erlangen Program: "Vergleichende Betrachtungen über neuere geometrische Forschungen".* (In Ostwald's Klassiker der exakten Wissenschaften, 253.) Leipzig, 1974.
Koenigsberger, Leo. *Hermann von Helmholtz.* New York, 1965 (1st edition, Oxford, 1906).
Kragemo, Helge Bergh: Ludvig Sylow. *Norsk Matematisk Tidsskrift,* 1933, pp. 73–123.
Kristianias historie, I-V. Kristiania, 1923–28.
Kronen, Torleiv. *Ut over grensene: norske vitenskapsmenn i Frankrike 1150–1940.* Oslo, 1985.
Lampe, Johan Fredrik. *Bergens Stifts Biskoper og Præster, I and II.* Kristiania, 1895, 1896.
Laache, S. *Die Anämie.* Universitesprogram. Christiania, 1883.
Leipziger mathematische Antrittsvorlesungen. Auswahl aus den Jahren 1869–1922. Leipzig, 1987.
Lindstrøm, Tom. *Kalkulus.* Oslo, 1995.
Lützen, Jesper. The Mathematical Correspondence between Julius Petersen and Ludvig Sylow. *Amphora.* Festschrift for Hans Wussing. Basel, 1992, pp. 439–467.
Løvland, Jørgen and Lars Knutson Liestøl. *Syn og Segn,* 1913, pp. 1–9.
Laache, S. *Die Anämie.* Universitetsprogram, Christiania, 1883.
Marthinsen, G. Wilhelm Ostwald. *Nature* Nr. 5, 1954, pp. 144–159.
Monrad, Marcus Jacob. *Tankeretninger i den nyere Tid. Et kritisk Rundskue.* Christiania, 1874.
Myhre, Jan Eivind. Hovedstaden Christiania. *Oslo bys historie.* 3 Vols., Oslo, 1990.
Nerbøvik, Jostein. *Norsk historie 1870–1905.* Oslo, 1986.
Nielsen, Yngvar. *En Christianiensers Erindringer fra 1850- og 60 Aarene.* Kristiania, 1910.
Niemann, Walter. *Die Musik Skandinaviens.* Leipzig, 1906.
Nissen, Hartvig. *Afhandlinger vedrørende det høiere og lavere Skolevæsen.* Christiania, 1876.
Nordgaard, Ole. *Michael og Ossian Sars.* Kristiania, 1918.
Norske Studenter. Assembled & edited by P. Botten Hansen, Christiania, 1893.
Norske Universitets- og Skole-Annaler, 1870–1900.

Nørstebø, Sigurd. Frå Soga om andreeksamen ved Universitetet i Oslo. *Det Kgl. Norske Videnskabs Selskabs Skrifter* Nr. 5. Trondheim, 1962.

Olsen, Richard. *Beretning om den høiere kommunale Skole i Moss, 1810–1885,* in Inbydelsesskrift til den offentlige examen i juni og juli 1886 ved Moss kommunale middelskole [Invitation to the public examination of June and July 1886 at Moss Municipality Middle School]. Kristiania, 1886.

Onsager, Nils. *Av Andvakes eldste historie.* Oslo, 1939.

Ording, Fr. *Det lærde Holland.* Printed for Den Norske Historiske Forening, Oslo, 1927.

Os, Edvard. *Eid og Hornindal.* Published by Jacob Aaland. Oslo, 1953.

Økland, Fridthjof. *Michael Sars. Et minneskrift.* Oslo, 1925.

Østvedt, Einar. *Dyre Vaa.* Oslo, 1962.

Øverås, Asbjørn, A. E. Erichsen og Johan Due. *Trondheim Katedralskoles historie.* Trondheim, 1952.

Piene, Kay. Matematikkens stilling i den høiere skole i Norge efetr 1800. *Norsk Matematisk Tidsskrift,* 1937, pp. 52–68 and 1938, pp. 33–58.

Platou, Fredrik. Universitetets gymnastikksal. *St. Hallvard* 4/1957, pp. 184–191.

Poincaré, Henri. *La science et l'hypothése.* Paris, 1902. (*Vetenskapen och Hypoteserna.* Stockholm, 1911.)

Poincaré, Henri. *Mathematics and Science: Last Essays.* New York, 1963.

Richards, Joan L. *Mathematical Visions.* San Diego, 1988.

Riddervold, Astri. *Drikkeskikker.* Oslo, 1997.

Ringdal, Nils Johan. *Moss bys historie.* Moss, 1989

Rowe, David E. Felix Kleins Erlangen Antrittsrede. *Historia Mathematica,* 12, 1984, pp. 123–141.

Rowe, David E. Hilbert and the Göttingen Mathemaical Tradition. OSIRIS, 2nd series, New York, 1989, pp. 186–213.

Rudeng, Erik. *William Nygaard.* 1865–1952. Oslo, 1992.

Røynstrand, Åse. Då Ludvig M. Lindemann skreiv opp Ola Glomstulen på Mo Prestegard. *Tokke Historielag Årsskrift,* 1995.

Sars, Ernst. *Samlede værker,* 4 Vols., Kristiania, 1912.

Scriba, C. J. Friedrich Engel (1861–1941), Mathematiker. *Giessener Gelehrte in der ersten Hälfte des 20. Jahrhunderts.* Marburg, 1982, pp. 212-223.

Seip, Anne-Lise. *Forspill til parlamentarisme.* Oslo, 1968.

Seip, Anne Lise. Nasjonen bygges 1830–1870. *Norges Historie.* 8 Vols. Oslo, 1997.

Seip, Jens Arup. *Et regime foran undergangen.* Oslo, 1945.

Seip, Jens Arup. *Ole Jacob Broch og hans samtid.* Oslo, 1971.

Seip, Åsmund Arup. *Lektorene. Professjon, organisasjon og politikk 1890–1980.* FAFO-Rapport Nr. 108. Oslo, 1990.

Skolem, Thoralf Albert. Ludvig Sylow og hans videnskabelige arbeider. *Norsk Matematisk Tidsskrift,* 1919, pp. 1–13.

Slagstad, Rune. *De nasjonale strateger.* Oslo, 1998.

Smith, David E. and Jekutheil Ginsburg. *A History of Mathematics in America before 1900.* Chicago, 1934.

Sogeskrift for Eid, årgang 1994, 1997.

Stenseth, Bodil. *En norsk elite. Nasjonsbyggerne på Lysaker 1890–1940.* Oslo, 1993.
Stenseth, Bodil. *Eilert Sundt og det Norge han fant.* Oslo, 2000
Strømholm, Per. Norsk vitenskapshistorie: En bibliografi for naturvitenskap. Oslo, 1975.
Struik, D. *Matematikkens historie.* København, 1966.
Stubhaug, Arild. *Et foranskutt lyn. Niels Henrik Abel og hans tid.* Oslo, 1996.
Sunde, Hjalmar. *Sør-Trøndelag infanteriregiment.* Trondheim, 1984.
Sylow, Ludvig. Forelæsninger over algebraisk Lignings og Substitutions theorie. Bent Birkeland, ed. *Vitenskapshistorisk skriftserie,* Nr. 1, Mat. inst., University of Oslo, 1997.
Sørensen, Øystein. *1880-årene: 10 år som rystet Norge.* Oslo, 1984.
Sørensen, Øystein, ed. *Jakten på det norske.* Oslo, 1998.
Tobies, Renate and Fritz König: *Felix Klein.* Leipzig, 1981.
Ullrich, Peter. Über die Wichtigkeit einer lesbaren Handschrift. *DMV* 4/199, pp. 39–42.
Vigander, Haakon. Norske studenters gymnastikk- og fekteforening, 1882–1932 (Manuscript.)
Vinje, Finn-Erik. Hvordan snakket Henrik Ibsen. Om talemålet på 1800-tallet. *P-2 Akademiet Vol. O,* Oslo, 1999, pp. 149–167.
Visitaser i Eid og Hornindal 1807–1990. Samla av Erling Tomasgard.
Vogt, Jørgen Herman. *Optegnelser om sit Liv og sin Embedsvirksomhed 1784–1846,* published by Den Norske Historiske Forening. Christiania, 1871.
Wallem, Fredrik B. *Det norske Studentersamfund gjennem hundrede Aar. I–II.* Kristiania, 1916.
Weil, André. *The Apprenticeship of a Mathematician.* Boston, 1992.
Weyl, Hermann. Centenary Lectures delivered by C. N. Yang, R. Penrose, A. Borel at the ETH Zürich. Berlin, Heidelberg, New York, 1986.
Wussing, Hans. *Die Genesis des abstrakten Gruppenbegriffes.* Berlin, 1969.

Index of Names

$\varphi_p s + \psi_q t + \ldots$

3 Gl. 5^{t} O. ... El. von r s t.

$W(x\,y\,z\,p\ldots$

$\Pi(x\,y\,z\,p\,q\,r\ldots$

$\dfrac{\partial \Pi}{\partial r} \ldots$ Gl. [illegible]

$y' + a + a_1 y + a_2 y^2 + \ldots$

$a_k' = A_k(a \ldots a \ldots (x)) \ldots$

$\rho\,r\,x$	1	
φ'.	2	1
3	3	2
3	4	3
3	5	4

3 ans 5^{t} O.

$\Omega(x\,y\,z\;\varphi(x)\;\psi(y)) = 0$

$\Omega_x' + \Omega_z' \ldots + \Omega_1\,\varphi' = 0$

$\Omega_{..}\;\varphi''\,\Omega_{..}\,\psi''\;\Omega_{..}\,($

$\Omega_x + \Omega_1 \varphi' = 0 \quad \Omega_y + \ldots$

$\Omega_{xx} + 2\Omega_{x1}\varphi' + \Omega_{11}\varphi'^2 + \Omega_1 \varphi''$

~~$x\,y\,z$~~	~~φ~~	
$p\,q$	φ'	φ
$r\,s\,t$	φ''	
...		

$\Omega(x\,y\,z\,\varphi) = \psi(y)$

$\Omega_x + \Omega_z p + \Omega_\varphi \varphi' = 0$

$\Omega_{xx} \ldots \Omega_{xy} \ldots \Omega_{y1} \ldots$

$\Omega(x\,y\,z\;\varphi(x)\;\psi(y)\;\chi(z\ldots$

~~$W(x\,y\,z\,p\,q\,r\,s\,t$~~ $\chi(z)$ [illegible]